Heat and Mass Transfer

Series Editors: D. Mewes and F. Mayinger

L. P. Yarin · G. Hetsroni

Combustion of Two-Phase Reactive Media

With 249 Figures

Springer

Series Editors

Prof. Dr.-Ing. Dieter Mewes
Universität Hannover
Institut für Verfahrenstechnik
Callinstr. 36
30167 Hannover, Germany

Prof. em. Dr.-Ing. E.h. Franz Mayinger
Technische Universität München
Lehrstuhl für Thermodynamik
Boltzmannstr. 15
85 Garching, Germany

Authors

Prof. Dr. L. P. Yarin
Prof. Dr. G. Hetsroni
Technion City
Faculty of Mechanical Engineering
32 000 Haifa, Israel

ISBN 3-540-40339-6 Springer-Verlag Berlin Heidelberg NewYork

Cataloging-in-Publication Data applied for
Bibliographic information published by Die Deutsche Bibliothek
Die Deutsche Bibliothek lists this publication in the Deutsche Nationalbibliografie; detailed bibliographic data is available in the Internet at <http://dnb.ddb.de>

Springer-Verlag Berlin Heidelberg New York
Springer-Verlag is a part of Springer Science+Business Media
springeronline.com

Typesetting: Camera-ready by author
Cover-Design: deblik, Berlin
Printed on acid free paper 62/3020/kk - 5 4 3 2 1 0

Preface

The subject of the book is combustion of two-phase reactive media. The study of this problem is of utmost importance for the understanding of complex phenomena characteristic of burning of solid and liquid fuels. Improvements in industrial furnaces, internal combustion and jet engines, as well as in high temperature combustion reactors, directly depend on progress in the field of combustion of two-phase media.

Combustion of two-phase media is accompanied by a number of physico-chemical processes, such as decomposition of fossil fuels, melting and swelling of solid particles, droplet evaporation, intensive heat release by chemical reactions, etc. The interphase and particle-to-particle interactions affect significantly the hydrodynamical flow structure, the intensity of heat and mass transfer and, as a result, the general characteristics of combustion of two-phase media. Two-phase combustion of porous matrixes when gaseous oxidizer filters through them adds extra features to the combustion process.

A multiplicity of phenomena characteristic of the combustion of two-phase media determine the contents of the book. We consider only a number of principal problems related to the processes developing under conditions of phase changes, intensive heat release and strong interaction of dispersed and continuous phases. These include the problems of heating, ignition and combustion of a single particle, droplet and bubble, combustion wave propagation in two-phase media, as well as the thermal regimes of combustion reactors.

To a certain degree, the choice of the problems considered is subjective. It was determined by the problems which, in our opinion, are of significant scientific interest, as well as illustrate the specific approaches characteristic of the theory of combustion of two-phase media. In the book we also used the results of our own investigations of a number of problems of two-phase flows, as well as the materials of graduated courses given by the authors at the Technion – Israel Institute of Technology.

The book consists of three parts. They deal with the dynamics of a single particle (Chaps. 1.1 – 1.5), combustion wave propagation in two-phase reactive media (Chaps. 2.6 – 2.9) and thermal regimes of high temperature combustion reactors (Chaps. 3.10 – 3.11).

A number of comprehensive monographs dealing with mechanics of two-phase flows have been published in the last decades. Numerous articles concerning various aspects of this problem are also published in the current literature.

However, as far as we know, there are practically no special monographs devoted to combustion of two-phase media. We have attempted to fill this gap.

The book is not an undergraduate text, although it could be used for a graduate-level course. It is intended for professional scientists and engineers interested in the combustion of two-phase reactive media.

Acknowledgements

We would like to express our gratitude to our colleagues Professor S. Haber, Professor A. Solan, Professor M. Sokolov, Professor G.S. Sukhov, Dr. E. Wacholder, Dr. A.L. Genkin, Dr. A.N. Guzhiev, Dr. P.L. Gussika, Dr. A.D. Lebedev, Dr. V.N. Likhachev, Dr. V.N. Pushkin, and Dr. N.N. Soldatkina for their participation in joint investigations of a number of problems considered in this book.

We are especially grateful and deeply indebted to Professor A.L Yarin for many valuable discussions and comments made after reading the manuscript.

Special thanks are directed to Mr. E. Goldberg and Mrs. E.L Yarin for correction of the text, as well as to Mrs. E. Pogrebnyak, Mrs. M. Schreier and Mrs. A. Rosen for their help in preparation of the manuscript.

We also thank our wives Nelly Sakharov and Ruth Hetsroni for their encouragement and patience while the manuscript was being developed.

During the work on the book we were recipients of grants from the Committee of the Council of Higher Education, the Israel Academy of Sciences and the Humanities. L.P. Yarin was also partially supported by the Center of Absorption (State of Israel) and Israel Council for Higher Education.

Contents

Introduction

0.1 General overview

The usage of fire became one of the most remarkable events in the history of humanity. It determined the face of modern civilization. The development of heat-power stations, automobile and railway transportations, central heating systems, aviation and astronautics, etc. would not have been possible without traditional sources of energy that are based on fossil fuels.

Despite the fact that the combustion process has been used for thousand years, the true mechanism of this process became clear only in the first half of the 20th century when a number of fundamental researches concerning kinetics of chemical reactions, theory of thermal explosion and flame propagation, combustion of non-premixed gases, turbulent combustion and stability of flame were performed. These researches laid the foundation of the modern theory of combustion which involves a wide class of problems associated with combustion of various types of fuels, as well as their application in energetics and chemical technology.

Over a long period of time the attention of investigators was concentrated on the studies of combustion of homogeneous media. Starting in the 1950s combustion of heterogeneous media, in particular the combustion of two-phase systems, became the subject of systematic theoretical and experimental investigations. The growing interest in this problem was related to the elaboration of gas turbines, jet engines, as well as to the improvement of industrial furnaces working on solid and liquid fuels. It was also related to the problems associated with environmental protection. At that time the important features of combustion of two-phase systems were revealed and the simplest models of this process were suggested. It was shown that the combustion of two-phase media possesses certain peculiarities related to the interplay of the strong dependence of chemical reaction rate on temperature with dynamic and thermal phase interaction. This process essentially depends not only on physico-chemical properties of reactants and the thermodynamic state of the system, but also on its structural characteristics. In order to describe such complex phenomena, an approach which combines the methods of the classical theory of combustion with the methods of the theory of multiphase flows is called for. The application of such an approach to combustion of two-phase media is the main topic of the present book. It contains eleven chapters which deal with burning of isolated particles, combustion wave propagation in two-phase reactive systems, as well as with the thermal regimes of high temperature combustion reactions.

0.2 Scope and Contents of Part 1

The first part of the book deals with the behavior of a single particle in a fluid flow. It is concerned with the basics of dynamics of a reactive particle in two-phase flows. The reason that a separate part of the book is devoted to the dynamics of a single particle stems from the following: (i) it is important for understanding the mechanism of a number of microprocesses determining the particle heating, ignition and burning, (ii) it is the basis of the theory of combustion of two-phase media, and (iii) it is directly applicable to many cases in engineering and technology.

Part 1 includes five chapters related to particle drag, its heating, ignition and burning, as well as to particle/particle and particle/turbulence interaction.

Chapter 1.1 deals with the drag of a solid particle, a droplet and a bubble. The classical results of the theory of hydrodynamic resistance and a number of specific problems related to the particle drag under conditions of intensive heat and mass transfer are discussed. Effects of flame in the vicinity of particle on its drag are also considered.

In Chapt. 1.2 the problems of particle heating, devolatilization and vaporization are considered. The basic correlations used to calculate the heat transfer coefficient are presented in this chapter. The focus is on the thermal regime of a particle of a fossil fuel under the conditions of decomposition of solid material and filtration of volatiles through the solid matrix. Some results related to vaporization of one- and multicomponent droplets are also discussed.

The concepts of particle ignition and combustion are considered in Chapt. 1.3. The mechanisms of homogeneous and heterogeneous ignition, the peculiarities of ignition and combustion of coal and metallic particles, as well as the models of combustion of one- and multicomponent droplets are treated in this chapter.

Some problems concerning collective effects in dense particle clouds are briefly considered in Chapt. 1.4.

Chapter 1.5 deals with turbulence intensity in dilute two-phase flows. The effect of particle size distribution and of the concentration of a heavy admixture on turbulence intensity is considered. Turbulence modulation in two-phase flows loaded with fine or coarse particles, as well as temperature fluctuations in two-phase flows are studied in detail in this chapter. The theoretical analysis of these phenomena is based on the mixing-length theory, modified to account for the peculiarities of viscous interaction of particles with carrier fluid and for the effect the admixture inertia. A brief discussion of the effect of temperature fluctuations on the average rate of chemical reaction is also presented in Chapt. 1.5.

0.3 Contents of Part 2

In the second part of the book a number of problems arising in relation to wave propagation in two-phase media are considered. In this part the attention is focused on the study of the mechanism of flame propagation in media loaded with reactive particles, droplets of liquid fuel, or bubbles filled with a gaseous oxidizer.

Chapter 2.6 deals with the combustion wave propagation in gas–particle systems. The characteristics of flame propagation under the conditions of conductive or radiant heat transfer are considered.

Combustion wave propagation in bubble media with low and high volumetric content of gaseous oxidizer are discussed in Chapt. 2.7.

Chapter 2.8 is devoted to the flame propagation under the conditions of filtration combustion of porous media.

Chapter 2.9 describes in brief some questions of the theory of turbulent heterogeneous flames.

0.4 Contents of Part 3

The third part of the book treats some specific, technologically important problems characteristic of high temperature combustion reactors. It contains two chapters, in which the results of the study of the ideally stirred and displacement reactors are presented.

The model of an ideally stirred gas–liquid reactor is considered in Chapt. 3.10. This model takes into account the principal features of the gas–liquid reactive systems, e.g. the multistage character of the process, as well as its dependence on the physico-chemical and structural properties of the two-phase media. The theoretical description of the process is based on the thermal theory of combustion and on a model of two-phase media as an interpenetrating continuum. In this approximation the heat and mass balance equations for the actual and space-averaged parameters are obtained. The dimensional analysis of the system of the governing equations is used to reduce the number of the dimensionless groups describing the processes in the gas–liquid ideally stirred combustion reactors. The general characteristics of the gas–droplet, bubble and jet gas–liquid ideally stirred combustion reactors are discussed. The features of the ideally stirred combustion reactors are considered in relation to the change of their regimes, as well as to the variation of the physico-chemical and structural characteristics of the gas–liquid media. The thermal regimes of the gas–solid particle ideally stirred combustion reactor are also considered in Chapt. 3.10.

Chapter 3.11 is devoted to the displacement combustion reactors. Two characteristic types of such reactors (bubble and filtration) are considered in this chapter. The kinematic balance method is used for the analysis of the thermal regimes of both types of the displacement reactors discussed. Special attention is paid to the states of the filtration reactor: its stationary regimes, stability of the stationary states, hysteresis phenomena, and regimes with self-sustained oscillations.

1

Dynamics of a single particle

1.1 Drag of solid particles, droplets and bubbles

1.1.1 Basic relations

Background. The drag on a particle moving in a viscous fluid remains the greatest challenge of modern hydrodynamics. This knowledge is essential for numerous applications in engineering, in particular for the calculations of the atomized fuel distribution in combustion chambers, as well as for heating, ignition and combustion of a single particle and a spray.

In spite of the fact that the first investigations of particle drag were carried out a long time ago (Newton's experiment in 1710), systematic study of this problem began only in the second half of the 19th century after the pioneering work of (Stokes 1851).

Stokes' work dealt with the drag on a spherical particle moving with a constant velocity in a viscous incompressible fluid. By solving the Navier–Stokes equations for the creeping flow at the low Reynolds numbers ($\mathrm{Re} \ll 1$), he found the relation between the drag force exerted on the particle by the fluid and the particle diameter, its velocity, and the physical properties of the fluid.

Beginning with the classical work of Stokes, the attention of investigators was concentrated on the particle drag at Re=O(1) (Oseen 1910, 1927, Goldstein 1938, Proudman and Pearson 1957, etc.). At the same time, a number of experimental and theoretical studies of the drag under a variety of conditions and modes of particle motion were performed. Particle rotation, acceleration, vaporization and the flow characteristics (turbulence intensity, the existence of the velocity, pressure and temperature gradients) were accounted for. At present, detailed data on drag of inert and reactive particles moving in incompressible and compressible fluids in laminar and turbulent flows are available. Comprehensive reviews of these results were given in a number of surveys and monographs dealing with the problems of applied hydrodynamics and theory of multiphase flows (Happel and Brenner 1983, Clift et al. 1978, Hetsroni 1982, Soo 1990, Kim and Karrila 1991, Crow et al. 1998, Sadhal et al. 1997). Below we consider briefly some of the most important results found in the studies of drag on the inert and reactive particles.

A. Rigid particle: dimensional analysis. We begin with a fine spherical particle moving rectilinearly in an infinite viscous, incompressible, fluid. First, we consider the physical parameters characteristic of in the present problem. Taking into account the drag force resulting from the inertial and viscous forces, it is possible to assume that the particle drag is determined by the four dimensional parameters

$$f_d = \Phi(\rho, \mu, d, v) \tag{1.1}$$

where f_d is the drag force, ρ and μ are the density and viscosity of fluid; d and v are the particle diameter and velocity.

In the case of particle motion with a constant speed it is customary to express the particle drag in terms of the drag coefficient C_d

$$C_d = \frac{4f_d}{\pi d^2(\frac{1}{2}\rho u^2)} \tag{1.2}$$

where $u = -v$ is the velocity of the undisturbed fluid in the frame associated with the particle. Accordingly, we rearrange the functional equation (1.1) to a dimensionless form. For this purpose, we assume that the product Π of certain powers of the dimensional parameters of the problem is dimensionless

$$\Pi = f_d^{\alpha}\rho^{\beta}\mu^{\gamma}d^{\omega}v^{\varepsilon} \tag{1.3}$$

where α, β, γ, ω and ε are the (yet unknown) exponents.

Bearing in mind that in terms of the three fundamental units of mechanics M, L, and T (mass, length, and time), the parameters of the present problem have the following dimensionalities

$$f_d:[LMT^{-2}];\quad \rho:[ML^{-3}];\quad \mu:[L^{-1}MT^{-1}];\quad d:[L];\quad v:[LT^{-1}] \tag{1.4}$$

we obtain three equations for determining the five unknown exponents in Eq. (1.3)

$$\begin{aligned} &\alpha - 3\beta - \gamma + \omega + \varepsilon = 0 \\ &\alpha + \beta + \gamma = 0 \\ &2\alpha + \gamma + \varepsilon = 0 \end{aligned} \tag{1.5}$$

Accordingly, the number of the exponents, which can be chosen arbitrarily, equals two. Assuming these exponents are A and B (A and B are arbitrary constants; $A \neq B$), we find the dimensionless groups of the problem as per

$$C'_d = \frac{f_d}{\rho u^2 d^2}, \qquad \mathrm{Re} = \frac{ud}{\nu} \tag{1.6}$$

Here ν is the kinematic viscosity of the fluid, and Re is the Reynolds number.

Equation (1.1) then reads

$$F(C'_d, \mathrm{Re}) = 0 \tag{1.7}$$

where F is a dimensionless function of the two dimensionless numbers. Solving Eq. (1.7) for C'_d , we obtain

$$C_d = F_1(\mathrm{Re}) \tag{1.8}$$

where $C_d = C'_d\ (8/\pi)$ is the drag coefficient.

The correlation (1.8) indicates that the drag coefficient of a spherical particle depends only on one dimensionless group: the Reynolds number.

General correlations. The dimensional analysis can provide only a functional relationship between dimensionless groups. To find a specific form of the dependence (1.8), it is necessary to solve the problem of the flow of viscous fluid around the particle. Consider a spherical particle. For incompressible fluid this problem reduces to integrating of the following system of the Navier–Stokes and continuity equations

$$\rho(\mathbf{v}\cdot\nabla)\mathbf{v} = -\nabla P + \mu\nabla^2\mathbf{v} \tag{1.9}$$

$$\nabla\cdot\mathbf{v} = 0 \tag{1.10}$$

where P is the pressure, and the boldface symbols represent vector quantities

The term $\rho(\mathbf{v}\cdot\nabla)\mathbf{v}$ on the left-hand side of Eq. (1.9) is negligible at low Reynolds numbers. The problem is thus simplified considerably and reduces to integration of the set of linear Stokes equations

$$\nabla P - \mu\nabla^2\mathbf{v} = 0; \qquad \nabla\cdot\mathbf{v} = 0 \tag{1.11}$$

It is emphasized, however, that such a simplification even in the case of $\mathrm{Re} \ll 1$ is valid only in a relatively thin fluid layer immediately adjoined the particle surface (Landau and Lifschitz 1959).

Taking into account that

$$\nabla\times(\nabla P) = 0 \tag{1.12}$$

$$\nabla^2 v = \nabla(\nabla\cdot\mathbf{v}) - \nabla\times(\nabla\times\mathbf{v}) \tag{1.13}$$

we eliminate the pressure from the first equation of the system (1.11) and reduce it to the following form

$$\nabla \times [\nabla(\nabla \cdot \mathbf{v}) - \nabla \times (\nabla \times \mathbf{v})] = 0 \tag{1.14}$$

Since for an incompressible fluid $\nabla \cdot \mathbf{v} = 0$, we obtain

$$\nabla \times [\nabla \times (\nabla \times \mathbf{v})] = 0 \tag{1.15}$$

Solution of the linear equation (1.15) should satisfy the non slip boundary condition at the particle surface in the frame of reference associated with the particle. It should also approach a uniform flow field far enough from the particle. The solution is given by

$$\overline{v}_r = (1 - \frac{3}{2}\overline{r}^{-1} + \frac{1}{2}\overline{r}^{-3})\cos\theta \tag{1.16}$$

$$\overline{v}_\theta = (1 - \frac{3}{4}\overline{r}^{-1} + \frac{1}{4}\overline{r}^{-3})\sin\theta \tag{1.17}$$

where $\overline{v}_r = v_r/u$, $v_0 = v_0/u$ are the velocity components in the spherical system of coordinates r, θ, φ associated with the particle center; $\overline{r} = r/r_o$, r_o is the particle radius, and u is the velocity far from the particle (Fig. 1.1a).

The calculation of the flow field around a spherical particle at $Re \ll 1$ (the Stokes problem) is described in detail in a number of monographs dealing with theoretical hydrodynamics (Landau and Lifschitz 1959, Batchelor 1967).

A spherical particle in a flowing viscous fluid experiences pressure and viscous forces. Projection of the net force onto x axis yields the particle drag. The projection of the net force acting on an elementary area of the particle surface $dS = 2\pi r_o^2 \sin\theta d\theta$ (Fig. 1.1b) equals

$$df_d = (\tau_s \sin\theta + P_s \cos\theta)dS \tag{1.18}$$

where τ_s is the tangent viscous traction on the particle surface, P_s is pressure. Then the drag force is given by

$$f_d = 2\pi r_o^2 \left\{ \int_0^\pi \tau_s \sin^2\theta d\theta + \int_0^\pi P_s \cos\theta \sin\theta d\theta \right\} \tag{1.19}$$

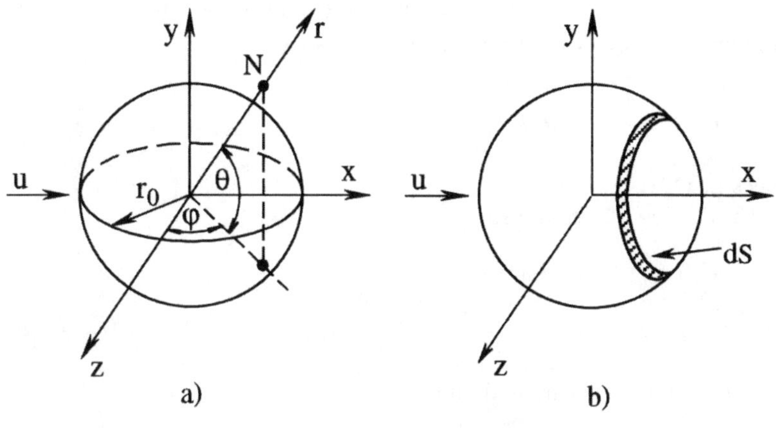

Fig. 1.1 The scheme of the flow about a spherical particle: a) the reference frame associated with the particle center; b) an elementary area on the particle surface

From Eqs. (1.16), (1.17) and (1.11) we find

$$\tau_s = -\frac{3}{2}\frac{\mu u}{r_o}\sin\theta \tag{1.20}$$

$$P_s = -\frac{3}{2}\frac{\mu u}{r_o}\cos\theta + P_\infty \tag{1.21}$$

where P_∞ is the ambient pressure.

Evaluating the integrals on the right-hand side of Eq. (1.19), we obtain the following Stokes expression for the drag force

$$f_d = f_f + f_p = 3\pi\mu u d \tag{1.22}$$

where $f_f = 2\pi\mu u d$ and $f_p = \pi\mu u d$ are the contributions to the total particle drag of the viscous friction and pressure, respectively.

In accordance with the definition (1.2) the drag coefficient of a spherical particle is given by

$$C_d = \frac{24}{Re} \tag{1.23}$$

The Stokes formulae (1.22), (1.23) are valid only for low Reynolds numbers. The deviation of the predicted values of C_d from experimental data on the drag coefficient does not exceed 2% at $Re \sim 0.24$, and 20% at $Re \sim 0.75$.

Oseen (1910) improved the Stokes theory by taking into account the inertia term in Eq. (1.9). Taking into account that the velocity of the fluid far from the particle is close to the velocity of the undisturbed flow, the term $\rho(\mathbf{v}\cdot\nabla)\mathbf{v}$ can be approximated by $\rho(\mathbf{u}\cdot\nabla)\mathbf{v}$. Then the linear equation

$$\rho(\mathbf{u}\cdot\nabla)\mathbf{v} = -\nabla P + \mu\nabla^2\mathbf{v} \tag{1.24}$$

replaces Eq.(1.9)

The drag coefficient given by Oseen's theory is

$$C_d = \frac{24}{Re}(1+\frac{3}{16}Re) \tag{1.25}$$

The latter formula is accurate for $Re \lesssim 1$

The next approximation for the drag coefficient was obtained by Goldstien (1938) in the form of the Reynolds number series

$$C_d = \frac{24}{Re}(1+\frac{3}{16}Re+\frac{19}{1280}Re^2+....) \tag{1.26}$$

Proudman and Pearson (1957), using the method of the matched asymptotic expansions, found that logarithmic terms should enter the series. The Eq.(1.26) is replaced by the following expression for the drag coefficient

$$C_d = \frac{24}{Re}\left\{1+\frac{3}{16}Re+\frac{9}{160}Re^2\ln\frac{1}{2}Re+O(Re^2)\right\} \tag{1.27}$$

which is valid up to $Re \approx 4$

The analysis of Proudman and Pearson was extended by Chester and Breach (1969) who extended the series (1.27) up to the term of order $Re^3 \ln Re$.

Dennis and Walker (1971) calculated numerically the drag coefficient of a spherical particle for Reynolds numbers in the range 0.1 to 40. At low Re their results appear to be close to the theoretical predictions of Goldstein (1938), Proudman and Pearson (1957) and Chester and Breach (1969).

Expressions like (1.23) and (1.25) – (1.27) allow for prediction of the drag coefficient for the simplest types of flow without vortex shading from the particle surface. In this case the Reynolds number is low and the drag coefficient varies inversely with the Re, the Stokes law, or sufficiently close to this law.

At high Reynolds numbers the character of the dependence $C_d(Re)$ changes qualitatively. In the range $750 < Re < 3.5 \cdot 10^5$ the drag coefficient is close to a constant equal to 0.445 (Newton's law). At higher Re it decreases sharply (Fig. 1.2) (Schlichting 1960). This results from the change of the flow structure at

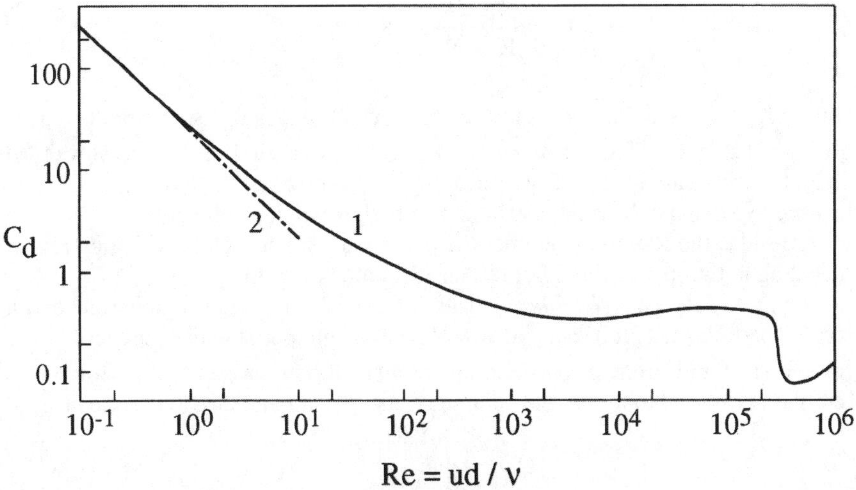

Fig. 1.2 The drag coefficient of a spherical particle: curve 1, the dependence $C_d(Re)$; curve 2, the Stokes law curve. Reprinted from Schlichting (1960), with permission

Table 1.1 Coefficients of Eq. (1.29) Reprinted from Morsi and Alexander (1972), with permission

Re	K_1	K_2	K_3
$0–10^{-1}$	24	0	0
$10^{-1}–10^{0}$	22.73	0.0903	3.69
$10^{0}–10^{1}$	29.1667	– 3.8889	1.222
$10^{1}–10^{2}$	46.5	– 116.67	0.6167
$10^{2}–10^{3}$	98.33	– 2778	0.3644
$10^{3}–5\cdot10^{3}$	148.62	$-4.75\cdot10^{4}$	0.357
$5\cdot10^{4}$	– 490.546	$57.87\cdot10^{4}$	0.46
$10^{4}–5\cdot10^{4}$	– 1662.5	$5.4167\cdot10^{6}$	0.511181

higher Re, when transition to turbulence happens in the boundary layer, which leads to flow reattachment to the surface. Calculation of C_d at high enough values of the Reynolds numbers is a highly complicated problem. The alternative is to use empirical formulae of acceptable accuracy available in the literature: (Soo 1990, Crowe et al. 1998). A good approximation of the experimental data at Re < 800 is given by the formula of Schiller and Naumann (1933)

$$C_d = \frac{24}{Re}(1 + 0.15\,Re^{0.687}) \tag{1.28}$$

Its accuracy is about 5%.

Morsi and Alexander (1972) suggested a three–term correlation to calculate drag coefficient at $0 < Re < 5\cdot10^4$,

$$C_d = \frac{K_1}{Re} + \frac{K_2}{Re^2} + K_3 \tag{1.29}$$

where K_1, K_2 and K_3 are constants for several ranges of the Reynolds number given in Table 1.1. The accuracy of Eq. (1.29) is about 1–2%. Putman's (1961) and Clift and Gauvin's (1970) correlations (see Crowe et al. 1998, pp. 71–72) can be used to calculate the drag coefficient at high Reynolds numbers.

Besides the Reynolds number there are many other factors which affect particle drag in two-phase flow. Let us consider some of them.

Effect of the particle shape. The difference in the drag of spherical and irregular particles is significant. At low Reynolds numbers the dependences $C_d(Re)$ for spherical and irregular particles are similar: in both cases the drag force is directly proportional to the particle diameter, its velocity and the fluid viscosity.

$$f_d = A\mu ud \tag{1.30}$$

where A is a dimensionless constant depending on the particle shape.

In spite of the large variety in the shape, the numerical coefficient A is of the same order of magnitude for different particles. For a thin disk oriented downstream or normal to the flow A equals 5.34 and 8, respectively, whereas for a spherical particle it equals 3π (Boothroyd 1971).

The effect of the particle shape is noticeable only at high Reynolds numbers. In this case the drag coefficient depends on two dimensionless parameters: the Reynolds number and the shape factor (Boothroyd 1971).

$$C_d = C_d(Re, \phi) \tag{1.31}$$

where $\phi = S_{eq}/S$ is the shape factor; S_{eq} is the surface area of some "equivalent" sphere of the same volume and S is the actual surface area. The surface area and diameter of the equivalent sphere are $S_{eq} = \pi^{1/3}(6V)^{2/3}$ and $d_{eq} = (6V/\pi)^{1/3}$, V is the particle volume; the Reynolds number is based on the equivalent diameter.

The dependence (1.31) is presented in Fig. 1.3 as a family of curves $C_d(Re)$ corresponding to different values of ϕ. It is seen that the drag coefficient of irregular particles is larger than the one for a comparable spherical particle. The difference increases significantly as Re increases.

Effect of the particle rotation. Rotation of solid particles and droplets results in a lift force and effects the combustion process significantly. Some aspects of this problem were considered by Pearlman and Sohrab (1991) in the context of turbulent spray combustion. The effect is noticeable at high angular rotation speed, when the lift force becomes commensurable with the other forces acting on the particle. This modifies particles trajectories, as well as their distribution in the furnace space (Genkin et al. 1981). As a result, the rates of heating and vaporization of atomized fuel change (Eastop 1973, Lozinski and Matalon 1992).

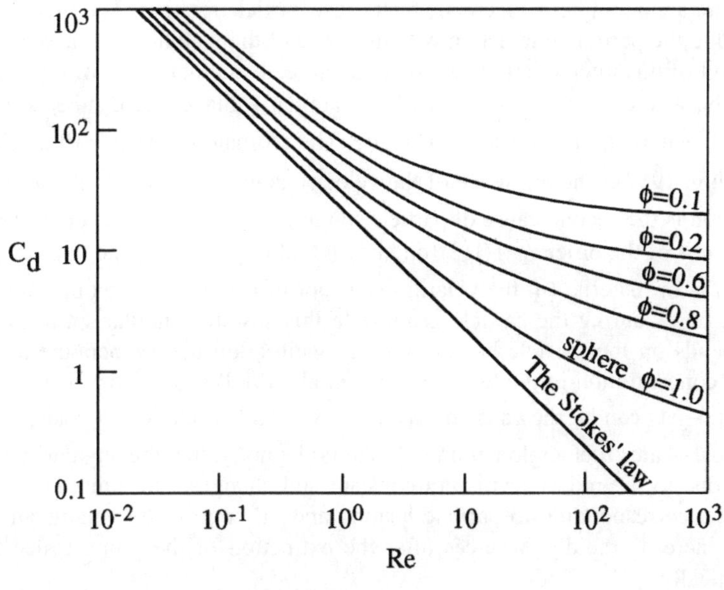

Fig. 1.3 The drag coefficient of irregular particles. Reprinted from Boothroyd (1971), with permission

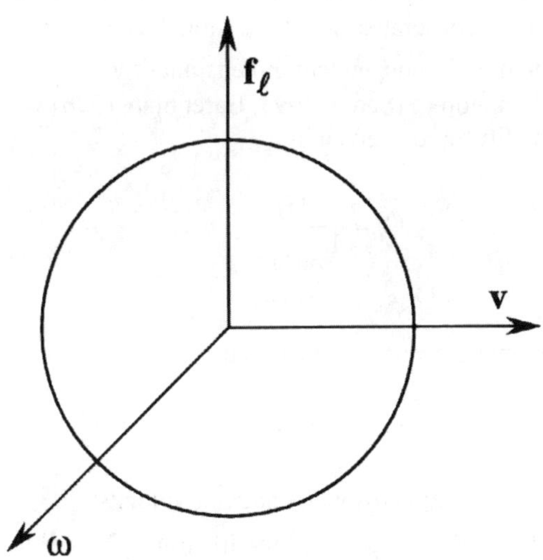

Fig. 1.4 Lift force on a spherical particle rotating in fluid. On the sketch **v**, **ω** and $\mathbf{f}_\ell$ are the vectors of the particle velocity, the angular speed of particle rotation and of the lift force, respectively

There are a number of factors which cause particle rotation. The first of these is related to the particle interaction with the walls of the feeding pipes and nozzles. As a result of this interaction, the particles acquire an angular speed of rotation ω. At $d \sim \delta$, where δ is the thickness of the laminar sublayer, ω can be estimated as $u_*^2/2\nu$, where u_* is the friction velocity. The estimate yields $\omega \sim 10^5 s^{-1}$ (Yarin and Sukhov 1987). The existence of the velocity gradient du/dy in the near-wall flow region is the second cause of particle spinning. In this case the angular speed of rotation is of the order $\omega \sim (1/2)(du/dy)$ (Boothroyd 1971). The third factor is the ejection of material (of the volatiles or vapor of water contained in solid fuel) from the pores during the particle heating. In this case the angular speed of rotation depends on the particle heating rate, as well as on the temperature and the oxidizer concentration in the furnace. Kang et al. (1988) showed that volatile (vapor) micro–jets can be the cause of fast ($\omega \sim 3 \cdot 10^4 s^{-1}$) rotation of coal particles and of coal–water fuel agglomeration. It should be noted that the angular speed of rotation due to the micro–jets is not constant and changes with time. During the first stage, corresponding to particle heating and burning of the volatile/air mixture, ω increase rapidly, whereas after the extinction of the flame it decreases monotonically.

Particle rotation is the cause of the Magnus lift force $\mathbf{f}_\ell$ which is directed normally to the plane formed by the vectors $\mathbf{v}$ and $\boldsymbol{\omega}$ (Fig. 1.4). The magnitude of this force depends on the physical properties of the fluid and on the particle diameter, as well as on its velocity $\mathbf{v}$ (relative to the fluid) and the angular speed $\boldsymbol{\omega}$. The given problem incorporates six dimensional characteristics: $\rho, \mu, d, u = -v$ and ω, three of them with independent dimensionalities. Then, in accordance with the Buckingham Π theorem (Sedov 1993, Barenblatt 1996) we find that the coefficient based on the lift force is given by

$$C_\ell = \frac{4 f_\ell}{\frac{1}{2}\rho u^2 \pi d^2} \tag{1.32}$$

where two dimensionless groups are involved

$$C_\ell = \Phi_1(\mathrm{Re}, \gamma) \tag{1.33}$$

Here $\gamma = \omega d/2u$ is the dimensionless angular velocity.

Rubinow and Keller (1961) determined the lift force $\mathbf{f}_\ell$ and the torque $\mathbf{T}_\omega$ for the low translational and rotational Reynolds numbers $\mathrm{Re} = vd/\nu$ and $\mathrm{Re}_\omega = \omega d^2/\nu$, as per

$$\mathbf{f}_\ell = \pi r^3 \rho (\boldsymbol{\omega} \times \mathbf{v})[1 + O(\mathrm{Re})] \tag{1.34}$$

$$\mathbf{T}_\omega = -8\pi r^3 \boldsymbol{\omega}[1+O(Re)] \tag{1.35}$$

At $Re \gg 1$ and $Re_\omega \gg 1$ the lift force is given by Goldshtik (1972) as

$$\mathbf{f}_\ell = \frac{8}{3}\pi r^3 \rho(\mathbf{v} \times \boldsymbol{\omega}) \tag{1.36}$$

Comparison of Eq. (1.34) with Eq. (1.22) yields the following estimate for the ratio of the Magnus to the Stokes force

$$\frac{f_\ell}{f_d} \sim Re_\omega \tag{1.37}$$

The lift force acting on a particle moving in shear flow was estimated by Saffman (1965, 1968). In the case of the velocity gradient du/dy and relatively low Reynolds number it is given by

$$f'_\ell = 6.46\mu|\mathbf{v}|\frac{r_o^2}{\nu^{\frac{1}{2}}}\left(\frac{du}{dy}\right)^{\frac{1}{2}} \tag{1.38}$$

where u is the flow velocity, r_o is the particle radius.

The ratio of this force to the Stokes force has the order of magnitude

$$\frac{f'_\ell}{f_d} \sim Re'_\omega \tag{1.39}$$

where

$$Re'_\omega = \frac{\omega' r_o^2}{\nu}; \quad \omega' = \frac{\nu^{\frac{1}{2}}}{r_o}\left(\frac{du}{dy}\right)^{\frac{1}{2}}$$

A number of important results regarding the Saffman's lift force were obtained by Dandy and Dwyer (1990), McLaughlin (1991), Auton (1987), and Mei (1992). In particular, Dandy and Dwyer (1990) showed that at a fixed shear rate the lift and drag coefficients for a spherical particle, normalized versus uniform flow, are approximately constant over a range $40 \le Re \le 100$ whereas C_d and C_ℓ increase sharply with Reynolds number (at $Re < 10$).

The experimental data on the lift force of spinning spherical particles were obtained in several investigations (Maccoll 1928, Davies 1949, Barkla and Auchterlonie 1971, etc.). Measurements by Barkla and Auchterlonie (1971) and

Tusji et al. (1985) showed that the lift coefficient of a spinning particle is practically independent of Re at high Reynolds numbers. In the interval 500< Re <3000 the lift coefficient is determined by the magnitude of the dimensionless angular speed of the particle $\gamma = \omega d/2u$. The dependence of $C_\ell(\gamma)$ can be presented as per

$$C_\ell = C \cdot \gamma \tag{1.40}$$

where the constant C equals 0.09 ± 0.02 for $\gamma > 5$, 0.16 ± 0.04 for $2 < \gamma < 4$, and 0.04 ± 0.01 for $\gamma < 0.7$.

The lift coefficient of rotating particles in the range of the intermediate Reynolds numbers was studied experimentally by Oesterle and Dinh (1998). Their measurements showed that in the range $10 < \mathrm{Re} < 140$ the lift coefficient depends on two dimensionless parameters: on the Reynolds number and on the dimensionless angular speed. The dependences of the C_ℓ on γ are shown Fig. 1.5 for several ranges of the Reynolds number. For $10 \le \mathrm{Re} \le 40$ (Fig. 1.5a) C_ℓ is proportional to γ, in the intermediate range there is a less pronounced trend (Fig. 1.5b), while at higher Re (Fig. 1.5c) the lift coefficient does not depend on γ.

The lift coefficient for the whole range 10 < Re < 140 can also be estimated from the empirical relation proposed by Oesterle and Dinh (1998)

$$C_\ell = 0.45 + (2\gamma - 0.45)\exp(-0.075\gamma^{0.4}\,\mathrm{Re}^{0.7}) \tag{1.41}$$

Comparison of this relation with the measurements of Oesterle and Dinh, as well as with the data of Barkla and Auchterlonie (1971) and Tsuji et al. (1985), is presented in Fig.1.6. It is seen that Eq. (1.41) correlates fairly well with the experimental data.

Effect of gravity and pressure gradient. Consider the forces acting on particles due to gravity and the pressure gradient. The gravity force is given by

$$\mathbf{f}_g = \mathbf{g}\int_V \rho_p dV \tag{1.42}$$

where $\mathbf{g}$ is the acceleration due to gravity, ρ_p and V are the particle density and volume.

When ρ_p is constant, Eq. (1.42) yields

$$\mathbf{f}_g = \frac{g\rho_p \pi d^3}{6} \tag{1.43}$$

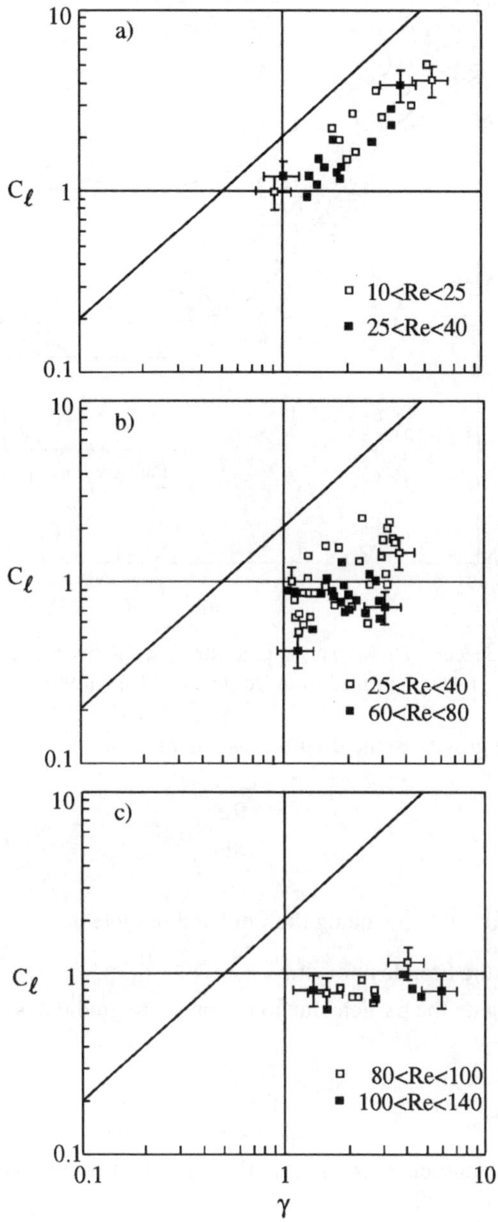

Fig. 1.5 The lift coefficient as a function of the dimensionless angular speed γ (solid lines correspond to the Rubinov and Keller results). Reprinted from Oesterle and Dinh (1998), with permission

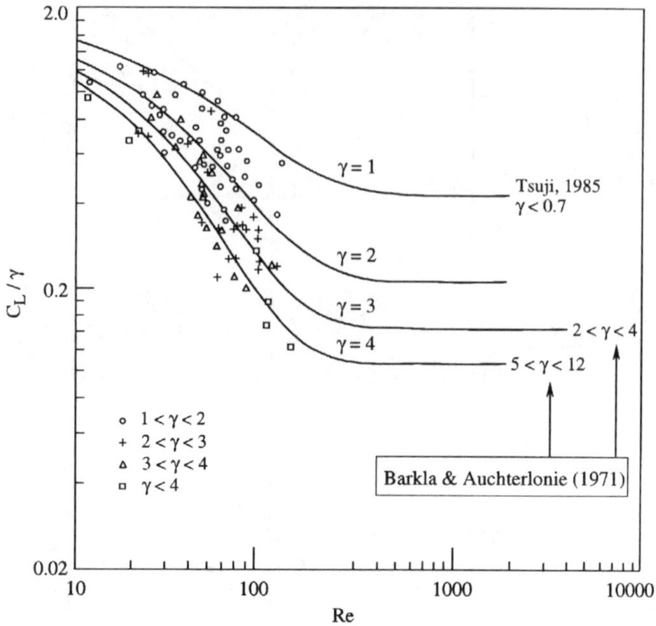

Fig. 1.6 The dependence $C_\ell(\mathrm{Re},\gamma)/\gamma$ on the Reynolds number. The solid lines correspond to solution of Eq. (1.41). Reprinted from Oesterle and Dinh (1998), with permission

The ratio of the gravity to the drag forces equals

$$\frac{f_g}{f_d} = \frac{\rho_2}{\rho_1}\frac{\mathrm{Re}}{18\mathrm{Fr}} \tag{1.44}$$

where ρ_1 and $\rho_2 = \rho_p$ being the fluid and particle density, respectively; $\mathrm{Fr} = u^2/gd$ is the Froude number.

The force acting on the particle due to the pressure gradient is

$$\mathbf{f}_p = -\int_S P\,\mathbf{n}\,dS \tag{1.45}$$

where S is the particle surface, **n** is the outer unit normal vector on the particle surface.

Using the Gauss theorem, Eq. (1.45) can be transformed to the following form

$$\mathbf{f}_p = -\int_V \nabla P dV \tag{1.46}$$

When the pressure gradient is constant, Eq. (1.46) yields

$$\mathbf{f}_p = -\nabla P \cdot \frac{\pi d^3}{6} \tag{1.47}$$

Substitution of the hydrostatic pressure $\nabla P = -\rho g \mathbf{e}_y$ ($\mathbf{e}_y$ being the unit vector of the vertical, y, direction), yields the force magnitude.

$$f_p = \frac{\rho g \pi d^3}{6} \tag{1.48}$$

The buoyancy force $\mathbf{f}_b$ acting on a particle immersed in fluid is

$$\mathbf{f}_b = \mathbf{g}\frac{\pi d^3}{6} \cdot \rho_2 \left(1 - \frac{\rho_1}{\rho_2}\right) \tag{1.49}$$

The ratio of the buoyancy to the Stokes force is given by

$$\frac{f_b}{f_d} = \frac{\mathrm{Re}}{18\,\mathrm{Fr}}\frac{\rho_2}{\rho_1}\left(1 - \frac{\rho_1}{\rho_2}\right) \tag{1.50}$$

Bearing in mind that $\rho_2/\rho_1 >> 1$ and $\rho_1/\rho << 1$ for particle/gas and bubble/liquid media, respectively, we obtain the following estimates

$$\frac{f_b}{f_d} \sim \frac{\mathrm{Re}}{18\mathrm{Fr}}\frac{\rho_2}{\rho_1} \tag{1.51}$$

for dust flow, and

$$\frac{f_b}{f_d} \sim \frac{\mathrm{Re}}{18\,\mathrm{Fr}} \tag{1.52}$$

for bubbly flow.

Effect of particle–fluid temperature difference. The temperatures of the particle and the surrounding fluid enter the dependence of the drag force on the characteristic parameters of the problem

$$f_d = \Phi(\rho_\infty, \mu_\infty, d, \quad v_\infty, T_\infty, T_p) \tag{1.53}$$

where the subscripts p and ∞ refer to the particle and ambient parameters.

Accordingly, the drag coefficient is given by

$$C_d = F(Re, \psi) \tag{1.54}$$

where $\psi = T_p / T_\infty$.

The temperature difference is due to variation of the physical properties of the fluid within the boundary layer. It affects the drag force, as was studied by Fendell et al. (1968), Kassoy et al. (1966), and Dwyer (1989). Kassoy et al. evaluated the drag coefficient on a spherical particle moving in high-velocity flow of a perfect gas. Assuming a linear dependence of the viscosity and thermal conductivity on the temperature, as well as constant specific heat and Prandtl number, they derived the following relation

$$C_d = \frac{24}{Re}\left(\frac{16C}{3K} - \frac{K}{3}\right) \tag{1.55}$$

where $K = \Omega(\Omega + 2)$, and C is some tabulated function of the parameter $\Omega = (\psi - 1)$ given in Table 1.2.

According to Eq. (1.55) the particle drag coefficient increases almost linearly with Ω. When Ω is of the order of unity, the drag coefficient at low Reynolds number increases up to 70% over the isothermal value corresponding to the Stokes drag.

Effect of particle acceleration. Unsteady motion leads to additional forces applied by the fluid to the particle. They are related to acceleration of the surrounding fluid (virtual mass force), as well as with the viscous effects due to delay in flow development as the velocity changes with time (Basset force).

Equations for unsteady motion of spherical particles in stagnant incompressible fluid were derived by Basset (1961), Boussinesq (1903) and Oseen (1910, 1927). Recently a number of modified model equations describing

Table 1.2 Coefficients in Eq. (1.55) Reprinted from Kassoy et al. (1966), with permission

Ω	$C(\Omega)$	Ω	$C(\Omega)$
0.095	$0.419474 \cdot 10^{-1}$	0.789	0.897344
0.183	$0.925079 \cdot 10^{-1}$	0.844	$0.102325 \cdot 10^{1}$
0.265	0.151344	0.897	$0.115615 \cdot 10^{1}$
0.342	0.2181198	0.949	$0.129603 \cdot 10^{1}$
0.414	0.292868	1.00	$0.144283 \cdot 10^{1}$
0.483	0.375190	1.049	$0.159648 \cdot 10^{1}$
0.549	0.465022	1.098	$0.175696 \cdot 10^{1}$
0.612	0.562255	1.145	$0.192416 \cdot 10^{1}$
0.673	0.666783	1.191	$0.209802 \cdot 10^{1}$
0.733	0.778508	1.236	$0.22785 \cdot 10^{1}$

unsteady particle motion in stationary or weakly fluctuating flows were proposed (Mei 1994, Mei et al. 1991, Berlemont et al. 1990, Maxey and Riley 1983, Chang and Maxey 1994, 1995). These modifications are dealing mostly with the history term in the drag force and, in particular, account for the effect of the initial velocity difference between the fluid and the particle. Comparison of the predictions of various model equations with the numerical solutions of the full Navier–Stokes equations was recently carried out by Kim et al. (1998). Based on their results, Kim et al. (1998) proposed a new correlation for the particle drag.

At the same time there are a number of semi-empirical expressions for accelerated particle motion. Odar and Hamilton (1964) proposed the following equation for the force exerted by a viscous fluid on a sphere which is accelerating arbitrarily and moving rectilinearly in otherwise quiet fluid.

$$-\mathbf{F} = \frac{1}{2}C_d \pi r^2 |\mathbf{v}|\mathbf{v} + C_m \frac{4}{3}\pi r^3 \rho \frac{d\mathbf{v}}{dt} + C_H r^2 (\pi\mu\rho)^{\frac{1}{2}} \int_0^t \frac{d\mathbf{v}/dt}{(t-t')} dt' \tag{1.56}$$

where $\mathbf{F}$ is the force exerted by a viscous fluid on a particle, $C_d = C_d(Re)$ is the Stokes drag coefficient, $C_m(A_c)$ and $C_H(A_c)$ are the mass and history coefficients depending solely on the acceleration parameter $A_c = v^2/((dv/dt)d)$.

The dependences $C_m(A_c)$ and $C_H(A_c)$ were found experimentally by Odar (1966). They are

$$C_m = 1.05 - \frac{0.066}{A_c^2 + 0.12} \tag{1.57}$$

$$C_H = 2.88 + \frac{3.12}{(A_c + 1)^3} \tag{1.58}$$

Magnaudet et al. (1995), using numerical calculation data on flow past a spherical particle, found that the total drag coefficient of an accelerated particle may be given (for $0.1 \le Re \le 300$) by the expression

$$C_{d.AC} = C_d(1 + 0.25A_c^{-1}) + 0.55A_c^{-1} \tag{1.59}$$

where $C_{d.AC}$ is the drag coefficient of the accelerated particle, and $C_d = C_d(Re)$.

Another approach to calculate of the drag force of an accelerated particle was adopted by Karanfilian and Kotas (1978). It consists in expressing the drag force as

$$f_d = C_{d.AC}\rho_1 \frac{v^2}{2} \cdot S_f \tag{1.60}$$

where the drag coefficient $C_{d.AC}$ depends on two dimensionless groups: the Reynolds number, and the acceleration parameter A_c; S_f is the particle cross-section.

The dependence $C_{d.AC}(Re, A_c)$ reads

$$C_{d.AC} = C_d(Re)[1 + A_c^{-1}]^n \tag{1.61}$$

where $n = 1.2 \pm 0.03$.

The expression (1.61) gives a reasonably good agreement with experiments at $10^2 \le Re < 10^4$, and $0 < A_c < 1.05$ (Fig. 1.7).

The general equation for motion of a spherical particle (without rotation), the Basset–Boussinesq–Oseen equation has the following form

$$m\frac{d\mathbf{v}}{dt} = \mathbf{f}_d + \mathbf{f}_p + \mathbf{f}_m + \mathbf{f}_B + \mathbf{f}_g \tag{1.62}$$

where $\mathbf{f}_m$ and $\mathbf{f}_B$ are the inertial force related to the virtual mass, and the Basset forces, given respectively by

$$\mathbf{f}_m = \frac{2}{3}\pi r^3 \rho_1 \frac{\partial}{\partial t}(\mathbf{u} - \mathbf{v}) \tag{1.63}$$

$$\mathbf{f_B} = 6r\sqrt{\pi\rho_1 u}\int_0^t \frac{\partial(\mathbf{u} - \mathbf{v})/\partial t}{(t - t')^{\frac{1}{2}}} dt' \tag{1.64}$$

Here $\mathbf{v}$ and $\mathbf{u}$ are the particle velocity and the velocity of the ambient medium, m is the particle mass.

The effect of the virtual mass and of the Basset forces becomes insignificant at large particle/fluid density ratios, as well as at low frequencies of the stream oscillations (Hjelmfelt and Mockros 1966, Rudinger 1980, Voir and Michaelides 1994). The ratio of the inertial force and of the Basset force to the Stokes drag force is given by

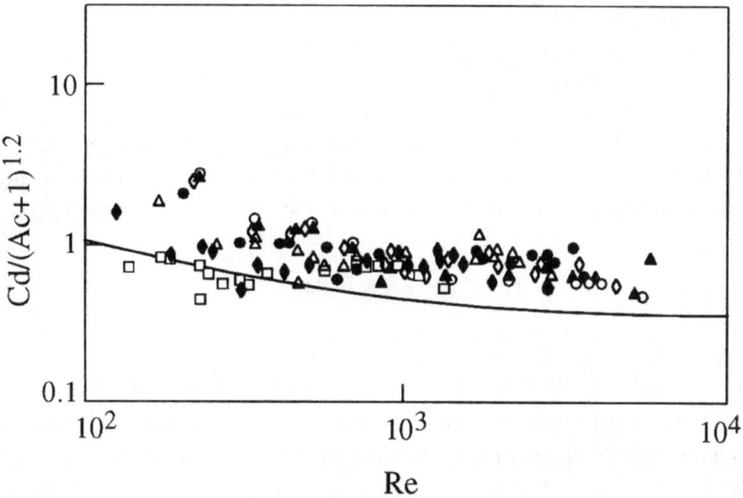

Fig. 1.7 The drag coefficient versus the Reynolds number. Reprinted from Karanfilian and Kotas (1978), with permission. ○, A_c=0.1; Δ, A_c=0.2; ◊, A_c=0.5; ●, A_c=1.0, ▲, A_c=2.2; ♦, A_c=5.2; □, A_c=10.5; The solid line shows the drag coefficients C_d in steady motion at: A_c=0

$$\frac{f_m}{f_d} = \frac{1}{9}\frac{r^2}{\nu t_*}\frac{1}{(\bar{u}-\bar{v})}\frac{\partial}{\partial \bar{t}}(\bar{u}-\bar{v}) \tag{1.65}$$

$$\frac{f_B}{f_d} = \frac{1}{\pi^{\frac{1}{2}}}\left(\frac{r^2}{\nu t_*}\right)\frac{1}{(\bar{u}-\bar{v})}\int_0^{\bar{t}}\frac{\partial}{\partial \bar{t}}(\bar{u}-\bar{v})\frac{d\bar{t}'}{(\bar{t}-\bar{t}')^{\frac{1}{2}}} \tag{1.66}$$

where the dimensionless parameters are defined as $\bar{u} = u/u_*$, $\bar{v} = v/u_*$, $\bar{t} = t/t_*$, $\bar{t}' = t'/t_*$, and u_* and t_* being characteristic scales of velocity and time.

Using the particle relaxation time $t_r = \left(d^2(\rho_2/\rho_1)\right)/18\nu$ as a characteristic time, we obtain the following estimate for the non-dimensional group $r^2/(\nu t_*)$:

$$\frac{r^2}{\nu t_*} = 4.5\frac{\rho_1}{\rho_2} \tag{1.67}$$

Since the order of magnitude of the ratio ρ_1/ρ_2 in gas/particle mixtures is about 10^{-3}, the effect of the virtual mass and of the Basset forces is negligible for these types of two-phase flows.

Effect of free stream turbulence. The particle drag coefficient in turbulent flows is modified by a number of factors accounting for the characteristics of the average and fluctuating fields (spectrum of turbulence energy, scales of turbulent vortices, etc.). In this case it is possible to assume that the drag force is given by

$$f_d = \Phi(\rho, \mu, d, u, \tilde{u}', \ell) \tag{1.68}$$

where $\tilde{u}' = \sqrt{u'^2}$ is the root–mean square of turbulent fluctuations of the carrier fluid, and ℓ is the turbulence length scale (it is implied that $\tilde{u}'$ and ℓ are some characteristic values of the turbulent fluctuations and of the scale of turbulence).

In this approximation the problem involves seven dimensional parameters, three of them with independent dimensions. Accordingly, we obtain

$$C_d = F_1(\mathrm{Re}, T_u, \overline{\ell}) \tag{1.69}$$

where $T_u = \tilde{u}'/u$ is the dimensionless turbulence intensity, and $\overline{\ell} = \ell/d$ is the dimensionless scale of turbulence.

At given $\overline{\ell}$ the effect of $\tilde{u}'$ on the drag coefficient depends on the Reynolds number. At sufficiently high Re, close to the transition to turbulence in the particle boundary layer, an increase in turbulence intensity of the free stream is accompanied by a decrease in C_d. This results from the shift of the separation point towards the rear stagnation point. Over a range of low Reynolds numbers the drag coefficient slightly increase with $\tilde{u}'$ of the free stream. This effect is due to the intensified viscous dissipation. The effect of the turbulence scale on the drag coefficient depends on the ℓ/d ratio. At $(\ell/d) \ll 1$ the effect of turbulence is to transport in the particle a non-steady flow field, whereas at $(\ell/d) > 1$ the drag coefficient increases with ℓ/d.

The above-mentioned effects were studied by Torobin and Gauvin (1961), Clamen and Gauvin (1969), Uhlherr and Sinclair (1970). The effect of the free stream turbulence on the drag coefficient of spherical liquid droplets (at low Weber numbers) was investigated by Warnica et al. (1994). The cumulative results on the effect of turbulence on the particle drag (the data by Torobin and Gauvin 1961, Clamen and Gauvin 1969, Uhlherr and Sinclair 1970, Warnica et al. 1994, Zarin and Nicholls 1971, and Rudoff and Bachalo 1988) are presented in Fig. 1.8. A brief survey of a number of works on the effect of the free stream turbulence on

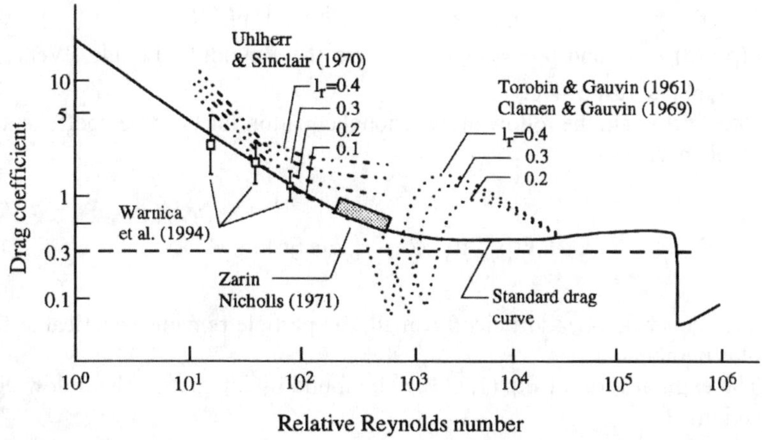

Fig. 1.8 The effect of the free stream turbulence on the particle drag. Reprinted from Crowe et al. (1998), with permission.

the particle drag can be found in the monographs by Soo (1990) and Crowe et al. (1998).

B. Deformable particles. In distinction from the rigid particles, the drag of droplets and bubbles depends not only on the velocity distribution in the outer flow but also on the inner fluid circulation. The particle deformation is controlled by the equilibrium between the surface tension and the hydrodynamic forces arising as a result of particle/fluid interaction.

To find the dimensionless groups determining the drag coefficient of deformable particles, we follow the standard procedure of the dimensional analysis outlined above. First we write the characteristic parameters of the problem. They are

$$f_d, \rho_1, \rho_2, \mu_1, \mu_2, d, v, g, \sigma_t, \Delta\rho \tag{1.70}$$

where σ_t is the surface tension, $\Delta\rho = |\rho_1 - \rho_2|$, and subscripts 1, 2 refer to outer and inner flow.

The problem is characterized by ten dimensional parameters, three of them with independent dimensionality. In accordance with the Buckingham Π theorem, the number of the dimensionless groups is seven

$$C_d, \rho_{2.1}, \mu_{2.1}, \mathrm{Re}, \mathrm{Fr}, \mathrm{We}, \mathrm{Eo} \tag{1.71}$$

where $\rho_{2.1} = \rho_2/\rho_1$, $\mu_{2.1} = \mu_2/\mu_1$, $\mathrm{Re} = (v_1 d_1)/\nu_1$, $\mathrm{Fr} = v_1^2/(gd)$, $\mathrm{We} = (\rho_1 v_1^2 d)/\sigma_t$, and $\mathrm{Eo} = \Delta\rho g d^2/\sigma_t$, are the Reynolds, Froude, Weber and Eötvös numbers, respectively.

Thus we obtain the following functional equation for the drag coefficient of a deformable particle

$$C_d = F(\rho_{2.1}, \mu_{2.1}, \mathrm{Re}, \mathrm{Fr}, \mathrm{We}, \mathrm{Eo}) \tag{1.72}$$

With the Weber number sufficiently small, the particle remains spherical at finite Reynolds numbers.

The explicit form of Eq. (1.72) can be found by integrating the following set of equations

$$\rho^{(i)}(v^{(i)} \cdot \nabla)v^{(i)} = -\nabla p^{(i)} + \mu^{(i)}\nabla^2 v^{(i)} + \rho^{(i)} g^{(i)} \tag{1.73}$$

$$\nabla \cdot \mathbf{v}^{(i)} = 0, \qquad i = 1,2 \tag{1.74}$$

The solutions are subject to the following boundary conditions: (i) uniformity of the flow field at infinity, (ii) continuity of tangential component of the viscous-tension tensor at the particle surface, and a jump of the normal stresses related to surface tension, (iii) equality to zero of the normal velocity components and continuity of the tangential ones at the interphase surface when the latter is assumed to be the undeformable, (iv) finite velocity everywhere inside the particle.

The problem (1.73), (1.74) was solved for a creeping flow about a spherical drop by Hadamard (1911) and Rybczynski (1911). They derived the following relation for the drag force acting on a drop moving in a uniform fluid

$$f_d = 3\pi\mu_1 v_1 d \frac{2\mu_1 + 3\mu_2}{3(\mu_1 + \mu_2)} \tag{1.75}$$

Accordingly, the drag coefficient reads

$$C_d = \frac{16}{\mathrm{Re}} \frac{1 + \frac{3}{2}\mu_{2.1}}{1 + \mu_{2.1}} \tag{1.76}$$

Equation (1.76) acquires the simplest forms corresponding to the drag on a rigid particle or a bubble at $\mu_{2.1} \to \infty$ and $\mu_{2.1} \to 0$,

$$C_d = \frac{24}{Re}, \quad C_d = \frac{16}{Re} \tag{1.77}$$

Note that the drag coefficient on a spherical particle with internal circulation does not depend on the density ratio of the inner and outer fluids, because the creeping flow approximation was employed.

The effect of the parameters $\rho_{2.1}$ and $\mu_{2.1}$ on the particle drag at low but finite Reynolds numbers was investigated by Oliver and Chung (1985). They showed that the drag coefficient is given by

$$C_d = C_{d.c} + 0.0126(C_{d.c}Re)^2 \tag{1.78}$$

where $C_{d.c}$ is the drag coefficient for creeping flow,(Re=0). It is seen that at low (but finite) Reynolds numbers the ratio $\rho_{2.1}$ still does not affect the drag coefficient.

The structure of the outer and inner flows changes qualitatively at high Reynolds numbers when thin boundary layers forms on both sides of the interface. This enables one to subdivide the flow field into three characteristic regions (the external, the intermediate and the internal ones) in which potential, boundary–layer like and circulatory flows occur.

Moore (1963) and Harper and Moore (1968) derived a relation for the drag of deformable particles using the boundary layer approximation. The drag coefficient of a spherical gas bubble moving in liquid medium is (Moore 1963)

$$C_d = \frac{48}{Re}\left(1 - \frac{2.21}{Re^{\frac{1}{2}}} + O(Re^{-\frac{5}{6}})\right) \tag{1.79}$$

The second term in Eq. (1.79) accounts for the dissipation due to the boundary layer and the wake behind the bubble. Omission of this term leads to Levich's relation $C_d = 48/Re$ (Levich 1962) obtained by calculation of viscous dissipation using an imaginary an irrotational flow field about bubble.

The relation (1.79) was verified by Brabston and Keller (1975) versus the results of computations of viscous flow past a bubble at $0.1 < Re < 200$. They found a fairly good agreement between the theoretical and numerical predictions at sufficiently high Reynolds numbers; the discrepancy does not exceed 5% at $Re = 60$ and steeply decreases as Re increases.

At comparable densities of the particle and the surrounding fluid the drag coefficient of a liquid droplet depends only on the Reynolds number and the viscosity ratio (Harper and Moore 1968)

$$C_d = \frac{48}{\mathrm{Re}}(1 + \frac{3}{2}\mu_{2.1}) \tag{1.80}$$

Effect of the particle deformation. Distortion of the particle shape manifests itself at finite Weber numbers (Moore 1959, 1965, Taylor and Acrivos 1964). As the Weber number increases, drops acquire first the form of an oblate spheroid and subsequently an umbrella-like geometry. At small deformations the particle surface resembles a spheroid. Adopting rectangular axis OX,OY,OZ with O at the center of the particle and OZ parallel to the undisturbed flow we write the equation for the particle surface as follows

$$\frac{(x^2 + y^2)}{a^2} + \frac{z^2}{b^2} = 1 \tag{1.81}$$

where a and b are the transverse and longitudinal semi-axes $(a \geq b)$.

The ratio of the semi–axes $\chi = a/b$ depends on the Weber number as

$$4\chi^{-\frac{4}{3}} \frac{(\chi^3 + \chi - 2)[\chi^2 \sec\chi^{-1} - (\chi^2 - 1)]^2}{(\chi^2 - 1)} = \mathrm{We} \tag{1.82}$$

The relation (1.82) reduces to a linear dependence $\chi(\mathrm{We})$ as χ tends to unity

$$\chi = 1 + \frac{9}{64}\mathrm{We} \tag{1.83}$$

Thus, the deviation from the spherical shape becomes negligible at $\mathrm{We} < 0.1$.

The drag coefficient of a distorted bubble moving in a liquid with low viscosity is given by Moore(1965) as

$$C_d = \frac{48}{\mathrm{Re}} G(\chi) \left\{ 1 + \frac{H(\chi)}{\mathrm{Re}^{\frac{1}{2}}} + O(\mathrm{Re}^{-\frac{1}{2}}) \right\} \tag{1.84}$$

where

$$G(\chi)=\frac{1}{3}\frac{\chi^{\frac{4}{3}}(\chi^2-1)^{\frac{2}{3}}\left[(\chi^2-1)^{\frac{1}{2}}-(2-\chi^2)\sec^{-1}\chi\right]}{\left[\chi^2\sec^{-1}\chi-(\chi^3-1)^{\frac{1}{2}}\right]^2}$$

and $H(\chi)$ is an increasing function of $\chi:-2\le H(\chi)\le 8$ at $1\le\chi\le 4$.

Recalling that $G(\chi)$ increases steeply with χ, it is not difficult to see that the particle deformation is accompanied by an increase on its drag coefficient.

Taylor and Acrivos (1964) derived a relation for C_d accounting for the particle deformation, as well as for the difference in density and viscosity between the outer and the inner fluid

$$C_d=C_{d.c}\left\{\begin{aligned}&1+\frac{Re}{8}\left(\frac{3\mu_{2.1}+2}{\mu_{2.1}+1}\right)+\frac{1}{40}\left(\frac{3\mu_{2.1}+2}{\mu_{2.1}+1}\right)^2 Re^2\ln Re+\\&+\frac{\lambda We}{5(\mu_{2.1}+1)}\frac{(3\mu_{2.1}^2-\mu_{2.1}+8)}{(3\mu_{2.1}+2)}\end{aligned}\right\} \tag{1.85}$$

where

$$\lambda=\frac{1}{4(\mu_{2.1}+1)^3}\left\{(\frac{81}{80}\mu_{2.1}^3+\frac{57}{20}\mu_{2.1}^2+\frac{103}{40}\mu_{2.1}+\frac{3}{4})-\frac{\gamma-1}{12}(\mu_{2.1}+1)\right\}$$

γ is the ratio of the density of interior to that of the exterior fluids, $C_{d.c}$ is the drag coefficient for the creeping flow at $Re\to 0$.

The last term in expression (1.85) represent the effect of the deformation on O(We). The relation (1.85) becomes significantly simpler for the case $\mu_{2.1}\to 0$ corresponding to a bubble moving in a liquid

$$C_d=C_{d.c}\left(1+\frac{1}{4}Re+\frac{1}{10}Re^2\ln Re+4.5\lambda We\right) \tag{1.86}$$

Oliver and Chung (1987) proposed the following relation for the drag coefficient of a deformable particle at low Reynolds numbers (0<Re<2)

$$C_d=C_{d.c}\left\{1+\frac{0.40}{16}\cdot Re\left(\frac{3\chi+2}{\chi+1}\right)\right\} \tag{1.87}$$

Drag in shear flow. This problem was examined by Legendre and Magnaudet (1998). The drag coefficient on the bubble was estimated on the basis of the numerical calculations over wide ranges of Reynolds number $(0.1 \le \mathrm{Re} \le 500)$ and of the dimensionless shear rate Sr $(0 \le \mathrm{Sr} \le 1)$. It was found that the drag coefficient varies slightly with Sr.

$$C_d = C_{d.u}(1+0.55\mathrm{Sr}^2) \tag{1.88}$$

where $C_{d.u}$ is the drag coefficient in the uniform flow, Sr is defined as $\mathrm{Sr} = (2r\varepsilon)/U_0$, U_0 and ε are the parameters of the velocity distribution in y direction: $U = (U_0 + \varepsilon y)\mathbf{e}_x$, $\mathbf{e}_x$ is the unit vector of the x direction.

Kurose and Komori (1999) carried out a detailed numerical study of the drag and lift forces acting on a rigid spherical particle in linear shear flow. The calculations, covering the ranges of the particle Reynolds number $0 < \mathrm{Re} \le 500$ and of the shear rate $0 < \mathrm{Sr} \le 0.4$, revealed some characteristics of such flows. In particular, it was shown that a change in the particle Reynolds number leads to changes in the lift force and its direction: at $\mathrm{Re} < 60$ the lift force is directed from the low-fluid-velocity side to the high-fluid-velocity side, whereas at $\mathrm{Re} > 60$ the direction is reversed. Accordingly, the lift coefficient has a negative value at Re > 60. The latter is attributed to flow separation behind the particle.

The migration of deformable drop in shear flow was considered by Haber and Hetsroni (1971), Wohl and Rubinov (1974) and Chan and Leal (1979). These works were reviewed in depth by Leal (1980).

1.1.2 Effect of vaporization

Qualitative consideration. Consider the effect of vaporization (volatilization) on particle drag. Release of vapor from the particle surface is accompanied by thickening of the boundary layer, reduction of the velocity gradient and, accordingly, by decrease of the drag force. Moreover, a non uniform velocity distribution over the interface leads to the appearance of some additional forces and also affects the drag on the particle. To illustrate this effect, consider the momentum flux through the particle surface when it vaporizes in a stagnant atmosphere or in a gas flow (Fig. 1.9).

Assume that the Grashof number is sufficiently small for the effect of the buoyancy forces to be negligible

$$\mathrm{Gr} = \frac{g\beta\Delta T d^3}{\nu^2} << 1 \tag{1.89}$$

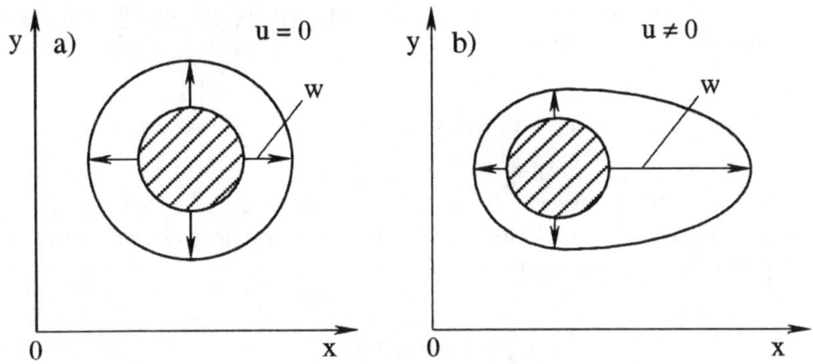

Fig. 1.9 Velocity distribution over a droplet vaporizing in a stagnant atmosphere a) and in a gas flow b)

where β is the thermal expansion coefficient, ΔT is the temperature difference between the particle and the surrounding fluid, ν is the kinematic viscosity, d is the particle diameter.

Under these conditions the radial flow pattern near a droplet (particle) vaporizing in a stagnant atmosphere is symmetric about the median plane normal to the flow direction.

Calculate the total momentum flux around the droplet surface

$$f_e = \int_S \rho_S \, \mathbf{w}\mathbf{w}' \, dS \tag{1.90}$$

where ρ_S is the vapor density, $\mathbf{w}$ the vapor velocity relative to the inertia system of coordinates, $\mathbf{w}' = (\dot{r} + w)$, $\dot{r}$ is the regression rate of the droplet surface, S=**n**S. Recalling that $\mathbf{w}$ and $\mathbf{w}'$ do not depend on θ, in the spherical symmetric case we find

$$\mathbf{f}_e = \rho_s w w' \int_S d\mathbf{S} = 0 \tag{1.91}$$

Therefore in this case the momentum flux through the interface does not contribute to the particle drag.

In contrast to the preceding case, when the droplet vaporizes in a gas flow, the integral in Eq. (1.90) does not vanish due to the dissimilarity of the velocity distribution in the front and rear sections of the particle surface.

The contribution of f_e to the particle drag depends on a number of different factors, reflecting the characteristics of heat and mass transfer. It is determined in the end by the velocity of the undisturbed flow u and the vapor velocity at the particle surface **w**. Thus we can assume that the drag of a vaporizing particle depends on six dimensional parameters:

$$f_d = \Phi(\rho, \mu, d, u, w, D) \tag{1.92}$$

where D is the diffusivity.

In accordance with Buckingham's Π theorem, the drag coefficient is determined by two dimensionless groups

$$C_d = F(Re, \overline{w}) \tag{1.93}$$

where $\overline{w} = (wr_0)/D$, and r_o is the initial droplet radius.

Choosing as characteristic the initial vapor velocity during vaporization of a droplet in a stagnant atmosphere (Yang 1993)

$$\overline{w} = \ln(1+B) \tag{1.94}$$

we transform Eq. (1.93) into

$$C_d = \varphi(Re, B) \tag{1.95}$$

where $B = (c_p(T_\infty - T_s))/q_e$ and $B = (c_p(T_\infty - T_s) + (C_{o,\infty} q)/\sigma)/q_e$ is the Spalding transfer number for non burning and burning droplet, c_p is the specific heat of gas–vapor mixture, q and q_e are the heat of combustion and latent heat of vaporization, $C_{o,\infty}$ the ambient oxygen concentration, σ the stoichiometric oxidizer-to-fuel mass ratio, T_s and T_∞ are the droplet surface temperature that corresponds to saturated pressure under conditions of thermodynamic equilibrium and ambient temperature.

One should note that the drag of a vaporizing drop depends also on the viscosity ratio of the inner and outer fluids. This ratio determines the intensity of the circulation inside the drop, the conditions at the interface gas–vapor, flow separation at its surface and, accordingly, the total drag (Gal-Or and Yaron 1973).

A. Theoretical analysis and quantitative estimates. The drag coefficient on a vaporizing surface can be estimated in the framework of boundary layer theory, as well as of the similarity theory and dimensional analysis. Spalding (1954) and Arpaci (1997) considered this problem for laminar and turbulent flames developing over a flat plate. It is noteworthy that in both cases the dependence of the drag coefficient on the Reynolds and Spalding numbers is

$$C_d = C_{d.B=0} F(B) \tag{1.96}$$

where $C_{d.B=0}(Re)$ is the drag coefficient for a non vaporizing surface and $F(B)$ is some decreasing function of the Spalding number (different for laminar and turbulent flows).

Thus, in non gradient flows mass release from a vaporizing surface leads to reduction of the drag coefficient.

The problem of the drag on a vaporizing droplet reduces to solving the Navier–Stokes equations for compressible fluid flow with a given (or calculated) radial velocity distribution on the droplet surface. Continuity of the tangential velocity and shear stress at the interface should be accounted for.

Muggia (1956) estimated the effect of vaporization on the drag of a spherical particle at low Reynolds numbers. Within the framework of the Oseen approximation for the flow about the particle he derived the relation

$$C_d = C_{d.St} \left\{ \frac{2+B}{2+\frac{3}{4}B-\frac{3}{16}Re(2+B)} - \frac{B}{3}\frac{(4+Re)}{(2+B)} \right\} \tag{1.97}$$

where $C_{d.St}$ is the Stokes drag coefficient.

The dependencies $\bar{C}_d = C_d / C_{d.St}$ as a function of the Spalding number for two values of Re are presented in Fig. 1.10. There is a significant difference between drag of non vaporizing ($B = 0$) and vaporizing ($B > 0$) particles. The increase in the Spalding number is accompanied by a steep decrease of the drag coefficient. This effect manifests itself very clearly at $Re<1$ where mass transfer plays an important role.

In the past the drag of vaporizing droplets was a subject to several theoretical investigations aimed at improving understanding of its effect on vaporizing and burning droplets.

Renksizbulut and Yuen (1983 a,b) carried out a numerical and experimental investigation of a droplet vaporizing in a high temperature air stream at $5<Re<2\cdot10^3$. They showed that the droplet drag coefficient can be presented as the product of two functions $C_d = F(B)\psi(Re)$. One of them depends on the Spalding transfer number B and the other on the Reynolds number Re. The first of these functions accounts for the reduction in the total drag due to the blowing effect related to vaporization. The form of the dependence $C_d(B)$ stays the same for intermediate Reynolds numbers $20<Re<100$ (Renksizbulut and Haywood 1988).

Fendall et al. (1966) estimated the drag on a burning droplet by using the "thin flame theory". They showed that asymmetric vaporization tends to increase the drag coefficient in comparison with the Stokes drag.

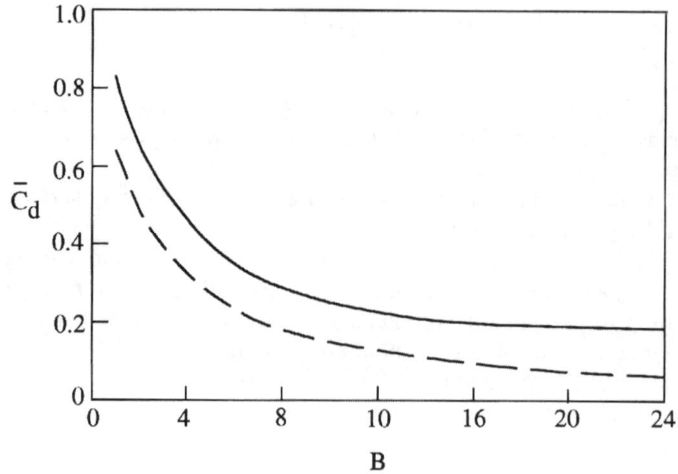

Fig. 1.10 The drag coefficient of a vaporizing droplet. The dashed line is for $Re \ll 1$; the solid line is for Re=1

The effect of non uniform radial velocity distribution on the drag of a spherical droplet was investigated in Sadhal and Ayyaswamy (1983). Assuming that the radial velocity distribution over the droplet surface is given as

$$v_r = A_0 + a_1(\theta) \tag{1.98}$$

they obtained the following relation for the drag due to the viscous stress, pressure and momentum flux at the interface

$$f_d = -\frac{4\pi}{3}\left\{3A_{00} + \frac{C}{2A_{00}}\right\} \tag{1.99}$$

where $a_1(\theta) = a_{01} + a_{11}\cos\theta$; A_0, a_{01} and a_{11} are assumed to be given parameters: $A_0 > 0$, $a_{01} > 0$, $a_{11} < 0$ for the vaporization effect; $A_0 > 0$, $a_{01} > 0$, $a_{11} > 0$ for the radial velocity distribution with maximum values of v_r at the rear stagnation point (burning in the wake), $\mu_{1.2} = \mu_1/\mu_2$, v_∞ is the velocity of the undisturbed flow

$$A_{00} = \frac{A_0 r_o}{\mu_1};\ A_{11} = \frac{a_{11}}{v_\infty};\ C = \frac{-\{(3+2\mu_{1.2}) - A_{11}(1+2\mu_{1.2})\}A_{00}^2}{\left\{-1+\frac{1}{6}(3+2\mu_{1.2})A_{00}^2 + (1+A_{00}-\frac{1}{3}\mu_{1.2}A_{00}^2)\exp(-A_{00})\right\}}.$$

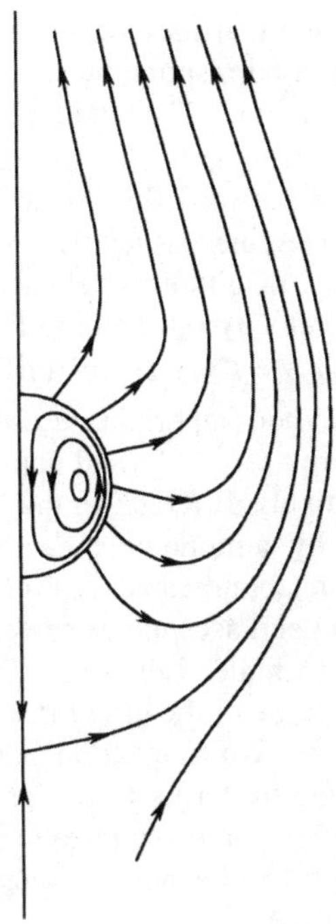

Fig. 1.11 Streamlines of the flow inside and outside the vaporizing droplet. Reprinted from Sadhal and Ayyaswamy (1983), with permission (the parameter values: ε =0.5, μ_{12} =0.1, A_{00}=2, A_{01}=0.5, A_{11}=–0.25)

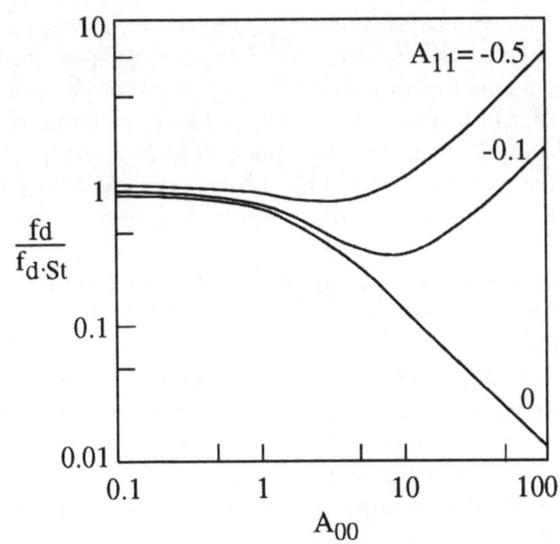

Fig. 1.12 The normalized drag force versus radial velocity. Reprinted from Sadhal and Ayyaswamy (1983), with permission (evaporative droplet μ_{12} = 0, A_{00}>0)

The streamlines of the flows inside and outside the vaporizing droplet are shown in Fig. 1.11. These data correspond to the case where the radial velocity distribution has a maximum at $\theta = \pi$. The circulation of the fluid inside the droplet (recalling Hill's spherical vortex) is due to the liquid–gas interaction at the interface. The streamlines of the outward flow issue from the droplet surface and then turn downstream. The streamlines issued from the front part of the drop contain a specific point at which the longitudinal velocity component changes its sign.

The effect of the radial velocity on the drag force is illustrated in Fig. 1.12 where the dependencies $f_d/f_{d.st} = \Phi(A_{00})$ for different values of A_{11} are presented; $f_{d.St}$ corresponds to the non vaporizing droplet. It is emphasized that the curves corresponding to uniform $(A_{11} = 0)$ and non uniform $(A_{11} < 0)$ radial velocity distributions are qualitatively different. In the former case an increase of radial velocity is accompanied by a monotonic decrease of the drag force, while in the latter the behavior is more complicated. In Fig. 1.12 it is seen that the curves have two branches (descending and ascending) corresponding to low and high radial velocities, respectively. At smaller values of A_{00} an increase of the radial velocity is accompanied by a decrease of the drag force, due to reduction of the pressure drop from the front to the rear stagnation point as a consequence of the negative contribution of the inertia terms to the pressure distribution. At sufficiently high values of A_{00} the drag force begins to increase due to the non-uniform radial velocity distribution. Its effect becomes dominant at high values of A_{00} and leads to increase of the drag with A_{00}.

In both cases A_{11}=–0.5 and A_{11}=–0.1 the curves $f_d/f_{d.st}(A_{00})$ have a characteristic minimum. From the minimum point the domains where thickening of the boundary layer due to the momentum flux at the interface is a dominant begin. The effect of variable density manifests itself in a reduction of the overall drag as compared to the case with constant density (Gogos et al. 1986).

The time variation of the drag coefficient of a vaporizing droplet slowly moving in a gas medium was studied by Gogos and Ayyaswamy (1988). The overall drag coefficient increases very weakly during the initial stage of the process when the droplet radius change does not exceed 50%. At times comparable to the droplet lifetime a steep increase in C_d is observed. (Fig. 1.13). The dependences of the friction, vaporization and pressure drag components depend on time similarly to the total drag coefficient C_d.

B. Experimental studies. During the last 50 years a number of experimental investigations on the drag on a single liquid fuel droplet burning in an oxidizing atmosphere were undertaken. The techniques used were: (1) the captive drop method, (ii) the supporting sphere technique, and (iii) the forced drop technique. The measurements were carried out over a wide range of liquid fuels, droplet diameters, velocities, temperatures and pressures and concentrations of the gaseous oxidizer. That made it possible to reproduce regimes corresponding to various types of flame (attached and envelope) and estimate their influence on the drag force.

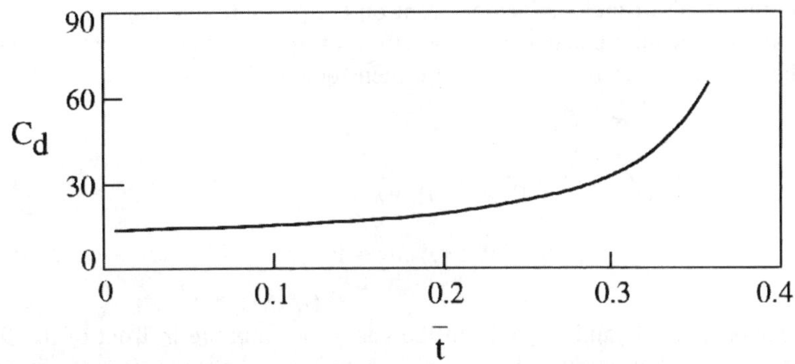

Fig. 1.13 The drag coefficient versus time. Time is rendered dimensionless by the droplet time life. Reprinted from Gogos and Ayyaswamy (1988), with permission

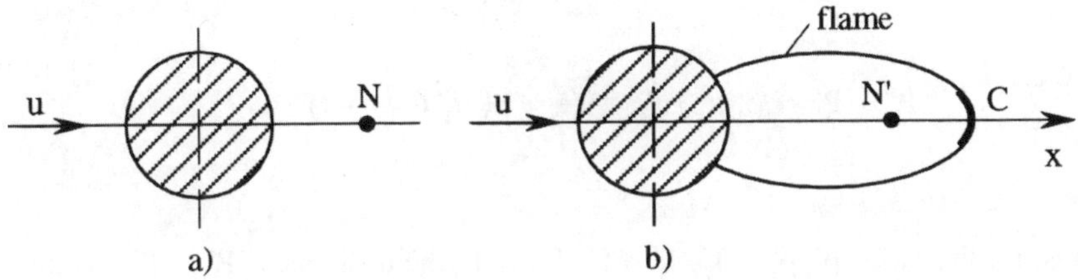

Fig. 1.14 Droplet in a gas flow: a) without flame, b) with an attached flame

Effect of attached and envelope flames. Ignition of reactive mixture of fuel vapor and oxidizer is accompanied by emergence of a flame front in the vicinity of the droplet. Depending on the flow conditions, two different types of the flame occur: (i) the envelope flame, and (ii) the attached flame (Fig. 1.14). The first of them is characteristic of the droplet burning in a stagnant medium, whereas the second one is for of burning of coarse droplets in a gas flow with high enough relative velocities. Khudyakov (1947), using the supporting sphere technique, measured the drag force of non burning and burning spheres of diameter 20 mm and 40 mm, at $\mathrm{Re} = (1 \text{ to } 11)10^3$. He found that the drag coefficient of a non-burning droplet is substantially higher than that of a burning one. Thus, the existence of a flame (in the given case an attached one) leads to reduction of the drag coefficient.

The above effect is due to the influence of the combustion front on the pressure in the wake of the droplet. It may be illustrated as follows. Consider a spherical non-burning and a burning droplet in a gas flow as shown in Fig. 1.14. Consider the pressure at points N and N' at the flow axis far from the droplet surface. In the first case the pressure at point N is close to the ambient pressure $P_1 \simeq P_\infty$, while in second case the pressure at point N' does not equal P_∞. To estimate the effect of the combustion front on the pressure distribution, we use the mass and momentum balance equations for an infinitely thin flame front. Assuming that the flow near the axis x is one-dimensional and the central sector c of the flame is flat, we have

$$\rho_1' u_1' = \rho_2' u_2' \tag{1.100}$$

$$\rho_1' u_1'^2 + P_1' = \rho_2' u_2'^2 + P_2' \tag{1.101}$$

where subscripts 1 and 2 refer to the gas–vapor mixture in front of the flame and the combustion products beyond it, respectively. Primes indicate parameters of the burning droplet.

Combining Eqs. (1.100) and (1.101), we obtain the following estimate for the pressure difference between the two sides of the combustion front:

$$P_1' - P_2' = \rho_1' u_1' (u_2' - u_1') = \rho_1' u_1'^2 (\frac{T_2'}{T_1'} - 1) \tag{1.102}$$

Thus, for the ratio $\rho_1'/\rho_2' = T_2'/T_1'$, $(T_2'/T')_1 > 1$ the difference $P_1' - P_2'$ is positive. Bearing in mind that the pressure at point N equals $P_1 \approx P_\infty$ and the pressure at point N' equals $P_1' > P_2' \approx P_\infty$, we see that $P_1' > P_1$. Thus, existence of an attached flame leads to an increase of the pressure in the wake region of a burning droplet and accordingly to a decrease of its drag.

The drag of burning droplets with envelope and attached flames was also studied by Eisenklam and Arunachalam (1996). Their experiments were carried out with freely falling droplets of pentane, heptane and benzene at $100 \le Re \le 400$. The envelope flame was produced by preliminary heating of the droplet and oxygen enrichment of the mixture in the igniter, whereas the attached flame was produced by ignition of cold droplets immediately as they entered the cross-section of the combustion chamber. It was found that the drag coefficient is lower in the case of the attached flame than in the case of the envelope flame. Comparison of results related to the drag of burning and non burning drops indicates that they differ significantly. A similar effect was noted by Bolt and Saad (1957). They described a considerable difference between the terminal velocities of free falling burning droplets and of solid non-burning spheres of the same size.

The above data by Eisenklam and Arunachalam (1966) show that the deviation of the drag coefficient of burning droplets from the standard curve $C_d(Re)$ depends on the type of flame, as well as on the Reynolds number. Thus, for an envelope flame it is higher and for an attached flame lower than the standard value.

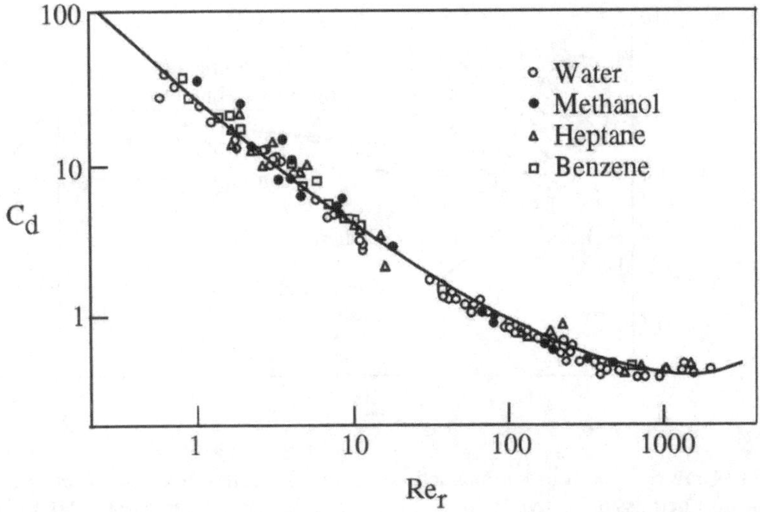

Fig. 1.15 The drag coefficient of droplets in air as a function of the modified Reynolds number Re_r. Reprinted from Yuen and Chen (1976), with permission $Re_r = Re(\mu(T_\infty)/\mu(T_r))$ $T_r = T_p - (T_\infty - T_p)/3$

Effect of variable physical properties of fluid. The above comparison of the results related to the drag of burning and non-burning droplets does not account for the effect of variable physical properties of the fluid in the boundary layer. Various attempts were made to include this effect by modifying the Reynolds number Re_r [reference to the mean conditions 1/3, rule $Re_r = Re(\mu(T_\infty)/\mu(T_r))$, $T_r = T_p + (T_\infty - T_p)/3$ and the drag coefficient by the mass transfer number (Eisenklam and Arunachalam 1966, Eisenklam et al. 1967, Yuen and Chen 1976, Natarajan 1973]. Yuen and Chen (1976) showed that a fairly good correlation between the experimental data and the standard $C_d(Re)$ curve can be achieved by assuming that the characteristic density and viscosity are determined by the ambient and 1/3 rule temperatures, respectively (Fig. 1.15).

Effect of acceleration. Ingebo (1957) investigated the drag of droplets accelerating in a cloud. The measurements show that experimental data on the drag coefficient of water, trichloroethyl and isooctane droplets are grouped near a single curve

$$C_d = 27\,Re^{-0.84} \tag{1.103}$$

Comparison of the data for accelerating droplets with their steady–state counterparts shows that acceleration leads to reduction of C_d.

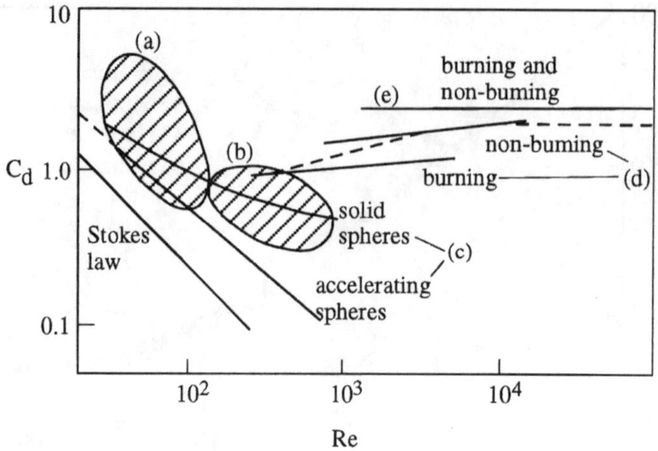

Fig. 1.16 Drag coefficient of burning and non-burning droplets. Reprinted from Williams (1973), with permission. Curves a, b, c, d and e are the data by Eisenklam, Arunachalam and Weston; Natarajan and Brzustowski; Ingebo; Rabin; Schallenmuller and Lawhead; Iaarsma and Derkson, respectively

Briffa (1981) studied experimentally the transient drag of droplets in a spray. The measurements show that deceleration leads to a lower drag coefficient compared with the steady state value. The following empirical relation was suggested

$$C_d = 0.13 + 7.63\,(\mathrm{Re} + 12.8)^{-0.82} \tag{1.104}$$

Effect of pressure. Data on the effect of pressure are scarce. Ingebo (1962) estimated the drag of ethanol droplets for a pressure range of approximately $(0.5–6.5)\cdot 10^5$ Pa, and proposed the relation

$$C_d = \mathrm{const}\left\{\mathrm{Re}^{0.84}\left(\frac{g\ell_{(*)}}{c_{(*)}^2}\right)^{0.2}\right\}^{-1} \tag{1.105}$$

where $\ell_{(*)}$ and $c_{(*)}$ are the mean free path and the root-mean-square velocity of the molecules.

Taking into account that $\ell_{(*)} = (\pi\sqrt{2}\delta_{(*)}^2 n_{(*)})^{-1}$, $c_{(*)} = (3R_g T)^{1/2}$ and $P = kn_{(*)}T$ (Kuchling 1980) ($\delta_{(*)}$ and $n_{(*)}$ being the molecule diameter and numerical concentration, k and R_g the Boltzmann and gas constants), we find that the drag coefficient is proportional to $P^{-0.64}$ in the pressure range $(0.5$ to $6.5)\cdot 10^5$ Pa.

Data from several investigations concerning the drag of burning and non-burning droplets over a wide range of Reynolds numbers were compared by Williams (1973). The results (Fig. 1.16) indicate a weak dependence of the drag coefficient on Re at high Reynolds numbers and a steep variation at low Re.

1.1.3 Effect of combustion

Three main factors determine the drag of burning solid particles: (i) the mass transfer due to the combustion process, (ii) a large temperature difference between the particle and ambient medium, (iii) surface and volumetric reactions on the particle and in its boundary layer.

Depending on the relationship between the rates of surface and volumetric reaction different combustion mechanisms lead to different temperature distributions over the particle surface. For example, when a surface reaction (heterogeneous process) is dominant, the point with the maximum temperature is located at the particle surface, whereas in the case of volumetric reaction in a gaseous mixture (homogeneous process) it is located in the flame enveloping the particle.

The influence of the above factors on the particle drag differs widely. For example, mass transfer leads to thickening of the boundary layer and reduction of the drag coefficient, whereas temperature difference between particle and medium affects mainly the variation of the physical properties of the fluid.

The effect of burning (in fact, mass transfer) on the drag of a solid spherical particle suspended or accelerating in a gas stream was studied by Crowe et al. (1963). It was shown that the drag coefficient of a burning particle decreases when the mass transfer intensity and the acceleration increase. This tendency manifests itself at sufficiently high ratios of the mass flux from the particle surface to that of undisturbed flow, $f > 0.025$ and the acceleration parameter $A_c > 10^2$. The theoretical drag coefficient versus Reynolds number for a burning particle is shown in Fig. 1.17. It is seen that the drag coefficient decreases with increasing parameter f. In contrast with that, measurements show that the drag coefficient of burning powder particles is larger than that of non burning ones. This contradiction is probably due to the over simplification of the theoretical model of Crowe et al. (1963), which does not account for a number of peculiarities of the given phenomenon.

Ogasawara et al. (1967) studied the drag of burning cylindrical and spherical model particles in air flow under conditions corresponding to the formation of an attached flame in the particle wake. In these experiments model particles with porous walls through which liquid fuel was supplied were used. The drag of these burning bodies was measured. The measurements showed that there is a significant difference between the drag in the case with and without a flame, reaching 30% for cylindrical model particles and 40% for spherical ones. It is noteworthy that the drag ratio of a particle with a flame to one without a flame is independent of the Reynolds number in the $1000 < Re < 6000$ interval (Fig. 1.18).

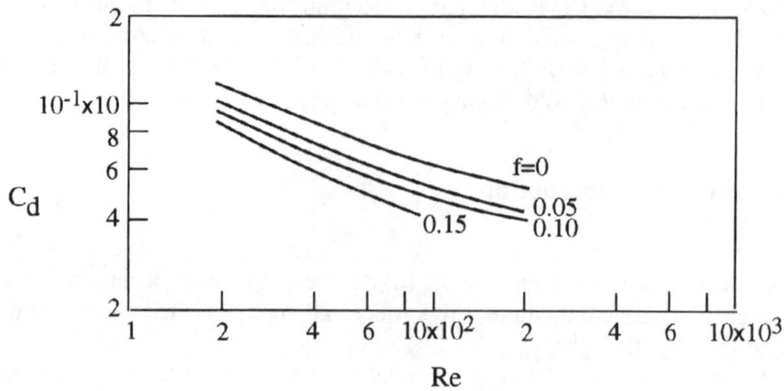

Fig. 1.17 The drag coefficient versus Reynolds number for a burning particle. Reprinted from Crowe et al. (1963), with permission. f is the ratio of mass flux from the particle surface to that in the free stream

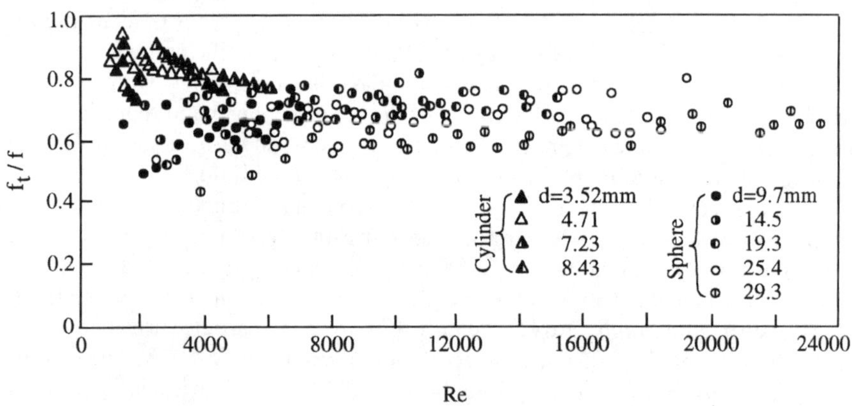

Fig. 1.18 The dependence $f_f/f = \Phi(\mathrm{Re})$ Reprinted from Ogasawara et al. (1967), with permission. f_f and f are drag on particles with and without flame, respectively

The presence of a flame affects directly the gas density behind the particle, as well as the velocity and temperature distribution in the cross-section of the particle wake (Figs. 1.19 and 1.20). To estimate these effects Ogasawara et al. (1967) applied the momentum theorem to the control surface enclosing the particle (Fig. 1.21). Assuming that the pressure in the remote cross-sections of the wake is close to the ambient one, the authors obtained the following expression for the drag force of the particle with and without a flame

$$f_f = \rho_\infty u_\infty^2 \ell^{1-K} d^{1+K} \int_{-\infty}^{+\infty} \bar{\rho}_f \bar{u}_f (1 - \bar{u}_{f'}) \bar{y}^K d\bar{y} \tag{1.106}$$

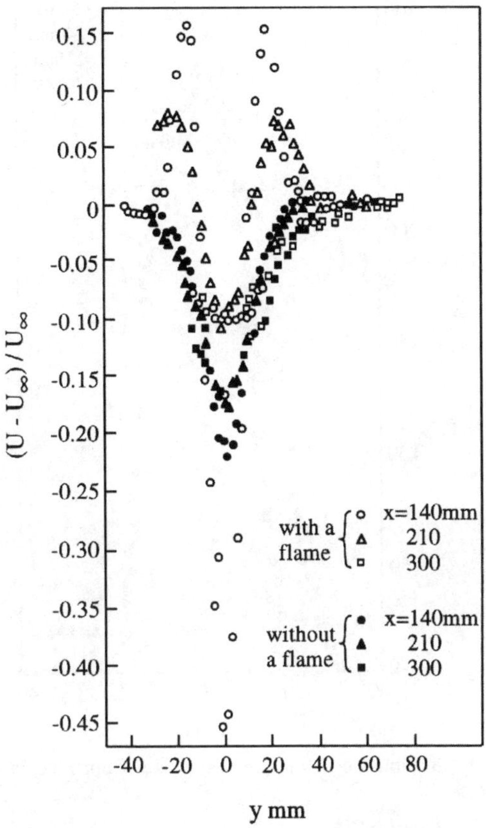

Fig. 1.19 Velocity distribution in the particle wake. Reprinted from Ogasawara et al. (1967), with permission

$$f = \rho_\infty u_\infty^2 \ell^{1-K} d^{1+K} \int_{-\infty}^{+\infty} \bar{u}(1-\bar{u})\bar{y}^K d\bar{y} \tag{1.107}$$

Here f is the drag force acting on the particle, $\bar{u} = u/u_\infty$, $\bar{\rho} = \rho/\rho_\infty$, $\bar{y} = y/y_\infty$; d is the diameter of a spherical or a cylindrical particle; ℓ is the length of a cylindrical particle, $K = 0;1$ refers to a cylindrical and spherical particle; the subscript f refers to a particle with a flame.

Replacing the actual temperature in the cross-section of the wake by the mean temperature T_m and assuming that according to the equation of state of ideal gas $\rho_m = (\rho_\infty T_\infty)/T_m$, we find

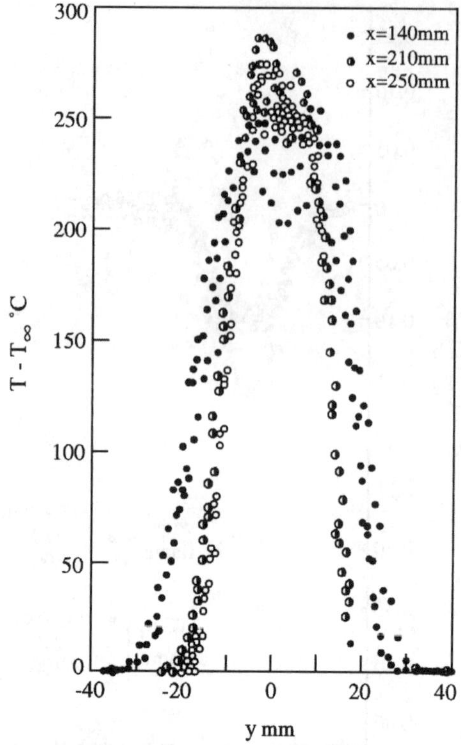

Fig. 1.20 Temperature distribution in the particle wake. Reprinted from Ogasawara et al. (1967), with permission

$$\frac{f_f}{f} = \theta^{-1} \frac{\int_{-\infty}^{+\infty} \bar{u}_f (1-\bar{u}_f) \bar{y}^K d\bar{y}}{\int_{-\infty}^{+\infty} \bar{u}(1-\bar{u}) \bar{y}^K dy} \tag{1.108}$$

where $\theta = (T_m - T_\infty)/T_\infty$.

The calculations show that ratio f_f/f changes slightly with Re. In the interval $2 \cdot 10^3 < Re \le 2.4 \cdot 10^4$ it equals 0.7 and 0.6 for a cylindrical and spherical particle, respectively.

Makino (1992) applied another approach in studying the drag coefficient of burning particles. In the context of pulverized–coal combustion, he calculated the drag coefficient of the carbon particles in compressible low–Re flow, using the matched asymptotic expansion method. The model accounts for variation of the physical properties with temperature as well as for heat losses due to radiation.

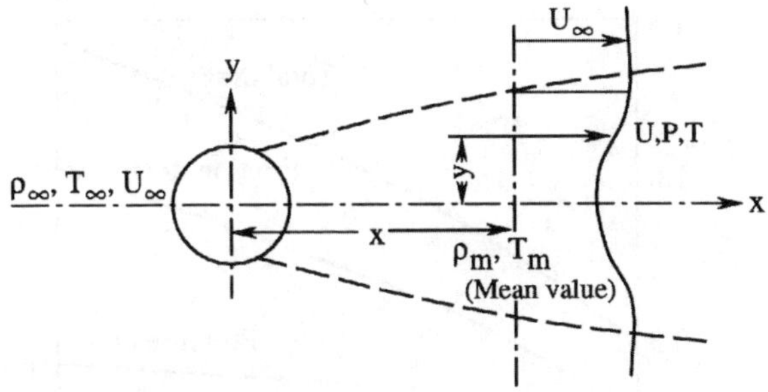

Fig. 1.21 The scheme of flow behind a particle

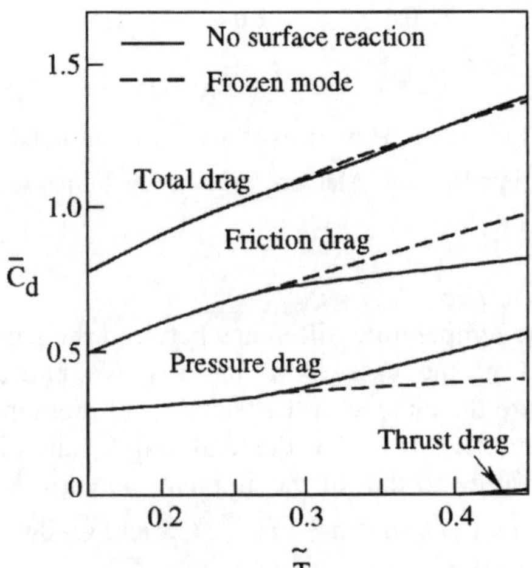

Fig. 1.22 Drag coefficient versus the particle temperature. Reprinted from Makino (1992), with permission

Makino's solution shows that the overall drag coefficient of the burning particle, as well as its friction, pressure and thrust components of the drag coefficient, increase with the particle temperature T_p (Fig. 1.22). At low temperatures ejection from the surface is very low and the drag coefficient difference between burning and non burning particles is negligible. The latter indicates that the chemical reaction does not directly affect the particle drag, only its temperature. The situation changes at high temperatures, when the mass transfer from the particle surface sharply increases and the friction and pressure components of the drag coefficient corresponding to a burning and a non burning particle become different. At high temperatures the friction component of the drag coefficient of a burning particle is lower and the pressure component of the drag coefficient is higher than for a non burning one. As a result, the overall drag coefficient is practically independent of the mass transfer intensity and is determined by the particle temperature.

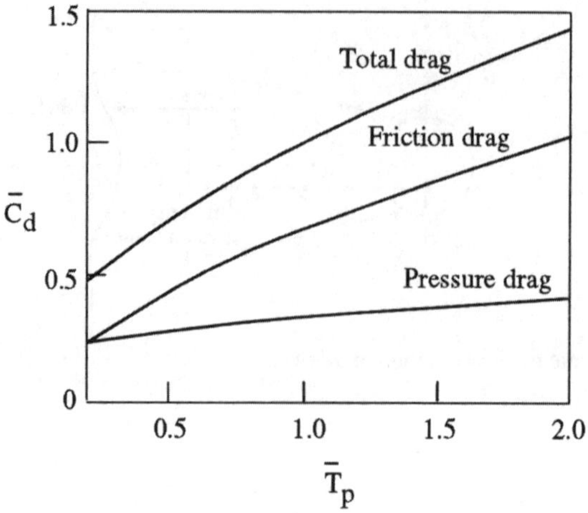

Fig. 1.23 The effect of the temperature ratio of particle to surrounding medium, for a non-reactive sphere. Reprinted from Makino (1992), with permission. $\overline{C}_d = C_d/C_{d.st}$, $\overline{T}_p = T_p/T_\infty$

The effect of the temperature difference between the particle surface and the surrounding medium in the case of a non reactive particle is illustrated in Fig. 1.23. In this figure the change of the friction and pressure components of the drag coefficient (C_f and C_p), as well as the total drag C_d, are plotted against the ratio of particle temperature to that of the ambient medium. It is seen that the increase of T_p/T_∞ results in an increase of C_f, C_p, and C_d due to the change in the physical properties of the fluid with temperature.

Comparison of the theoretical and experimental results for the drag coefficient of a burning particle is presented in Fig. 1.24 (data by Niksa et al. 1984 and Mitchell 1986). It is seen that the Makino's model agrees fairly well with experimental data.

A number of experimental investigations of the drag of a burning particle were undertaken in the former Soviet Union in the period 1950–85 (Babii and Kuvaev 1986). Measurements on carbon, coal, coke and charcoal particles were carried out over wide ranges of particle diameter, oxygen concentration and initial and ambient temperatures: $0.1<d<15$ mm, $0.21 < C_{o\infty} < 100\%$, $300 < T_{po} < 1{,}400\,\text{K}$, $300 < T_\infty < 600\,\text{K}$. Under these conditions the particle temperature changes in the interval $1{,}300< T_p< 2{,}670$ K.

The results of these experiments are presented in Fig. 1.25. They show that the drag coefficient of a burning particle is larger than that of a nonburning one, the difference being of 1 to 1.5 orders of magnitude at Re ~ 10 and decreasing as Re increases. At sufficiently high Reynolds numbers (Re ~ 10^3) corresponding to development of turbulent flow, the drag coefficients are of the same order.

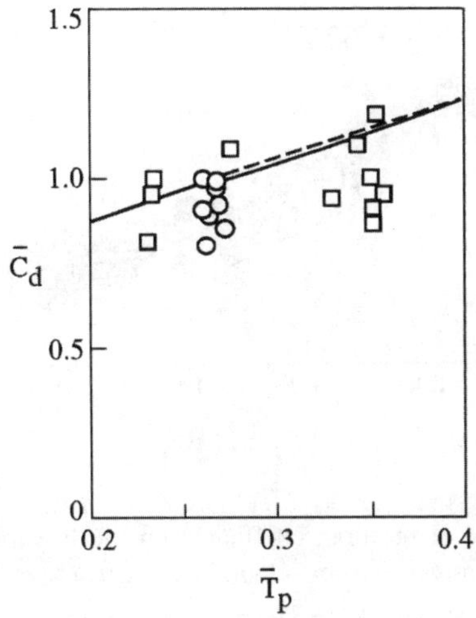

Fig. 1.24 The drag coefficient of a burning carbon particle. Reprinted from Makino (1992), with permission. Solid curve shows Makino's result for a burning carbon particle; dashed curve shows Makino's result for a non reactive sphere. □, data by Niksa et al., reprinted from Niksa et al. (1984), with permission, ○, data by Mitchell, reprinted from Mitchell (1986), with permission. $\bar{C}_d = C_d/C_{d.st}$, $\bar{T}_p = T_p / T_\infty$

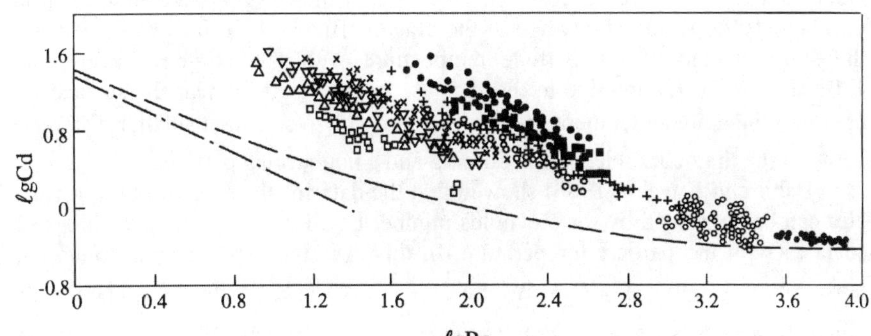

Fig. 1.25 The dependence of the drag coefficient on the Reynolds number is defined by ambient parameters. Reprinted from Babii and Kuvaev (1986), with permission. ●, $C_{o\infty}$=0.23, T_∞ = 293 K; carbon, ○, 0.23< $C_{o\infty}$ <0.49, T_∞ =293 K, carbon charcoal

●, $C_{o\infty}$ =0.98, T_∞ =293 K, anthracite, ■, $C_{o\infty}$ =1.0, T_∞ =293 K, anthracite, ∇, $C_{o\infty}$ =0.23, T_∞ =623 K, anthracite, Δ, $C_{o\infty}$ =0.23, T_∞ =903 K, anthracite, □, $C_{o\infty}$ =0.23, T_∞ =1083 K, anthracite, +, $C_{o\infty}$ =0.23, T_∞ =293 K, coke, ×, $C_{o\infty}$ =0.23, T_∞ =573 K, coke

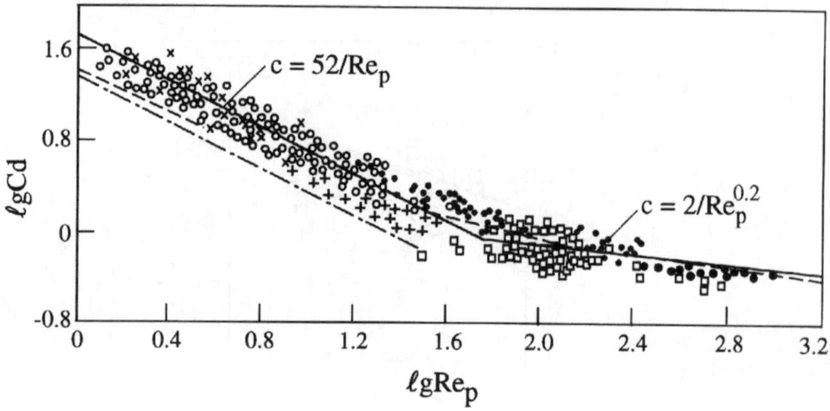

Fig. 1.26 The dependence of the drag coefficient on the Reynolds number based on the particle temperature Reprinted from Babii and Kuvaev (1986), with permission. ●, $C_{0\infty}$ =0.23, T_∞ =293 K; carbon, •, 0.23 < $C_{0\infty}$ < 0.49, T_∞ =293 K; carbon, charcoal, ϕ, non-reactive particle, O, $C_{0\infty}$ =0.23, T_∞ =293 K, anthracite, +, $C_{0\infty}$ = 0.23, T_∞ =293 K, coke, ×, $C_{0\infty}$ =0.23, T_∞ =553 K, coke

The oxygen concentration in a gaseous mixture as well as the ambient temperature affect the particle drag. The increase of the oxygen concentration from 0.21% to 100% results in change of the drag coefficient by a factor of 2.5 due to the steep change of the particle temperature with the oxygen concentration 1300< T_p < 2670 K) and due to the increase in gas viscosity near the surface. Increase of the ambient temperature at a constant $C_{0\infty}$ results in a smaller difference between the drag coefficients of a burning and a nonburning particle.

Babii and Kuvaev (1986) showed that the data for the drag of burning particles can be correlated by the Reynolds number based on the properties values associated with the particle temperature. In this case the experimental points collapse onto a curve $C_d(Re_p)$, where $Re_p = (ud)/\nu(T_p)$ with $\nu(T_p)$ being the kinematic viscosity corresponding to the particle temperature (Fig. 1.26). This dependence can be expressed as

$$C_d = \frac{52}{Re_p} \tag{1.109}$$

for $1 < Re_p < 50$, and as

$$C_d = \frac{2}{Re_p^{0.4}} \tag{1.110}$$

for $50 < Re_p < 700$.

It should be noted that Eqs. (1.109), (1.110), like other empirical correlations, are applicable only under conditions close to those of the experiments.

References

Arpaci VS (1997) Microscales of turbulence heat and mass transfer Correlations. Gordon and Breach, Amsterdam

Auton TR (1987) The lift force on a spherical body in a rotational flow. J. Fluid Mech. 183: 199–218

Babii VI, Kuvaev JaF (1986) Combustion of coal dust and coal dust flame calculation. (in Russian) Energoatomizdat, Moscow

Barenblatt GI (1996) Scaling, self-similarity and intermediate asymptotics. Cambridge University Press, Cambridge

Barkla HM, Auchterlonie LJ (1971) The Magnus or Robins effect on rotating spheres. J. Fluid Mech. 47: 437–447

Basset AB (1961) A treatise on hydrodynamics, vol. 2. Dover, New York

Batchelor GK (1967) An introduction to fluid dynamics. Cambridge University Press, Cambridge

Berlemont A, Desjonqueres P, Gouesbet G (1990) Particle Lagrangian simulation in turbulent flows. Int. J. Multiphase Flow 16: 19–34

Bolt JA, and Saad MA (1957) Combustion rates of freely falling fuel drops in a hot atmosphere. The Sixth Symposium (International) on Combustion. The Combustion Institute, Reinhold, New York, pp. 717–725

Boothroyd RG (1971) Following gas-solids suspensions. Charman and Hall, London

Boussinesq J (1903) Theorie analytique de la chaleur. L'Ecole Polytechnique, Paris

Brabston DC, Keller HB (1975) Viscous flows past spherical gas bubbles. J. Fluid Mech. 69: 179–189

Briffa FEJ (1981). Transient drag in sprays. The Eighteenth Symposium (International) on Combustion. The Combustion Institute, Pittsburgh, Pa., pp. 307–319

Chan PC-H, Leal LG (1979) The motion of a deformable drop in a second-order fluid. J. Fluid Mech. 92: 131–170

Chang EJ, Maxey MR (1994) Unsteady flow about a sphere at low to moderate Reynolds number. Part 1. Oscillatory motion. J. Fluid Mech. 277: 347–379

Chang EJ, Maxey MR (1995) Unsteady flow about a sphere at low to moderate Reynolds number. Part 2. Accelerated motion. J. Fluid Mech. 303: 133–153

Chester W, Breach DR (with an Appendix by Proudman I.). (1969) On the flow past a sphere at low Reynolds number. J. Fluid Mech. 37: 751–760

Clamen A, Gauvin WH (1969) Effect of turbulence on the drag coefficient of spheres in a supercritical flow regime. AIChE J. 15: 184–189

Clift R, Grace JR, Weber ME (1978) Bubbles, drops and particles. Academic Press. New York

Crowe CT, Nicholls JA, Morrison RB (1963) Drag coefficients of inert and burning particles accelerating in gas streams. The Ninth Symposium (International) on Combustion. The Combustion Institute. Academic Press, New York, 395–406

Crowe CT, Sommerfeld M, Tsuji Y (1998) Multiphase flows with droplets and particles. CRC Press, New York

Dandy DS, Dwyer HA (1990) A sphere in shear flow at finite Reynolds number: effect of shear on particle lift, drag and heat transfer. J. Fluid Mech. 216: 381–410

Davies JM (1949) The aerodynamics of golf balls. J. Appl. Phys. 20: 821–828

Dennis SCR, Walker JDA (1971) Calculation of the steady flow past a sphere at low and moderate Reynolds numbers. J. Fluid Mech. 48: 771–789

Dwyer HA (1989) Calculations of droplet dynamics in high temperature environments. Prog. Energ. Combust. Sci. 15: 131–158

Eastop TD (1973) The influence of rotation on the heat transfer from a sphere to an air stream. Int. J. Heat Mass Transfer 16: 1954–1957

Eisenkalm, P., and Arunachalam, S.A. 1966. The drag resistance of burning drops. Combust. Flame, 10: 171–181

Eisenklam P, Arunachalam SA, Weston JA (1967) Evaporation rates and drag resistance of burning drops. The Eleventh Symposium (International) on Combustion. The Combustion Institute, Pittsburgh, Pa., pp. 715–728

Fendell FE, Coats DE, Smith EB (1968) Compressible slow viscous flow past a vaporizing droplet. AIAA J. 6: 1953–1960

Fendell FE, Sprankle ML, Dodson DS (1966). Thin-flame theory for a fuel droplet in slow viscous flow. J. Fluid Mech. 26: 267–280

Gal-Or B, Yaron I (1973) Diffusion drag upon slowly evaporating droplets. Phys. Fluids. 16: 1826–1829

Genkin AL, Gnatyuk TA, Yarin LP (1981) Distribution of concentration of particles polydispersed material in technological cyclone chambers. Theor. Found. Chem. Technol. 5: 787–791 (in Russian)

Gogos G, Ayyaswamy PS (1988) A model for the evaporation of a slowly moving droplet. Combust. Flame. 74: 111–129

Gogos G, Sadhal SS, Ayyaswamy PS, Sundararajan T (1986) Thin-flame theory for the combustion of a moving liquid drop: effects due to variable density. J. Fluid Mech. 171: 121–144

Goldshtik MA (1972) The elementary theory of pulverized layer. Zh. Prikl. Mekh. Tekhn. Fiz. 6: 106–112 (in Russian)

Goldstein S (1938) Modern developments in fluid dynamics. Oxford University Press, Oxford

Haber S, Hetsroni G (1971) The dynamics of a deformable drop suspended in an unbounded Stoks flow. J. Fluid Mech. 49: 257–277

Hadamard JS (1911) Mouvement permanent lend d'une sphere liquid et visqueuse dans un liquid visqueux. C.R. Acad. Sci. Paris 152: 1735–1738

Happel J, Brenner H (1983) Low Reynolds number hydrodynamics. Martinus Nijhoff, The Hague

Harper JF, Moore DW (1968) The motion of a spherical liquid drop at high Reynolds number. J. Fluid Mech. 32: 367–391

Hetsroni G (1982) Handbook of multiphase systems. Hemisphere, New York

Hjelmfelt AT, Mockros LF (1966) Motion of discrete particles in a turbulent fluid. Appl. Sci. Res. 16: 149–164

Ingebo RD (1957) Atomization, acceleration, and vaporization of liquid fuels. The Sixth Symposium (International) on Combustion. The Combustion Institute. Reinhold, New York, pp. 684–687

Ingebo RD (1962) Heat-transfer and drag coefficients for ethanol drops in a rocket chamber burning ethanol and liquid oxygen. The Eighth Symposium (International) on Combustion. The Combustion Institute. Williams and Wilkins, Baltimore, pp. 1104–1113

Kang SW, Sarofim AF, Beer JM (1988) Particle rotation in coal combustion: statistical experimental and theoretical studies. The Twenty-second Symposium (International) on Combustion. The Combustion Institute, Pittsburgh, Pa., pp. 145–152.

Karanfilian SK, Kotas TJ (1978) Drag on a sphere in unsteady motion in a liquid at rest. J. Fluid Mech. 87: 85–96

Kassoy DR, Adamson TCJr, Messiter AF (1966). Compressible low Reynolds number, flow around a sphere. Phys. Fluids. 9: 671–681

Khudyakov GN (1947) On the combustion of liquid droplets in flight. Isvest. Akad. Nauk SSSR. Otdel. Tekn. Nauk. 4: 508–513 (in Russian)

Kim I, Elghobashi S, Sirignano WA (1998) On the equation for spherical- particle motion: effect of Reynolds and acceleration numbers. J. Fluid Mech. 367: 221–253

Kim S, Karrila SJ (1991) Microhydrodynamics: Principles and Selected Applications. Butterworth – Heinemann, Boston

Kuchling H (1980) Nachschlagebucher fur Grundlagenfacher Physik. VERB Fachbuchverlag, Leipzig

Kurose R, Komori S (1999) Drag and lift forces on a rotating sphere in a linear shear flow. J. Fluid Mech. 384: 183–206

Landau LD, Lifschitz EM (1959) Fluid mechanics, 2nd edn. Pergamon, New York

Leal LG (1980) Particle motions in a viscous fluid. In: Van Dyke M, Wehausen JV Lumley JL (eds), Annu. Rev. Fluid Mech. 12: 435–476

Legendre D, Magnaudet J (1998) The lift force on a spherical bubble in a viscous linear shear flow. J. Fluid Mech. 368: 81–126

Levich VG (1949) Motion of gaseous bubbles with high Reynolds numbers Zh. Eksper. Teoret. Fiz. 19: 18–24. (in Russian), 1962. Physicochemical hydrodynamics. Prentice-Hall, Englewood Cliffs, NJ

Lozinski D, Matalon M (1992) Vapaorization of a spinning fuel droplet. The Twenty-fourth Symposium (International) on Combustion. The Combustion Institute, Pittsburgh, Pa., pp. 1483–1491

Maccoll JH (1928) Aerodynamics of a spinning sphere. J.R. Aero. Soc. 32: 777–798

Magnaudet J, Rivero M, Fabre J (1995) Accelerated flows past a rigid sphere or a spherical bubble. Part 1. Steady straining flow. J. Fluid Mech. 284: 97–135

Makino A (1992) Drag coefficient of a slowly moving carbon particle undergoing combustion. Combust. Sci. Techol. 81: 169–192

Maxey MR, Riley JJ (1983) Equation of motion for a small rigid sphere in a nonuniform flow. Phys. Fluids. 26: 863–889

McLaughlin JB (1991) Inertial migration of small sphere in linear shear flows. J. Fluid Mech. 224: 261–274

Mei R (1992) An approximate expression for the shear lift force on a spherical particle at finite Reynolds number. Int. J. Multiphase Flow 18: 145–147

Mei R (1994) Flow due to an oscillating sphere and an expression for unsteady drag on the sphere at finite Reynolds number. J. Fluid Mech. 270: 133–174

Mei R, Lawrence CJ, Adrian RJ (1991) Unsteady drag on a sphere at finite Reynolds number with small fluctuations in the free-stream velocity. J. Fluid Mech. 233: 613–631

Mitchell RE (1986) Experimentally determined overall burning rates of coal chars. In: The spring meeting of the west states section. The Combustion Institute, Banff, Canada. April 27–30

Moore DW (1959) The rise of a gas bubble in viscous liquid. J. Fluid Mech. 6: 113–130

Moore DW (1963) The boundary layer on a spherical gas bubble. J. Fluid Mech. 16: 161–176

Moore DW (1965) The velocity of rise of distorted gas bubbles in a liquid of small viscosity. J. Fluid Mech. 23: 749–766

Morsi SA, Alexander AJ (1972) An investigation of particle trajectories in two-phase flow systems. J. Fluid Mech. 55: 193–208

Muggia A (1956) Speed of evaporation and drag coefficient for a droplet in a gas stream. Aerotecnica Roma 36: 127–131

Natarajan R (1973) Experimental drag coefficients for evaporating and burning drops at elevated pressure. Combust. Flame 20: 199–209

Niksa S, Mitchell RE, Hencken KP, Tichenor DA (1984) Optically determined temperatures, sizes, and velocities of individual carbon particles under typical combustion conditions. Combust. Flame 60: 183–193

Odar F (1966) Verification of the proposed equation for calculation of the forces on a sphere accelerating in a viscous fluid. J. Fluid Mech. 25: 591–592

Odar F, Hamilton WS (1964) Forces on a sphere accelerating in viscous fluid. J. Fluid Mech. 18: 302–314

Oesterle B, Dinh TB (1998) Experiments on the lift of a spinning sphere in a range of intermediate Reynolds numbers. Exp. Fluid 25: 16–22

Ogasawara M, Adachi T, Yashiki T (1967) Study on the drag of cylinder and sphere with flames supported in air stream. Bull. JSME. 10: 825–832

Oliver DLP, Chung JN (1985) Steady flows inside and around a fluid sphere at low Reynolds numbers. J. Fluid Mech. 154: 215–230

Oliver DLR, Chung JN (1987) Flow about a fluid sphere at low to moderate Reynolds numbers. J. Fluid Mech. 177: 1–18

Oseen CW a. (1910). Über die Stoke's Formel, und über eine verwendte Aufgabe in der Hydrodynamik. Ark. Math. Astronom. Fys. 6: 29, 1–20; b (1927) Hydrodynamik. Akademische Verlagsgesellschaft Leipzig

Pearlman HG, Sohrab SH (1991) The role of droplet rotation in turbulent spray combustion modeling. Combust. Sci. Technol. 76: 321–334

Proudman J, Pearson JRA (1957) Expansions at small Reynolds numbers for the flow past a sphere and circular cylinder. J. Fluid Mech. 2: 237–367

Renksizbulut M, Haywood RJ (1988) Transient droplet evaporation with variable properties and internal circulation at intermediate Reynolds numbers. Int. J. Multiphase Flow 14: 189–202

Renksizbulut M, Yuen MC (1983a). Numerical study of droplet evaporation in a high temperature stream. J. Heat Transfer 105: 389–397

Renksizbulut M, Yuen MC (1983b) Experimental study of droplet evaporation in a high temperature air stream. J. Heat Transfer 105: 384–388

Rubinow SI, Keller JB (1961) The transverse force on a spinning sphere moving in a viscous fluid. J. Fluid Mech. 11: 447–459

Rudinger G (1980) Fundamentals of gas-particle flow. Handbook of powder technology vol.2. Elsevier, Amsterdam

Rudoff RR, Bachalo WD (1988) Measurements of droplet drag coefficient in polydispersed turbulent field. AIAA Paper 88 – 0235

Rybczynski W (1911) Uber die fortschreitende Bewegung einer flussingen Kugel in einem zahen Medium. Bull. Inst. Acad. Sci. Cracovie. ser A. 40–46

Sadhal SS, Ayyaswamy PS (1983). Flow past a liquid drop with a large non-uniform radial velocity. J. Fluid Mech. 133: 65–81

Sadhal SS, Ayyaswamy PS, Chung JN (1997) Transport phenomena with drops and bubbles. Springer, New York

Saffman PG (1965) The lift on a small sphere in a slow shear flow. J. Fluid Mech. 22: 385–400

Saffman PG (1968) Corrigendum to "The lift on a small sphere in a slow shear flow". J. Fluid Mech. 31: 624

Schiller L, Naumann A (1933) Uber die grundlegenden Berechnungen bei der Schwerkraftaufbereitung. Ver. Deut. Ind. 77: 318–320

Schlichting H (1960) Boundary layer theory. McGraw Hill, New York

Sedov LI (1993) Similarity and dimensional methods in mechanics, 10th edn., Boca Raton, Fla CRC Pres.

Soo SL (1990) Multiphase fluid dynamics. Science Press and Gower Technical, Beijing

Spalding DB (1954) Mass transfer in laminar flow. Proc. R. Soc. London, Ser. A. 221: 78–88

Stokes GG (1851) On the effect of internal friction of fluids on the motion of pendulums. Trans Cambridge Philos. Soc. 9: 8–106

Taylor TD, Acrivos AS (1964) On the deformation and drag of a falling viscous drop at low Reynolds number. J. Fluid Mech. 18: 466–476

Torobin LB, Gauvin WH (1961) The drag coefficient of single spheres moving in steady and accelerated motion in a turbulent fluid. AIChE J. 7: 615–619

Tsuji Y, Morikawa Y, Mizuno O (1985) Experimental measurement of the Mangus force on a rotating sphere at low Reynolds numbers. Trans. ASME, J. Fluids Eng. 107: 484–488

Uhlherr PHT, Sinclair CG (1970) The effect of free stream turbulence on the drag coefficient of spheres. Proc. Chemea. A conference convened by Australian National Committee of the Institution of Chemical Engineers and the Australian Academy of Science. Chatwood, Australia: Butterworths of Australia and the Institution of Chemical Engineers 70: 1–13

Voir DJ, Michaelides EE (1994) The effect of the history term on the motion of rigid spheres in a viscous fluid. Int. J. Multiphase Flow 20: 547–556

Warnica WD, Renksizbulut M, Strong AB (1994) Drag coefficients of spherical liquid droplets. Part 2. Turbulent gaseous fields. Exp. Fluids 18: 265–276

Williams A (1973) Combustion of droplets of liquid fuels: a review. Combust. Flame 21:1–31

Wohl PR, Rubinov SI (1974) The transfer force on a drop in an unbounded parabolic flow. J. Fluid. Mech. 62: 185–207

Yang JC (1993) Heterogeneous combustion. Environmental implications of combustion processes. In: Puri IK (ed) CBC Press, London 97–137

Yarin LP, Sukhov GS (1987) Fundamentals of combustion theory of two-phase media. Energoatomizdat, Leningrad (in Russian).

Yuen MC, Chen LW (1976) On drag of evaporating liquid droplets. Combust. Sci. Technol. 14: 147–154

Zarin NA, Nicholls JA (1971) Sphere drag in solid rockets non-continuum and turbulence effects. Combust. Sci. Technol. 3: 273–285

1.2 Heat and mass transfer

1.2.1 Heat and mass transfer coefficients

Dimensional analysis. Consider steady heat transfer from a spherical particle immersed in unbounded flow of a thermally conductive incompressible fluid. We will assume that the effect of buoyancy force is negligible, so that the heat transfer process occurs under the conditions corresponding to solely forced convection. As the first step to study this problem, we consider the functional dependence of the heat flux at the particle surface on the characteristic parameters. It is possible to assume that the flux depends on physical properties of the fluid, the relative velocity of the particle, its diameter, as well as on the temperature difference between the particle and the surrounding fluid

$$q = \Phi(\rho, \mu, d, u, \lambda, c_p, \Delta T) \tag{2.1}$$

Here q is the heat flux at the particle surface, ρ, μ, λ and c_p are the density, viscosity, thermal conductivity and specific heat of fluid, ΔT is the temperature difference between the particle and the fluid.

In accordance with the approach which was used to study the particle drag (see Chapt. 1.1), we assume that the product of the powers of the characteristic parameters of the problem is dimensionless. Then we have

$$\Pi_t = q^\alpha \rho^\beta \mu^\gamma d^\delta u^\varepsilon \lambda^\xi c_p^\eta \Delta T^\chi \tag{2.2}$$

where $\alpha, \beta, \gamma, \delta, \varepsilon, \xi, \eta, \chi$ are unknown exponents; Π_t is a dimensionless parameter.

Take into account that the dimensional parameters $q, \rho, \mu, d, u, \lambda, c_p$ and ΔT measured in units containing M, L, T,Q, θ (Q and θ are units of heat and temperature), namely:

$$\begin{aligned} &q:[QT^{-1}L^{-2}];\quad \rho:[ML^{-3}];\quad \mu:[L^{-1}MT^{-1}];\quad d:[L];\quad v:[LT^{-1}]; \\ &\lambda:[QT^{-1}\theta^{-1}L^{-1}]; c_p:[Q\theta^{-1}M^{-1}];\quad \Delta T:[\theta] \end{aligned} \tag{2.3}$$

Then we obtain five algebraic equations to determine the eight unknown powers in Eq. (2.2)

$$\begin{aligned} &\beta+\gamma-\eta=0 \\ &-2\alpha-3\beta-\gamma+\delta+\varepsilon-\xi=0 \\ &\alpha+\gamma+\varepsilon+\xi=0 \\ &\alpha+\xi+\eta=0 \\ &-\xi-\eta+\chi=0 \end{aligned} \tag{2.4}$$

Thus, the number of the exponents whose values may be chosen arbitrarily equals three. Using the procedure described in Chapt. 1.1, we find the dimensionless groups of the problem:

$$\mathrm{Nu}=\frac{hd}{\lambda}, \quad \mathrm{Re}=\frac{ud}{\nu}; \quad \mathrm{Pr}=\frac{\nu}{\alpha} \tag{2.5}$$

where $h = q/\Delta T$ is the heat transfer coefficient, $\alpha = \lambda/(\rho c_p)$ is the thermal diffusivity; Nu, Re and Pr are the Nusselt, Reynolds and Prandtl numbers, respectively. Equation (2.1) yields

$$F(\mathrm{Nu}, \mathrm{Re}, \mathrm{Pr}) = 0 \tag{2.6}$$

Solving Eq. (2.6) for Nu, we obtain the functional equation for the dimensionless heat transfer coefficient

$$\mathrm{Nu} = F(\mathrm{Re}, \mathrm{Pr}) \tag{2.7}$$

The explicit form of the dependence (2.7) is determined by solving the Navier–Stokes and energy equations. For flows of incompressible fluids with constant properties the momentum and the continuity equations can be integrated independently from the energy one. Then the energy equation should be solved with known velocity distribution

$$(\mathbf{v}\cdot\nabla)T = \alpha\nabla^2 T \tag{2.8}$$

where $\mathbf{v}$ is a known function of coordinates, in the steady state case.

The boundary conditions for Eq. (2.8) are:

$$r = r_s \qquad -\lambda \frac{\partial T}{\partial r} = h(T_s - T_\infty) \tag{2.9}$$

$$r \to \infty \qquad T \to T_\infty$$

Here subscripts s and ∞ refer to the particle surface and the ambient medium far from the particle, respectively.

In the simpliest case corresponding to motionless fluid at rest ($v = 0$) the solution of the problem (2.8)–(2.9) yields

$$\mathrm{Nu} = 2 \tag{2.10}$$

Correlations for the Nusselt (Sherwood) number. For creeping flows expressions for the Nusselt number are typically presented in the form of a series of the Peclet number. For example, Acrivos and Taylor (1962) using the method of matched asymptotic expansion obtained the following analytical expression for the Nusselt number at $\mathrm{Re} \ll 1$, $\mathrm{Pr} \ll 1$

$$\mathrm{Nu} = 2 + \mathrm{Pe} + \mathrm{Pe}^2 \ln \mathrm{Pe} + 0.829\,\mathrm{Pe}^2 + \frac{1}{2}\mathrm{Pe}^3 \ln \mathrm{Pe} \tag{2.11}$$

where $\mathrm{Pe} = \mathrm{PrRe}$ is the Peclet number.

Rimmer (1968) derived a relation for the Nusselt number which is valid for $\mathrm{Re} \ll 1$ and $\mathrm{Pr} \sim 1$. Using the approach by Proudman and Pearson (1957) to describe the flow field he obtained the following result

$$\mathrm{Nu} = 2 + \mathrm{Pe} + \mathrm{Pe}^2 \ln\mathrm{Pe} + f(\mathrm{Pr})\mathrm{Pe}^2 \tag{2.12}$$

where

$$f(\mathrm{Pr}) = \frac{1}{2}\{(2\,\mathrm{Pr}^2 - \mathrm{Pr} - 4\gamma - \frac{213}{40}) + 2(\mathrm{Pr}^3 - 3\,\mathrm{Pr} - 2)\ln \mathrm{Pr} - 2(\mathrm{Pr}+1)^2(\mathrm{Pr}-2)\ln(\mathrm{Pr}+1)\}$$

where $\gamma = 0.5772156...$ is the Euler number. The values of the Nusselt number calculated via Eq. (2.11) and Eq. (2.12) are close to each other: the difference does not exceed 5% at $0 < \mathrm{Re} < 1$.

Acrivos and Goddard (1965) found the asymptotic Nusselt number for small and moderate Reynolds number and $\mathrm{Pe} \to \infty$.

$$\mathrm{Nu} = 2\mathrm{Pe}^{\frac{1}{3}}\left\{0.6245 + 0.461\mathrm{Pe}^{\frac{1}{3}} + O(\mathrm{Pe}) + O(\mathrm{Pe}^{-\frac{1}{3}})\right\} \tag{2.13}$$

A numerical study of the heat transfer from a spherical particle (0.1<Re<40, Pr~0.11) was performed by Dennis et al. (1973) who used the full Navier–Stokes equations. The comparison of these results with the Rimmer's theory shows that they are almost identical at $\mathrm{Re} \ll 1$.

A number of empirical correlations have been proposed for evaluation of the Nusselt number in a wide range of the Reynolds and Prandtl numbers (see Soo 1990). In particular, we mention the correlation valid in the range $1<\mathrm{Re}<7\cdot10^4$, $0.6<\mathrm{Pr}<400$

$$\mathrm{Nu} = 2 + 0.459\,\mathrm{Re}^{0.55}\mathrm{Pr}^{0.33} \tag{2.14}$$

In the case of mass transfer temperature is replaced by concentration of liquid vapor or a gas. The mass flux depends of the values of the mass transfer coefficient h_m replaeing h. Also for the dimensionless mass transfer coefficient the Sherwood number replaces the Nusselt number.

Heat and mass transfer from droplets or bubbles depend not only on the physical properties of the carrier fluid and relative velocity but also on circulatory flow inside the droplets and their deformation. For a spherical droplet settling with a constant velocity in a continuous fluid ($\mathrm{Re} \ll 1$) Levich (1962) derived the following expression for the Sherwood number

$$\mathrm{Sh} = \frac{2}{\sqrt{6\pi}}\mathrm{Pe}_*^{\frac{1}{2}}\left(\frac{\mu_1}{\mu_1+\mu_2}\right)^{\frac{1}{2}} \tag{2.15}$$

where $\mathrm{Sh} = (h_m r_o)/D$ is the Sherwood number, r_o is the particle radius, h_m is the mass transfer coefficient, the diffusional Peclet number $\mathrm{Pe}_* = \mathrm{Sc}\cdot\mathrm{Re}$, $S_c = \nu/D$ is the Schmidt number, μ_1 and μ_2 are the viscosities of the continuous and dispersed phases, respectively, ν and D are the kinematic viscosity and mass diffusivity.

The effect of the internal circulation and droplet deformation on the mass transfer was considered in detail by Gupalo et al. (1985). For a nondeformable droplet they obtain in the case of large Pe_*

$$\mathrm{Sh} = \sqrt{\frac{2}{3\pi}}\left\{\frac{1}{\mu_{2.1}+1} + \frac{3\mu_{2.1}+2}{\mu_{2.1}+1}\frac{\mathrm{Re}}{8}\right\}^{\frac{1}{2}} \mathrm{Pe}_*^{\frac{1}{2}} + 0.41\left(\frac{3}{4}\mu_{2.1}+1\right) + \phi(\mu_{2.1}, \mathrm{Re}) \tag{2.16}$$

where $\mu_{2.1} = \mu_2/\mu_1$, $\phi(\mu_{2.1}, \mathrm{Re})$ is a tabulated function which changes in the range $0 < \phi(\mu_{2.1}, \mathrm{Re}) < 0.2$ at $0 < \mathrm{Re} < 10$ and $\mu_{2.1}\mathrm{Pe}_*^{\frac{1}{2}} < O(1)$ at Re=O(1). The value of $\phi(\mu_{2.1}, \mathrm{Re})$ increase with $\mathrm{Re} - [\phi(\mu_{2.1}, 10) = 0.32]$.

Mass transfer from a deformable droplet at large Pe_* is described as follows. In the case of small Re and We

$$\mathrm{Sh} = \left[1 + \delta \mathrm{We}\frac{3}{2}\frac{\mu_{2.1}-1}{\mu_{2.1}+1}\right]\mathrm{Sh}_{sp} \tag{2.17}$$

where $\mathrm{Sh}_{sp} = (2\,\mathrm{Pe}_*)/(3\pi(\mu_{2.1}+1))$ is the Sherwood number for a spherical particle at the same conditions, $\mathrm{We} = (\rho u^2 r_o)/\sigma_t$ is the Weber number, σ_t is the surface tension,

$$\delta = \frac{1}{4(\mu_{2.1}+1)^3}\left\{\left(\frac{81}{80}\mu_{2.1}^3 + \frac{57}{20}\mu_{2.1}^2 + \frac{103}{40}\mu_{2.1} + \frac{3}{4}\right) - \frac{\rho_{2.1}-1}{12}(\mu_{2.1}+1)\right\}, \quad \rho_{2.1} = \frac{\rho_2}{\rho_1}.$$

At $\mu_{2.1} < 1$ (small viscosity of the dispersed phase) droplet deformation leads to a decrease in Sh, whereas at $\mu_{2.1} > 1$ it leads to an increase in Sh.

For bubbles $(\mu_{2.1} \to 0, \quad \rho_{2.1} \to 0)$ the Sherwood number is given by

$$\mathrm{Sh} = (1 - \frac{1}{2}\mathrm{We})\mathrm{Sh}_{sp} \tag{2.18}$$

For $\mathrm{Re} \to \infty$, as in the potential flow about a bubble,

$$\mathrm{Sh}_\infty = \sqrt{\frac{2}{\pi}}\Omega(\chi)\chi^{-\frac{1}{3}}\mathrm{Pe}_*^{\frac{1}{2}} \tag{2.19}$$

where

$$\Omega(\chi) = \sqrt{\frac{2}{3}}\frac{(\chi^2-1)^{\frac{3}{4}}}{\chi^{\frac{2}{3}}}\left[\arctan(\chi^2-1)^{\frac{1}{2}} - \frac{(\chi^2-1)}{\chi^2}\right]^{-\frac{1}{2}}$$

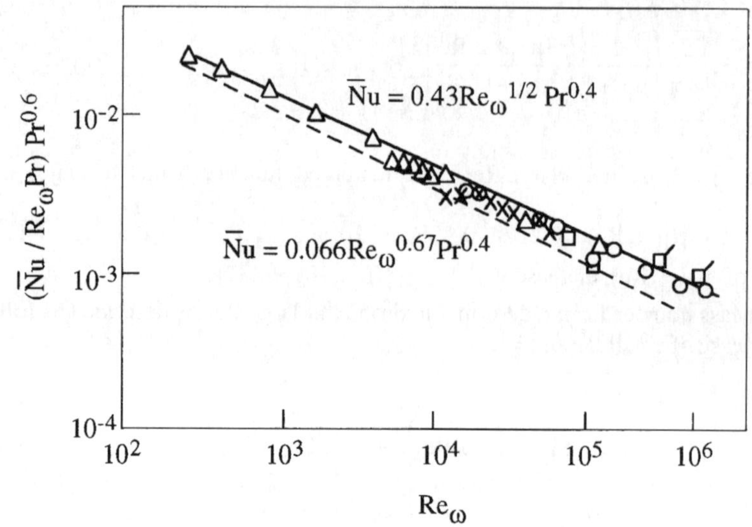

Fig. 2.1 Convective heat transfer from a rotating sphere. Reprinted from Kreith (1968), with permission. Δ, cooling in oil (Pr=217, d=2 in), O, cooling in water (Pr=4.52, d=2 in), ⊗, heating in water (Pr=4.52, d=2 in), □, cooling in water (Pr=4.52, d = 6 in), □́, cooling in water in a small tank (Pr=4.52, d=6 in), ×, cooling in air (Pr = 0.72, d = 6 in)

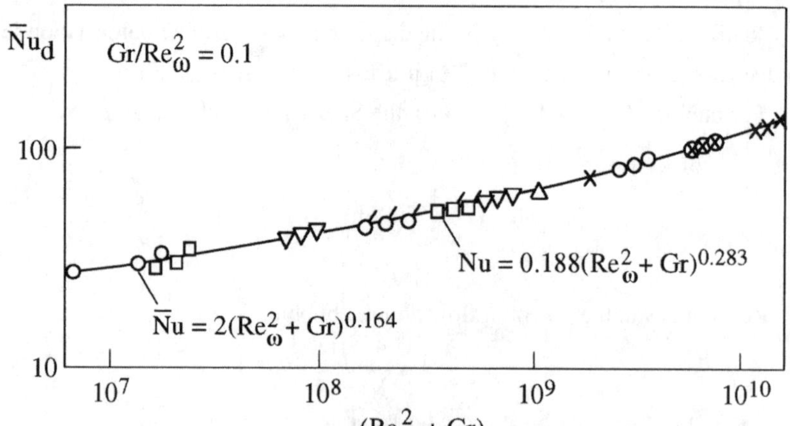

Fig. 2.2 Experimental data for convective heat transfer from a rotating sphere. Reprinted from Kreith (1968), with permission. O, natural convection; □, ○́, ⌀, ▽, ◇, ▷, △, × , mixed convection

and χ is the ratio of the large to small semiaxes of spheroid which approximates the bubble shape.

The numerical calculations of Gupalo et al. (1985) show that at large but finite Reynolds numbers and $0 < \mathrm{We} \le 1.5$ the Sherwood number average over the droplet surface can be presented as follows

$$\mathrm{Sh} = \sqrt{\frac{2}{3}}\left\{(1-\mathrm{Re}^{-\frac{1}{2}}) + (0.003 - 0.2\,\mathrm{Re}^{-\frac{1}{2}})\mathrm{We}\right\}\mathrm{Pe}_*^{\frac{1}{2}} \tag{2.20}$$

It should be noted that for practical calculations of Sh in the range of the Reynolds and Schmidt numbers $1 < \mathrm{Re} < 10^4$, $0.6 < \mathrm{Sc} < 400$, empirical correlations similar to Eq. (2.14) are used.

The effect of particle rotation on the Nusselt number. Convective heat transfer of a spinning sphere in a moveless or flowing fluid was studied by a number of investigators during the last 150 years (Kreith 1968). The results obtained show that the effect of rotation on heat transfer of a spherical particle is due to the influence of the centrifugal force on fluid around the particle. The particle rotation about the axis through its poles promotes secondary currents that are directed towards the particle poles and outwards from the equatorial region. The superposition of the secondary currents on the main flow (due to the inertial and buoyancy and pressure forces) determines the hydrodynamical structure of the flow and its evolution, as well as the general characteristics of heat and mass transfer.

Therefore the intensity of the heat transfer on a spinning spherical particle is determined by the following three nondimensional parameters accounting for the effect of the inertial, buoyancy and centrifugal forces:

$$\mathrm{Re} = \frac{u_\infty d}{\nu}, \quad \mathrm{Gr} = \beta g (T_s - T_\infty)\frac{d^3}{\nu^2}, \quad \mathrm{Re}_\omega = \frac{\omega d^2}{\nu} \tag{2.21}$$

Here u_∞ is the velocity of the undisturbed flow, ω is the angular speed of rotation, β is the thermal expansion coefficient, g is the gravity acceleration.

Depending on the relative importance these parameters, different types of heat transfer can be realized (natural convective, mixed or predominantly forced types).

To describe the heat and mass transfer of a spinning spherical particle, typically approaches based on the boundary layer theory are used (Schlichting 1960). A summary of various methods of calculation of the heat and mass transfer in rotating systems, as well as the generalized data of a number of experimental investigations is contained in the monograph by Dorfman (1963) and in the survey by Kreith (1968) (application of the boundary layer theory for studying the convective heat transfer over rotating bodies is also discussed in Dorfman and Serazeldinov (1965), Banks (1965), Chao and Greif (1974), Lee et al. (1978)).

Some experimental data on the heat transfer from a spinning spherical particle to air for natural, mixed and forced convection are presented in Figs. 2.1 and 2.2 (the data by Nordle and Kreith 1961 and Kreith et al. 1963). On these graphs the empirical correlations for the average Nusselt number are also plotted. Under the conditions of forced convection the heat transfer coefficient depends on the particle rotational Reynolds number Re_ω. The dependence on Re_ω is different for ranges of low and high values of Re_ω. It may be expressed as

$$\overline{Nu} = 0.43\, Re_\omega^{0.5}\, Pr^{0.4} \tag{2.22}$$

for $Re_\omega < 5 \cdot 10^5$, and

$$\overline{Nu} = 0.066\, Re_\omega^{0.67}\, Pr^{0.4} \tag{2.23}$$

for $Re_\omega > 5 \cdot 10^5$ (cf. Fig. 2.1).

The difference of the exponents of Re_ω in Eqs. (2.22) and (2.23) is due to the effect of the intensive equatorial current at high spinning speed. In the case of mixed convection, the Nusselt number depends on Re_ω and Gr. This dependence is shown in Fig. 2.2 in the form of $\overline{Nu}$ as a function of the complex $(Re_\omega^2 + Gr)$. It is described by the following correlation

$$\overline{Nu} = 2(Re_\omega^2 + Gr)^{0.164} \tag{2.24}$$

For $(Re_\omega^2 + Gr) < 10^8$, and by

$$\overline{Nu} = 0.188(Re_\omega^2 + Gr)^{0.283} \tag{2.25}$$

for $(Re_\omega^2 + Gr) > 10^8$ (cf. Fig. 2.2)

The empirical correlations for the average Nusselt number of a spherical particle rotating in an air at rest or in an air stream were suggested by Eastop (1973). In the case of air at rest

$$\overline{Nu} = 0.353\, Re_\omega^{0.5} \tag{2.26}$$

whereas for a particle rotating in an air stream

$$\overline{\mathrm{Nu}} = 0.288\,\mathrm{Re}_{\omega}^{0.6}\left\{1 + 0.167\left(\frac{\mathrm{Re}_{\omega}}{\mathrm{Re}} - 0.54\right)\right\} \tag{2.27}$$

The latter is valid at $(\mathrm{Re}_{\omega}/\mathrm{Re}) > 0.54$.

The comparison of a number of theoretical predictions with experimental data on heat transfer of a spherical particle at forced convection was done by Hussaini and Sastry (1976). They showed that all the dependences $\overline{\mathrm{Nu}}(\mathrm{Re}_{\omega}, \mathrm{Pr})$ proposed by a number of authors are given by $\overline{\mathrm{Nu}} = A\,\mathrm{Re}_{\omega}^{0.5}\,\mathrm{Pr}^{0.4}$, with only the factor A being different. For example, according to Banks (1965), Dorfman and Mironova (1970), Dorfman and Serazetdinov (1965) and Hussaini and Sastry (1976) the factor A equals 0.25; 0.252; 0.276; 0.291, respectively, whereas the experimental value of A is 0.373.

Heat transfer characteristics for the case of laminar natural, mixed, and forced convection about a spinning spherical particle were considered in Chao and Greif (1974), Suwono (1980), Rajasekaran and Palekar (1985) and Tieng and Yan (1993). Chao and Greif (1974) solved the problem of heat transfer of a rotating isothermal sphere with and without a uniform stream along the axis of rotation. They derived the following relation for the local Nusselt number

$$\mathrm{Nu}\left(\frac{r_o^2\omega}{\nu}\right)^{-\frac{1}{2}}\mathrm{Pr}^{-\frac{1}{3}} = 1.108\begin{Bmatrix}1 - 0.2068\phi^2 - 0.0056\phi^4 - \\ -0.0024\phi^6 - \ldots\end{Bmatrix}\left(\frac{\partial\theta}{\partial\eta}\right)_{\eta=0} \tag{2.28}$$

where $\phi = x/r_o$, x is the arclength coordinate along the generatrix, r_o is the sphere radius, $\theta = (T - T_{\infty})/(T_s - T_{\infty})$ is the dimensionless temperature, T_s and T_{∞} are the temperatures of the particle surface and of the undisturbed fluid, respectively, $\eta = zg(x)$, z is the coordinate normal to the surface, g(x) is the function of x which is found during solution of the problem, $-(\partial\theta/\partial\eta)_{\eta=0}$ is a function of ϕ and Pr. It tends to 1.1199 as $\mathrm{Pr} \to \infty$.

Since $(-\partial\theta/\partial\eta)_{\eta=0}$ is constant as $\mathrm{Pr} \to \infty$, the limiting value of the local Nusselt number is proportional to $\mathrm{Pr}^{1/3}$. The change of the heat transfer coefficient over an isothermal spherical particle rotating in a quiescent medium and in a fluid flow is illustrated in Figs. 2.3 and 2.4. In both cases the maximum of the heat transfer coefficient corresponds to the stagnation point $\phi = 0$. For $\phi > 0$ the Nusselt number gradually decreases with ϕ. The effect of rotation on the heat transfer coefficient is characterized by the rotation parameter $\lambda_{\omega} = ((2r\omega)/(3u_{\infty}))^2 = (\mathrm{Re}_{\omega}/(3\,\mathrm{Re}_{\infty}))^2$, where $\mathrm{Re}_{\infty} = (u_{\infty}d)/\nu$. An increase of

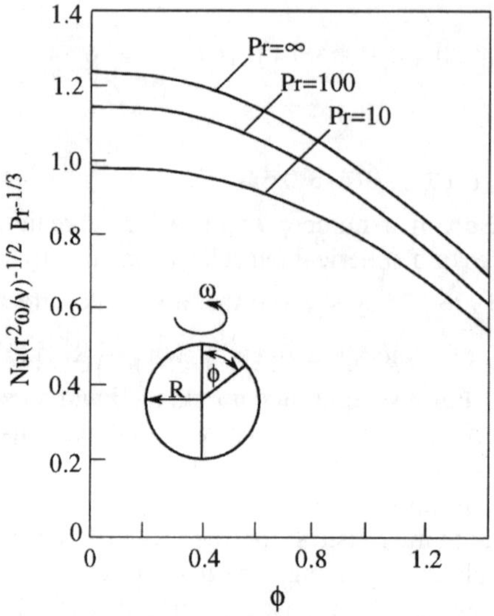

Fig. 2.3 Local Nusselt number over an isothermal sphere rotating in a quiescent fluid. Reprinted from Chao and Greif (1974), with permission

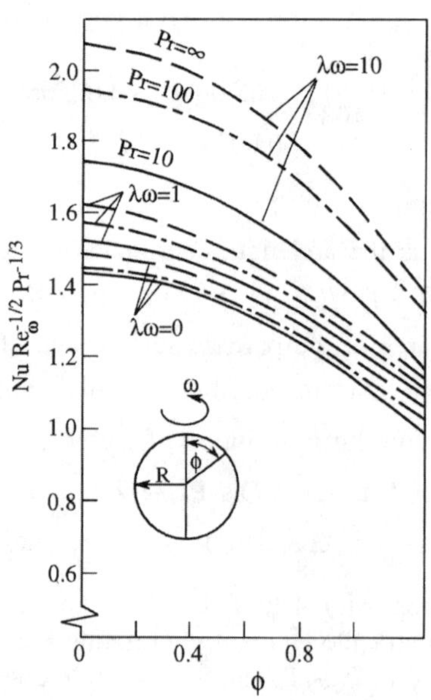

Fig. 2.4 Local Nusselt number over an isothermal sphere rotating in a fluid flow. Reprinted from Chao and Greif (1974), with permission. - - - -, Pr = ∞ ; — · —, Pr = 100; ——, Pr = 10

particle rotation speed promotes secondary flow and thus leads to an increase in the heat transfer coefficient.

The theoretical analysis of the heat transfer characteristics under the conditions corresponding to a joint effect of buoyancy and centrifugal forces was given by Rajasekaran and Palekar (1985). In this work the numerical calculations for the flow about a rotating sphere were performed for two kinds of thermal conditions on the particle surface: (i) uniform wall temperature, and (ii) uniform surface heat flux. It was found that the local Nusselt number increased with the buoyancy. For an equivalent buoyancy effect, the Nusselt number is larger in case (ii) than in case (i). The ratio Nu_f/Nu_T (Nu_f and Nu_T are the Nusselt numbers for constant heat flux and constant wall temperature, respectively) increases with increasing of the rotation speed.

A detailed experimental study of the heat transfer of a spinning particle at $0 < Re_\omega < 33320$ was carried out by Tieng and Yan (1993). They used holographic interferometry to observe the flow structure in the vicinity of a spherical particle under the conditions of interaction of the centrifugal and buoyancy currents. The measurements of the heat transfer were performed for conditions corresponding to laminar mixed and forced convection. It was found that the buoyancy effect is dominant at $Re_\omega < 10^3$ when the intensity of the heat transfer is determined by natural convection. In the range $10^3 < Re_\omega < 9\cdot 10^3$ a sharp change of the flow structure is observed. It manifests itself in formation of a turbulent zone and jet eruption which affect the heat transfer process. At $Re_\omega > 9\cdot 10^3$ the forced convection is dominant. In this case the contribution of the jet eruption to the heat transfer is essential.

For calculations of the average Nusselt number for mixed convection, Tieng and Yan (1993) suggested the following relation which is valid at $Pr = 0.71$:

$$\overline{Nu}^3 = \overline{Nu_N}^3 + \overline{Nu_F}^3 \tag{2.29}$$

where $\overline{Nu}$, $\overline{Nu_N}$ and $\overline{Nu_F}$ are the overall Nusselt number and the average Nusselt numbers for the natural and forced convection, respectively, and are given by

$$\overline{Nu_N} = 2 + 0.392 Gr^{0.31} \tag{2.30}$$

at $1 < Gr > 10^5$

$$\overline{Nu_F} = 2 + 0.175 Re_\omega^{0.583} \tag{2.31}$$

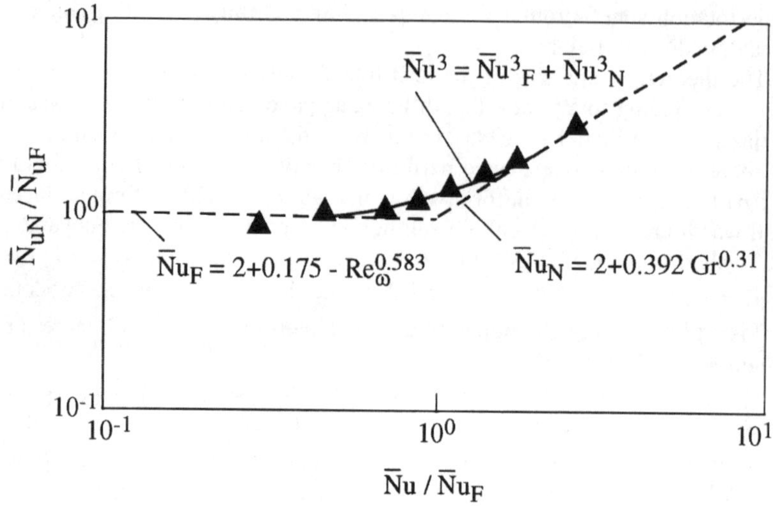

Fig. 2.5 The dependence for $\overline{Nu} / \overline{Nu}_F$ Reprinted from Tieng and Yan (1993), with permission. ----, asymptotes; ▲, data of Tieng and Yan (1993)

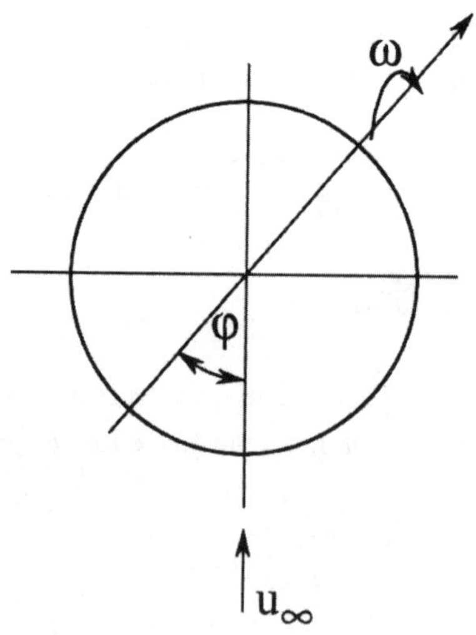

Fig. 2.6 Scheme of the flow

at $Gr=0$, $10< Re_{\omega} <10^4$, Pr=0.71. Fig.ure 2.5 shows that the relations (2.29)–(2.31) agree fairly well with the experimental data in the range $2\cdot10^{-1} <(\overline{Nu}/\overline{Nu}_F)<3$.

The laminar mixed convection about an isothermal spinning sphere in a flow directed arbitrarily to the axis of rotation was studied by Palec and Daguenet (1987) (Fig. 2.6). In these experiments the regimes close to forced and mixed convection were realized. The effect of the particle rotation on the mass transfer coefficient is illustrated in Fig. 2.7. An increase of the rotation parameter λ_{ω} leads to a noticeable increase of the average Sherwood number. At the same time, the effect of the inclination of the particle axis is very weak as Fig. 2.8 shows. The data show that for φ in the range of 0 to 45° the value of the $\overline{Nu}\,Re_{\infty}^{-\frac{1}{2}}$ changes only by 2%.

The effect of injection and suction on the heat transfer characteristics of rotating spherical particles was considered by Lien et al. (1986) and Hatzikonstantinou (1990). They showed that the heat transfer of a spinning particle depends essentially on the rate of injection/suction. The local Nusselt number increases with suction, while it decreases with injection.

Effect of temperature difference. A finite temperatures difference between the particle and surrounding fluid is the cause of changes of the physical properties of the fluid within the particle boundary layer. Kassoy et al. (1966) estimated the rate of the heat transfer from a spherical particle in compressible low Reynolds number flow. Using the method of matched asymptotic expansions, they calculated the Nusselt number for small $\left((T_p-T_{\infty})/T_{\infty}=O(Re),\quad Re\ll1\right)$ and moderate $\left((T_p-T_{\infty})/T_{\infty}=O(1)\right)$ temperature differences. For flows with small Mach numbers they obtained the following relations for the heat transfer coefficient

$$Nu=2+Re\,Pr\left(1-\frac{\psi-1}{Re\,Pr}\right)+O\left(Re^3\ln Re)\right) \tag{2.32}$$

for $\psi=1+O(Re)$, and

$$Nu=\frac{1+\psi}{2\psi}\left[2+Re\,Pr+O(Re^2)\right] \tag{2.33}$$

for $\psi=1+O(1)$. Here $\psi=T_p/T_{\infty}$. In both cases, an increase in the particle temperature is accompanied by a decrease in the heat transfer coefficient. The rate of decrease in the Nusselt number is different for small and moderate values of the ratio $(T_p-T_{\infty})/T_{\infty}$. In the former case $dNu/d\psi=-1$, whereas in the latter case

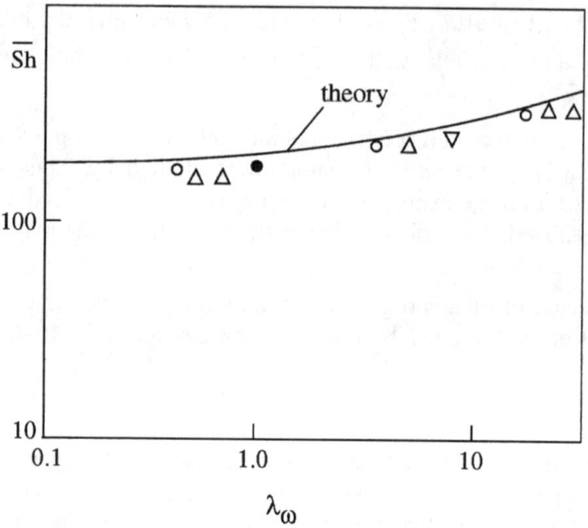

Fig. 2.7 The effect of particle rotation on the mass transfer coefficient Reprinted from Palec and Daguent (1987), with permission. $\overline{\mathrm{Sh}}$ is the average Sherwood number. ∇, $\varphi = 0°$; ▲, $\psi = 10°$; O, $\varphi = 15°$; ●, $\varphi = 20°$; $\mathrm{Gr}/\mathrm{Re}_{\infty}^{2} = 2750$

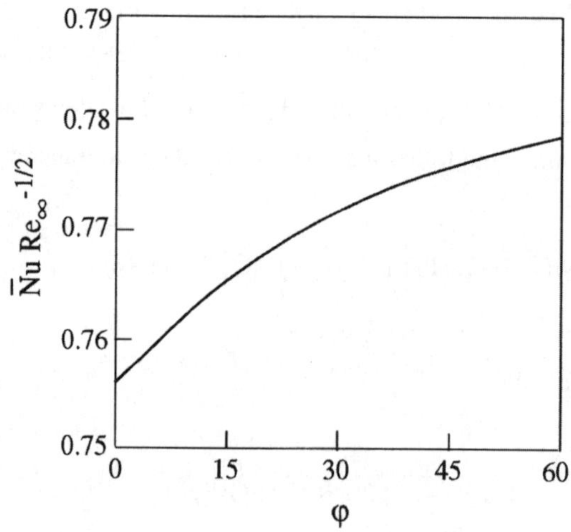

Fig. 2.8 The average Nusselt number versus φ. Reprinted from Palec and Daguent (1987), with permission. $\mathrm{Pr} = 1$, $\mathrm{B} = 1$, $\mathrm{Gr}/\mathrm{Re} = 0$

$dNu/d\psi = -\left[2 + Re\,Pr + O(Re^2)\right]/24$. At a sufficiently high particle temperature $(\psi \sim 2)$, the derivative $d\,Nu/d\,\psi = -1/2$, i.e. it approaches one-half of the value corresponding to heating of a small particle.

Heat transfer to a spherical particle immersed in a high temperature flow was studied numerically by Sayegh and Gauvin (1979). The calculations showed that at $0<Re<50, 0.25 < \psi < 1$ the general patterns of fluid and temperature fields (the streamlines and isotherms in the vicinity the particle, the vortical structure near the rear part of the particle) do not change significantly with ψ. At the same time, a number of flow characteristics, such as the velocity and temperature distributions near the particle, fluid recirculation in the wake, etc., undergo noticeable changes in comparison with the isothermal flow. This leads to changes in both particle drag and heat transfer coefficient. The latter is expressed as

$$Nu = 2fo + 0.473\,Pr^m\,Re_{0.19}^{0.552} \tag{2.34}$$

where Nu is the overall Nusselt number, $m = 0.78 \cdot Re_{0.19}^{-0.145}$, $2fo = (2(1-\psi^{1+x}))/((1+x)(1-\psi)\psi^x)$, x is the value of the exponent of T, in the viscosity and thermal conductivity λ; the reference temperature for λ is the particle temperature dependence, and that for ν is $T_{0.19} = T_p + 0.19\,(T_\infty - T_p)$. Accordingly, the Nusselt number is defined by the particle temperature, whereas the Reynolds number $Re_{0.19}$ is defined at $T_{0.19}$.

It is emphasized that usage of the parameter values based on the mean film temperature leads to errors as high as 20% for the Nusselt number.

From the correlation (2.34) it follows that the heat transfer coefficient increases with particle cooling.

Turbulence effects. Free stream turbulence tends to enhance heat transfer from a particle to the surrounding fluid because of the effect of the eddies penetrating from the external flow into the particle boundary layer. This disturbs the near-wall flow, facilitating transition to turbulence and shifting the separation point of the boundary layer downstream. The characteristic parameters determining heat transfer from a spherical particle immersed in a turbulent flow are

$$h,\ \rho,\ \mu,\ v,\ d,\ \lambda,\ c_p,\ \tilde{v}',\ \ell \tag{2.35}$$

where v' and ℓ are the velocity fluctuation ($\tilde{v}' = \sqrt{v'^2}$ is the root mean square of the turbulence fluctuations) in the carrier fluid and integral scale of turbulence, respectively.

The dimensional analysis yields the Nusselt number depending on four dimensionless groups which account for the parameters of the average and fluctuating motions

$$\mathrm{Nu} = \mathrm{F}(\mathrm{Re}, \mathrm{Pr}, \mathrm{T_u}, \bar{\ell}) \tag{2.36}$$

where $T_u = \sqrt{\overline{v'^2}}/v$ is the turbulence intensity, $\bar{\ell} = \ell/d$ is the dimensionless scale of turbulence.

The relation (2.36) shows that the contribution of turbulence to the heat transfer is determined by two dimensionless parameters accounting for the turbulence intensity and the size of turbulent eddies.

Depending on the values of these parameters, different conditions of the heat transfer may be realized. When $\bar{\ell} \gg 1$, the particle experiences a time-dependent flow, whereas when $\bar{\ell} \ll 1$ the flow becomes quasi-steady. In the latter case the heat transfer coefficient depends on the Reynolds and the Prandtl numbers, as well as on the turbulence intensity T_u (Kestin 1966)

$$\mathrm{Nu} = \mathrm{F}(\mathrm{Re}, \mathrm{Pr}, \mathrm{T_u}) \tag{2.37}$$

In particular for a spherical particle immersed in the turbulent flow $(2\cdot 10^3 < \mathrm{Re} < 6.5\cdot 10^4,\ 1 < T_u < 17\%)$ Eq. (2.37) takes the form (Lavender and Pei 1967)

$$\mathrm{Nu} = 2 + \mathrm{f}(\mathrm{Re_T})\mathrm{Re}^{0.5} \tag{2.38}$$

where $\mathrm{Re_T}=T_u\mathrm{Re}$ is the turbulent Reynolds number; $\mathrm{f}(\mathrm{ReT_u})=0.629\,\mathrm{Re_T^{0.035}}$ for $\mathrm{Re_T} < 10^3$ and $\mathrm{f}(\mathrm{Re_T}) = 0.145\,\mathrm{Re_T^{0.25}}$ for $\mathrm{Re_T}>10^3$.

The difference in the dependences $\mathrm{f}(\mathrm{Re_T})$ corresponding to small and large values of $\mathrm{Re_T}$ are related to the peculiarities of flows about a particle at low and high turbulence intensity. The influence of the free stream turbulence on the particle boundary layer is weak enough at low T_u. When turbulent disturbances reach some critical value, laminar/turbulent transition in the particle boundary layer occurs. It is accompanied by significant changes in the flow structure, reducing dramatically the particle drag. The measurements by Lavender and Pei (1967) showed that the critical value of the turbulent Reynolds number corresponding to the transition is about 10^3 (Fig. 2.9).

Raithby and Eckert (1968) studied the effect of both turbulence intensity and scale on the heat transfer from a spherical particle. These experiments confirmed the previous results and showed that the free stream turbulence affects essentially the heat transfer coefficient. For example, an increase in the turbulence intensity of 5% causes an increase in the Nusselt number of 7.5 and 17.5% for $\mathrm{Re}=3.6\cdot 10^3$ and $5\cdot 10^4$, respectively. It was also shown that the heat transfer rate decreases with $\bar{\ell}$ (cf. Fig. 2.10). This effect is more pronounced at high values of the Reynolds numbers and turbulent intensity.

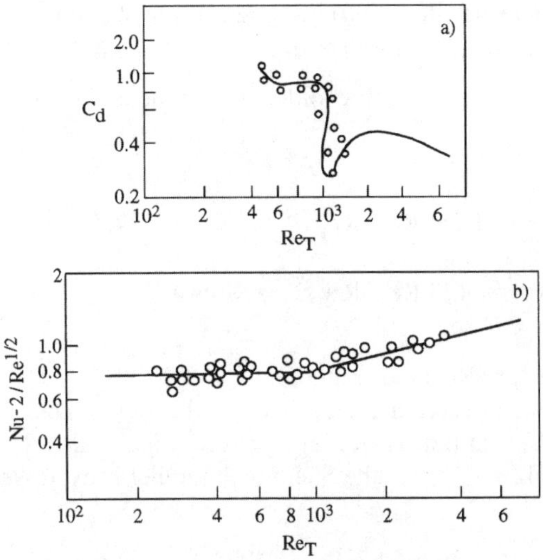

Fig. 2.9 The effect of the free stream turbulence on the drag and the heat transfer coefficient of a spherical particle. Reprinted from Lavender and Pei (1967), with permission. a) Dependence of the drag coefficient on the turbulent Reynolds number. b) Dependence of the Nusselt number on the turbulent Reynolds number

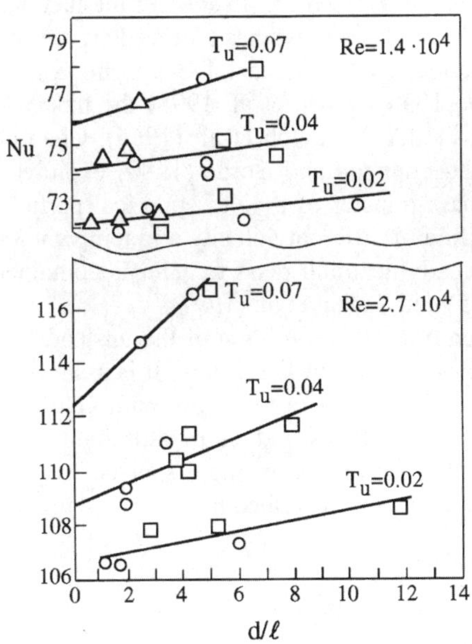

Fig. 2.10 The effect of the turbulence scale on the heat transfer coefficient Reprinted from Raithby and Eckert (1968), with permission. □, d=5.08 cm; ○, d=2.54 cm; Δ , d=1.27 cm.

Gostowski and Costello (1990) investigated the heat transfer from a spherical particle in a turbulent flow with $1 < T_u < 36\%$. For the low ($Re_T < 7 \cdot 10^3$) and high ($Re_T > 7 \cdot 10^3$) turbulent Reynolds number the authors obtained the following relations

$$Nu = 1.255\, Re^{0.5}\, Re_T^{0.0214} \quad \text{at } Re_T < 7 \cdot 40^3 \tag{2.39}$$

$$Nu = 1.128\, Re^{0.5}\, R\, e_T^{0.283} \quad \text{at } Re_T > 7 \cdot 70^3 \tag{2.40}$$

Here Re is the ordinary Reynolds number.

The correlations describing the effect of the free stream turbulence on the mass transfer coefficient have a similar structure (Sandoval-Robles et al. 1981); in particular, for $0.042 < T_u < 0.3$ the Sherwood number is expressed as

$$Sh = 6.82\ Re^{0.559} \cdot T_u^{\ 0.069} \tag{2.41}$$

where $T_u = (\overline{v_r'^2} + \overline{v_\varphi'^2} + \overline{v_\theta'^2})^{1/2} / (3v)$, v_r', v_φ' and v_θ' are the components of the fluctuation velocity in r, φ and θ directions, respectively.

Unsteady heat and mass transfer. In spite of the fact that the problem of heat and mass transfer in multiphase flows has received significant attention for a long time and many theoretical and experimental results are available (Boothroyd 1971, Clift et al. 1978, Soo 1990, Crowe et al. 1998) the unsteady counterpart of this problem is much less studied. Levich et al. (1965), Chao (1969), Konopliv and Sparrow (1972), and Abramzon and Borde (1980) explored some aspects of the unsteady heat and mass transfer of a solid spherical particle or a drop suddenly subject to a flow of fluid of different velocity and temperature. A survey of the results concerning unsteady heat and mass transfer is contained in the monographs of Gupalo et al. (1985) and Sadhal et al. (1997).

Below we sketch briefly the problem of the unsteady heat and mass transfer from spherical particles or drops in a gas flow. It is assumed that a fine particle (a dispersed phase) is suddenly subject to a flow with velocity $v_{1.0}$ and temperature $T_{1.0}$. Let the initial particle velocity and temperature be $v_{2.0}$ and $T_{2.0}$. As a result of the dynamic and thermal interaction of the dispersed and continuous phases, unsteady heat transfer occurs. It is described by the following system of equations

$$\frac{\partial \mathbf{v}_i}{\partial t} + (\mathbf{v}_i \cdot \nabla)\mathbf{v}_i = -\frac{1}{\rho_i} \nabla P_i + \nu_i \nabla^2 \mathbf{v}_i \tag{2.42}$$

$$\nabla \cdot \mathbf{v}_i = 0 \tag{2.43}$$

$$\frac{\partial T_i}{\partial t} + (\mathbf{v}_i \cdot \nabla) T_i = \alpha_i \nabla^2 T_i \tag{2.44}$$

where $\rho, \mathbf{v}$ and T are the density, velocity and temperature (bold denotes vectors), $\mathbf{v}$ has components v_r, v_φ and v_θ (r, φ and θ are the spherical coordinates associated with the particle center), P is the pressure, ν and α are the kinematic viscosity and thermal diffusivity, subscripts 1 and 2 refer to the continuous and dispersed phases, respectively; Eqs. (2.42)–(2.44) are written for ρ_i, ν_i and α_i constant.

In the frame of the model of the incompressible fluid with constant transfer coefficients the momentum equation can be integrated independently from the energy equation. This enables us to assume that the velocity distribution inside and outside of the particle is known. It is determined by the Reynolds numbers of the internal (in the case of a drop) and external flows and the ratio of the fluid viscosities.

Then the problem of the unsteady heat transfer can be posed. To solve it, Eq. (2.44) should be integrated subject the following initial and boundary conditions

$$t = 0: \quad \begin{cases} 0 \le r < r_s & T_2 = T_{2.0} \\ r_s < r \le \infty & T_1 = T_{1.0} \end{cases} \tag{2.45}$$

$$t > 0: \quad \begin{cases} r = 0 & \left(\dfrac{\partial T_2}{\partial r}\right)_{r=0} = 0 \\ r = r_s & T_{1.s} = T_{2.s}; \quad \lambda_1 \left(\dfrac{\partial T_1}{\partial r}\right)_{r=r_s} = \lambda_2 \left(\dfrac{\partial T_2}{\partial r}\right)_{r=r_s} \\ r \to \infty & T_1 \to T_{1.0} \end{cases} \tag{2.46}$$

where λ_i are the thermal conductivities, subscripts 0 and s refer to the initial state and the particle surface, respectively.

The parameters are rendered dimensionless, as follows: $\overline{r} = r/r_s$, $\overline{v}_i/v_*$ (v_* is characteristic velocity scale and $\overline{T}_i = (T_i - T_{1.0})/(T_{2.0} - T_{1.0})$ Then Eq. (2.44) and the conditions (2.45) and (2.46) take the form

$$\frac{\partial \overline{T}_i}{\partial \mathrm{Fo}_i} + \mathrm{Pe}_i (\overline{v}_i \cdot \nabla) \overline{T}_i = \nabla^2 \overline{T}_i \tag{2.47}$$

and

$$\mathrm{Fo}_i = 0: \quad \begin{cases} 0 \le \bar{r} < 1 & \bar{T}_2 = 1 \\ 1 < \bar{r} \le \infty & \bar{T}_1 = 0 \end{cases} \tag{2.48}$$

$$\mathrm{Fo}_i > 0: \quad \begin{cases} \bar{r} = 0 & \left(\dfrac{\partial \bar{T}_2}{\partial r}\right)_{\bar{r}=0} = 0 \\ \bar{r} = 1 & \bar{T}_{1.s} = \bar{T}_{2.s}; \quad \left(\dfrac{\partial \bar{T}_1}{\partial r}\right)_{\bar{r}=1} = \lambda_{2.1}\left(\dfrac{\partial \bar{T}_2}{\partial \bar{r}}\right)_{\bar{r}=1} \\ \bar{r} \to \infty & \bar{T}_1 \to 0 \end{cases} \tag{2.49}$$

where $\lambda_{2.1} = \lambda_2/\lambda_1$, $\mathrm{Pe}_i = (v_* d)/\alpha_i$ is the Peclet number, $\mathrm{Fo}_i = (t\alpha_i)/d^2$ is the Fourier number.

The intensity of the unsteady heat transfer is characterized by the heat transfer coefficient, which can be defined as $h = (\lambda_1/(T_{1.0} - T_{1.s}))(\partial T_1/\partial r)_{r=r_s}$. The Nusselt number is $\mathrm{Nu} = (h \cdot d)/\lambda_1$. Taking into account that $\mathrm{Fo}_1 = \mathrm{Fo}_2\alpha_{1.2}$ and $\mathrm{Re}_1 = \mathrm{Re}_2\, \nu_{2.1}$, $\mathrm{Pe}_i = \mathrm{Pe}_2\alpha_{2.1}$, $(\alpha_{1.2} \cdot \alpha_{2.1} = 1)$ we can write the functional dependence of the Nusselt number on the dimensionless parameters of the problem

$$\mathrm{Nu} = F(\mathrm{Re}_i, \mathrm{Pe}_i, \mathrm{Fo}_i, \lambda_{2.1}, \nu_{2.1}, \alpha_{2.1}) \tag{2.50}$$

where the Reynolds, Peclet and Fourier numbers are based on the parameters of the continuous or dispersed phase.

In order to find a particular form of the dependence (2.50), it is necessary to find a solution of Eq. (2.47) subject to the conditions (2.48) and (2.49). For this aim, approximate approaches based on various problem simplifications are used. These include assumptions about the physical properties of the dispersed and continuous phases, as well about the operating parameters determining hydrodynamic and thermal structures of the flow.

Let us consider some of these simplifications. First, we estimate the ratio of the thermal diffusivities of the continuous and dispersed phases. At a moderate pressure corresponding to sub-critical conditions, gas densities are much smaller than the solid (liquid) densities. Therefore $\rho_{2.1}$ is of order 10^3. Since the product $\lambda_{1.2}\, c_{p2.1}$ is of order 1 (Reid et al. 1987), the ratio $\alpha_{1.2} = \rho_{2.1}\lambda_{1.2}c_{p2.1} >> 1$. Accordingly, the characteristic time of the unsteady thermal processes in the gas phase is much shorter than that in the dispersed phase. This enables us to use the quasi-steady approximation to calculate the temperature field in the carrier fluid. In this case the system of the governing equations takes the following form

$$\mathrm{Pe}_1(\overline{\mathbf{v}}_1 \cdot \nabla)\overline{T}_1 = \nabla^2 \overline{T}_1 \tag{2.51}$$

$$\frac{\partial \overline{T}_2}{\partial F_{02}} + \mathrm{Pe}_2(\overline{\mathbf{v}}_2 \cdot \nabla)\overline{T}_2 = \nabla^2 T_2 \tag{2.52}$$

An additional significant simplifications in the theoretical description of the unsteady heat transfer is related to the assumption that the effect of fluid motion is negligible, which corresponds to the purely conductive case: Pe = 0. Such a process was considered by Cooper (1972). In the special case when the thermal properties of the dispersed and continuous phases are identical $(\alpha_1 = \alpha_2 = \alpha;\ \mathrm{Fo}_1 = \mathrm{Fo}_2 = \mathrm{Fo})$ the average Nusselt number is expressed as

$$\overline{\mathrm{Nu}} = \pi^{-\frac{1}{2}}\mathrm{Fo}^{\frac{1}{2}}\left\{\left[\exp(-\frac{1}{\mathrm{Fo}})-1\right]+\frac{1}{2}\mathrm{Fo}^{-1}\left[1+\exp\left(-\frac{1}{\mathrm{Fo}}\right)\right]\right\} \tag{2.53}$$

where $\overline{\mathrm{Nu}} = \frac{Q(t)\, r_o}{4\pi r^2 \lambda}(T_{1.0} - T_{2.0})^{-1}$ is the average (over the particle surface) Nusselt number, Q(t) is the interfacial heat flux, $\mathrm{Fo} = (t\alpha)/r_o^2$ is the Fourier number, r_o is the particle radius.

In the limits corresponding to small and large times $(\mathrm{Fo} \ll 1$ and $\mathrm{Fo} \gg 1)$ Eq. (2.53) simplifies further. Then the average Nusselt number is given by

$$\overline{\mathrm{Nu}} = \pi^{-\frac{1}{2}}[\frac{1}{2}\mathrm{Fo}^{-\frac{1}{2}} - \mathrm{Fo}^{\frac{1}{2}}] \simeq \frac{1}{2}\pi^{-\frac{1}{2}}\mathrm{Fo}^{-\frac{1}{2}} \tag{2.54}$$

for $\mathrm{Fo} \ll 1$, and by

$$\overline{\mathrm{Nu}} = \pi^{-\frac{1}{2}}[\frac{1}{2} - \mathrm{Fo}^{-\frac{1}{2}} + \mathrm{Fo}^{-1}] \simeq \frac{1}{2}\pi^{-\frac{1}{2}} \tag{2.55}$$

for $\mathrm{Fo} \gg 1$.

In the general case $(\lambda_1 \neq \lambda_2;\ \alpha_1 \neq \alpha_2)$ the Nusselt number is expressed as

$$\overline{\mathrm{Nu}} = \frac{2}{\pi}\lambda_1\lambda_2(\alpha_1\alpha_2)^{\frac{1}{2}}\int_0^\infty \frac{\exp(-\xi^2 \mathrm{Fo}_2)(\xi\cos\xi - \sin\xi)^2 d\xi}{[\alpha_2(\lambda_1^2\xi^2\sin^2\xi) + \alpha_1\{\lambda_2(\xi\cos\xi - \sin\xi) + \lambda_1\sin\xi\}^2]} \tag{2.56}$$

Unsteady heat and mass transfer at finite Peclet numbers was analysed by Levich (1962), Ruckenstein (1965), Oliver and Chung (1986), and Konopliv and Sparrow (1972). Ruckenstein (1965) studied the unsteady mass transfer between a drop and a continuous phase for two types of flow: (i) for flows with small Reynolds numbers when the Rybczynski–Hadamard solution (1911, 1911) for the flow field is valid, (ii) for the potential flows. In the case when $\mathrm{Re} \ll 1$, Ruckenstein (1965) derived the following relation for the Sherwood number:

$$\mathrm{Sh}_2 = \left(\frac{\mathrm{Pe}_2}{\pi}\right)^{\frac{1}{2}} \left[\mathrm{H}\left(\frac{\mathrm{D}_2}{\mathrm{D}_1}\right)^{\frac{1}{2}} + 1 \right]^{-1} \psi(\varepsilon) \tag{2.57}$$

where $\mathrm{Sh}_2 = \frac{2h_m r_o}{D_2}$ is the Sherwood number, $\mathrm{Pe}_2 = \frac{1}{2}\left(\frac{\mu_1}{\mu_1+\mu_2}\right)\frac{r_o u}{D_2}$ is the Peclet number, μ is viscosity, $\varepsilon = \frac{1}{2}\frac{ut}{r_o}\frac{\mu_1}{\mu_1+\mu_2}$, u is the translational velocity of the drop, H is the equilibrium dissolution constant, and $\psi(\varepsilon)$ is given by

$$\psi(\varepsilon) = \int_0^\pi \sin^3\theta \left\{ \cos\theta - \frac{1}{3}\cos^3\theta - \frac{1-\tan^2\frac{\theta}{2}e^{2\varepsilon}}{1+\tan^2\frac{\theta}{2}\cdot e^{2\varepsilon}} + \frac{1}{3}\left(\frac{1-\tan^2\frac{\theta}{2}e^{2\varepsilon}}{1+\tan^2\frac{\theta}{2}e^{2\varepsilon}}\right)^3 \right\}^{-\frac{1}{2}} d\theta$$

In the case when $D_1 \gg D_2$ and the rate of mass transfer is determined by diffusion in the dispersed phase, Eq. (2.57) reduces to the relation

$$\mathrm{Sh}_2 = \left(\frac{\mathrm{Pe}_2}{\pi}\right)^{\frac{1}{2}} \psi(\varepsilon) \tag{2.58}$$

For short time intervals $(\mathrm{Fo} \to 0)$ the following estimation of the Sherwood number derives from Eq. (2.58)

$$\mathrm{Sh}_2 = \frac{2}{\sqrt{\pi}}\mathrm{Fo}_2^{-1/2} \tag{2.59}$$

For long time $(\varepsilon \gg 1)$, Eq. (2.58) yields

$$Sh_2 = \frac{4}{3}\left(\frac{Pe_2}{\pi}\right)^{\frac{1}{2}} \tag{2.60}$$

When $D_2 \gg D_1$ and the rate of mass transfer is determined by diffusion in a continuous phase, Eq. (2.57) reduces (at Le=1) to

$$Sh_1 = \left(\frac{Pe_1}{\pi}\right)^{\frac{1}{2}} \psi(\varepsilon) \tag{2.61}$$

where Le is the Lewis number. For long times $(\varepsilon \gg 1)$, the latter expression takes the form

$$Sh_1 = \frac{4}{\sqrt{3\pi}}(Pe_1)^{\frac{1}{2}} \tag{2.62}$$

In the case of the potential flow about a drop the Sherwood numbers for the continuous and dispersed phases are given by the following relation

$$Sh_i = \left(\frac{Pe_i^*}{\pi}\right)\left[H\left(\tilde{D}_i\right)^{\frac{1}{2}} + 1\right]^{-1} \psi(\varepsilon^*) \tag{2.63}$$

where $\tilde{D}_1 = D_1/D_2$, $\tilde{D}_2 = D_2/D_1$, $\varepsilon^* = (3tu)/(2r_o)$, $Pe_i^* = 3Pe_i/4$. At llong time ($\varepsilon^* \gg 1$), corresponding to the steady state, Eq. (2.63) yields

$$Sh_i = \frac{2}{\sqrt{\pi}}\left(\frac{2r_o u}{D_i}\right)^{\frac{1}{2}} \tag{2.64}$$

Konopliv and Sparrow (1972) studied the heat transfer from a spherical particle under the assumption that the velocity field is Stokesian and the temperature field is of the boundary layer type, which corresponds to large values of the Prandtl number. In the case of a constant surface temperature they obtained the following relation for the average Nusselt number

$$\left(\frac{3}{4}\mathrm{Pe}\right)^{-\frac{1}{2}}\overline{\mathrm{Nu}}=\frac{2}{\sqrt{\pi\tau}}+\frac{19}{48}\frac{\tau}{\Gamma(7/2)}-\frac{125687}{15380}\frac{\tau^{\frac{11}{2}}}{\Gamma(13/2)}+\ldots \tag{2.65}$$

where $\tau=((3\mathrm{Pe})/4)^{2/3}\,\mathrm{Fo}$. The first term on the right hand side in Eq. (2.65) results from pure conduction, while the other terms are associated with convection. As can be seen from Eq. (2.65), conduction is the dominant mode at small times. At $\tau \ll 1$ Eq. (2.65) yields

$$\overline{\mathrm{Nu}}=\left(\frac{3}{4}\mathrm{Pe}\right)^{\frac{1}{3}}\mathrm{Fo}^{-\frac{1}{2}} \tag{2.66}$$

Unsteady heat transfer under the conditions of large Reynolds and Peclet numbers was investigated by Chao (1969). Assuming that the flow is inviscocid and there exists a thin boundary layer the following relation for the average Nusselt number was obtained

$$\overline{\mathrm{Nu}}=\frac{2\,\mathrm{Pe}^{\frac{1}{2}}}{\sqrt{\pi}(1+a)}\,I\left(\frac{ut}{r_o}\right) \tag{2.67}$$

where $a=\lambda_{1.2}\rho_{1.2}c_{1.2}$, and the function $I(ut/r_o)$ is given by

$$I=\frac{\sqrt{3}}{4}\int_0^{\pi}\frac{\sin^2\theta\,d\theta}{\left[(f-\cos\theta)-\frac{1}{3}(f^2-\cos^2\theta)\right]^{\frac{1}{2}}},\qquad f=\frac{1-\frac{1-\cos\theta}{1+\cos\theta}\exp(-3\frac{ut}{r_o})}{1+\frac{1-\cos\theta}{1+\cos\theta}\exp(-3\frac{ut}{r_o})}$$

For short and long times the function $I(ut/r_o)$ reduces to

$$I\to\frac{1}{2}\left(\frac{ut}{r_o}\right)^{-\frac{1}{2}} \tag{2.68}$$

while for $(ut/r_o)\gg 1$, and

$$I\to 1 \quad \text{at} \quad t\to\infty \tag{2.69}$$

Accordingly Eq. (2.67) takes the form

$$\overline{Nu} = \frac{\sqrt{2}}{(1+a)} Fo^{-\frac{1}{2}} \tag{2.70}$$

at $(ut/r_o) \ll 1$ and

$$\overline{Nu} = \frac{2Pe^{\frac{1}{2}}}{\sqrt{\pi}(1+a)} \tag{2.71}$$

at $(ut/r_o) \gg 1$.

Unsteady heat transfer at $0 < Pe < 10^3$ was studied numerically by Abramzon and Borde (1980), Oliver and Chung (1986), and Abramzon and Elata (1984). It was shown that in all the cases studies (corresponding to different values of the density, specific heat and diffusivity of continuous and dispersed phases) the heat transfer coefficient decreases with Fo at short times and approaches an asymptotic value at large values of the Fourier number. This shows that the transient effects are important in calculation of the particle heat transfer only at the early stages of particle heating or cooling and devolatilization (vaporization).

Heat transfer in natural convection. Natural convection from a heated sphere was a subject of the investigations of Fendell (1968), Chiang et al. (1964), Lin and Chao (1974), Dennis et al. (1973), Potter and Riley (1980), Geoola and Cornish (1981, 1982), and Riley (1986). At present there are a number of correlations allowing estimation of the values of the drag and heat transfer coefficients in a wide range of the Grashof and the Prandtl numbers (Geoola and Cornish 1981, Dudek et al. 1988, Jafarpur and Yovanovich 1992).

Geoola and Cornish (1981) proposed the following correlation for the average Nusselt number based on the results of numerical simulations of the heat transfer by natural convection from a heated spherical particle

$$Nu = 2 + 0.39\,(Ra)^{0.42} \tag{2.72}$$

for $0.05 \le Gr < 50$; $Pr = 0.72$, and

$$Nu = 2 + 0.75(Ra)^{0.25} \tag{2.73}$$

for $0.36 \le Ra < 12500$; $50 \le Gr \le 12500$; $0.72 \le Pr \le 100$. Here $Ra = GrPr$ is the Rayleigh number.

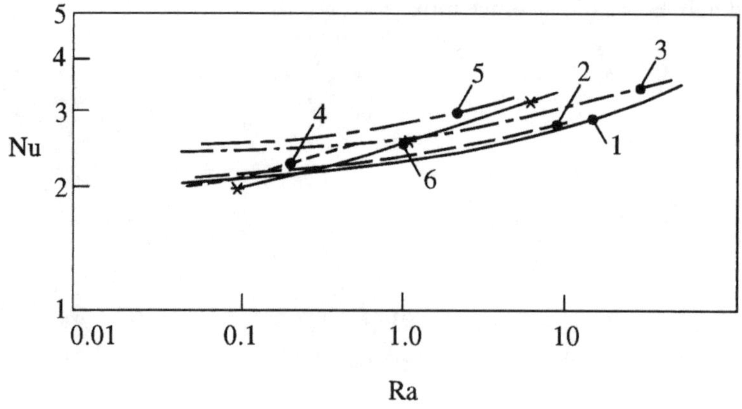

Fig. 2.11 The Nusselt number versus the Rayleigh number. Reprinted from Geoola and Cornish (1981), with permission. Curve 1 corresponds to Eq. (2.72), curve 2 corresponds the data of Hossain (1966), curve 3, data of Mathers et al. (1957), curve 4, data of Tsubouch and Sato (1960), curve 5, data of Yuge (1960), curve 6, data of Jafarup and Yovonovich (1992).

The correlation (2.72) is compared in Fig. 2.11 to the experimental data of Yuge (1960), Tsubouchi and Sato (1960), Mathers et al. (1957) and Hossain (1966), as well as to the theoretical relation of Jafarpur and Yovanovich (1992). It is seen that Eq. (2.72) underestimates the values of the Nusselt number. The latter, probably, results from the difference in the conditions of the experiments from those assumed theoretically, as Eq. (2.72) does not take into account the influence of the additional heat losses due to conduction and radiation, etc. (Geoola and Cornish 1981), while in the experiments some of these factors could be of importance.

An approximate analytical solution of the problem of heat transfer from a sphere due to of natural convection was obtained by Jafarpur and Yovanovich (1992). Using the method based on the linearization of the energy equation, they derived a relation for the average Nusselt number which is valid in the range $0 \le \mathrm{Ra} < 10^8$ at any Pr.

$$\mathrm{Nu} = 2 + \frac{0.600\mathrm{Ra}^{\frac{1}{4}}}{\left[1+\left(\frac{0.5}{\mathrm{Pr}}\right)^{\frac{9}{16}}\right]^{\frac{4}{9}}} \tag{2.74}$$

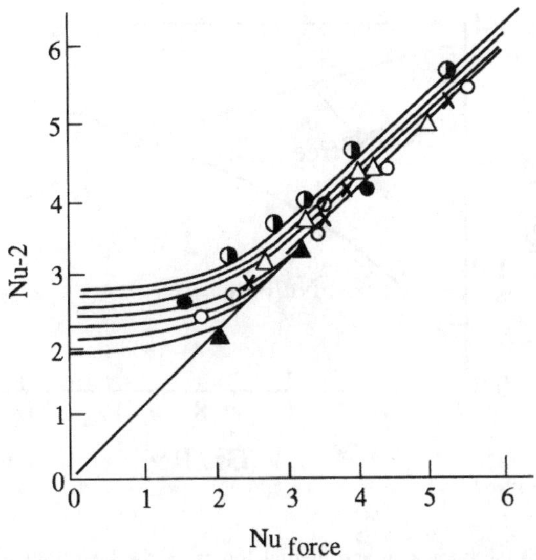

Fig. 2.12 The dependence (Nu – 2) on Nu_{force}. Reprinted from Yuge (1960), with permission. ◐, Gr=2.56; Δ, Gr=2.49; ×, Gr=2.34; ⌀, N_g=2,24; ●, Gr=2.09; O, Gr=1.49, ▲, Gr=1.75.

In the limiting cases corresponding to small or large Prandtl numbers, Eq. (2.74) reduces to

$$Nu = 2 + 0.714\, Ra^{\frac{1}{4}}\, Pr^{\frac{1}{4}} \tag{2.75}$$

for $Pr \to 0$, and to

$$Nu = 2 + 0.600\, Ra^{\frac{1}{4}} \tag{2.76}$$

for $Pr \to \infty$. Equation (2.76) slightly overestimates the Nusselt number at large Ra and Gr. The difference between the values of the Nusselt number calculated from Eqs. (2.72), (2.73) and Eq. (2.74) does not exceed 15% in the range $0 \le Gr \le 50$ and increases up to 25% at Gr = 12,500.

Heat transfer in mixed convection. There are a number of experimental and theoretical investigations dealing with mixed convection from a spherical particle (Yuge 1960, Klyachko 1963, Acrivos 1966, Hieber and Gebhart 1969, Chen and Mucoglu 1977, Wong et al. 1986, Nguyen et al. 1993).

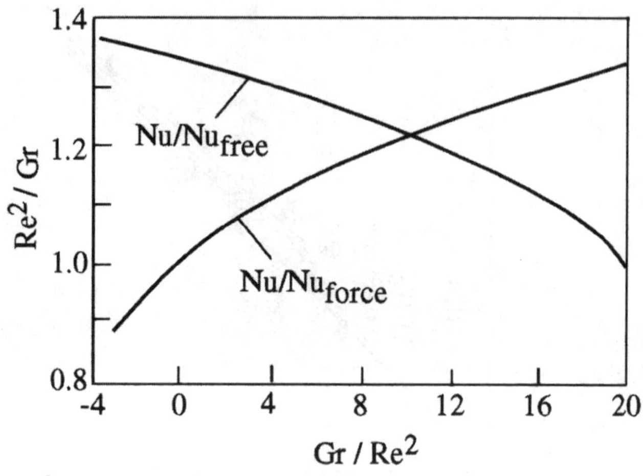

Fig. 2.13 Effect of the buoyancy and inertial forces on the heat transfer at the stagnation point of a sphere. Reprinted from Chen and Mucoglu (1977), with permission. Pr = 0.72. Nu_{free} denotes the Nusselt number for pure natural convection

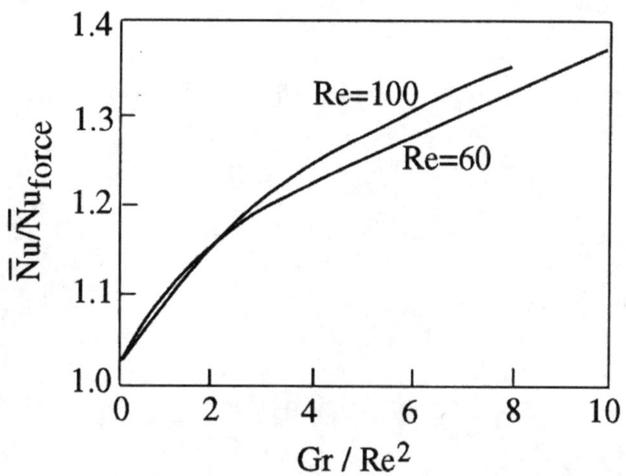

Fig. 2.14 The effect of Gr/Re^2 on the ratio of the average Nusselt number for mixed convection $\overline{Nu}$ to the average Nusselt number for the forced convection. Reprinted from Wong et al. (1986), with permission

Yuge (1960) studied heat transfer from a sphere immersed in air flow (Pr = 0.715). The measurements were carried out in the range of the Reynolds numbers from 1 to $1.44 \cdot 10^5$ and of the Grashof numbers from 1 to 10^5. Under these conditions the regimes of natural, mixed and forced convection occur. His experimental data are presented in Fig. 2.12 on the parametrical plane ($\mathrm{Nu}-2$)–$\mathrm{Nu}_{\mathrm{force}}$ as a family of curves corresponding to several constant values of the Grashof number. Here $\mathrm{Nu}_{\mathrm{force}}$ is the Nusselt number for the forced convection. In such a presentation of the measurements results the line passing through the origin at the angle of 45° corresponds to pure forced convection, whereas deviations of the other curves from this line characterize the natural convection effect.

Hieber and Gebhart (1969) explored the effect of gravity on the heat transfer from a heated sphere that is maintained at a steady uniform temperature in vertical flow. Using the method of the matched asymptotic expansions, they estimated the effect of the buoyancy force on the velocity, temperature and pressure fields, as well as the contribution of natural convection in the heat transfer. For mixed convection at small Reynolds and Grashof numbers and Pr = 1, Hieber and Gebhart (1969) derived the following expression for the Nusselt number

$$\mathrm{Nu} = 1 + \mathrm{Nu}^{\mathrm{F}}(\mathrm{Re}) + \mathrm{Nu}^{\mathrm{B}}(\mathrm{Re},\ \varepsilon) \tag{2.77}$$

where Nu^{F} is given by Rimmer (1968) for the forced convection (see Eq. (2.12)) and N^{B} reads

$$\mathrm{Nu}^{\mathrm{B}} = (2\ln 2 - 1)\varepsilon\,\mathrm{Re} + \varepsilon\,\mathrm{Re}^2 \ln \mathrm{Re} + O(\varepsilon\,\mathrm{Re}^2) + O(\varepsilon^2\,\mathrm{Re}) \tag{2.78}$$

In Eqs. (2.77) and (2.78) the Grashof, Reynolds and Nusselt numbers are based on the sphere radius, and $\varepsilon = \mathrm{Gr}/\mathrm{Re}^2$. The first term on the right hand side of Eq. (2.77) corresponds to the pure condition in fluid at rest, the second to the effect of the forced convection, and the third accounts for the effect of natural convection.

Heat transfer due to mixed convection determines not only the value of the ratio of the buoyancy force to inertial $\mathrm{Gr}/\mathrm{Re}^2$ but also the direction of the buoyancy force. Chen and Mucoglu (1977) studied mixed convection in the framework of the numerical study of convective heat transfer for different directions of the buoyancy force. The calculations showed that under the conditions of the mixed convection, the Nusselt number increases with increasing buoyancy forces in the case of flow in the direction of buoyancy. In the case of flow against the buoyancy direction, the Nusselt number decreases with increasing buoyancy forces. The effect of the buoyancy forces on the heat transfer becomes significant for $\mathrm{Gr}/\mathrm{Re}^2 > 1.65$ and $\mathrm{Gr}/\mathrm{Re}^2 < -1.33$ (cf. Fig. 2.13).

A detailed study of the mixed heat transfer from an isothermal spherical particle immersed in fluid flow was performed by Wong et al. (1986) and Nguyen et al. (1993). They solved numerically the full Navier–Stokes and energy equations by the finite element method. Some results of their calculations are shown in Fig. 2.14. It is seen that the Nusselt number corresponding to mixed convection significantly exceeds the Nusselt number for forced convection. This difference increases when the ratio $\mathrm{Gr}/\mathrm{Re}^2$ increases. The Reynolds number only slightly affects the value of the ratio $\overline{\mathrm{Nu}}/\overline{\mathrm{Nu}}_{\mathrm{force}}$. The absolute value of the Nusselt number depends on the direction of the buoyancy force relative to the direction of the main flow. It increases in the cocurrent flow case and decreases for the countercurrent case when the flow is directed opposite to the buoyancy force. The change of the Nusselt number does not exceed 17% for $(\mathrm{Gr}/\mathrm{Re}^2) \le 40$ When the buoyancy force and main flow have the same direction, their effect is less pronounced than when they have opposite directions.

Heat (mass) transfer in shear flow. The effect of the velocity gradient on the heat (mass) transfer from a particle was considered by Frankel and Acrivos (1968), Acrivos (1971, 1980), Batchelor (1979, 1980) and Gupalo et al. (1985).

Frankel and Acrivos (1968) derived the following relation for the Nusselt number which is valid for low Reynolds and Peclet numbers

$$\mathrm{Nu} = 2 + \frac{0.9104}{(2\pi)^{\frac{1}{2}}} \cdot \mathrm{Pe}^{\frac{1}{2}} \tag{2.79}$$

where $\mathrm{Pe} = (r_o^2 \gamma)/\alpha$ is the Peclet number, γ is the shear rate at infinity, α is the thermal diffusivity.

As can be seen from Eq. (2.79), an increase in the velocity gradient is accompanied by an increase in the intensity of the heat transfer. At high values Pe the Nusselt number approaches to 8.9 (Acrivos 1971) Heat transfer from a particle of an arbitrary shape was studied by Acrivos (1980) and Batchelor (1979). It was shown that the Nusselt number of such a particle is determined by two dimensionless groups accounting for the intensity of heat transfer in the case without shear, as well as for the value of the shear rate. Batchelor (1980) estimated the mass transfer rate from a spherical particle immersed in turbulent flow. Assuming that the flow field in the vicinity of the particle can be presented as a superposition of a translating motion and shear, he obtained the following relation for the Sherwood number

$$\mathrm{Sh} = 0.55 \left(\frac{r_o^2 \varepsilon^{\frac{1}{2}}}{D \nu^{\frac{1}{2}}} \right)^{\frac{1}{3}} \tag{2.80}$$

Here ε is the mean dissipation rate, and Sh is the Sherwood number based on the particle radius. Eq. (2.80) is valid for low $\mathrm{Re}_\varepsilon = (r_o^2 \varepsilon^{1/2})/\nu^{3/2}$ and high $\mathrm{Pe}_\varepsilon = (r_o^2 \varepsilon^{1/2})/(D\nu^{3/2})$.

Gupalo et al. (1985) proposed a simple interpolation formula for the Sherwood number in the whole range of the Peclet number Pe_ε

$$Sh = 1 + 0.55 Pe_\varepsilon^{\frac{1}{3}} \tag{2.81}$$

Here the Sherwood number is based on the particle radius.

Effect of chemical reactions. A chemical reaction that proceeds on the surface of a particle is typical for combustion in two-phase reactive flows. The effect of the heterogeneous reactions on the heat and mass transfer mostly depends on the ratio of the rates of the chemical reaction and diffusion. Depending on the value of the ratio different regimes of the process exist, namely the kinetic regime, corresponding to a finite rate of reaction, or the diffusion regime corresponding to an infinite rate of reaction and, accordingly, to a full absorption of a substance at the particle surface. Thus, the problem of the heat and mass transfer from reactive particles involves a number of parameters accounting for the characteristics of chemical reactions (the activation energy, the order of reaction, etc.), flow hydrodynamics (the Reynolds number, the shear rate), as well as for the physical properties of fluid (the Prandtl and the Schmidt numbers). The analytical solutions of the problem on heat (mass) transfer from a reactive particle are based on the assumptions that $Re \ll 1$ and the Peclet number is small or very large. That allows one to use the approximations of the creeping flow or of the diffusion boundary layer for the description of the flow field around a reactive particle.

The mass transfer from a reactive spherical particle was studied in the creeping flow approximation by Taylor (1963). Later on the theory of the heat (mass) transfer at small and very large Peclet numbers was extended by Polyanin and Sergeev (1980), Gupalo and Rayzantsev (1974), and Polyanin (1982,1984). These results are summarized in the monograph by Gupalo et al. (1985). Following these investigations, we illustrate the effect of the chemical reaction on the particle mass (heat) transfer using the examples corresponding to the diffusion and kinetic regimes of the process.

At low Peclet number the Sherwood number corresponding to an isothermal reaction of the first order is given by

$$Sh = k^* + \frac{1}{2}k^*(Pe + Pe^2 \ln Pe) + \frac{1}{2}k^{*2}F(k^*, Sc)Pe^2 + \frac{1}{4}k^{*3}Pe^3 \ln Pe + O(Pe^3) \tag{2.82}$$

where Sc is the Schmidt number, $k^* = k(1+k)^{-1}$, $k = k_s r_o C_\infty^{n-1} D^{-1}$, k_s is the pre-exponential in the Arrhenius law, n is the order of reaction, C_∞ is the concentration far from the particle and the function $F(k^*, Sc)$ is determined by the following relation

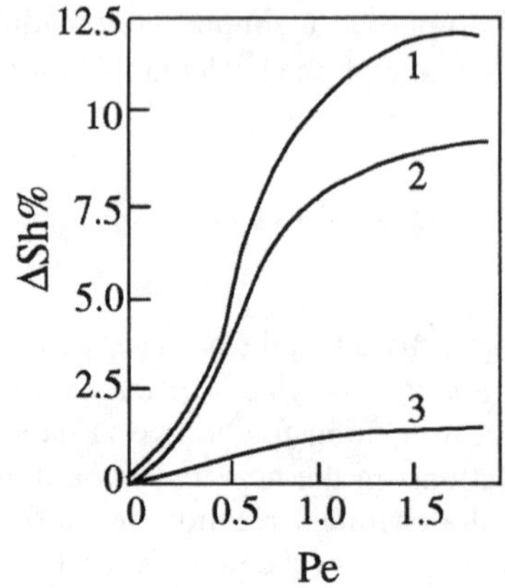

Fig. 2.15 The dependence of the Sherwood number increment on Pe for the diffusion regime of heterogeneous reaction. Reprinted from Gupalo et al. (1985). Curve 1, Sc=0.7, curve 2, Sc=1, curve 3, Sc=10

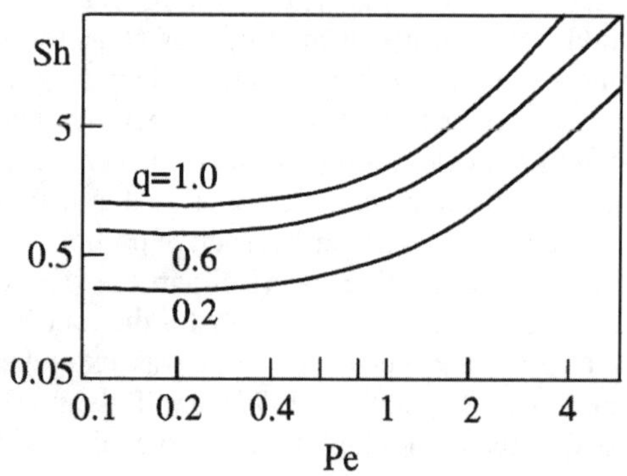

Fig. 2.16 The dependence of the mean Sherwood number on Pe for different regimes of heterogeneous reaction. Reprinted from Gupalo et al. (1985). Sc=1

$$F(k^*,Sc) = -\frac{119}{80} + \gamma + \frac{1}{2}k^* - \frac{3}{32}(2-k^*)^{-1} + \frac{Sc^2}{2} - \\ -\frac{Sc}{4} + (Sc+1)^2(\frac{Sc}{2}-1)\ln(1+Sc^{-1}) \quad (2.83)$$

Here γ is the Euler number.

In the particular case corresponding to an infinitely large rate of chemical reaction (the diffusion regime of the process), Eq. (2.82) reduces to

$$\mathrm{Sh} = 1 + \frac{1}{2}(\mathrm{Pe} + \mathrm{Pe}^2 \ln \mathrm{Pe}) + \frac{1}{2}F(1,\mathrm{Sc})\mathrm{Pe}^2 + \frac{1}{4}\mathrm{Pe}^2 \ln \mathrm{Pe} + O(\mathrm{Pe}^3) \tag{2.84}$$

where

$$F(1,\mathrm{Sc}) = -\frac{173}{160} + \gamma + \frac{\mathrm{Sc}^2}{2} - \frac{\mathrm{Sc}}{4} - (\mathrm{Sc}+1)^2\left(\frac{\mathrm{Sc}}{2} - 1\right)\ln(1+\mathrm{Sc}^{-1})$$

Equation (2.84) is identical to the Rimmer relation (Eq. 2.12) for the Nusselt number with Sh, $\mathrm{Pe} = (ud)/D$ and Sc standing instead of Nu, $\mathrm{Pe} = (ud)/\alpha$ and Pr, respectively.

The effect of the physical properties and velocity of fluid on the mass transfer of a reactive particle is illustrated in Figs. 2.15 and 2.16. It is seen that the increase of the Sherwood number $(\Delta \mathrm{Sh} = (\mathrm{Sh}/\mathrm{Sh}_\infty) - 1$, Sh_∞ is the Sherwood number for $k \to \infty)$ with Pe achieves 10% at $\mathrm{Sc} \sim 0.7$ and $\mathrm{Pe} \sim 1$. This effect diminishes with increasing Schmidt number.

The mass (heat) transfer under non isothermal conditions is of importance for the combustion theory. In the case when the rate of reaction W is determined by the Arrhenius law, $W(C,T) = k_s C \exp(-E/RT)$, the Sherwood number is given by

$$\mathrm{Sh} = k\left(1 - \frac{\mathrm{Sh}}{\mathrm{Sh}_\infty}\right)\exp\left\{-\left[\theta_\infty\left(1 - H\,\frac{\mathrm{Sh}}{\mathrm{Sh}_\infty}\right)\right]^{-1}\right\} \tag{2.85}$$

where $\theta_\infty = (RT_\infty)/E$ $H = q(DC_\infty)/(\lambda T_\infty)$, q is the heat released in the reaction.

Depending on the values of the parameters θ_∞ and k, Eq. (2.85) has one, two or three roots. In the last case they correspond to different thermal regimes of the process: the low, the intermediate and the high temperature regimes. This implies that the intensity of the mass (heat) transfer is different in the case of particle heating and combustion.

To estimate the intensity of the mass (heat) transfer in the intermediate range of the Peclet number, the following relations for the Sherwood number can be used

$$\mathrm{Sh} = 1 + \frac{0.5\mathrm{Pe}}{1 + 0.8\mathrm{Pe}^{\frac{1}{2}}} \tag{2.86}$$

for a solid particles, and

$$\mathrm{Sh} = 1 + \frac{0.5\mathrm{Pe}}{1 + 1.1(1+\mu_{2.1})^{\frac{1}{2}} \mathrm{Pe}^{\frac{1}{2}}} \tag{2.87}$$

for droplets and bubbles. The deviation of Eq. (2.86) from the results of the numerical calculations does not exceed 10%; for Eq. (2.87) the deviation is about 12% for $\mu_{2.1} = 0$ (a bubble) and increases up to 16% for $\mu_{2.1} \to \infty$ (a solid particle).

1.2.2 Particle heating

The pattern of the process. A. Particles of solid fuels. Particle heating is a basic phenomenon characteristic of a number of the processes related to the combustion of solid fuels. These include such processes as drying, devolatilization, ignition, etc. The rate of particle heating in the combustion chambers depends on the physico-chemical properties of solid fuel, particle shape and size, as well as on the conditions of heat and mass transfer with the environment.

Solid fuels possess a complicated structure because of the existence of numerous micropores and cracks penetrating a solid matrix, Fig. 2.17. The voids significantly enlarge the accessible surface of solid reagent and promote transport of gaseous oxidizer inside the particle, as well as removal of volatiles. At the same time, solid fuels are multicomponent substances containing solid reagent (also in some cases, oxidizer), as well as inert matter and moisture. Components of solid fuels include a number of elements and chemical compounds. For example, the composition of such typical organic fuels as coals includes carbon, hydrogen, oxygen, sulfur and nitrogen (Smith and Smoot 1990). The mass percentage of these elements varies as: C–65–95%, H_2– 2–7%, O_2 up to 25%, S up to 10% and N_2–1–2%. Most coals also contain alkaline metals such as calcium, sodium and magnum. The moisture content in coals varies from 2% to 20%. Also much higher contents, up to 70%, have been recorded for some lignite coals.

An increase of the particles temperature during heating is accompanied by decomposition of solid matter and formation of volatiles. The latter include a mixture of carbon monooxide and dioxide, water vapor, nitrogen, hydrocarbon gases such as CH_4, C_2H_4, C_2H_6, etc. During devolatilization of coals hydrocarbon liquids and tar also emerge.

Thus, heating leads to a transformation of a homogenous particle into a multiphase agglomerate of solid, liquid and gaseous components. The evolution of such an agglomerate depends on a number of factors relating to the properties of all phases and their interaction with each other.

The behavior of heated particles depends significantly on the conditions of volatile removal. They are different for various stages of particle heating. During the early stage of the process, when the particle temperature is much less than its melting temperature, volatile removal occurs by filtration through the porous matrix towards the particle surface. In contrast, the other mechanism of volatile

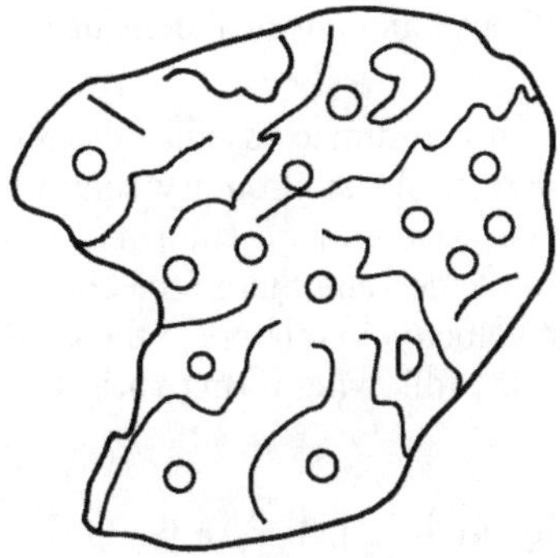

Fig. 2.17 Solid fuel structure

removal occur when the particle temperature exceeds the melting point. In this case volatile bubbles arise in the liquid phase contained inside of a particle. Displacement of these bubbles determines the rate of devolatilization. It should be noted that the existence of volatile bubbles also affects the particle's structure. Particles can swell and crack due to the volatile's pressure. These effects are due to the thermal expansion of the bubbles leading to deformation of the solid matrix.

B. Droplets of liquid fuels. The temperature of droplets of liquid fuels is mostly governed by the boiling temperatures of their components. Accordingly, the heating process of droplets of monocomponent fuels can be presented schematically as two successive stages corresponding to the preliminary heating of liquid and its vaporization. During the first stage of the process, the droplet temperature increases from an initial value T_0 up to the boiling temperature, T_B. For the second stage of the process, the droplet temperature remains practically constant up to complete vaporization of the liquid. Heating of droplets of multicomponent fuels is a multistaged process. In the first stage the droplet temperature increases up to the boiling temperature of the most volatile component. Subsequently, the temperature successively increases up to the boiling temperatures of the less volatile components of the fuel.

Energy balance equation. Consider a heated particle as a two-phase medium containing the initial matter (a solid phase) and products of its decomposition (the gaseous phase). The densities of the gaseous and solid phases and their volumetric contents are $\rho_1^o;\rho_2^o$ and $\xi_1;\xi_2$, respectively (here and hereinafter subscripts 1 and 2 refer to gaseous and solid phases).

In order to describe the heat and mass transfer of a heated particle, we use the model of interpenetrating and interacting continua (Nigmatulin 1991). In accordance with this model a particle is treated as an aggregate of the gaseous and solid continua with the effective densities

$$\rho_i = \rho_i^o \xi_i \tag{2.88}$$

Here ρ_i and ρ_i^0 are the effective and physical densities of the i-th phase, ξ_i is the volumetric content.

Choosing the system of coordinate associated with the solid matrix, the energy balance equation is derived for an arbitrary control volume V inside the particle. The volume should be large enough compared to the heterogeneity size of the particle matter. Taking into account that in the chosen coordinate system volatiles move with a filtration velocity v_f relative to the solid matrix, we can write the energy balance equation in the following form (Yarin and Sukhov 1987)

$$\int_V \frac{\partial}{\partial t}(\rho_i h_i)\,dV = -\int_S \mathbf{q}_i \cdot \mathbf{n}\,dS - \varepsilon_i \int_S \rho_l h_l \mathbf{v}_f \cdot \mathbf{n}\,dS + \int_V Q_i dV + \int_V H_i dV \tag{2.89}$$

where h is the enthalpy, q is the specific heat flux, Q_i is the rate of the i-th phase enthalpy change due to phase transition, H_i is the rate of the i-th phase enthalpy change due to the heat exchange with the other phases of the heterogeneous mixture, ε_i is a factor equal to 1 or 0 for the gaseous and solid phases, respectively.

The filtration velocity in the third term on the right hand side in Eq. (2.89) is defined as

$$v_f = \frac{1}{\rho_l}\sum_s \rho_s v_s \tag{2.90}$$

where subscript s corresponds to the components of the gaseous phase.

The term on the left hand side in Eq. (2.89) accounts for the change of the enthalpy of the i-th phase in volume V. The first term on right hand side in Eq. (2.89) accounts for the heat flux through the surface S of the control volume V; the second one describes the convective heat transfer due to volatile filtration through the solid matrix. It is assumed that different components move with the same speed $\mathbf{v}_f$. The third and the fourth terms account for the rate of change of the i-th phase enthalpy due to the phase transition and interfacial heat transfer, respectively.

We transform Eq. (2.89) by using the Gauss theorem. As result, we obtain

$$\frac{\partial}{\partial t}(\rho_i h_i) = -\nabla \cdot \mathbf{q}_i - \varepsilon_i \nabla \cdot (\rho_l h_l \mathbf{v}_f) + Q_i + H_i \tag{2.91}$$

The term Q_i in Eq. (2.91) can be presented as the sum of two components accounting for heat release (absorption) due to phase conversion (melting, evaporation, etc.) Q_i' and heat release due to chemical reaction between the solid reagent and gaseous oxidizer Q_i''.

$$Q_i = Q_i' + Q_i'' \tag{2.92}$$

Typically it is assumed that the heat of chemical reactions concentrates in the final products. Accordingly, we will assume that

$$Q_{in}'' = 0, \qquad Q_{fi}'' \neq 0 \tag{2.93}$$

where subscripts in and fi refer to the initial reagents and final products.

The term H_I in Eq. (2.91) can be split in three components accounting for (i) the heat transfer from a substance undergoing phase conversion, (ii) the contact heat transfer due to thermal non equilibrium of the phases, (iii) the heat transfer due to radiation.

$$H_i = H_i' + H_i'' + H_i''' \tag{2.94}$$

where

$$H_i' = h_i \frac{\partial \rho_i}{\partial t}, \quad \left(H_1' = \sum_s h_s \left(\frac{\partial \rho_s}{\partial t} + \nabla \cdot (\rho_s v_s)\right) \right), \quad H_i'' = \sum_j H_{ij}''$$

subscripts i and j refer to gaseous and solid (reactive and inert components and final products) phases. Each term of the sum $\sum_j H_{ij}$ expresses the intensity of the contact heat transfer between the i-th and j-th phases.

In order to calculate the intensity of the contact heat transfer between the i-th and the j-th phases, we should use relation

$$H_{ij}'' = h_{ij} s_{ij} (T_i - T_j) \tag{2.95}$$

where h_{ij} is the heat transfer coefficient depending on the physical properties of the interacting phases, as well as on the shape of solid inclusions and relative motion of the phases, s_{ij} is the specific area of the interfacial.

The intensity of the interphase heat transfer due to radiation can be estimated as

$$H_i''' = \tilde{\sigma} \sum_j \varepsilon_i \, \varepsilon_j s_{ij}^R (T_i^4 - T_j^4) \tag{2.96}$$

where $\tilde{\sigma}$ is the Stefan–Bolzmann constant, ε_i and ε_j are the emissivity of the phases, superscript R corresponds to the radiative heat transfer.

At moderate values of pressure, gas in the particle pores is transparent, so that $H_i'''=0$. Accordingly, the heat transport due to radiation occurs in the porous medium directly between the solid elements through the surfaces. It is emphasized that porous media possess huge interfacial area (Essenhich 1977). This allows the assumption that s_{ij}^R is infinitely large. As a result, the problem may be significantly simplified. Indeed, when $s_{ij}^R \to \infty$ the heat transfer of finite intensity is possible only at $(T_i - T_j) \to 0$, i.e. under the conditions corresponding to the thermal equilibrium: $T_i = T_j = T$. The one - temperature approximation makes it possible to exclude from consideration a number of parameters $(h_{ij}, s_{ij}, s_{ij}^R$, etc.$)$, which cannot be estimated directly for porous media with complicated internal structure. Then solutions of the problems on particle heating, devolatilization and combustion can be formulated in the frame of the one-temperature model. They are based on an energy equation for a heterogeneous medium as a whole (Yarin and Sukhov 1987).

In order to derive the energy equation of a heterogeneous medium as a whole, a sum of Eqs. (2.91) for different values of i should be considered. Taking into account that $\sum_i H_i' = 0$ due to energy conservation at the interphase, the result of summation reads

$$\frac{\partial}{\partial t}(\sum_i \rho_i h_i) + \nabla \cdot (\rho_1 h_1 \mathbf{v}_f) = -\nabla \cdot (\sum_i \mathbf{q}_i) + Q \tag{2.97}$$

where $Q = Q' + Q''$, Q' and Q'' are the total rates of heat release due to phase conversion and chemical reaction, respectively.

Heat fluxes $\mathbf{q}_i$ in Eq. (2.97) can be presented formally as a generalized Fourier law

$$\mathbf{q}_i = -\lambda_i \nabla T \tag{2.98}$$

where $\lambda_i = \lambda_i^c + \lambda_i^R$, λ_i^c and λ_i^R are the thermal conductivity coefficients corresponding to conductive and radiative heat transfer.

Then the first term in right hand side of Eq. (2.98) takes the following form.

$$-\nabla \cdot (\sum_i \mathbf{q}_i) = \nabla \cdot (\lambda \nabla T) \tag{2.99}$$

where $\lambda = \sum_i (\lambda_i^c + \lambda_i^R)$ is the macroscopic thermal conductivity coefficient.

The macroscopic thermal conductivity λ depends on the structural characteristics of the porous medium, on the thermal conductivity of each phase, as well as on the particle temperature. A relatively strong dependence $\lambda(T)$ which is observed at high temperature $(T > 1{,}300\ K)$ shows that the radiative heat transfer plays a dominant role under these conditions. At $T > 1{,}300$ K the following correlation for λ can be used

$$\lambda = \tilde{\sigma}\varepsilon d\chi_p T^3 \tag{2.100}$$

where χ_p is the dimensionless factor depending on porous shape.

When convective heat transfer is dominant (in the low temperature domain) λ can be estimated using the relation

$$\lambda = 0.026[0.25\lg\frac{d}{0.06} + 0.1(d - 0.06) + 5.91\rho + 0{,}72\rho^2 + 1] \tag{2.101}$$

which is valid at $T \sim 300$ K. Here d [mm] is the scale of heterogeneity, $\rho\left[kg/m^3\right]$ is the density of the substance.

Using Eq. (2.99), Eq. (2.97) takes the form

$$\frac{\partial}{\partial t}\left(\sum_i \rho_i h_i\right) + \nabla\cdot(\rho_1 h_1 \mathbf{v}_f) = \nabla\cdot(\lambda\nabla T) + Q \tag{2.102}$$

Taking into account that

$$\nabla\cdot(\rho_1 h_1 \mathbf{v}_f) = h_1\nabla\cdot(\rho_1\mathbf{v}_f) + \rho_1\mathbf{v}_f\cdot\nabla h_1 \tag{2.103}$$

we transform the total energy balance to the form

$$\frac{\partial}{\partial t}\left(\sum_i \rho_i h_i\right) + h_1\nabla\cdot(\rho_1\mathbf{v}_f) + \rho_1\mathbf{v}_f\cdot\nabla h_1 = \nabla\cdot(\lambda\nabla T) + Q \tag{2.104}$$

Mass and momentum balance equations. Consider the balance of a gaseous substance in a chosen control volume V surrounded by surface S. Taking into account that the mass in the volume varies due to the mass flux through the surface

S, as well as due to transformation inside V, we can write the mass balance equation in the following form:

$$\int_V \frac{\partial \rho_s}{\partial t} dV = -\int_S \rho_s \mathbf{v}_s \cdot \mathbf{n} dS + \int_V G_s dV \tag{2.105}$$

where G_s is the specific rate of change of the s-th component due to chemical conversion (phase transition and chemical reactions).

Using the Gauss theorem we arrive at

$$\int_V \left(\frac{\partial \rho_s}{\partial t} + \nabla \cdot (\rho_s \mathbf{v}_s) - G_s \right) dV = 0 \tag{2.106}$$

Since, the control volume V is arbitrary, Eq. (2.106) reduces

$$\frac{\partial \rho_s}{\partial t} + \nabla \cdot (\rho_s \mathbf{v}_s) = G_s \tag{2.107}$$

Summation over s yields the mass balance equation for the gaseous phase as a whole

$$\frac{\partial \rho_1}{\partial t} + \nabla(\rho_1 \mathbf{v}_f) = G_1 \tag{1.108}$$

where $G_1 = \sum_s G_s$.

Now we consider the mass balance equation for the solid phase. Since, in the chosen system of coordinates the solid matter is motionless, the change of its mass within the control volume V is determined only by phase conversion. Therefore, the mass balance equation of the solid matter is

$$\int_V \left(\frac{\partial \rho_2}{\partial t} - G_2 \right) dV = 0 \tag{2.109}$$

where G_2 is the specific rate of solid phase transformation due to chemical conversion.

Equation (2.109) yields

$$\frac{\partial \rho_2}{\partial t} = G_2 \tag{2.110}$$

By virtue of the mass conservation, G_1 and G_2 are related as

$$G_1 + G_2 = 0 \tag{2.111}$$

Combining Eqs. (2.108) and (2.110) yields

$$\frac{\partial \rho}{\partial t} + \nabla \cdot (\rho_1 \mathbf{v}_f) = 0 \tag{2.112}$$

where $\rho = \rho_1 + \rho_2$.

Flows of multicomponent gaseous mixtures are highly complicated, since each of the components possesses a different relative velocity w_s defined as

$$\mathbf{w}_s = \mathbf{v}_s - \mathbf{v}_f \tag{2.113}$$

Taking into account Eq. (2.113), we transform Eq. (2.107) to the form

$$C_s\left(\frac{\partial \rho_1}{\partial t} + \nabla \cdot (\rho_1 \mathbf{v}_f)\right) + \rho_1 \frac{\partial C_s}{\partial t} + \rho_1 (\mathbf{v}_f \cdot \nabla) C_s = -\nabla \cdot (\rho_s \mathbf{w}_s) + G_s \tag{2.114}$$

where $C_s = \rho_s / \rho$ is the concentration of the s-th component. Bearing in mind Eq. (2.108), Eq. (2.114) reduces to

$$\rho_1 \frac{\partial C_s}{\partial t} + \rho_1 (\mathbf{v}_f \cdot \nabla) C_s = -\nabla \cdot (\rho_s \mathbf{w}_s) + G_s - C_s \sum_s G_s \tag{2.115}$$

The diffusion mass flux in Eq. (2.115) is determined by the Fick law

$$\rho_s \mathbf{w}_s = -\rho D_s \nabla(C_s) \tag{2.116}$$

Here D_s is the macroscopic diffusion coefficient of the s-th component.

Consider the momentum balance equation for the gaseous phase. In the integral form this equation reads

$$\int_V \frac{\partial \rho_1 \mathbf{v_f}}{\partial t} dV = -\int_S \rho_1 v_{f_n} v_{f_n} \mathbf{n} dS + \int_S \mathbf{P}_n dS + \int_V \rho_1 \mathbf{g} dV + + \int_V \mathbf{F_{fr}} dV + \int_V \mathbf{F_p} dV \tag{2.117}$$

Here $v_{f_n} = \mathbf{v_f} \cdot \mathbf{n}$ is the normal component of the filtration velocity, $\mathbf{P_n} = -\mathbf{n}P$.

The first term on the right in Eq. (2.117) accounts for the momentum flux through the surface S, and the second one expresses the effect of the surface forces P_n related to pressure and viscosity stress in the gaseous phase. The third term accounts for the effect of gravity (**g** is gravity acceleration). The fourth and the fifth terms account for the resultant drag force $\mathbf{F_{fr}}$ and the force $\mathbf{F_p}$ acting on gas from the particle side due to nonuniform macroscopic pressure from the solid phases. It is expressed as

$$\mathbf{F_p} = \sum_k \mathbf{P_k} \tag{2.118}$$

where subscript $k = 1,2,3,...$ refers to solid components.

At low density and filtration velocity of the gaseous phase, it is possible to neglect the effect of gravity and inertia forces, as well as the viscosity stresses in Eq. (2.117) compared to the friction forces acting at the developed interface surface inside the particle. Bearing in mind Eq. (2.118) we transform Eq. (2.117). Using the Gauss theorem, we can present Eq. (2.117) (after all is neglected) in the following reduced form

$$\mathbf{F_{fr}} = \nabla P \tag{2.119}$$

where $P = P_1 + \sum_k P_k \simeq P_1$ is the pressure of the mixture of effective continua.

The friction force can be presented in the form

$$\mathbf{F_{fr}} = -\chi_* \rho_1 \nu_1 \mathbf{v}_f \tag{2.120}$$

where χ_* is a correction coefficient, and ρ_1 and ν_1 are the density and the kinematic viscosity of fluid.

The coefficient χ_* accounts for the dependence of the friction drag force on the geometry of the porous medium. The coefficient k that is inversely proportional to χ_* is called the permeability of the porous medium. From the dimensional arguments it is seen that k is proportional to the square of a characteristic size of the pores d_{ef}

$$k = k_o\, d_{ef}^2 \qquad (2.121)$$

where k_o is a dimensionless coefficient depending on the porosity of the medium and the tortuosity of the pores.

There are a number of correlations for calculations of the permeability of porous media. For example, the Carman–Kozeny formula $k = (m^3 d_{ef}{}^2)/(120(1-m)^2)$, where m is porosity. They are based on the representation of a real porous medium with arbitrary void distribution by an ideal medium of a regular structure. These correlations allow for an estimate of the order of magnitude of the permeability of porous media.

Denoting

$$k_f = k(\rho_1 \nu_1)^{-1} \qquad (2.122)$$

we present Eq. (2.119) in the form of the Darcy law

$$\mathbf{v_f} = -\, k_f \nabla P \qquad (2.123)$$

where k_f is the filtration coefficient.

Additional correlations. The system of Eqs. (2.97), (2.108), (2.110) and (2.135) should be supplemented by an equation of state of the gas, for example, by that for the ideal gas

$$P_1 = \rho_1^o R_g T_1 \qquad (2.124)$$

as well as by a macrokinetic law for the rate of chemical reaction, and by the correlations determining the dependence of the physical properties on temperature; here R_g is a gas constant of a specific gas.

Special cases. Heating of coal particles undergoing pyrolysis. The solution of the problem on particle heating is significantly simplified when the effect of the mass transfer is negligible compared to the effect of conversion of solid material is accompanied by the formation of condensed products only. This corresponds to an early stage of pyrolysis of coal particles. According to Phuoc and Mathur (1991) this process may be schematically presented as follows. Assume that a spherical coal particle is instantaneously immersed in a hot gaseous medium. As a consequence of the heat transfer with the environment, a non-uniform temperature field is formed inside the particle. The temperature of the solid material has a maximum at the particle surface and a minimum at its center. It is also assumed that process of phase conversion is activated. This means that there is a specific temperature (pyrolysis temperature T_{py}), at which the transition of the initial material into a pyrolysis product occurs. Accordingly, when the surface temperature reaches T_{py}

at the particle surface, an infinitely thin pyrolysis front is formed. During the particle heating this front propagates inwards in the particle. The pyrolysis front subdivides the particle in two domains: the external one (the pyrolysis layer) and the internal one (the initial material).

Taking into account that in this case $\mathbf{v_f} = 0$, $Q = 0$, we consider Eq. (2.104) for the external and internal domains. Assuming that λ_i and c_{pi} are constant, $h_i = c_{pi} T_i$ and $\xi_i = 1$ in each of the domains, we reduce the problem to the solution of the following set of equations:

$$\frac{\partial T_1}{\partial t} = \alpha_i \nabla^2 T_1 \tag{2.125}$$

$$\frac{\partial T_2}{\partial t} = \alpha_2 \nabla^2 T_2 \tag{2.126}$$

where $\alpha = \lambda/(\rho^\circ c_p)$ is the thermal diffusivity; subscripts 1 and 2 refer to the initial material and the pyrolysis product.

The initial and boundary conditions of the problem arc

$$t = 0 \quad T_1 = T_2 = T_\infty \tag{2.127}$$

$$t>0 \quad \begin{cases} r=0: & \dfrac{\partial T_1}{\partial r} = 0 \\ r = r_f: & \lambda_1 \left(\dfrac{\partial T_1}{\partial r}\right) - \lambda_2 \left(\dfrac{\partial T_2}{\partial r}\right) = Q_{py} \\ r = r_s: & \lambda_1 \dfrac{\partial T_1}{\partial r} = \lambda_2 \dfrac{\partial T_2}{\partial r} = I - \left[h(T_s - T_\infty) + \tilde{\sigma}\varepsilon(T_s^4 - T_\infty^4)\right] \end{cases} \tag{2.128}$$

where $Q_{py} = \rho_2 q_{py} (dr_f/dt) = q_{py} W_{py}$, W_{py} is the rate of matter conversion, q_{py} is the latent heat of pyrolysis, r_f the radius corresponding to the pyrolysis front, $I = I_0 \varepsilon_0$, I_0 is the radiative heat flux from an external source of radiation.

The numerical solution of the system of Eqs. (2.125) and (2.126) found in Phuoc and Mathur (1991) showed that two different scenarios of the process can be realized depending on the ratio of the characteristic time of particle pyrolysis ($t_{py} = r_s/v_{py}$, v_{py} is the speed of propagation of the pyrolysis front) and the time that is needed to start oxidation of the particle surface t_{ox}. When $t_{py}/t_{ox} < 1$ (low I, small r_s and high λ_1), the particle is totally pyrolyzed before the surface temperature reaches the temperature at which an intensive oxidation of the solid material begins. In the second case corresponding to $t_{py}/t_{ox} > 1$ the particle

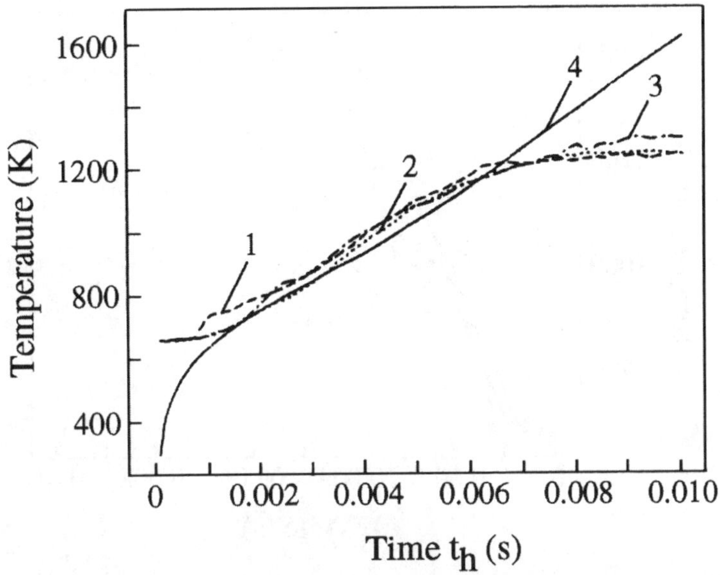

Fig. 2.18 Comparison of the predicted and measured temperature histories of carbon spheres. Reprinted from Maloney et al. (1991), with permission. Pulse duration was fixed at 10 ms. Curve 1, 135–μm diameter and 1,100-W/cm^2 pulse intensity; curve 2, 140–μm diameter and 1,160-W/cm^2 pulse intensity; curve 3, 135–μm diameter and 1,160-W/cm^2 pulse; curve 4, to the model prediction for a 135–μm diameter particle, and 1,160 -W/cm^2 pulse intensity

oxidation begins under the conditions when a narrow pyrolysis layer is formed close to the particle surface.

Temperature histories of carbon and coal particles during the early stage of heating and devolatilization were studied by Maloney et al. (1991). Experiments were carried out with particles of diameters $d = 115–140\,\mu m$ which were rapidly heated up by means of a pulse laser with the energy flux $(500–1000)[W/cm^2]$ and heating rate about 10^5 $[K/s]$. The results for the carbon spheres are plotted in Fig. 2.18. It is seen that the temperature of the carbon particles increases practically linearly during the initial stage of heating ($1 < t_h < 6.5$ [ms]) when the particle temperature increases from $T_p = 850$ K to $T_p = 1{,}200$ K. Latter on ($t_h >$ 6.5 [ms]) the rate of particle heating decreases significantly.

The application of the energy equation in the form of Eqs. (2.125) and (2.126) for carbon and coal particles was based on a number of simplifying assumptions. The comparison of the experimental data with the theoretical predictions in Fig. 2.18 shows that there is a fairly good agreement between the measured and the calculated particle temperatures within the range $1 < t_h < 6.5$ [ms]

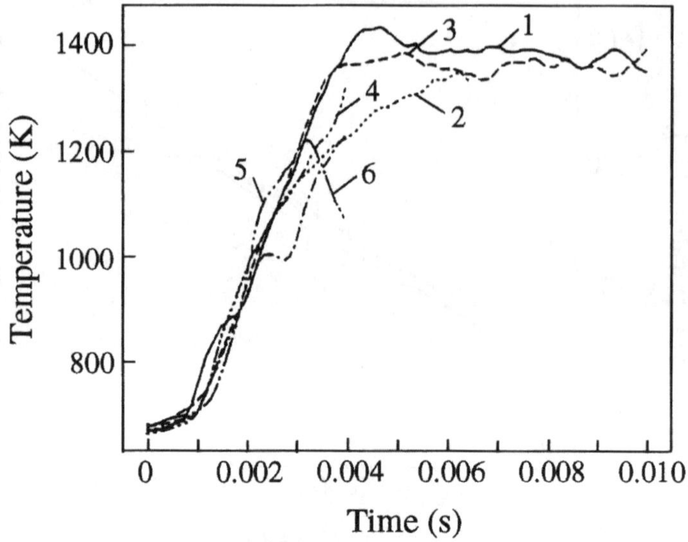

Fig. 2.19 Measured temperature records for six particles of PSOC 1451 coal. Reprinted from Maloney et al. (1991), with permission. Curve 1, 127–μm diameter and 1,100-W/cm^2 intensity, 10-ms pulse duration; curve 2, 115–μm diameter and 1,040-W/cm^2 intensity, 10-ms pulse duration; curve 3, 110–μm diameter and 1,050-W/cm^2 intensity, 10-ms pulse duration ; curve 4, 110–μm diameter and 1,050-W/cm^2 intensity, 6-ms pulse duration; curve 5, 120–μm diameter and 1,040 W/cm^2 intensity, 4-ms pulse duration; curve 6, 115–μm diameter and 1,040-W/cm^2 intensity, 3-ms pulse duration

(the deviation does not exceed 50 K). Therefore, simplified theory appears to be quite satisfactory. A closer inspection, however, reveals drawbacks of this theory at high temperatures when such processes as volatilization, particle fragmentation, and attenuation of particle emission resulting from screening of the particle surface by a volatile aerosol cloud, etc. become important.

The data on heating of coal particles are shown in Fig. 2.19. There are two characteristic stages of the process. The first one corresponds to a steep growth of the particle temperature from its initial value to the maximum one. The behavior of the particle during this stage causes a number of additional phenomena, namely, particle rotation due to light-volatile ejection, particle swelling by up to 30%, etc. During the second, high temperature stage the particle temperature remains practically constant as a result of attenuation of the radiative flux by the volatile cloud and particle fragments.

The comparison of the predicted and experimental data on the temperature history of a coal particle is presented in Fig. 2.20. Unfortunately, it is not satisfactory: there is a significant difference in results even using a measured particle size. Thus, in spite of attracting such data the simplest model of particle heating is confined to qualitative analysis of the phenomenon.

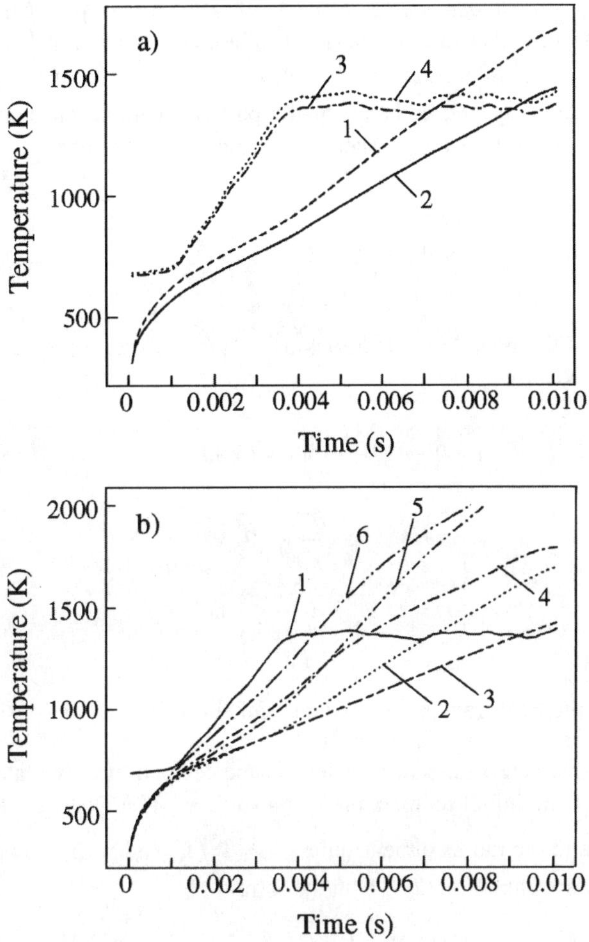

Fig. 2.20 Comparison of the temperature histories predicted and measured for PSOC 1451D coal particles. The coal particle of an initial diameter of 110–μm is heated by a 10-ms pulse of 1,040-W/cm^2 intensity. Reprinted from Maloney et al. (1991), with permission. a) Temperature versus particle emissivity ε: curve 1, measured temperature assuming ε =0.8; curve 2, predicted temperature assuming ε =0.8; curve 3, measured temperature assuming ε =1.0; curve 4, predicted temperature assuming ε =1.0. b) Calculations performed using the following property assumptions: curve 1, measured temperature history assuming $\varepsilon = 1.0$; curve 2, prediction using the measured size; curve 3, prediction using the measured size as the model input; curve 4, prediction using the measured size and a constant particle thermal conductivity; curve 5, prediction using the measured size and a constant particle c_p; curve 6, prediction using the measured size, constant particle thermal conductivity and c_p

Lumped-capacitance model. Consider heating of a spherical particle in a fluid flow. Assume that rates of phase transition and of chemical reactions inside the particle are negligible, so that Q = 0. Assume also $\mathbf{v}_f = 0$, $\rho_2 = \rho_2^{\circ}$ and $\lambda = \lambda_2$. Thermal conductivity and enthalpy of the particle material are considered to be constant $\lambda_2 = \text{const}$, $h_2 = c_2 T_2$. Then Eq. (2.104) takes the form

$$\rho_2^{\circ} c_2 \frac{\partial T_2}{\partial t} = \lambda_2 \nabla^2 T_2 \tag{2.129}$$

The initial and the boundary conditions of the problem for a pure convective heat transfer are given by

$$t = 0 \qquad T_2 = T_2(0,r) \tag{2.130}$$

$$t > 0 \quad \begin{cases} r = 0: & \dfrac{\partial T_2}{\partial r} = 0 \\ r = r_s: & \lambda_2 \left(\dfrac{\partial T_2}{\partial r}\right) = h(T_\infty - T_s) \end{cases} \tag{2.131}$$

where h is the heat transfer coefficient, and subscripts s and ∞ refer to the particle surface and the undisturbed fluid.

As the characteristic scales of length and temperature, we take the particle radius r_s and the initial temperature of its surface $T_s' = T_s(0, r_s)$. Rendering the temperature and the radius dimensionless $\tilde{T} = T_2/T_s'$ and $\tilde{r} = r/r_s$, we arrive at the second boundary condition (2.131) in the form

$$\left(\frac{\partial \tilde{T}}{\partial \tilde{r}}\right)_s = \text{Bi}(\tilde{T}_\infty - \tilde{T}_s) \tag{2.132}$$

where $\text{Bi} = (h r_s)/\lambda_2$ is the Biot number.

Taking into account the first condition (2.131), we estimate the temperature gradient inside the particle

$$0 \le \left(\frac{\partial \tilde{T}}{\partial \tilde{r}}\right) \le \left(\frac{\partial \tilde{T}}{\partial \tilde{r}}\right)_s \tag{2.133}$$

From the conditions (2.132) and (2.133) it follows that the particle temperature does not depend on $\tilde{r}$ at small Biot number. Then the temperature is a function of time only $\tilde{T}_2 = \tilde{T}_2(t)$. This makes it possible to model a particle as a homogeneous lump of matter with a uniform time-dependent temperature. The energy balance equation for such a lump is

$$\rho_2^o c_2 \frac{dT_2}{dt} = h\,s(T_\infty - T_s) \tag{2.134}$$

where T_2 is the particle temperature, T_2=T_s, s is the specific particle surface (the ratio of the surface area to the volume). In the case of the combined (convective and radiative) heat transfer boundary condition at the particle surface, Eq. (2.132) is replaced by

$$\left(\frac{\partial \tilde{T}}{\partial \tilde{r}}\right)_s = \mathrm{Bi}(\tilde{T}_\infty - \tilde{T}_s) + \mathrm{St}(\overline{T}_\infty^4 - \tilde{T}_s^4) \tag{2.135}$$

where $\mathrm{Bi} = (hr_s)/\lambda_2$ is the Biot number and $\mathrm{St} = (\tilde{\sigma}\varepsilon r_s T_s'^3)/\lambda_2$ the Stefan number.

The lumped model is valid when $\mathrm{St} \simeq \mathrm{Bi} \ll 1$. In this case the energy equation takes the form

$$\rho_2^o c_{p_2} \frac{dT_2}{dt} = hs(T_\infty - T_2) + \tilde{\sigma}\varepsilon s(T_\infty^4 - T_2{}^4) \tag{2.136}$$

The initial condition for Eqs. (2.134) and (2.136) is

$$t = 0, \qquad T_2 = T_0 \tag{2.137}$$

Equation (2.136) has a simple analytical solution in the case of the pure convective or pure radiative heat transfer. In the first case the solution is

$$\frac{\Delta T}{\Delta T_o} = 1 - \exp\left(-\frac{t}{t_*}\right) \tag{2.138}$$

where $\Delta T = T_2 - T_0$, $\Delta T_0 = T_\infty - T_o$, $t_* = \rho_{2.1} c_{2.1} (d^2/(12\alpha_1))$ is the thermal relaxation time at $Nu = 2$, Nu is the Nusselt number, $\rho_{2.1} = \rho_2/\rho_1$, $c_{2.1} = c_2/c_1$, and α_1 is the thermal diffusivity.

The thermal relaxation time determines the duration of the particle heating at which its excess dimensionless temperature $\Delta T/\Delta T_o$ decreases in time e. It strongly depends on the particle diameter and increases as d^2. The dynamic relaxation time of a spherical particle corresponding to velocity equilibration under the action of the Stokes drag force is (Soo 1990, Crow et al. 1998)

$$t = \rho_{2.1} \frac{d^2}{18\nu_1} \tag{2.139}$$

Then the ratio of the thermal relaxation time t_* to the dynamical time t is

$$\tau = \frac{3}{2} c_{2.1} \, Pr \tag{2.140}$$

where $Pr = \nu_1/\alpha_1$ is the Prandtl number, $\tau = t_*/t$.

The heat capacity of the majority of diatomic gases is about 1 $[kJ/kg \cdot K]$ (e.g. for O_2 0.915, for N_2 1.031, and for air $1.005[kJ/kg \cdot K]$. On the other hand the heat capacity of solid substance varies within the range 0.8–$8.0[kJ/kg \cdot K]$. However, the heat capacity of widespread solid fuels such as coals is about $1[kJ/kg \cdot K]$. Since the Prandtl number for gases is about 0.75, the ratio of the thermal to dynamic relaxation times τ for coal dust flow is O(1). In solid/liquid flows τ is of the order of O(Pr). Since the Prandtl number of liquids (excluding liquid metals) exceeds unity, the initial stage of particle heating occurs in these flows when the relative velocity of the phases is negligible. In contrast with this in coal dust flows where the dynamic and thermal relaxation times are close to each other, the effect of the particles, t motion on their heating is significan.

Integration of Eq. (2.136) in the case of a pure radiaive heat transfer yields

$$\ln \frac{(1+\bar{T})(1-\bar{T}_o)}{(1-\bar{T}_o)(1+\bar{T}_o)} - 2\arctan \frac{\bar{T}-\bar{T}_o}{(1+\bar{T}_o)} = \frac{t}{t_*^R} \tag{2.141}$$

where $\bar{T} = T_2/T_\infty$, $\bar{T}_o = T_o/T_\infty$, $t_*^R = d^2/\alpha_R$ is the thermal relaxation time for radiative heat transfer, $\alpha_R = \lambda_R/(\rho_2^o c_2)$, and $\lambda_R = \tilde{\sigma}\varepsilon \, T_\infty^3 d$.

It is seen that in the case of a pure radiative heat transfer, heating of particles does not depend on their motion.

The lumped-capacitance model can also be used for studying the particle temperature when phase transitions occur immediately on the particle surface, e.g. as in the cases of evaporation and condensation. The problem reduces to integration of Eq. (2.104) which takes the following form

$$\frac{\partial \rho_2 h_2}{\partial t} = \nabla \cdot (\lambda_2 \nabla T_2) \tag{2.142}$$

with the initial and boundary conditions

$$t = 0 \qquad T_2 = T_2(0, r) \tag{2.143}$$

$$t > 0 \quad \begin{cases} r = 0: & \left(\dfrac{\partial T_2}{\partial r}\right) = 0 \\ r = r_s: & \lambda_2 \left(\dfrac{\partial T_2}{\partial r}\right) = h(T_\infty - T_s) + \tilde{\sigma}\varepsilon(T_\infty^4 - T_s^4) + h_1 \dfrac{dm}{dt}\dfrac{1}{S} \end{cases} \tag{2.144}$$

where $dm/dt = -h_m S \rho_* (C_s - C_\infty)$, h_m is the mass transfer coefficient, h_l is the gaseous phase enthalpy, C_S is the concentration of the gaseous substance formed at the solid surface, ρ_* is the characteristic density of the gaseous substance.

When the Biot and Stefan numbers are sufficiently small and the dimensionless group $(h_l h_m \rho_* r_o)/\lambda_2 \ll 1$ (r_o is the initial particle radius), the temperature gradient inside the particle is negligible. Using the lumped-capacitance model ($T_2 = T_s$) we can present the energy equation as

$$\frac{d\rho_2 h_2}{dt} = hs(T_\infty - T_2) + \tilde{\sigma}\varepsilon s(T_\infty^4 - T_2^4) + h_1 \frac{d\rho_2}{dt} \tag{2.145}$$

or

$$\rho_2 \frac{dh_2}{dt} = hs(T_\infty - T_2) + \tilde{\sigma}\varepsilon s(T_\infty^4 - T_2^4) + (h_1 - h_2)\frac{d\rho_2}{dt} \tag{2.146}$$

Taking into account that $\rho_2 V = m_2$, $h_1 - h_2 = q_e$, where V is the particle volume and q_e is the latent heat of the phase transition process we finally obtain

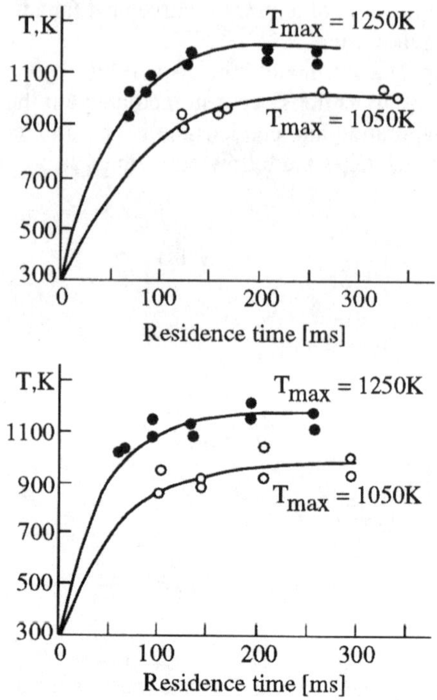

Fig. 2.21. Comparison of the measured and calculated particle temperature as a function of the residence time for 1,250 K (•) and 1,050 K (○) gas temperatures. Reprinted from Fletcher (1989), with permission. a) For 115–μm diameter particles. b) For 6–μm diameter particles

$$m_2 \frac{dh_2}{dt} = hS(T_\infty - T_2) + \varepsilon s(T_\infty^4 - T_2^{\ 4}) + q_e \frac{dm_2}{dt} \tag{2.147}$$

Fletcher (1989) used the lumped-capacitance model to study heating and devolatilization of particles of bituminous coals with high volatiles content in nitrogen at gas temperatures of 1,050–1,250 K. To improve the agreement between the theoretical and experimental results he introduced in the energy balance equation a correction factor accounting for the effect of the mass transfer on the heat transfer coefficient. To estimate the mass of volatiles released from the particle surface, as well as the particle swelling, the one-step Arrhenius kinetic law and a linear dependence of the particle diameter on volatiles content were used. The calculations showed that measured rate of coal particle heating is higher than the predicted one. However, with an appropriate choice of the initial value of the thermal relaxation time, it is possible to reach a fairly good agreement between the calculation and experiment (Fig. 2.21). It is emphasized that the lumped-capacitance model allows for the estimation of a number of important characteristics of such complex phenomena as mass loss, particle swelling, etc.

1.2.3 Devolatilization

Background. Devolatilization is an elementary step in coal combustion. It consists of thermal decomposition of coal and escape of volatiles from particles of solid fuel to ambient space. This process involves a number of successive and parallel stages such as heating of coal particles, their pyrolysis, swelling and melting of solid material, filtration of volatiles through the porous matrix or migration of bubbles inside the particles, etc. Accordingly, the devolatilization rate of coal particles depends on the composition of the parent coal, kinetics of pyrolysis reactions, physical properties of solid and gaseous phases, as well as on the conditions of particle/fluid interaction. These problems were tackled in numerous of investigations carried out during the last decades. At present, a number of semi-empirical models for rapid coal devolatilization have been proposed (Badzioch and Hawksley 1970, Anthony and Howard 1976, Niksa 1986, Niksa and Kerstein 1986, Niksa et al. 1987, Unger and Suuberg 1981). The models account for the general feature of the devolatilization process. To calculate the devolatilization rate of different types of coals it is necessary to determine the kinetic constants of the pyrolysis reaction (Beck and Hayhurst 1990, Solomon et al. 1992). The results of these studies were summarized by Anthony and Howard 1976, Gavalas 1981, Saxena 1990, and Solomon et al. 1992). Reffering the reader to these surveys, we shall not discuss its actual structure of coal and the mechanism of chemical reactions, its thermoplastic behavior, etc, but describe briefly the results concerning the formal kinetics of pyrolysis, as well as the influence of some physical factors on the devolatilization rate.

Phenomenological models. These models are based on some hypotheses concerning the dependence of the devolatilization rate on the current volatile content in the coal and on the particle temperature. The simpliest hypothesis consists in the assumption that the real process of coal pyrolysis may be simulated by a single global first-order reaction occurring throughout the coal particle (Badzoich and Hawksley 1970)

$$\frac{dV}{dt} = k(V^* - V) \tag{2.148}$$

Here V is the total volatiles evolved up to time t, V^* is the limiting value of V at $t \to \infty$, and k corresponds to the rate constant of the pyrolysis reaction.

The dependence of the devolatilization rate on temperature is contained in the constant k which is expressed by the Arrhenius law

$$k = k_o \exp\left(-\frac{E}{RT}\right) \tag{2.149}$$

where k_o is the pre-exponential factor, E is the activation energy, and R is the universal gas constant.

The model of Badzoich and Hawksley (1970) accounts for the effect of both principal factors determining non isothermal devolatilization, namely: V and T. At the same time it does not consider an actual pyrolysis reaction. Therefore, quantitative agreement between the theoretical predictions and experiments is typically achieved only within a narrow temperature range. In fact, the model (2.148)–(2.149) is valid only for small variation of temperature, in a situation close to isothermal.

Wiser et al. (1981) simulated the coal devoltilization by the n-th order reaction

$$\frac{dV}{dt} = k(V^* - V)^n \tag{2.150}$$

The magnitude of the parameter n is found by comparing Eq. (2.150) with experimental data on non isothermal devolatilization. Obviously, the efficiency of such a model depends on whether the parameter n is constant over a wide range of the particle temperatures. The experimental data show that the value of n depends on the temperature level. Therefore, for an early stage of coal devolatilization (in the low temperature range) n = 2, whereas for the final stage of the process (in the high temperature range) n = 1. This circumstance, as well as the dependence of the parameter n on the type of coal makes the use of the model (2.150) difficult.

A better fit of the theoretical results to experiment can be achieved with the help of the multireaction models (Pitt 1962, Anthony et al. 1975, Kobayashi et al. 1976). In accordance with these models, the actual process of coal decomposition is simulated by a number of pseudo-reactions of the first order. Each of these reactions is characterized by its own kinetic constants, which are chosen so that the given pseudo-reaction approaches the actual one within a certain temperature interval.

Kobayashi et al. (1976) proposed a model of coal devolatilization involving two competing volatile-producing reactions. The process of coal devolatilization is assumed to occur as per:

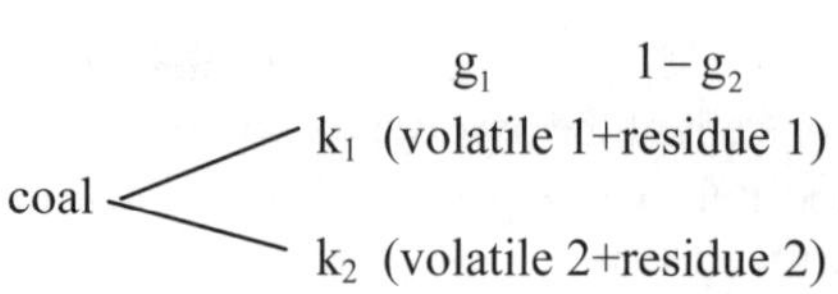

where k_1 and k_2 are the kinetic rate constants, and g_1 and g_2 are the mass fraction of coal transformed to volatiles via reactions 1 and 2. The kinetic rate constants are expressed via the Arrhenius law

$$k_i = k_{oi} \exp\left(-\frac{E_i}{RT}\right) \tag{2.151}$$

The magnitudes of the pre-exponental factor k_{oi}, and of the activation energy E_i are chosen such that the first reaction is dominant at low temperatures, whereas the second one is dominant at high temperatures. The model of two competing reactions model agrees fairly good with experiments on devolatilization of bituminous coals (Kobayashi et al. 1976, Ubhayakar et al. 1976).

Anthony et al. (1975) modeled the thermal decomposition of coal by numerous independent chemical reactions that occur more or less simultaneously. Each of them possesses its own kinetic constant different from the others. In the frame of this approach, the rate of devolatilization is expressed as

$$\frac{dV_i}{dt} = k_i (V_i^* - V) \tag{2.152}$$

Integration of Eq. (2.152) yields

$$V_i^* - V_i = V_i^* \exp(-\int_0^\infty k_i dt) \tag{2.153}$$

In order to simplify the problem the authors assumed that the pre-exponential factor k_{oi} in the form of the Arrhenius law (2.151) similar for all the reactions, which differ only in their activation energies. Besides that, a number of these reactions is assumed to be large enough to permit E to be expressed as a continuously distributed function $f(E) = dF/dE$, where F represents the fraction of the potential loss of volatiles V^* which have an activation energy less than E. Then dV^* can be written as

$$dV^* = V^* dF = V^* f(E) dE \tag{2.154}$$

with

$$\int_0^\infty f(E)\, dE = 1$$

Summation of the contributions of all the reactions yields the total amount of the volatile material yet unreleased

$$V^* - V = V^* \int_0^{\infty} \exp\left(-\int_0^{\infty} k(E)\, f(E)\, dE \right) \tag{2.155}$$

To evaluate the integral on the right hand side of Eq. (2.155) it is necessary to define the function $f(E)$. It may be taken as the Gaussian distribution function

$$f(E) = \frac{1}{\sigma(2\pi)^{1/2}} \exp\left\{ \frac{(E - E_o)^2}{2\sigma^2} \right\} \tag{2.156}$$

where E_0 and σ are the mean activation energy and the standard deviation.

Then Eq. (2.155) takes the form

$$V^* - V = \frac{V^*}{\sigma(2\pi)^{1/2}} \int_{-\infty}^{\infty} \exp\left(-\int_0^{t} k(E)\, dt \right) \exp\left\{ \frac{(E - E_o)^2}{2\sigma^2} \right\} dE \tag{2.157}$$

Since $k(E) = k_o \exp(-E/RT)$, Eq. (2.157) is transformed to the form

$$V^* - V = \frac{V^*}{\sigma(2\pi)^{1/2}} \int_{-\infty}^{\infty} \exp\left\{ -k_o \int_0^{t} \exp\left(-\frac{E}{RT} \right) dt \right\} \exp\left\{ \frac{(E - E_o)^2}{2\sigma^2} \right\} dE \tag{2.158}$$

The comparison of the calculated and experimental data on devolatilization of lignite and bituminous coals is presented in Fig. 2.22. It shows that the model of Anthony et al. (1975) agrees fairly well with the measurements. Ragland and Yang (1989) studied combustion of coarse coal particles ($5.5 < d_p < 9.9$ mm) in a convective high temperature flow. The measurements carried out at $63<Re_p<126$, $900\,K < T_{\infty} < 1{,}200\,K$, $4.5\%<C_{O_2}<21\%$ showed that the predictions of the theory of Anthony et al. (1975) agree well with the experimental data for the following values of the kinetic constants: $k_0 = 2.91 \cdot 10^8\, s^{-1}$, $E = 154.6$ [kJ/mol], $\sigma = 17.51$ [kJ/mol]

Influence of physical factors. As was mentioned earlier, heating of coal particles is accompanied by a significant change in the particle properties because of chemical conversion of solid material, as well as due to its swelling and melting. A coal particle comprrises a solid body which is penetrated by numerous pores

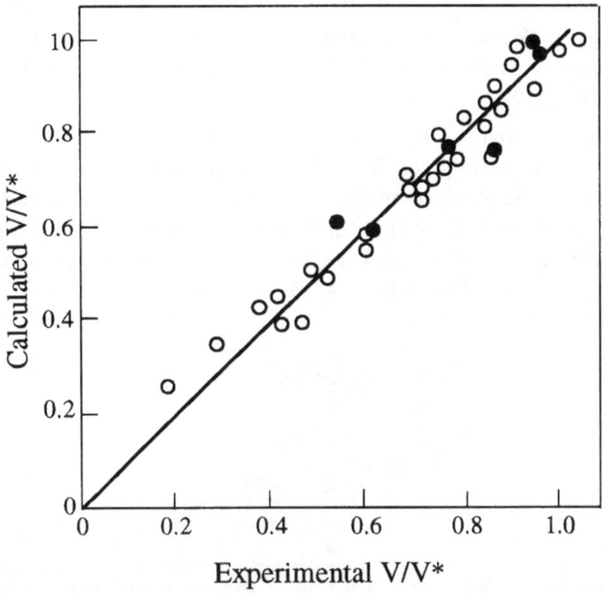

Fig. 2.22 Particle devolatilization. Comparison of the theoretical and experimental results. Reprinted from (Anthony et al. (1975), with permission. O the data for the lignite coals at all pressure, ● data for bituminous coal at 60 atm, the straight line of a unit slope fits the data fairly well.

and cracks. Gan et at. (1972) showed that coal particles have a wide pore size distribution, from micropores with $d<0.0012$ mm to macropores with $d>0.03$ mm. The change of the pore sizes affects both heat and mass transfer, and as a result, the devolatilization rate. Swelling of coal particles is characterized by the so-called "swelling index" which accounts for an increase in particle volume when it is heated (the index varies in the range of 3.5–9). The particles of plastic coals undergoing rapid heating are subdivided into six groups depending on their swelling (Saxena 1990). These include a wide spectrum of particles of different types of coal beginning with those that keep their volume virtually constant during heating and devolatilization, and ending up with particles that disintegrate.

Swelling of coal particles depends not only on the coal rank but also on the rate of heating, as well as on the particle temperature and residence time. Gale et al. (1995) showed that swelling and porosity of the particles of plastic coals increase or decrease depending on the heating rate. At low rate of particle heating swelling gradually increases with dT/dt, whereas at high rate of heating it decreases with dT/dt. The transition from an increasing swelling to a decreasing one occurs at $dT/dt \sim 5\cdot 10^3\,[K/s]$. For swelling coals a maximum porosity is reached near this heating rate. Note, that particle swelling and porosity correlate better with the maximum particle heating rate than with the particle temperature.

The effect of the heating rate on swelling and porosity of coal particles is illustrated by the following data: swelling ratio and porosity decrease by 35% and 15%, respectively, when the maximum particle heating rate increases from $2\cdot 10^4$ $[\mathrm{K/s}]$ to $7\cdot 10^4 [\mathrm{K/s}]$.

A strong dependence of the physical properties of coal particles on temperature determines the general approach for studying the devolatilization process. It consists in the simultaneous solution of the plastic and kinetic problems under the conditions of rapid particle heating and intensive interparticle mass transfer. The problem is significantly simplified for fine non plastic particles. In this case it is reduced to the solution of the following set of equations

$$\frac{dV}{dt} = k(V^* - V) \tag{2.159}$$

$$\rho c_p \frac{dT}{dt} = hs(T_\infty - T) \tag{2.160}$$

where $k = k_0 \exp(-E/RT)$, h is the heat transfer coefficient, s is the particle specific surface, c_p is the specific heat, ρ is the particle density, T_∞ is the ambient temperature. Equations (1.59) and (2.160) employ the lumped-capacitance model of heat transfer and the simplest kinetic model of pyrolysis.

For the purpose of future analysis it is worthwhile to exclude time from Eqs. (2.159) and (1.260), which yields

$$\frac{dV}{dT} = \frac{k_0 \rho c_p}{hs} \frac{V^* - V}{T_\infty - T} \exp\left(-\frac{E}{RT}\right) \tag{2.161}$$

Restricting our consideration to the early stage of devolatiliziation, we use the Frank-Kamenetskii transformation of the exponent in Eq. (2.161) (Frank-Kamenetskii 1969)

$$\exp\left(-\frac{E}{RT}\right) \approx \exp\left(-\frac{E}{RT_0}\right) \cdot \exp\left(\frac{E}{RT_0^2}(T - T_0)\right) \tag{2.162}$$

where T_0 is the initial particle temperature.

The transformation (2.162) is valid in the case of high activation energy E, which is typical true. We introduce an excess dimensionless temperature

$$\theta = \frac{E}{RT_0^2}(T - T_0) \tag{2.163}$$

Then Eq. (2.161) takes the following form

$$\frac{dV}{V^* - V} = Se^* \frac{\exp(\theta)}{\theta_\infty - \theta} d\theta \tag{2.164}$$

where

$$\theta_\infty = \frac{E}{RT_o^2}(T_\infty - T_o), \quad Se^* = \frac{k_o \rho c_p}{hs} \exp\left(-\frac{E}{RT_o}\right)$$

The integral of Eq. (2.164) determines the value of V as a function of the current temperature

$$\left(1 - \frac{V}{V^*}\right) = C(\theta_\infty - \theta)^{Se} \exp\left\{-Se\left[\frac{\theta_\infty - \theta}{1 \cdot 1!} - \frac{(\theta_\infty - \theta)^2}{2 \cdot 2!} + \frac{(\theta_\infty - \theta)^3}{3 \cdot 3} - \ldots\right]\right\} \tag{2.165}$$

where $Se = [(k_o \rho c_p)/(hs)] \cdot \exp(-\ (1 - \beta\theta_\infty)/\beta\ )$ is the Semenov number, $\beta = (RT_o)/E$, and C is the integration constant. Taking into account that $V = 0$ at $\theta = 0$ we obtain C as

$$C = \left\{\theta_\infty^{Se} \exp\left[-Se\left(\frac{\theta_\infty}{1 \cdot 1!} - \frac{\theta_\infty^2}{2 \cdot 2!} + \frac{\theta_\infty^3}{3 \cdot 3!} - \ldots\right)\right]\right\}^{-1} \tag{2.166}$$

Substitution of Eq. (2.166) into Eq. (2.165) yields

$$\left(1 - \frac{V}{V^*}\right) = \left(\frac{\theta_\infty - \theta}{\theta_\infty}\right)^{Se} \exp\left\{-Se\left[\frac{(\theta_\infty - \theta) - \theta_\infty}{1 \cdot 1!} - \frac{(\theta_\infty - \theta)^2 - \theta_\infty^2}{2 \cdot 2!} + \frac{(\theta_\infty - \theta)^3 - \theta_\infty^3}{3 \cdot 3!} - \ldots\right]\right\} \tag{2.167}$$

Bearing in mind that $\theta \ll 1$ it is possible to omit the terms containing $\theta^2, \theta^3, \ldots, \theta^n$ Then we obtain the following approximate expression for V:

$$\left(1-\frac{V}{V^*}\right)=\left(\frac{\theta_\infty-\theta}{\theta_\infty}\right)^{Se}\exp(Se\,\theta) \tag{2.168}$$

Now we determine the rate of devolatilization of a fine particle during the early stage of the process. For this purpose we calculate the derivative $\frac{d}{dt}\left(\frac{V}{V^*}\right)_{t=0}$. Using Eq. (2.168), we arrive at

$$\frac{d}{dt}\left(\frac{V}{V^*}\right)_{t=0}=Se\frac{1-\theta_\infty}{\theta_\infty}\left(\frac{d\theta}{dt}\right)_{t=0} \tag{2.169}$$

To calculate $(d\theta/dt)_{t=0}$ we use the thermal balance equation (2.160) which yields

$$\left(\frac{d\theta}{dt}\right)_{t=0}=\frac{\theta_\infty}{t_*} \tag{2.170}$$

where t_* is the thermal relaxation time. Then Eq. (2.169) takes the following form:

$$\left(\frac{dV}{dt}\right)_{t=0}=V^*\frac{Se}{t_*}(1-\theta_\infty) \tag{2.171}$$

Taking into account that $Se/t_* = k_o\exp(-(1-\beta\theta_\infty)/\beta)$ we obtain

$$\left(\frac{dV}{dt}\right)_{t=0}=V^*k_o(1-\theta_\infty)\exp\left(-\frac{1-\beta\theta_\infty}{\beta}\right) \tag{2.172}$$

Since θ_∞ and β have the orders of 10^{-1} and 10^{-2}, respectively, Eq. (2.171) takes the following form

$$\left(\frac{dV}{dt}\right)_{t=0}=V^*k_o\exp\left(-\frac{1}{\beta}\right) \tag{2.173}$$

The relation (2.173) shows that rate of devolatilization of fine coal particles does not depend on particle size and heating rate. That means that this process is essentially chemically controlled.

The above results correspond to the simplest situation when the temperature distribution inside the non-plastic particle is uniform (the case of small Biot numbers). In contrast with that the temperature distribution in coarse particle is inhomogeneous. The existence of the temperature difference between the particle surface and its center affects significantly the rate of devolatilization and the time of completion of this process (Saxena 1990, Solomon et al. 1992, Gat 1986, Suuberg 1988). The devolatilization rate of coarse particles is less than that of the fine particles because of the existence of the internal temperature gradients, slower heating rates, mass transfer limitations, repolymerization reactions, etc. Under certain conditions, the heat and mass transfer processes become dominant in the coarse particle devolatilization. The character of the process is determined by the competition of the external and internal heat transfer and chemical reactions.

The characteristic time scales of these processes are

$$t_{ex} \sim \frac{d^2}{\alpha_g}, \quad t_{in} \sim \frac{d^2}{\alpha_p}, \quad t_R \sim k_o^{-1} \exp\left(\frac{RT}{E}\right) \tag{2.174}$$

where t_{ex}, t_{in} and t_R are the characteristic time scales of the external and internal heat transfer processes and chemical reaction, respectively, $\alpha_g = (\lambda/(\rho c_p))_g$, $\alpha_p = (\lambda/(\rho c_p))_p$ are the thermal diffusivities, and subscripts g and p refer to the carrier gas and particle.

The magnitudes of the ratio of the characteristic times allows for the estimation of the conditions under which the devolatilization process is controlled by heat transfer or by the kinetics of the pyrolysis reactions (Suuberg 1988).

A self-consistent way to study devolatilization of coarse particles consists in the combined analysis of the heat and mass transfer occurring in the background of the chemical reaction (Phuoc and Durbetaki 1989). The model of a coarse particle undergoing pyrolysis is presented in Fig. 2.23. In accordance with this model, the domain under consideration includes a spherical particle and a gaseous film around its surface. It is subdivided in three regions where (I) the interparticle mass transfer, (II) the diffusion of mass and energy, and (III) the particle–free stream interaction occur.

In the frame of the quasi-stationary approach, the problem of volatile transport is described by the following equations:

$$r^{-2}\frac{d}{dr}(r^2 m_v) = \omega_v \tag{2.175}$$

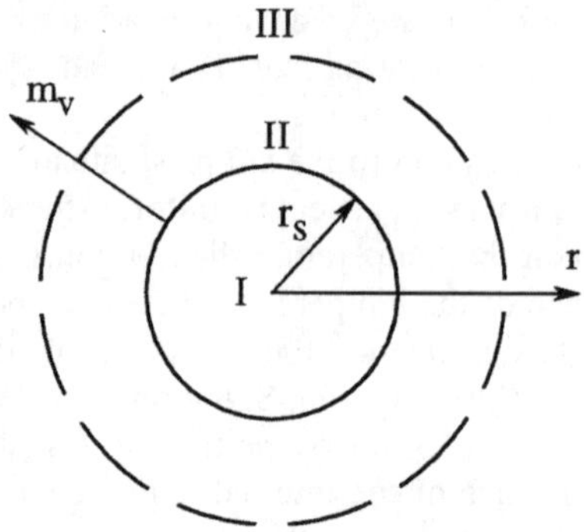

Fig. 2.23 Sketch of a devolatilizing coal particle undergoing pyrolysis. Reprinted from Phuoc and Durbetaki (1989), with permission. I, intraparticle mass transport. II, diffusion of mass and energy. III, free stream

$$r^{-2}\frac{d}{dr}(r^2 m_1) = 0 \tag{2.176}$$

for region I and

$$r^{-2}\frac{d}{dr}(r^2 m) = 0 \tag{2.177}$$

$$r^{-2}\frac{d}{dr}(r^2 m_i) = 0 \tag{2.178}$$

$$r^{-2}\frac{d}{dr}(r^2 Q) = 0 \tag{2.179}$$

for regions II and III.

In Eqs. (2.175)–(2.179) m is the total mass flow rate, $m_i = m_i C_i - (\lambda/c_p)(dC_i/dr)$ is the mass flow rates of the i-th species, C_i is the mass fraction of the i-th species, ω is the rate of volatiles production which is due to pyrolysis, $Q = mT - (\lambda/c_p)(dT/dr)$ is the heat flux, and subscripts v and 1 refer to volatiles and ambient gas, respectively. Spherical symmetry as assumed.

The system of the one-dimensional equations (2.175)–(2.179) describes an isothermal particle with constant physical properties of solid and gaseous materials, and equal binary diffusivities. The Lewis number (the ratio of the thermal and mass diffusivities) assumed to be one. The boundary conditions of the present problem are:

$$r = 0: \quad \left\{ \lim(r^2 m_i) = 0 \right.$$

$$r = r_s: \quad \begin{cases} m_{v.s} = m_{v.s} C_{v.s} - \rho D\left(\dfrac{dC_V}{dr}\right)_{r=r_s} \\ \lambda \left(\dfrac{dT}{dr}\right)_{r=r_s} = h(T_p - T_s), \quad T = T_s = T_p \\ C_{l.s} = 1 - C_{v.s} \end{cases} \tag{2.180}$$

$$r \to \infty: \quad \left\{ C_v = 0, \; C_l = 1, \quad T = T_\infty \right.$$

where subscript s corresponds to the particle surface.

To calculate the particle temperature during its pyrolysis, the lumped-capacitance model is used; the volatile production rates are calculated by means of Eq. (2.152), with the rate constant taken as in Anthony and Howard (1976).

The results of the computations carried out by Phuoc and Durbetaki (1989) revealed that both the heat and mass transfer affect the pyrolysis of coarse particles. In particular, they showed that an increase in the ambient temperature and a decrease in the particle size promote the devolatilization prosess, as well as overcome the diffusive transport within the particle. At a high rate of heat transport the weight loss is determined by the competition between pyrolysis and a secondary reaction, whereas at low heating rate the contribution of the internal mass transfer becomes more pronounced.

Heritzberg and Zlochower (1991) and Phuoc and Mathur (1991) developed another approach to the analysis of devolatilization of solid fuel. It is based on the assumption that decomposition of solid material occurs within a thin layer of solid substance which may be identified as the pyrolysis front. This front is initiated at the particle surface by the external heat flux and propagates into its interior (Fig. 2.24).

Heritzberg and Zlochower (1991) considered pyrolysis of a planar solid layer. They assumed that the rate of the pyrolysis front is linearly proportional to the net absorbed energy flux per unit area of the solid. The proportionality constant was assumed to be equal to the reciprocal of the total enthalpy change for the pyrolysis and devolatilization reactions. To calculate the rate of the pyrolysis front, the authors used the thermal balance equation which takes the following form in the frame of reference associated with the pyrolysis front

$$\rho u_f c_p(T) \frac{dT}{dx} = \frac{d}{dx}\left[\lambda(T) \frac{dT}{dx}\right] \tag{2.181}$$

Here u_f is the rate of pyrolysis front, and $c_p(T)$ and $\lambda(T)$ are the specific heat and thermal conductivity coefficient.

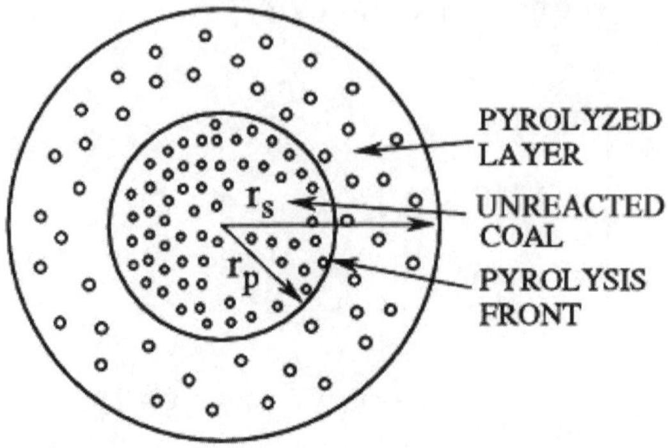

Fig. 2.24 Theoretical model. Reprinted from Phuoc and Mathur (1991), with permission

The boundary conditions for Eq. (2.181) are

$$\begin{aligned} x = 0 : \quad & I_{abs} = \lambda(T)\left(\frac{dT}{dx}\right)_s + \rho u_f \cdot q_e \\ x \to \infty : \quad & T(\infty) = T_0, \quad \left(\frac{dT}{dx}\right)_{x\to\infty} = 0 \end{aligned} \tag{2.182}$$

where q_e is the heat of vaporization, I_{abs} is the absorbed energy flux, T_0 is the undisturbed temperature.

Integration of Eq. (2.181) yields

$$\lambda(T_s)\left(\frac{dT}{dx}\right)_s = \rho u_f \int_{T_0}^{T_s} c_p(T)\, dT \tag{2.183}$$

where T_s is the surface decomposition temperature

Then using Eq. (2.183) and the first condition of (2.182) we find the mass rate of the pyrolysis wave

$$\rho u_f = I_{abs}\left[\int_{T_0}^{T_s} c_p(T)\, dT + q_e\right]^{-1} \tag{2.184}$$

Phuoc and Mathur (1991) considered the pyrolysis wave propagation in a spherical coal particle. They assumed that a particle with a uniform temperature distribution undergoes heating beginning from $t = 0$. At a certain moment of time $t = t_p$ the surface temperature reaches the temperature T_{pyr}. corresponding to coal pyrolysis. The latter leads to the formation of a pyrolysis front which propagates into the particle. Heat transfer occurs within the layers of unreacted coal and its char, whereas coal conversion proceeds at an infinitely thin pyrolysis front which subdivides the regions of unreacted coal and char. In accordance with this model, the evolution of the temperature field in the coal particle undergoing heating and devolatilization is described by Eqs. (2.125) and (2.126) with the boundary conditions (2.127) and (2.128). There are two different mechanisms of pyrolysis particle. At low radiant heat flux, small particle size and high thermal conductivity of unreacted coal, motion of the pyrolysis front is faster than the rise in the surface temperature, but slower than the rise in the temperature of the unrected core. That means that the coal particle is totally pyrolyzed before the surface temperature reaches the value when an intensive oxidation begins. The situation qualitatively changes for a coarse particle with low thermal conductivity. At high enough radiative flux the surface temperature reaches the value when an intensive oxidation begins before the pyrolysis front penetrates into the unreacted core.

Plastic particles. Devolatilization of plastic coals is more complicated than devolatilization of nonplastic ones. In this case the particle behavior during heating and devolatitlization determines the variation in particle porosity and size and such physical properties as the effective viscosity of the molten coal, its density, thermal conductivity, etc. Under these conditions the escape of volatiles occurs via bubbling, namely, by diffusion of volatiles through the molten coal to the surface of the bubbles located inside a particle and then by transfer of these bubbles to the particle surface.

Lee et al. (1977) investigated the structure of a coarse cylindrical particle of a plastic coal ($d = 1.9 \cdot 10^{-2}$ m $d = 5 \cdot 10^{-2}$ m high) at a fire-level heat flux. These experiments showed that in the coarse particles there are four characteristic regions with different states of the heated substance manifested by the density distribution (Fig. 2.25). Within the first low temperature region (25 °C < T < 410 °C) which is located far from the heated cross section there exists an unreacted coal. Due to the small mass loss, the bulk density here remains practically constant and close to the virgin coal density. At the temperature 410 °C < T < 450 °C corresponding to the second region a rapid process of thermal decomposition of coal into liquid occurs. As a result of the partial liquid vaporization, gas bubbles arise in the liquid. This leads to the formation of a two-phase system (metaplast) with the bulk density $\rho_{met} \sim 300\ [\mathrm{kg/m^3}]$. The densities of the melt and volatiles are $\rho_\ell \sim 900\ [\mathrm{kg/m^3}]$ and $\rho_v \sim 3\ [\mathrm{kg/m^3}]$, respectively. At the temperatures 450 °C<T<560 °C polymerization and condensation of the metaplast become significant. These processes are accompanied by the formation of coke and by an increase in the bulk density from that of the metaplast $\rho_{met} \sim 300\ [\mathrm{kg/m^3}]$ to the coke density $\rho_c \sim 850\ [\mathrm{kg/m^3}]$. Within the high temperature region 580 °C<T<800 °C most of volatiles are driven out.

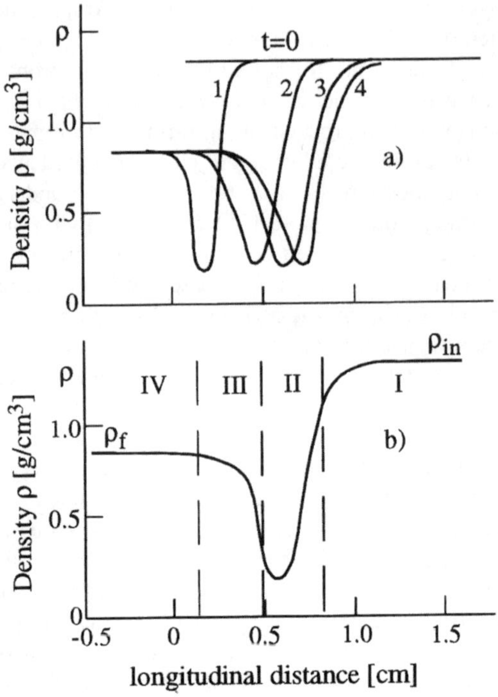

Fig. 2.25 Density distribution in the coarse coal particle. Reprinted from Lee et al. (1977), with permission. a) Curve 1, t=5 min; curve 2, t=10 min; curve 3, t=15 min, curve 4, t=20 min. b) $\rho_{in} = 1.33 \ \left[g/cm^3\right]$ is the density of unreacted coal, $\rho_f = 0.85 \ \left[g/cm^3\right]$ is the coke density

Devolatilization of plastic particles depends on several factors, such as the intensity of radiative heat flux, the ambient temperature, the turbulence of the carrier fluid, etc. Phuoc and Maloney (1988) studied pyrolysis of a single coal particle for the heat flux $(0.5\text{–}2.5)10^4 [W/m^2]$. Their observations showed that the devolatilization process includes three successive stages. The first of them corresponds to an insignificant particle heating to the temperature at which formation of light non-condensible volatiles begins. During this stage the particles are stationary and the mass loss is negligible. The heavy condensable volatiles are released in the next stage of devolatilization. This process is accompanied by the growth of bubbles, their migration through the plastic coal shell, bursting at the particle surface, etc. During the last stage of devolatilization bubbling at the particle surface ceases and tar evolution is ended. It is worthyof note that the process of devolatilization is accompanied by a number of phenomena such as particle rotation, condensation of heavy volatiles and formation of a dense cloud surrounding the particle, etc.

Additional phenomena were revealed by Costa et al. (1994) in their study of devolatilization of pulverized coal in a turbulent jet. The measurements carried out for a temperature variation within 1,680–1,870°C (heating rate of about $10^5\,[\mathrm{K/s}]$) showed that the rate of devolatilization increases with temperature and signficantly decreases with the particle size. These results are not unexpected and correlate qualitatively with the data of the above-mentioned works. However, the comparison of the measurement results performed in turbulent and non turbulent flows revealed a new effect, namely, the effect of turbulence on the devolatilization rate. The latter is probably due to the influence of turbulent fluctuations on the heat and mass transfer, as well as on the rate of pyrolysis.

1.2.4 Droplet evaporation

Qualitative description. Droplet evaporation has received significant attention and an extensive literature exists on this phenomenon. Many theoretical and experimental results concerning droplet evaporation and combustion in a stagnant gas or gas flow are summarized in the books by Spalding (1955), Khitrin (1957), Clift et al. (1978), Williams (1985), Kuo (1986), Glassman (1987), as well as in the reviews by Williams (1973), Faeth (1977, 1983), Law (1982) and Sirigano (1983). A comprehensive survey of the modern status of this problem is given in the monograph by Sirignano (1999).

Below we consider some characteristics of the evaporation of a single droplet. The overall processes which occur during the droplet lifetime are the following. A cold droplet that is injected into a hot gaseous medium undergoes an intensive heating by convective and radiative heat transfer. This process is accompanied by an increase of the droplet temperature, liquid vaporization and, accordingly, a decrease in the droplet mass and diameter.

The behavior of a droplet during the heating and evaporation is illustrated in Fig. 2.26, where the change of droplet surface temperature and its mass are depicted. At low temperatures close to the initial temperature T_0 the partial vapor pressure at the surface is very small. Accordingly, the vapor mass flux from the droplet is also relatively small. This allows one to neglect the mass loss during the first stage (I) of evaporation and to assume $m = m_0$. An increase in the temperature during the second stage (II) of the droplet heating leads to a significant increase in the rate of vaporization, so that the heat transfer occurs under the conditions of variable m and T. During the final stage (III) of the droplet heating, the temperature changes relatively slightly and tends to a saturated temperature T_S, since the latent heat of evaporation absorbs most of the energy supplied, whereas the droplet mass reduces to zero.

Formulation of the problem. Evaporation of a droplet in a high temperature gas flow is described using the system of the mass, momentum, energy and species conservation equations. For an incompressible fluid they are

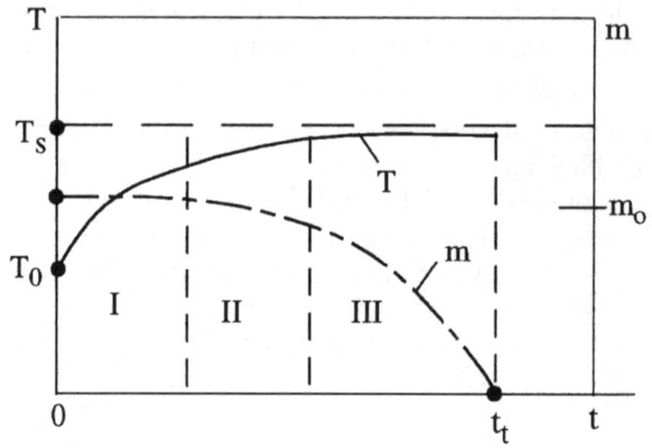

Fig. 2.26 The change of droplet temperature and vapor mass flux from the surface with time

$$\rho^{(\alpha)}\frac{\partial \mathbf{v}^{(\alpha)}}{\partial t}+\rho^{(\alpha)}(\mathbf{v}^{(\alpha)}\cdot\nabla)\mathbf{v}^{(\alpha)}=-\nabla P^{(\alpha)}+\mu^{(\alpha)}\nabla^{(2)}\mathbf{v}^{(\alpha)}+f^{(\alpha)} \tag{2.185}$$

$$\nabla\cdot\mathbf{v}^{(\alpha)}=0 \tag{2.186}$$

$$\rho^{(\alpha)}\frac{\partial h^{(\alpha)}}{\partial t}+\rho^{(\alpha)}(\mathbf{v}^{(\alpha)}\cdot\nabla)h^{(\alpha)}=\lambda^{(\alpha)}\nabla^2 T^{(\alpha)} \tag{2.187}$$

$$\rho^{(1)}\frac{\partial C^{(j)}}{\partial t}+\rho^{(1)}(\mathbf{v}^{(1)}\cdot\nabla)C^{(j)}=\rho^{(1)}D\nabla^2 C^{(j)} \tag{2.188}$$

where $\rho, \mathbf{v}, T$ and h are the density, the velocity, the temperature and the enthalpy (bold denotes vectors, $\mathbf{v}$ has the components u, v and w corresponding to the Cartesian axes, x, y and z, respectively), $C^{(j)}=\rho^{(j)}/\rho^{(1)}$ is the concentration of the j-th component of gaseous mixture, $\sum_{j=1}^{n}C^{(j)}=1$, $\mathbf{f}$ is a body force, D is the binary mass diffusion coefficient ($D^{(j)}$ is assumed to be equal for all the components), P is the pressure, λ and μ are the thermal conductivity and viscosity, respectively; superscripts $\alpha = 1; 2$ refer to vapor $(\alpha=1)$ and liquid $(\alpha=2)$, superscript $j=v$ and G refer to the vapor (v) and the ambient gas (G), respectively.

Consider the conditions at the droplet surface. They are expressed by the continuity of mass and thermal fluxes at the surface, and the balance of all the forces involved (Landau and Lifshiz 1959). The conditions of an evaporating droplet have the following form

$$\sum_{\alpha=1}^{2} \rho^{(\alpha)} \tilde{\mathbf{v}}^{(\alpha)} \cdot \mathbf{n}_i^{(\alpha)} = 0 \tag{2.189}$$

$$\sum_{\alpha=1}^{2} \left(\rho^{(\alpha)} \tilde{\mathbf{v}}^{(\alpha)} h^{(\alpha)} + \lambda^{(\alpha)} \nabla T^{(\alpha)} \right) \cdot \mathbf{n}_i^{(\alpha)} = 0 \tag{2.190}$$

$$\sum_{\alpha=1}^{2} (p^{(\alpha)} + \rho^{(\alpha)} \tilde{v}_i^{(\alpha)} \tilde{v}_i^{(\alpha)}) \mathbf{n}_i^{(\alpha)} = (\sigma_{ik}^{(2)} - \sigma_{ik}^{(1)}) \mathbf{n}_k + \sigma_t (r_1^{-1} + r_2^{-1}) \mathbf{n}_i^{(2)} + \nabla \sigma_t \tag{2.191}$$

where $\tilde{\mathbf{v}}^{(\alpha)} = \mathbf{v}^{(\alpha)} - \mathbf{v}_s$ is the velocity of the α- th phase relative to the surface $v_s = d\mathbf{r}_s/dt$, $\mathbf{r}_s$ is the current droplet radius, σ_t is the surface tension, σ_{ik} is the tensor of viscous tension, $\tilde{\mathbf{v}}^{(\alpha)} \mathbf{n}_i^{(\alpha)} = v_{i_n}$ and $\nabla T^{(\alpha)} \cdot \mathbf{n}_i^{(\alpha)}$ are normal components of the velocity vector and gradient at the interphase surface, r_1 and r_2 are the principal curvature radii of the interphase surface; $\mathbf{n}_i$ and $\mathbf{n}_k$ refer to the outward unit normal and the tangent directions; $\mathbf{n}_i^{(1)} = -\mathbf{n}_i^{(2)}$.

When the droplet surface is expressed by its radius vector $\mathbf{r} = \mathbf{r}(\xi, \eta)$ where ξ and η are the surface coordinates, the surface curvature K is given by Korn and Korn (1968)

$$K = r_1^{-1} + r_2^{-1} = \frac{EN - 2FM + GL}{EG - F^2} \tag{2.192}$$

where $E = \mathbf{r}_\xi \cdot \mathbf{r}_\xi$, $F = \mathbf{r}_\eta \cdot \mathbf{r}_\eta$, $G = \mathbf{r}_\xi \cdot \mathbf{r}_\eta$, $\mathbf{r}_\xi = \partial \mathbf{r}/\partial \xi$, $\mathbf{r}_\eta = \partial \mathbf{r}/\partial \eta$, $L = (\mathbf{r}_{\xi\xi}(\mathbf{r}_\xi \times \mathbf{r}_\eta))/|\mathbf{r}_\xi \times \mathbf{r}_\eta|$, $M = (\mathbf{r}_{\xi\eta}(\mathbf{r}_\xi \times \mathbf{r}_\eta))/|\mathbf{r}_\xi \times \mathbf{r}_\eta|$, $N = (\mathbf{r}_{\eta\eta}(\mathbf{r}_\xi \times \mathbf{r}_\eta))/|\mathbf{r}_\xi \times \mathbf{r}_\eta|$.
The vapor and liquid densities, pressure and temperature at the droplet surface are related by the following equation (Carey 1992):

$$\frac{1}{\rho^{(v)}} \frac{dP^{(v)}}{dT} + \frac{1}{\rho^{(2)}} \frac{dP^{(2)}}{dT} = \frac{q_e}{T} \tag{2.193}$$

where q_e is the latent heat of evaporation.

Equation (2.193) reduces to the Clapeyron–Clausius equation when the pressures in both phases are equal. For practical calculations of the saturation pressure of several liquids one uses the empirical equation by Antoine (Reid et al. 1987)

$$P = 0.001333224 \exp\left(A_* - \frac{B_*}{T + C_*}\right) \tag{2.194}$$

where P is given bar (1 bar = 10^{-5} [$\mathrm{N/m^2}$]), and T is in Kelvins; the values of the parameters A_*, B_*, and C_* for a number of liquids (methanol, ethanol, propanol 2, n-heptane, n-octane and n-decane) are presented in the book by Reid et al. (1987).

On a curved surface the vapor pressure $P^{(v)}$ deviates from the saturation pressure P_{sat} (the Kelvin effect). The deviation depends on the surface tension σ_t, the liquid density $\rho^{(2)}$, the gas constant R_g (in the present case R_g=R_v), the temperature T and the radius of curvature r. When $P^{(v)}$–$P_{sat}(T)$ is small compared to $2\sigma_t/r_2$ (which is mostly the case), $P^{(1)}$ can be approximated for most systems by the expression

$$P^{(v)} = P_{sat} \cdot \exp\left(-\frac{2\sigma_t}{R_v T \rho^{(2)} r_2}\right) \tag{2.195}$$

when the ratio $(2\sigma_t)/(R_v T \rho^{(2)} r_2) \ll 1$ it is possible to assume that $P^{(v)} = P_{sat}$. For example, for realistic conditions corresponding to the evaporation of water droplets: T=300 K, $\rho^{(2)} = 10^3$ [$\mathrm{kg/m^3}$], $R_v = 462$ [$\mathrm{J/kg \cdot K}$], $r_2 = 5 \cdot 10^{-4}\,\mathrm{m}$, $\sigma_t = 0.0727$ [N/m] the vapor pressure near the curved droplet surface equals $P^{(v)} = P_s \cdot \exp(-2 \cdot 10^{-6})$, i.e. $P^{(v)} \simeq P_{sat}$. In this case the vapor pressure at the interphase surface may be found from the Clapeyron–Clausius equation

$$\frac{dP^{(v)}}{dT^{(1)}} = \frac{q_e \rho^{(v)}}{T^{(1)}} \tag{2.196}$$

Assuming that the vapor is an ideal gas

$$P^{(v)} = R_v \rho^{(v)} T^{(1)} \tag{2.197}$$

and combining Eqs. (2.196) and Eq. (2.197), we arrive after the integration at the dependence of the vapor pressure on the temperature at the surface

$$P^{(v)} = \tilde{P} \cdot \exp\left(-\frac{q_e}{R_v T^{(l)}} \right) \tag{2.198}$$

where $\tilde{P} = P' \exp\left(q_e/(R_v T')\right)$, with P' and T' being some values of the pressure and temperature on the saturation line.

The d^2 law. The solution of Eqs. (2.185) and (2.188) with the conditions (2.189) and (2.191) pases great difficulties. Therefore, in studying droplet vaporization simplified models are widely used. They allow one to obtain simplified analytical solutions of the problem. The main assumptions which are used in the approximate analysis of droplet vaporization are the following:

1. The Reynolds number is based on the gas flow velocity and is much less than one and the Grashof number Gr= $(g\beta\Delta T d^3)/\nu^3 \ll 1$ (g, β and ΔT are the acceleration due to gravity, the thermal expansion coefficient, and the temperature difference between the droplet and the ambient gas). Under these conditions it is possible to neglect the effect of the forced and natural convection and to consider evaporation in a stagnant gas at infinity. In this case the velocity, temperature and concentration fields are spherically symmetric if the droplet is assumed to be spherical.
2. It is assumed that the characteristic time for changes in the gaseous phase is much shorter than in the liquid phase. This enables us to use the quasi-steady approach for the description of the processes in the gaseous phase.
3. The droplet temperature is uniform: constant or time dependent. This is essentially equivalent to the lumped-capacitance approximation introduced before.
4. Vapor near the surface is saturated (the thermodynamic equilibrium is reached at the surface almost immediately), and the Kelvin effect is negligible. The latter is certainly true for droplets of radii $r_s \geq 10^{-3}$ m.
5. Heat transfer occurs by conduction.
6. Internal fluid motion inside a droplet is absent, which also corresponds to the spherical symmetry assumed.

Using the above mentioned simplifications, consider droplet evaporation in a stagnant medium (say air). Assuming that liquid vapor and air are the only species over the droplet surface, we present the continuity equation and the species conservation equations as follows

$$\frac{1}{r^2}\frac{d}{dr}(\rho_1 v_1 r^2) = 0 \tag{2.199}$$

$$\rho_1 v_1 \frac{dC_j}{dr} - \frac{\rho_1 D}{r^2}\frac{d}{dr}\left(r^2 \frac{dC_j}{dr} \right) = 0 \tag{2.200}$$

where j= v, G corresponds to the vapor and ambient gas, respectively. The boundary conditions for Eqs. (2.199) and (2.200) are

$$r \to \infty \qquad C_v \to 0, \quad C_G \to 1 \tag{2.201}$$

$$r = r_s: \quad \begin{cases} \rho_1 v_1 C_G - \rho_1 D \dfrac{dC_G}{dr} = 0 \\ \rho_1 v_1 C_v - \rho_1 D \dfrac{dC_v}{dr} = \rho_1 v_1 \end{cases} \tag{2.202}$$

From Eq. (2.199) it follows that

$$\rho_1 v_1 r^2 = \rho_1 r_s^2 v_s = \frac{m}{4\pi} = \text{const} \tag{2.203}$$

where m is the vapor mass flux from the droplet surface.

Integration of Eq. (2.200) yields

$$\frac{dC_j}{dr} = \frac{K_1}{r^2} \exp\left(-\frac{\alpha}{r}\right) \tag{2.204}$$

$$C_j = \frac{K_1}{\alpha} \exp\left(-\frac{\alpha}{r}\right) + K_2 \tag{2.205}$$

where $\alpha = m/(4\pi D)$, K_1 and K_2 are constants. Taking into account that $C_j \to C_{j\infty}$ at $r \to \infty$, and $C_j = C_{js}$ at $r = r_s$, we find

$$K_1 = \alpha (C_{j\infty} - C_{js}) \left(1 - \exp\left(-\frac{\alpha}{r_s}\right)\right)^{-1} \tag{2.206}$$

$$K_2 = C_{j\infty} - \frac{K_1}{\alpha} \tag{2.207}$$

Express the derivative $(dC_v/dr)_s$ from the condition (2.202)

$$\left(\frac{dC_v}{dr}\right)_s = (C_{vs} - 1)\frac{\rho_1 v_1}{\rho_1 D} \tag{2.208}$$

From the relation (2.204) it follows that

$$\left(\frac{dC_v}{dr}\right)_s = \frac{K_1}{r_s^2}\exp\left(-\frac{\alpha}{r_s}\right) \tag{2.209}$$

Equating the right hand sides of Eqs. (2.208) and (2.209), we obtain

$$m = 4\pi\rho_1 D \ln(1+B)\cdot r_s \tag{2.210}$$

where $B = (C_{vs} - C_{v\infty})/(1 - C_{vs}) = (c_{p_1}(T_\infty - T_s))/q_e$ is the Spalding transfer number, c_{p_1} is the specific heat of the gaseous phase, T_∞ and T_s are the ambient and the droplet surface temperature.

The total mass flux from the evaporating droplet is equal to the changes of its mass. Accordingly we can write the mass balance equation as follows

$$-\frac{dM}{dt} = m = 4\pi\rho_2 D r_s \ln(1+B) \tag{2.211}$$

where $M = (4/3)\pi\rho_2 r_s^3$ is the current mass of the droplet.

Integration of Eq. (2.211) yields

$$d_s^2 = d_0^2 - kt \tag{2.212}$$

where d_0 is the initial droplet diameter, $k = [(8\rho_1 D)/\rho_2]\ln(1+B)$

Equation (2.212) is usually referred to as the d^2 law, which states that the square of the droplet diameter decreases linearly with time. Using Eq. (2.212), the total time of droplet vaporization corresponding to $d_s = 0$ is given by

$$t_t = \frac{d_0^2}{k} \tag{2.213}$$

Wood et al. (1960) generalized the d^2 law for vaporization of droplets of binary fluids. They showed that in this case the current droplet diameter may be determined by an equation identical to Eq. (2.212) in which the constant k is defined as follows:

$$k = \frac{8\lambda_1}{c_{p_1}\left(\sum C_{2i0}\rho_{2i}\right)} \ln\left\{1 + \frac{c_{pi}\left(T_\infty - T_s\right) + (C_{o.\infty}/\sigma)q}{\left(\sum C_{2i0}q_{ei}\right)}\right\} \tag{2.214}$$

where λ_1 and c_{p_1} are the gas phase thermal conductivity and specific heat, C_{2i0} is the initial mass fraction of the i-th component of liquid phase, ρ_{2i} is the density of the i-th component of liquid phase, C_{2io} is the initial concentration of i-th liquid fraction in fuel, $C_{0.\infty}$ is the ambient oxidation concentration, T_∞ and T_s are the ambient and the droplet surface temperature, respectively, q_{ei} is the specific heat of vaporization of the i-th component of the liquid phase, σ is the stiochiometric oxidizer/fuel mass ratio, q is the heat of combustion, and sujbscripts 1 and 2 correspond to the gaseous and liquid phases.

Characteristic times. The expression (2.213) determines the total time of droplet evaporation under the conditions when liquid temperature is constant. The duration of a real evaporation process when temperature is variable can be estimated by the simplest model of droplet heating. The model assumes that the process includes two successive stages: (i) heating at a variable temperature T_s and constant liquid mass M and (ii) evaporation at a constant temperature T_s and variable M.

To estimate the duration of these stages, we use the energy and the mass balance equations. First we determine the mass and heat transfer coefficients corresponding to evaporating droplet. Defining the Sherwood number as

$$Sh = -\frac{d}{C_{vs} - C_{v\infty}}\left(\frac{dC_v}{dr}\right)_s \tag{2.215}$$

and taking into account Eq. (2.208), we obtain

$$Sh = Sh_* \frac{\ln(1+B)}{B} \tag{2.216}$$

where $Sh = h_m/(\rho_1 D)$ is the Sherwood number, $Sh_* = 2$ is the Sherwood number corresponding to a nonvaporizing droplet, h_m is the mass transfer coefficient. At the Lewis number Le = 1, the Sherwood and the Nusselt numbers are equal to each other. Accordingly, we can write

$$\mathrm{Nu} = \mathrm{Nu}_* \frac{\ln(1+B)}{B} \tag{2.217}$$

where $\mathrm{Nu} = (hd)/\lambda$ is the Nusselt number, h is the heat transfer coefficient, and $\mathrm{Nu}_*=2$ is the Nusselt number for a nonvaporizing droplet.

In the first stage when the temperature varies, but the droplet mass is constant, the energy equation in the frame of the lumped-capacitance model is

$$M_0 c_p \frac{dT_s}{dt} = hS_0(T_\infty - T_s) \tag{2.218}$$

where M_0 and S_0 are the initial mass and surface of the droplet.

Integration Eq. (2.218) under the initial condition $t=0$, $T_s = T_0$, yields

$$t' = \frac{\rho_2 c_p d_0^2}{6\mathrm{Nu}_* \lambda_1} \ln \frac{T_\infty - T_s}{T_\infty - T_0} \tag{2.219}$$

where t' is the duration of the first stage of the process.

For the second stage of the process the thermal balance equation takes the form

$$-q_e \frac{dM}{dt} = hS(T_\infty - T_s) \tag{2.220}$$

where M and S are the current droplet mass and surface.

Integration of Eq. (2.220) yields

$$\Delta t = \frac{d_0^2 \rho_2 q_e}{6\mathrm{Nu}\lambda_1 (T_\infty - T_s)} \tag{2.221}$$

where $\Delta t = t'' - t'$ is the duration of the second stage of the process.

Then the ratio of the duration of the first stage of the process to the duration of the second one is

$$\tau = -\ln(1+B) \cdot \ln \frac{T_\infty - T_s}{T_\infty - T_0} \tag{2.222}$$

where $\tau = t'/\Delta t$

For small values of the Spalding numbers when $\ln(1+B) \approx B$ we have

$$\tau = -\frac{T_\infty - T_s}{q_e} c_p \ln\frac{T_\infty - T_s}{T_\infty - T_0} \tag{2.223}$$

An increase in the initial droplet temperature leads to a sharp decrease in τ if $T_0 \to T_\infty$. This means that at T_0 close to T_s it is possible to neglect the duration of the first stage of heating and to assume that $t_t = \Delta t$.

In the consideration above, the forced and natural convection effects, as well as the effects that are due to radiative heat transfer, droplet rotation, turbulence of the carrier fluid, etc. were not considered. They are discussed in the monograph by Sirignano (1999) (convective droplet vaporization, radiative heating of a droplet) and in Lozinski and Matalon (1992) (vaporization of a spinning droplet), Gokalp et al. (1992, 1994), Birouk et al. (1996) (turbulence effect on droplet vaporization), Burdukov and Nakoryakov (1965, 1967), Gopinath and Mills (1993), and Yarin et al. (1999) (effect of the acoustic field).

References

Abramzon B, Borde I (1980) Conjugate unsteady heat transfer from a droplet in creeping flow. AIChE J. 26: 536–544

Abramzon B, Elata C (1984) Unsteady heat transfer from a single sphere in Stokes flow. Int. J. Heat Mass Transfer 27: 687–695

Acrivos A (1966) On the combined effect of forced and free convection heat transfer in laminar boundary layer flows. Chem. Eng. Sci. 21: 343–352

Acrivos A (1971) Heat transfer at high Peclet number from a small sphere freely rotating in a simple shear field. J. Fluid Mech. 46: 233–240

Acrivos A (1980). Note on the rate of heat or mass transfer from a small particle freely suspended in a linear shear field. J. Fluid Mech. 98: 299–304

Acrivos A, Goddard JD (1965) Asymptotic expressions for laminar forced-convection heat and mass transfer. Part 1. Low speed flows. J. Fluid Mech. 23: 273–271

Acrivos A, Taylor TD (1962) Heat and mass transfer from single spheres in Stokes flow. Phys. Fluids. 5: 378–394

Anthony DB, Howard JB (1976) Coal devolatilization and hydrogasification. AIChE J. 22: 625–656

Anthony DB, Howard JB, Hottel HC, Meissner HP (1975) Rapid devolatilization of pulverized coal. The Fifteenth Symposium (International) on Combustion. The Combustion Institute, Pittsburgh, Pa., pp. 1303–1317

Badzioch S, Hawksley PGW (1970) Kinetics of thermal decomposition of pulverized coal particle. Ind. Eng. Chem. Process Desing Develop. 9: 521–530

Banks WHH (1965) The thermal laminar boundary layer on a rotating sphere. J. Appl. Math. Phys. 16: 780–788

Batchelor GK (1979) Mass transfer from a particle suspended in fluid with a steady linear ambient velocity distribution. J. Fluid Mech. 95: 369–400

Batchelor GK (1980) Mass transfer from small particle suspended in turbulent fluid. J. Fluid Mech. 98: 609–623

Beck NC, Hayhurst AN (1990) The early stages of the combustion of pulverized coal at high temperatures 1: the kinetics of devolatilization. Combust. Flame. 79: 47–74

Birouk M, Chauveau C, Sarh B, Quilgars X, Gokalp I (1996) Turbulence effects on the vaporization of monocomponent single droplets. Combust. Sci. Tech. 113: 413–428

Boothroyd RG (1971) Following gas-solids suspensions. Chapman and Hall. LTD. London

Burdukov AP, Nakoryakov VE (1965) On mass transfer in an acoustic field. J. Appl. Mech. Tech. Phys. 6: 51–55

Burdukov AP, Nakoryakov VE (1967) Effect of vibrations on mass transfer from a sphere at high Prandtl numbers. J. Appl. Mech. Tech. Phys. 8: 111–113

Carey VP (1992) Liquid–vapor phases-change phenomena. Hemisphere, Bristol, PA

Chao BT (1969) Transient heat and mass transfer to a translating droplet. ASME J. Heat Mass Transfer, 91: 273–281

Chao BT, Greif R (1974) Laminar forced convection over rotating bodies. Trans. ASME. 96: 463–466

Chen TS, Mucoglu A (1977) Analysis of mixed forced and free convection about a sphere. Int. J. Heat Mass Transf. 20: 867–875

Chiang T, Ossin A, Tien CL (1964) Laminar free convection from a sphere. Trans. ASME.J. Heat Transf. 86: 537–542

Clift R, Grace JR, Weber ME (1978) Bubbles, drops and particles. Academic Press, New York

Cooper F (1972) Heat transport from a sphere to an infinite medium. Int. J. Heat Mass Transfer. 20: 991–993

Costa M, Godoy S, Lockwood FC, Zhou J (1994) Initial stages of devolatilization of pulverized - coal in a turbulent jet. Combust. Flame. 96: 150–162

Crowe CT, Sommerfeld M, Tsuji Y (1998) Multiphase flows with droplets and particles. CRC Press, New York

Dennis SCR, Walker JDA, Hudson JD (1973) Heat transfer from a sphere at low Reynolds numbers. J. Fluid Mech. 60: 273–283

Dorfman LA (1963) Hydrodynamic resistance and the heat loss of rotating solids. Oliver and Boyd, Edinburgh and London

Dorfman LA, Mironova VA (1970) Solutions of equations for the thermal boundary layer ot a rotating axisymmetric surface. Int. J. Heat Mass Transfer. 13: 81–92

Dorfman LA, Serazetdinov AZ (1965) Laminar flow and heat transfer near rotating axisymmetric surface. Int. J. Heat Mass Transfer 8: 317–327

Dudek DR, Fletcher TH, Longwell JP, Sarofim AF (1988) Natural convection induced drag forces on spheres at low Grashof numbers: comparison of theory with experiment. Int. J. Heat Mass Transfer 31: 863–873

Eastop TD (1973) The influence of rotation on the heat transfer from a sphere to an air stream. Int. J. Heat Mass Transfer 16: 1954–1957

Essenhigh RH (1976) Combustion and flame propagation in coal systems: A Review. The Sixteenth Symposium (International) on Combustion. The Combustion Institute, Pittsburgh, Pa., pp. 353–374

Faeth GM (1977) Current status of droplet and liquid combustion. Prog. Energy Combust. Sci. 3: 191–224

Faeth GM (1983) Evaporation and combustion of sprays. Prog. Energy Combust. Sci. 9: 1–76

Fendell FE (1968) Laminar natural convection about an isothermally heated sphere at small Grashof number. J. Fluid Mech. 34: 163–176

Fletcher TH (1989) Time-resolved particle temperature and mass loss measurements of a bituminous coal during devolatilization. Combust. Flame. 78: 223–236

Frankel NA, Acrivos A (1968) Heat and mass transfer from small spheres and cylinders freely suspended in shear flow. Phys. Fluids. 11: 1913–1918

Frank-Kamenetskii DA (1969) Diffusion and heat transfer in chemical kinetics. 2nd edn., Plenum, New York.

Gale TK, Bartholomew CH, Fletcher TH (1995) Decreases in the swelling and porosity of bituminous coals during devolatilization at high heating rates. Combust. Flame. 100: 94–100

Gan H, Nadi SP, Walker PI (1972) Nature of the porosity in American coals. Fuel 51: 272–277

Gat N (1986) On internal temperature gradients in a pyrolysing coal particle. combust. Sci. Technol. 49: 297–303

Gavalas GR (1981) Coal pyrolysis. Elsevier, New York

Geoola F, Cornish ARH (1981) Numerical solution of steady-state free convective heat transfer from a solid sphere. Int. J. Heat Mass Transfer. 24: 1369–1379

Geoola F, Cornish ARH (1982) Numerical simulation of free convective heat transfer from a sphere. Int. J. Heat Mass Transfer 25: 1677–1687

Glassman I (1987) Combustion, 2nd edn. Academic Press, New York

Gokalp I, Chauveau C, Simon O, Chesneau X (1992) Mass transfer from liquid fuel drops in turbulent flow. Combust. Flame. 89: 286–298

Gokalp I, Chauveau C, Berrekam H, Ramos – Arroyo NA (1994) Vaporization of miscible binary fuel droplets under laminar and turbulent convective conditions. Atomization and Sprays. 4: 661–676

Gopinath A, Mills AF (1993) Convective heat transfer from a sphere due to acoustic streaming. Trans. ASME J. Heat Transfer 115: 332–341

Gostowski VJ, Costello FA (1970) The effect of free stream turbulence on the heat transfer from the stagnation point of a sphere. Int. J. Heat Mass Transfer. 13: 1382–1386

Gupalo YuP, Ryazantsev YuS (1974) Heat and mass transfer from a sphere with a chemical surface reaction in a laminar flow. Acta Astronaut. 1: 993–1005

Gupalo YuP, Polyanin AD, Ryazantsev YuS (1985) Mass heat transfer reactive particles with flow. (in Russian) Nauka, Moscow

Hadamard JS (1911) Mouvement permanent lent d'une sphere liquid et visqueuse dans une liquid visquese. C. R. Acad. Sci. Paris, 152: 1735–1738

Hatzikonstantinou P (1990). Unsteady mixed convection about a porous rotating sphere. Int. J. Heat Mass Transfer 33: 19–27

Heritzberg M, Zlochower IA (1991) Devolatilization wave structures and temperatures for the pyrolysis of polymethylmethacrylate, ammonium perchlorate, and coal combustion level heat fluxes. Combust. Flame 84: 15–37

Hieber CA, Gebhart B (1969) Mixed convection from a sphere at small Reynolds and Grashof numbers. J. Fluid Mech. 38: 137–159

Hossain MA (1966) Laminar free convection about an isothermal sphere at extremely small Grashof numbers. Ph.D. Thesis, Cornell University

Hussaini MY, Sastry MS (1976) The laminar compressible boundary layer on a rotating sphere with heat transfer. Trans. ASME. J. Heat Transfer 98: 533–535

Jafarpur K, Yovanovich MM (1992) Laminar free convection heat transfer from isothermal spheres: a new analytical method. Int. J. Heat Mass Tranfer 35: 2195–2201

Kassoy DR, Adamson TCJr, Messiter AF (1966) Compressible low Reynolds number flow around a sphere. Phys. Fluids 9: 671–681

Kestin J (1966) The effect of free-stream turbulence on heat transfer rates. In: Irvine TF, Hertnett (eds), Advances in heat transfer, vol. 3: pp. 1–32

Khitrin LN (1957) The physics of combustion. (in Russian) Moscow University, Moscow

Klyachko LS (1963) Heat transfer between a gas and a spherical surface with the combined action of free and forced convection. Trans. ASME. J. Heat Transfer 85C: 355–357

Kobayashi H, Howard JB, Sarofim AF (1976) Coal devolatilization at high temperatures. The Sixteenth Symposium (International) on Combustion. The Combustion Institute, Pittsburgh, Pa., pp. 411–425

Konopliv N, Sparrow EM (1972) Unsteady heat transfer and temperature for Stokesian flow about a sphere. ASME J. Heat Mass Transfer 94: 266–272

Korn GA, Korn TM (1968) Mathematical handbook. McGraw – Hill, New York

Kreith F (1968) Convective heat transfer in rotating systems. In: Irvine TF, Hertnett (eds), Advances in heat transfer. vol. 5: pp. 129–251

Kreith F, Roberts LG, Sullivan JA, Sinha SN (1963) Convection heat transfer and flow phenomena of rotating sphere. Int. J. Heat Mass Transfer 6: 881–895

Kuo KK (1986) Principles of combustion. Wiley, New York.

Landau LD, Lifshitz EM (1959) Fluid mechanics. 2nd edn. Pergamon, New York

Lavender WJ, Pei DCT (1967) The effect of fluid turbulence on the rate of heat transfer from spheres. Int. J. Heat Mass Transfer 10: 529–539

Law CK (1982) Recent advances in droplet vaporization and combustion. Prog. Energy Combust. Sci. 8: 171–201

Lee CK, Singer JM, Chaiken RF (1977) Coal pyrolysis at fire-level heat flux. Combust. Sci. Technol. 16: 205–213

Lee MH, Jeng DR, De Witt KJ (1978) Laminar boundary layer transfer over rotating bodies in forced flow. Trans. ASME, J. Heat Transfer 100: 497–502

Levich VG (1962) Physicochemical hydrodynamics. Prentice-Hall, Englewood Cliffs, NJ

Levich VG, Krylov VS, Vorotilin VP (1965) Towards the theory of nonstationary diffusion from a moving drop. Sov. Phys. Dokl. 161: 648–652

Lien F - S, Chen C - K, Cleaver JW (1986) Mixed and free convection over a rotating sphere with blowing and suction. Trans. ASME, J. Heat Transfer 108: 398–404

Lin FN, Chao BT (1974) Laminar free convection over two-dimensional and axisymmetric bodies of arbitrary countur. Trans. ASME. J. Heat Transfer 96: 435–442

Lozinski D, Matalon M (1992) Vaporization of a spinning fuel droplet. The Twenty-fourth Symposium (International) on Combustion. The Combustion Institute, Pittsburgh, Pa., pp. 1483–1491

Maloney DJ, Monazam ER, Woodruff SD, Lawson LO (1991) Measurements and analysis of temperature histories and size changes for single carbon and coal particles during the early stages of heating and devolatilization. Combust. Flame 84: 210–220

Mathers WG, Madden AJJr, Piret EL (1957) Simultaneous heat and mass transfer in free convection. Ind. Eng. Chem. 49: 961–968

Nguyen HD, Paik S, Chung JN (1993) Unsteady mixed convection heat transfer from a solid sphere: the conjugate problem. Int. J. Heat Mass Transfer 36: 4443–4453

Nigmatulin RI (1991) Dynamics of multiphase media. v.1 and 2 Hemisphere, London

Niksa S (1986). The distributed - energy chain model for rapid coal devolatilization kinetics. Part II: Transient weight loss correlations. Combust. Flame 66: 111–119

Niksa S, Kerstein AR (1986) The distributed - energy chain model for rapid coal devolatilization kinetics. Part 1: Formulation. Combust Flame 66: 95–109

Niksa S, Kerstein AR, Fletcher TH (1987) Predicting devolatilization at typical coal combustion conditions with the distributed - energy chain model. Combust. Flame 69: 221–228

Nordle RL, Kreith F (1961) Convective heat transfer from a rotating sphere. Int. Dev. in Heat Transfer. Am. Soc. Mech. Eng. New York, 461–467

Oliver DLP, Chung JN (1986) Conjugate unsteady heat transfer from a spherical droplet at low Reynolds numbers. Int. J. Heat Mass Transfer 29: 879–887

Palec GL, Daguenet M (1987). Laminar three-dimensional mixed convection about a rotating sphere in a stream. Int. J. Heat Mass Transfer 30: 1511–1523

Phuoc TX, Durbetaki P (1989) Heat and mass transfer analysis of a coal particle undergoing pyrolysis. Int. J. Heat Mass Transfer 30: 2331–2339

Phuoc TX, Maloney DI (1988) Laser pyrolysis of single coal particles in an electrodynamic balance. The Twnety-second Syumposium (International) on Combustion. The Combustion Institute, Pittsburgh Pa., pp. 125–134

Phuoc TX, Mathur MP (1991) Transient heating of coal particles undergoing pyrolysis. Combust. Flame 85: 380–388

Pitt GJ (1962) The kinetics of the evolution of volatile products from coal. Fuel 41: 267–274

Polyanin AD (1982) On nonisothermal chemical reaction at the particle surface in a laminar flow. Int. J. Heat Mass Transfer 25: 1031–1042

Polyanin AD (1984) An asymptotic analysis of some nonlinear boundary - value problems of convective mass and heat transfer of reacting particles with the flow. Int. J. Heat Mass Transfer 27: 163–189

Polyanin AD, Sergeev YuA (1980) Convective diffusion to a reacting particle in a fluid. Nonlinear surface reaction kinetics. Int. J. Heat Mass Transfer 23: 1171–1182

Potter JM, Riley N (1980) Free convection from a heated sphere at large Grashof number. J. Fluid Mech. 100: 769–783

Proudman I, Pearson JRA (1957) Expansions at small Reynolds numbers for the flow past a sphere and circular cylinder. J. Fluid Mech. 2: 237–269

Ragland KW, Yang JT (1989) Combustion of millimeter sized coal particles in convective flow. Combust. Flame 60: 285–297

Raithby GD, Eckert ERG (1968) The effect of turbulence parameters and support position on the heat transfer from spheres. Int. J. Heat Mass Transfer 11: 1233–1252

Rajasekaran R, Palekar MG (1985) Mixed convection about a rotating sphere. Int. J. Heat Mass Transfer 28: 956–968

Reid RC, Prausnitz JM, Poling BE (1987). The properties of gases and liquids. McGraw-Hill, New York

Riley N (1986) The heat transfer from a sphere in free convective flow. Comput. Fluids 14: 225–237

Rimmer PL (1968) Heat transfer from a sphere in a stream of small Reynolds number. J. Fluid Mech. 32: 1–7

Ruckenstein E (1967) Mass transfer between a single drop and a continuous phase. Int. J. Heat Mass Transfer 10: 1785–1792

Rybczynski W (1911) Uber die fortschreitende bewegung einer flussigen kugel in einem zahen medium. Bull. Inst. Acad. Sci. Cracovie ser. A. 40–46

Sadhal SS, Ayyaswamy PS, Chung JN (1997) Transport phenomena with drops and bubbles. Springer, New York

Sandoval-Robles JG, Delmas H, Couders JP (1981) Influence of turbulence on mass transfer between a liquid and solid sphere. AChIE J. 27: 819–823

Saxena SC (1990) Devolatilization and combustion characteristics of coal particles. Prog. Energy Combust. Sci. 16: 55–94

Sayegh NN, Gauvin WH (1979) Numerical analysis of variable property heat transfer to a single sphere in high temperature surroundings. AIChE J. 25: 522–534

Schlichting H (1960) Boundary layer theory. McGraw-Hill, New York

Sirignano WA (1983) Fuel droplet vaporization and spray combustion. Prog. Energy Combust. Sci. 9: 291–322

Sirignano WA (1999) Fluid dynamics and transport of droplets and sprays. Cambridge University Press, Cambridge

Smith KL, Smoot LD (1990) Characteristics of commonly – used U.S. coals – towards a set of standard research coals. Prog. Energy Combust. Sci. 16: 1–53

Solomon PR, Serio MA, Suuberg EM (1992) Coal pyrolysis: experiments, kinetic rates and mechanisms. Prog. Energy Combust. Sci. 18: 133–220

Soo SL (1990). Multiphase fluid dynamics. Science Press and Gower Technical, Beijing

Spalding DB (1955) Some fundamentals of combustion. Butterworths, London

Suuberg EM (1988) Significance of heat transport effects in determining coal pyrolysis rate. Energy Fuel 2: 593–595

Suwono A (1980) Buoyancy effects on flow and heat transfer on rotating axisymmetric round–nosed bodies. Int. J. Heat Mass Transfer 23: 819–831

Taylor TD (1963) Mass transfer from single spheres in Stokes flow with surface reactions. Int. J. Heat Mass Transfer 6: 993–994

Tieng SM, Yan AC (1993) Experimental investigation on convective heat transfer of heated spinning sphere. Int. J. Heat Mass Transfer 36: 599–610

Tsubouchi T, Sato S (1960) Heat transfer from fine wires and particles by natural convection. Res. Inst. High-Speed Mech., Tohoku University 12: 127–132

Ubhayakar SK, Stickler DB, von Posenberg CWJr, Gannon RE (1976) Rapdi devolatilization of pulverized coal in hot combustion gases. The Sixteenth Symposium (International) on Combustion. The Combustion Institute, Pittsburgh, Pa., pp. 427–436

Unger PE, Suuberg EM (1981) Modeling the devolatilization behavior of a softening bituminous coal. The Eighteenth Symposium (International) on Combustion. The Combustion Institute, Pittsburgh, Pa., pp. 1203–1211

Williams A (1973). Combustion of droplets of liquid fuels: A Review. Combust. Flame 21: 1–31

Williams FA (1985) Combustion theory. The fundamental theory of chemically reacting flow systems, 2nd edn. Benjamin/Cummings, Menlo Park, Calif.

Wiser WH, Hill GR, Kertamus NJ (1967) Kinetic study of the pyrolysis of high-volatile bituminous coal. Ind. Eng. Chem. Process Desing Develop. 6: 133–138

Wong K-L, Lee S-C, Chen C-K (1986) Finite element solution of laminar combined convection from a sphere. Trans. ASME. J. Heat Transfer 108: 860–865

Wood BJ, Wise H, Inami SH (1960) Heterogeneous combustion of multicomponent fuels. Combust. Flame 4: 235–242

Yarin LP, Sukhov GS (1987) Fundamentals of combustion theory of two-phase media. (in Russian) Energoatomizdat, Leningrad

Yarin AL, Brenn G, Kastner O, Rensink D, Tropea C (1999) Evaporation of acoustically levitated droplets. J. Fluid Mech. 399: 151–204

Yuge T (1960) Experiments on heat transfer from spheres including combined natural and forced convection. Trans. ASME. J. Heat Transfer 82: 214–220

1.3 Ignition and combustion of a single particle

1.3.1 Ignition of a coal particle

Mechanism of ignition. As has already been mentioned, coal particles undergoing an intensive heating evolve combustible volatiles which significantly affect the particles' behavior in a gas flow. Depending on the production of volatiles, various scenarios of particles ignition occur. For certain coal compositions, particle sizes, etc. two limiting regimes corresponding to pure heterogeneous or to pure homogeneous ignition can take place (Essenhigh et al. 1989). In the former case the primary ignition is due to a direct attack of oxidizer on solid matter, whereas in the latter case it is related to chemical reactions which proceed in a gaseous vapor–oxidizer mixture formed in the vicinity of the particle.

The effect of coal composition on ignition and combustion of coal particles is illustrated in Figs. 3.1 and 3.2. In these figures a snapshots from movie on the behavior of anthracite and lignite coal particles in a hot oxidizing medium are presented (data by Babii and Kuvaev 1986). The ignition of anthracite particles is heterogeneous. Such a process develops as follows. During the initial period ($0<t<2$ s) the particle temperature is low and its surface is dark. Then (at $t>2$ s) bright high temperature points appear at the particle surface. In the course of time the number of hot spots increases. Finally the heterogeneous reaction expands to the whole particle surface. The particle is burned out at about 7.781 s and is not seen any more.

The data presented in Fig. 3.1 correspond to the ignition of coal particles which evolve a low amount of volatiles (mass percentage of the burning substance $v\sim3\%$). The heterogeneous mechanism of ignition also sets in during of heating of coal particles with high content of volatiles. In this case the process of particle ignition is multi-stage. It involves the primary heterogeneous ignition of solid matter, a subsequent homogeneous ignition and combustion of volatiles in the gaseous flame enveloping a particle, and the secondary heterogeneous ignition of the solid matter (char) after the burning of volatiles has finished.

In contrast with the heterogeneous ignition in which the first stage of the process consists of ignition of solid matter, homogeneous ignition begins with ignition of the volatiles–oxidizer mixture in some domain, which is located in the vicinity of the particle surface. Then the gas flame enveloping the particle screens its surface and prevents oxidizer from penetrating through the particle surface (Fig. 3.2). In accordance with that, the particle temperature remains relatively low until the combustion of volatiles has closed. Then a direct attack of oxidizer on char takes place. That leads to the appearance of a number of centers of heterogeneous reaction (the bright spots on the particle surface) and to the transformation of the homogeneous process to the heterogeneous one.

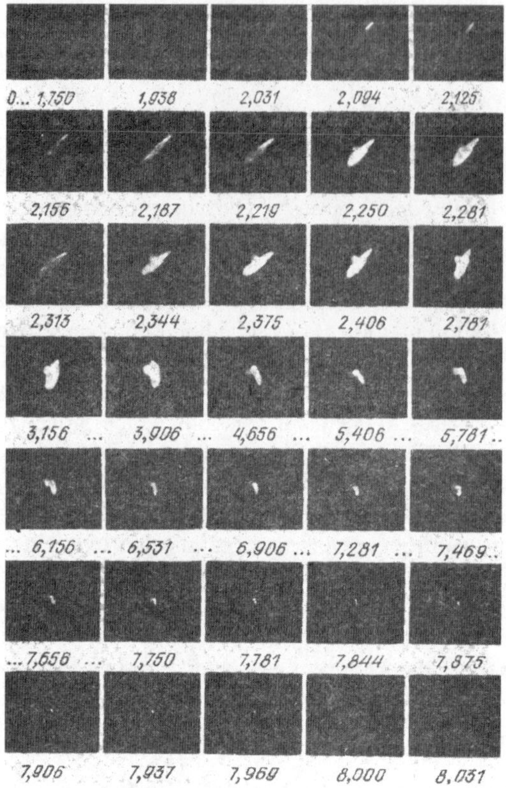

Fig. 3.1 Ignition and combustion of anthracite particle. ($T_\infty = 1{,}200$ K, $C_{O_2\infty} = 0.21\%$, t: time in seconds). Reprinted from Babii and Kuvaev (1986)

A realization of either of the regimes of particle ignition depends on the parameters determining the rates of particle heating and devolatilization. When the rate of particle heating is higher than the rate of devolatilization, the heterogeneous regime of ignition takes place. From the physical point of view this means that the primary heating of coal particles occurs under conditions when the increase in particle temperature overtakes pyrolysis of the solid substance. That ensures the access of oxidizer to the particle surface and its heterogeneous ignition. In the opposite case when the devolatilization rate exceeds the heating rate, the evolution of volatiles overtakes the increase in particle temperature. That creates favourable conditions for forming a gaseous reactive mixture in the vicinity of the particle and its homogeneous ignition. Since the ratio of the devolatilization rate to that of the heating is proportional to d^2, ignition of fine and coarse particles occurs, respectively, by the heterogeneous and homogeneous mechanism. A qualitative map of the ignition regimes of particles of different sizes is presented in Fig. 3.3. It is seen that there are three characteristic ranges of particle diameters, which correspond to the heterogeneous, hetero-homogeneous, and homogeneous ignition.

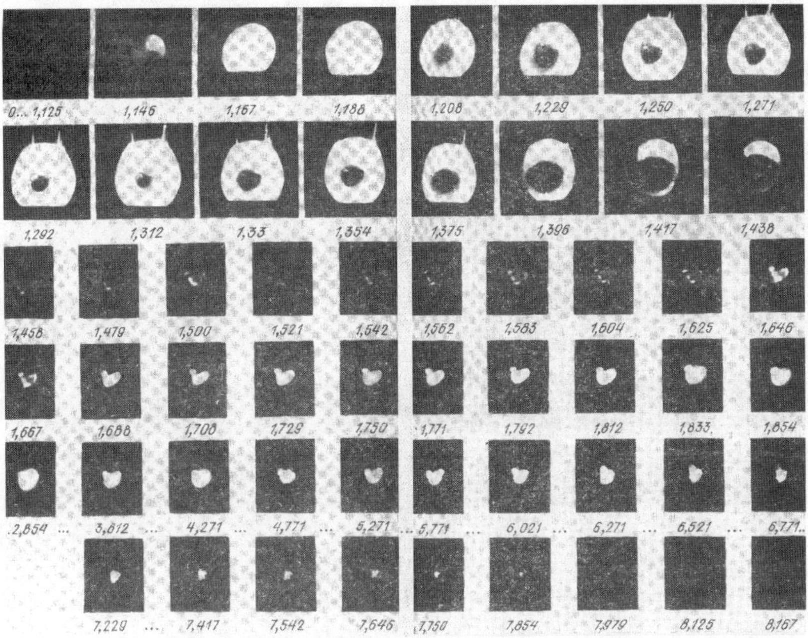

Fig. 3.2 Ignition and combustion of lignite coal particle ($T_\infty = 1,200$ K, $C_{O_{2,\infty}} = 0.21$, t: time in seconds). Reprinted from Babii and Kuvaev (1986)

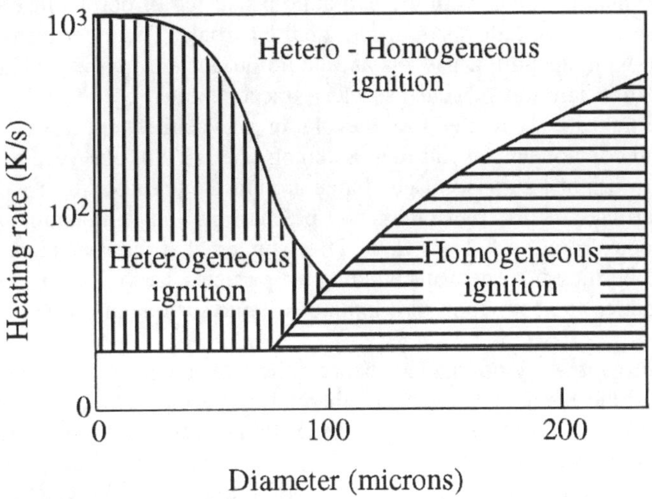

Fig. 3.3 Map of the ignition regimes as a function of the heating rate and particle size. Reprinted from Essenhigh et al. (1989), with permission

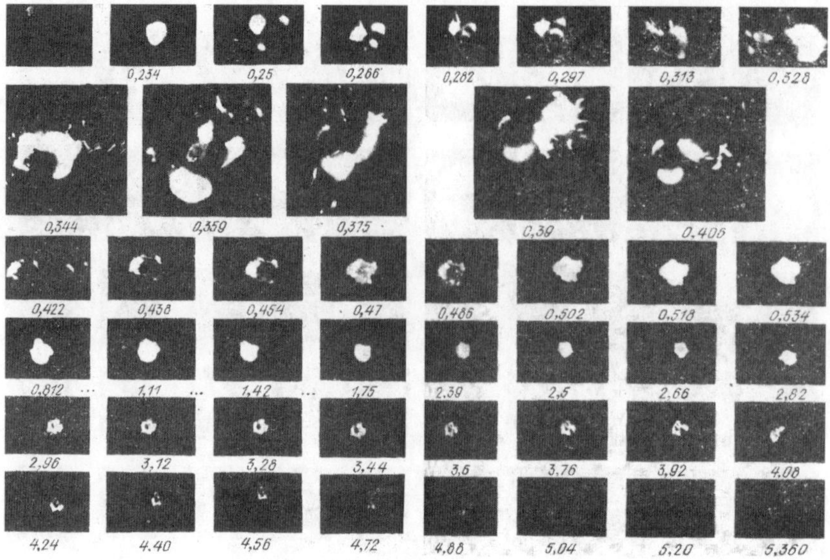

Fig. 3.4 Ignition and combustion of coal particle ($T_{\infty} = 1{,}400$ K, $C_{O_{2,\infty}} = 0.21$, t: time in seconds). Reprinted from Babii and Kuvaev (1986)

The mechanism which is described above does not account for a number of processes accompanying particle heating and ignition (swelling, melting, etc.). The change of the physical properties of a solid material at high temperature is accompanied by changes in the hydraulic resistance of porous medium and by an increase in the volatiles pressure inside the particle. The latter leads to particle explosion, with ejection of gas jets and liquid products of pyrolysis (Fig. 3.4). At the same time ignition of coal particles is accompanied by several phenomena, which are not directly related to the evolution of volatiles, and transformation of the mineral component of solid fuel. Solomon et al. (1990) observed a "secondary ignition" (a bright luminescent gas flame around the particle) during the ignition of coal particles of different ranks, and precharring and demineralizing of these coals introduced into a hot gas flow. They suggest that the secondary ignition is due to CO burning in a sheet surrounding the particles, as well as to fragmentation of solid matter and formation of a luminous flame surrounding fine particle fragments.

Heterogeneous ignition. The theory of heterogeneous ignition and extinction is based on the Semenov's theory of thermal explosion (1935). This theory was applied by Vulis (1961) and Frank-Kamenetskii (1969) in order to study heterogeneous ignition of coal particles. The authors used the thermal balance equation to analyze the thermal regime of combustion of a single particle when a first order heterogeneous reaction takes place.

In accordance with the theory of thermal explosion, the critical states of a system are determined as the unstable steady states. In the framework of this approach we discuss the ignition of carbon and coal particles. The following simplifications of the problem will be used: (i) the particle is spherical and its shape and size do not change during the particle heating and ignition, (ii) the physical properties of the solid and gaseous reagents are invariable, (iii) the gaseous medium does not absorb radiation, (iv) the Biot and the Stefan numbers are less than one.

The thermal balance equation in the of the framework lumped-capacitance heat transfer model is

$$mc_p \frac{dT_p}{dt} = Q_1 + Q_2 \tag{3.1}$$

where $Q_1 = qSC\,k_{ef}$ is the heat released; $Q_2 = hS(T_\infty - T_p) + \tilde{\sigma}\varepsilon S(T_w^4 - T_p^4)$ is the heat removed due to convection/conduction and radiation, m is the particle mass, c_p is the specific heat of the solid reagent, h is the heat transfer coefficient, $\tilde{\sigma}$ is the Stefan-Boltzmann constant, ε is the emissivity, q is heat of reaction, C is the oxidizer concentration in the environment, k_{ef} is the effective constant of the chemical reaction, T_p, T_∞ and T_w are the particle, the ambient and the furnace wall temperatures, respectively, S is the particle surface, and subscript p refers to the particle.

The effective constant of the chemical reaction, accounting for the existence of the kinetic as well as diffusion resistances, is expressed as

$$k_{ef} = \left(\frac{1}{h_m} + \frac{1}{k} \right)^{-1} \tag{3.2}$$

where $h_m = (ShD)/d$ is the mass transfer coefficient; Sh is the Sherwood number, D is the diffusion coefficient of oxygen in the oxygen–products mixture, d is the particle diameter, $k = k_o \exp(-E/RT)$ is the constant of the heterogeneous reaction, k_0 is the pre-exponential factor, and E and R are the activation energy and universal gas constant, respectively.

The initial condition for the problem is

$$t = 0, \quad T_p = T_{p.0} \tag{3.3}$$

Equation (3.1) is rendered dimensionless by the characteristic scales of time $t_* = (mc_pE)/(qCSk_oR)$ and temperature $T_* = E/R$. Using the dimensionless temperature and time $\theta = T/T_*$ and $\tau = t/t_*$ we reduce Eq. (3.1) to the following dimensionless equation in the case of a pure convective heat transfer.

$$\frac{d\theta_p}{d\tau} = \omega(\theta_\infty - \theta_p) + (\Omega + \exp(1/\theta_p))^{-1} \tag{3.4}$$

where $\omega = (hE)/(qCk_o R)$, $\Omega = k_0/h_m$ and $\theta_\infty = RT_\infty/E$. Its solution is subject to the initial condition

$$\tau = 0,\ \ \theta_p = \theta_{p.0} \tag{3.5}$$

In Eq. (3.4) it is assumed that the particle mass remains constant $(m \simeq m_0 = \text{const})$. The assumption holds only for the initial stage of the process, in particular for the ignition period, when changing of the particle mass is insignificant.

In the steady state $d\theta_p/dT = 0$, Eq. (3.4) reduces to

$$\omega(\theta_p - \theta_\infty) = (\Omega + \exp(1/\theta_p))^{-1} \tag{3.6}$$

The term on the left hand side in this equation expresses the heat losses due to the convective heat transfer from the particle to the carrier fluid, whereas the term on right hand side in Eq. (3.5) expresses the heat released by the chemical reaction that proceeds at the particle surface. The dependences $Q_I = (\Omega + \exp(1/\theta_p))^{-1}$ and $Q_{II} = \omega(\theta_p - \theta_\infty)$ are depicted on the parametrical plane $Q - \theta_p$ in Fig. 3.5. The intersection points of the curves $Q_I(\theta_p)$ and $Q_{II}(\theta_p)$ determine the steady states of the system. In general case there are three intersection points which correspond to the low, intermediate and high temperature states. It is not difficult to show that the intermediate state is unstable. Indeed, any perturbation increasing the particle temperature relative to that corresponding to point 2 $(\theta_p > \theta_{p2})$ is accompanied by an excess of the intensity of heat release over the intensity of heat removal. Therefore temperature θ_p should keep increasing and this point 2 is unstable. Also if due to a perturbation $\theta_p < \theta_{p2}$ the heat losses exceed the heat released, the temperature θ_p keeps decreasing and point 2 is also unstable.

Among the possible mutual locations of the curves $Q_I(\theta_p)$ and $Q_{II}(\theta_p)$ there are two particular positions at which these curves are tangential (Fig. 3.5b). The latter correspond to the critical states: ignition and extinction. As will be shown below, points I and E determine the values of the governing parameters corresponding to the particle ignition and extinction. In the general case the values of

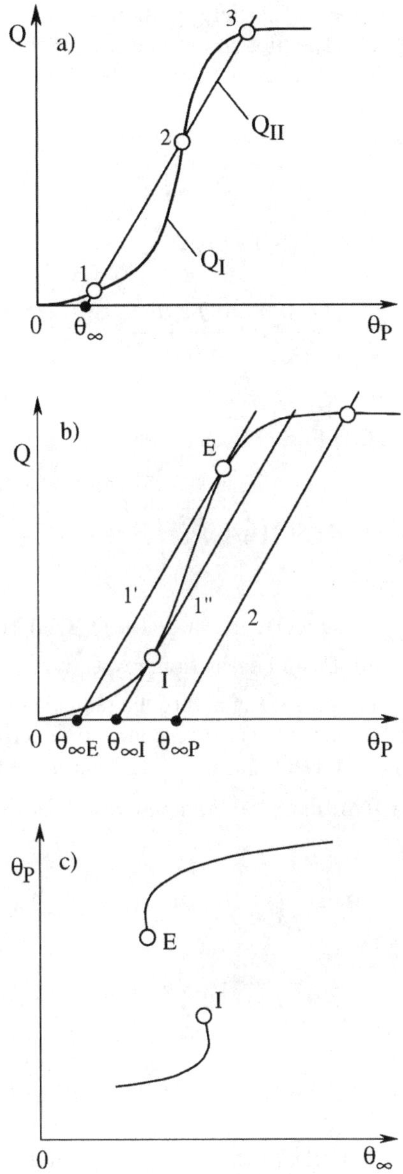

Fig. 3.5 The curves corresponding to the heat release Q_I and to the heat losses Q_{II}. a) Three intersection points: curve 1, the low temperature oxidation; curve 2, the intermediate (unstable) state; curve 3, the high temperature state. b) Possible mutual positions of the curves Q_I and Q_{II}. The points I and E correspond to the ignition and extinction. c) Change of the particle temperature θ_p with the ambient temperature θ_∞

the parameters corresponding to particle ignition and extinction do not coincide with each other (Fig. 3.5c). The plot shows that the dependence of the steady particle temperature on the ambient temperature (as well as on the other parameters) shows a characteristic hysteresis.

The critical states are determined by Semenov's conditions of tangency of the two curves

$$Q_I(\theta_p) = Q_{II}(\theta_p) \tag{3.7}$$

$$\frac{dQ_I(\theta_p)}{d\theta_p} = \frac{dQ_{II}(\theta p)}{d\theta_p} \tag{3.8}$$

The solution of the system of Eqs. (3.7) and (3.8) is

$$\theta_{p.cr} = 0.5\left(f_1 \pm \sqrt{f_1^2 - 4f_2}\right) \tag{3.9}$$

where $f_1 = (1+2\omega\Omega\theta_{\infty.cr})/(1+\omega\Omega)$, $f_2 = \theta_\omega((1+\theta_{\omega.cr}\Omega\omega)/(1+\Omega\omega))$, and $\theta_{\infty.cr} = \theta_{\infty I}$ and $\theta_{\infty E}$ for ignition and extinction, respectively.

In Eq. (3.9) minus and plus signs before the radical correspond to particle ignition and extinction, and subscript cr corresponds to the critical states. The product $\Omega\omega = h/(h_m qC) \approx c_p/(qC) \ll 1$ and $\theta_{\infty.cr} < 1$. Bearing in mind these estimates, we obtain from Eq. (3.10) the classical Semenov result for ignition

$$\theta_{pI} = \frac{1}{2}\left(1 - \sqrt{1 - 4\theta_{\infty I}}\right) \tag{3.10}$$

or

$$\theta_{pI} \simeq \theta_{\infty I} + \theta_{\infty I}^2 \tag{3.11}$$

Equations (3.10) and (3.11) relate the particle temperature at ignition θ_{pI} with to the minimal ambient temperature $\theta_{\infty I}$. The additional relation between these and $\theta_\infty = \theta_{\infty I}$ temperatures are given by Eq. (3.7) with $\theta_p = \theta_{pI}$ and $\theta_\infty = \theta_{\infty I}$. Therefore, the system of Eqs. (3.7) and (3.10) or (3.11) allows one to find θ_{pI} and $\theta_{\infty I}$.

The Semenov conditions (3.7), (3.8) determine the limiting values of the parameters (in particular, the minimum ambient temperature) at which the particle

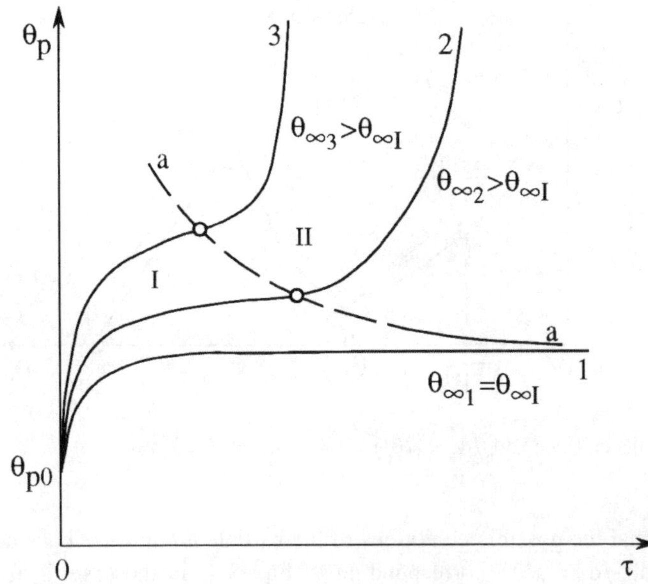

Fig. 3.6 Change in the particle temperature with time. Curve 1, ambient temperature $\theta_{\infty,1}$ equals the Semenov's ignition temperature $\theta_{\infty I}$; curves 2 and 3, the ambient temperatures $\theta_{\infty 2}$ and $\theta_{\infty 3}$ are higher than $\theta_{\infty I}$; curve a–a shows the locus of the inflection points

ignition occurs. At a higher ambient temperature θ_{∞} than the minimal one for ignition $\theta_{\infty I}$ (e.g. at $\theta_{\infty .2}$ in Fig. 3.5) the particle ignition takes place (Essenhigh et al. 1989).

The variation of the particle temperature θ_p with time for different values of θ_{∞} is depicted in Fig. 3.6. It is seen that the character of the dependence $\theta_p(\tau)$ depends essentially on the ambient temperature θ_{∞}. At $\theta_{\infty} \le \theta_{\infty I}$ the particle is not ignited at all and its temperature remains quite low. The situation changes qualitatively as θ_{∞} increases. When the ambient temperature exceeds the critical value $\theta_{\infty I}$ then a sharp increase in the particle temperature begins. In this case the curve $\theta_p(\tau)$ has an inflection point where $d^2\theta_p/d\tau^2 = 0$. The locus of the inflection points (curve a–a in Fig. 3.6) subdivides the parametrical plane $\theta_p - \tau$ into two characteristic domains in which the increment of the derivative $d\theta_p/d\tau$ decreases monotonically with θ_p (domain I) or increases with θ_p (domain II). An increase in the increment of the derivative $d\theta_p/d\tau$ promotes progressive particle heating and a sharp rise of the particle temperature corresponding to $d^2\theta_p/d\tau^2 > 0$. Therefore the curve a a″ in Fig. 3.6 can be considered as the boundary of particle ignition.

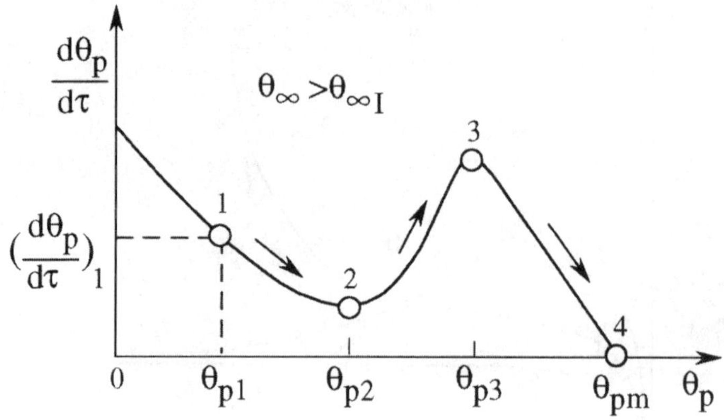

Fig. 3.7 The dependence $d\theta_p/d\tau = \varphi(\theta_p)$ for $\theta_\infty > \theta_{\infty I}$

To reveal the possible conditions of the particle ignition we consider the dependence $d\theta_p/d\tau = \varphi(\theta_p)$ corresponding to Eq. (3.4) in the case $\theta_\infty > \theta_{\infty I}$ (Fig. 3.7). Assume that the initial state of the system corresponds to point 1 where the particle temperature equals θ_{p1} and $(d\theta_p/d\tau) = (d\theta_p/d\tau)_1$. Since the derivative $(d\theta_p/d\tau) > 0$ within the range $\theta_{p1} \le \theta_p < \theta_{p2}$, the particle temperature increases monotonically in this temperature range. As a result, the point corresponding to the particle in Fig. 3.7 moves from point 1 to point 2 as is shown by the arrow. The state corresponding to point 2 is unstable. Any positive temperature perturbation leads to a progressive particle heating, since $d\theta_p/d\tau > 0$ for the branch 2,3 in Fig. 3.7. Such a process can be considered as particle ignition and point 2 can be considered as the ignition point. After the ignition the particle temperature increases up to a maximum temperature θ_{pm}. A subsequent temperature increase is impossible, since at $\theta_p > \theta_{pm}$ the derivative $(d\theta_p/d\tau)$ is negative.

At a ambient temperatures θ_∞ lower than that for Fig. 3.7, the curve $d\theta_p/d\tau = \varphi(\theta_p)$ is displaced down the vertical axis. At $\theta_\infty = \theta_{\infty I}$ the point 2 achieves the level $d\theta_p/d\tau = 0$ (Fig. 3.8a). This, indeed, corresponds to Semenov's conditions (3.7) and (3.8).

For $\theta_\infty < \theta_{\infty I}$, as in Fig. 3.8b, particles with the initial temperatures $\theta_p < \theta_{pI}$ cannot be ignited. Indeed, a particle with the initial temperature θ_{p_1} will be heated only to the temperature θ'_p (Fig. 3.8b). At the ambient temperature $\theta_\infty < \theta_{\infty I}$ particle ignition is possible only when its initial temperature is higher than the temperature θ''_p.

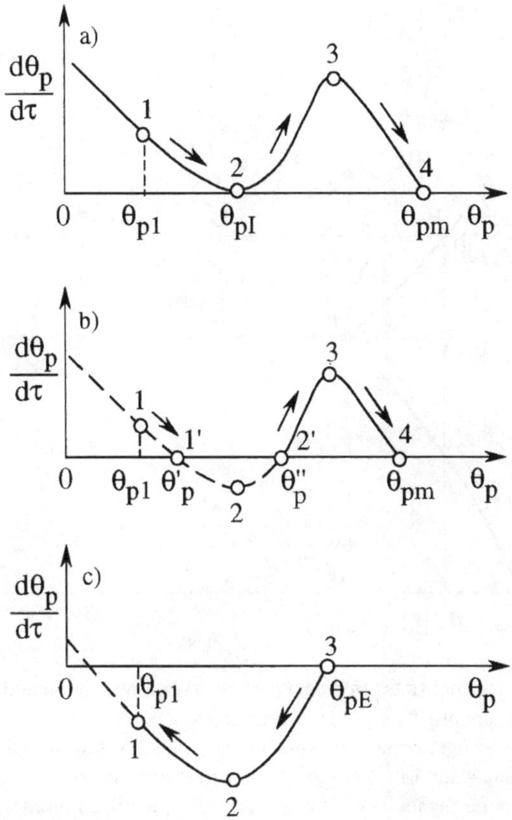

Fig. 3.8 The dependences $d\theta_p/d\tau = \varphi(\theta_p)$ for $\theta_\infty \leq \theta_{\infty I}$, a) $\theta_\infty = \theta_{\infty I}$; b), c) $\theta_\infty < \theta_{\infty I}$

At a sufficiently low ambient temperature, point 3 coincides with the level $d\theta_p/d\tau = 0$. This case corresponds to the second critical state (extinction) since any perturbation of the temperature leads to a transition from the state corresponding to point 3 to the state corresponding to point 1. It should be emphasized that the above consideration of the particle extinction in the frame of the present model is conditional since it does not account for the burning out of solid reagent. The latter is essential in the case of extinction.

Accordingly to the above consideration, the following conditions of particle ignition can be formulated:

$$\theta_\infty > \theta_{\infty I}, \qquad \frac{d^2\theta_p}{d\tau^2} = 0 \tag{3.12}$$

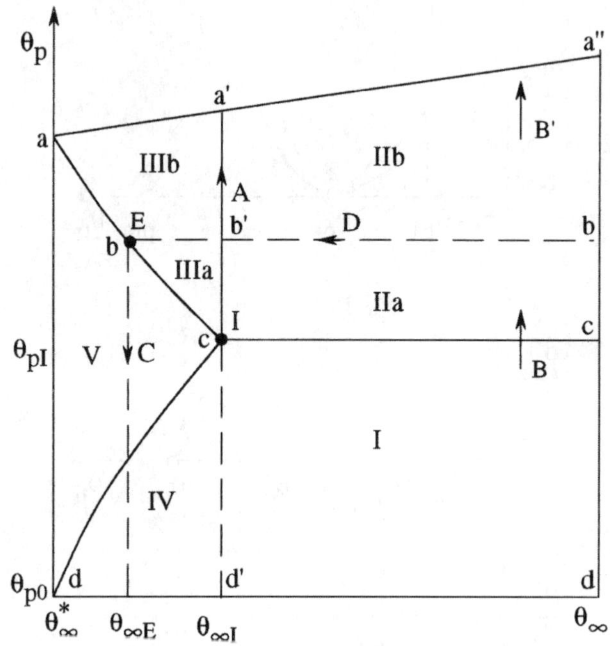

Fig. 3.9 Map of the thermal states of a coal particle at different initial and ambient temperatures. I, IV low temperature states; II transition from ignition to the high temperature state; III transition from the high temperature state to extinction; IV transition from the extinction to the low temperature state. Lines: a – a″ high temperature states, b – b transition from the intermediate state to the Semenov's extinction, c – c supercritical ignition, a – c extinction, c – d transition from initial state to the Semenov's ignition, at $\theta_\infty < \theta_{\infty I}$, d – d the initial state. The arrows A, B, C, and show the directions of transitions

$$\theta_\infty = \theta_{\infty I}, \quad \frac{d\theta_p}{d\tau} = 0, \quad \frac{d^2\theta_p}{d\tau^2} = 0 \tag{3.13}$$

$$\theta_\infty < \theta_{\infty I}; \quad \theta_p > \theta''_{p_2} \tag{3.14}$$

where θ'_{p_2} is the intermediate root of the equation

$$\omega(\theta_\infty - \theta_p) + (\Omega + \exp(1/\theta_p))^{-1} = 0 \tag{3.15}$$

The influence of the initial temperature on particle ignition was studied by Lisitsyn et al. (1971), Gurevich et al. (1971), and Klyachko and Kuntsev (1978).

The possible states of coal particles corresponding to various initial and ambient temperatures are illustrated on the $\theta_p - \theta_\infty$ plane in Fig. 3.9. There are five domains, which characterize the possible particle states. Domain I corresponds to a low temperature oxidation (line 1–2 in Figs. 3.7; 3.8a,b). This process can be unstable. It leads to an increase in the particle temperature from its initial value θ_{p0} up to the ignition temperature θ_{pI} if $\theta_\infty \geq \theta_{\infty I}$. Point I in Fig. 3.9 corresponds to Semenov's ignition, whereas the line c–c (at $\theta_\infty > \theta_{\infty I}$) corresponds to the supercritical ignition.

The unsteady particle heating after the ignition is characterized by the domain II. Within the subdomains IIa and IIb the process proceeds under the conditions $d^2\theta_p/d\tau^2 > 0$ (line 2–3 in Figs. 3.7 and 3.8 a,b) and $d^2\theta_p/d\tau^2 < 0$ (line 3–4 in Figs. 3.7 and 3.8 a,b). In both cases the process is accompanied by the particle heating and transition to the steady state at $\theta_p = \theta_m$. At a high enough initial temperature θ_{p0} ($\theta_{p0} > \theta_{pI}$) a decrease of the ambient temperature leads to a transition of the process into domain III and at the end to extinction E (line a–c). The transition from the high temperature state to the low temperature state (domain IV) corresponds to domain V.

In order to estimate the critical conditions of ignition and extinction we use the Frank-Kamenetskii approach (1969). Expanding the exponent in a Taylor series as per $\exp(-E/(RT)) \approx \exp(-E/(RT_\infty))\exp(\Theta)$, we reduce the steady - state thermal balance equation (3.4) with $d\theta_p/d\tau = 0$ to the following form

$$\delta = \Theta(\mu + \exp(-\Theta)) = f(\Theta) \tag{3.16}$$

Where $\delta = [(qEk_o)/(hRT_\infty^2)]\exp(-E/(RT_\infty))$, $\mu = (k_o/h_m)\exp(-E/(RT_\infty))$, and $\Theta = [E/(RT_\infty^2)](T_p - T_\infty)$ is the dimensionless temperature.

The shape of the function $f(\Theta)$ determines the character of the solution of Eq. (3.16), namely, the existence of a single or of a number of steady states (or critical conditions) of ignition and extinction. If $f(\Theta)$ is monotonic, there is a single solution, whereas if $f(\Theta)$ has an extremum, there are a number of solutions corresponding to different steady states. The extrema of the dependence $f(\Theta)$ determine the critical conditions of ignition and extinction. The condition $df(\Theta)/d\Theta = 0$ leads to the following relation for critical parameters

$$\mu = (\Theta - 1)\exp(-\Theta) \tag{3.17}$$

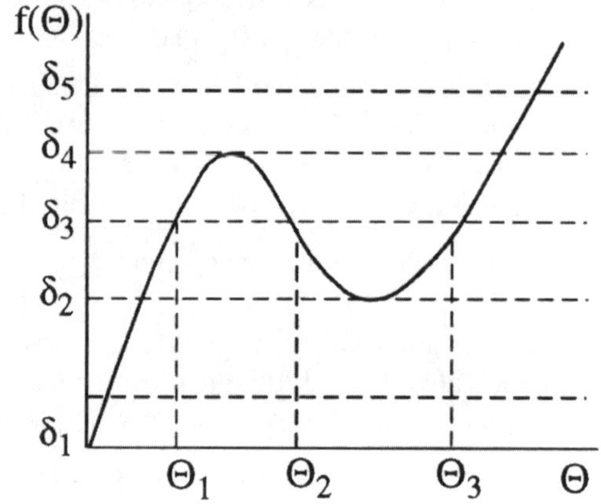

Fig. 3.10 The function $f(\Theta)$. Reprinted from Frank-Kamenetskii (1969)

The right hand side of Eq. (3.17) has a maximum at $\Theta = 2$ which corresponds to $\mu = 1/e^2$. When $\mu < (1/e^2)$, Eq. (3.17) has two solutions. On the other hand, when $\mu > (1/e^2)$ solutions of Eq. (3.17) are absent.

The function $f(\Theta)$ for $\mu < (1/e^2)$ is shown in Fig. 3.10. It is seen that different values of δ correspond to different solutions of the thermal balance equation. At small δ $(\delta = \delta_1)$ there is the only a single solution which corresponds to a weak particle heating (the low temperature regime). Large δ $(\delta = \delta_5)$ corresponds to the high temperature regime. At moderate values of δ $(\delta_2 < \delta < \delta_3)$ three solutions which correspond to the low, high and intermediate (unstable) states are possible.

Homogeneous ignition. The theory of homogeneous ignition of pyrolyzing particles is much less developed than the theory of heterogeneous ignition. There are a few models that allow one to estimate characteristics of homogeneous ignition (Annamalai and Durbetaki 1977, Gururajan et al. 1990, Du and Annamalai 1994, Annamalai and Ryan 1993). In accordance with these models the mechanism of ignition of pyrolyzing particle includes a number of successive stages such as heating, devolatilization etc. When the temperature of the coal particle reaches the pyrolysis level, thermal decomposition of the particle material occurs. As a consequence, volatiles are released and their mixing with the surrounding oxidizer in the vicinity of the particle leads to a reactive volatile–oxygen mixture. Homogeneous ignition of this mixture is possible if the volatile concentration in the gaseous mixture is flammable. When the volatile concentration is below the flammability level heterogeneous reaction proceeds on the particle surface. Thus, the ignition mode may change depending on the rates of the heat and mass transfer, as well as on the chemical reaction involved (Du and Annamalai 1994). Heterogeneous and homogeneous ignition correspond to the conditions (3.12) and (3.14), as well as to the existence of a temperature peak at some radial location where the homogeneous ignition occurs.

For calculating the critical states of combustion of coal particles it is important to know the kinetic constant of coals of various types. Fu and Zeng (1992) showed that the activation energy of any kind of coal is practically constant $(E \sim 152\ [kJ/mol])$, whereas the frequency factor varies with the coal type. The analysis of Fu and Zeng (1992) demonstrates that there is a universal relationship between the frequency factors of carbon and coal char particles. The ratio k_{0c}/k_{och} (k_{oc} and k_{och} are the frequency factor for carbon and coal char particle, respectively) determines the single dimensionless group $F_z = (v + w)^2 \times c \times 100$ which depends only on the proximate analysis of the parent coal; where c, v, and w are the carbon, the volatile and the intrinsic moisture contents (%).

The parameter F_z enables generalization of the data on the temperature of heterogeneous ignition of coal char particles and coals (Fu and Zhang 1992)

1.3.2 Droplet ignition

Qualitative pattern. Consider the behavior of a small droplet of liquid fuel with an initial temperature $T_{2.0}$ which is suddenly introduced into a hot oxidizing environment with temperature $T_{1.\infty}$. The existence of the finite temperature difference between the interacting phases leads to heat transfer from the carrier fluid to the droplet, its heating and vaporizing. The vapor of liquid fuel leaves the droplet surface and mixes with the oxidizer in the surrounding medium. As a result of mixing of the gaseous reactants, and a subsequent chemical reaction in the reactive mixture, heterogeneous temperature and concentration fields developing in the vicinity of the droplet.

During the initial stage of the process the temperature and the reactants, concentrations change in the range $T_{2.0} \le T \le T_{1.\infty}$, $C_{v\infty} \le C_v \le C_{vs}$, $(C_o = 1 - C_v - C_a)$ where C_v, C_o and C_a are the vapor, the oxidizer and the inert component concentrations, respectively, and subscripts s and ∞ are related to the droplet surface and the environment. The temperature and concentration profiles are monotonous and similar to the profiles for the regime of pure evaporation. In the subsequent periods of time the parameters, distribution in the mixing layer changes markedly (Fig. 3.11). The chemical reaction proceeding in the vapor–oxidizer mixture is accompanied by consumption of the initial reactants, formation of the combustion product, as well as by intensive heat release. Under these conditions the evolution of the temperature and concentration fields is determined by two factors: (i) hydrodynamics of the flow due to liquid evaporation, and (ii) the rate of chemical reaction in the vapor–oxidizer mixture. The contribution of each of these factors can be estimated by the ratio of the characteristic hydrodynamic time τ_h to the chemical reaction time τ_R, i.e. by the Damkohler number $Da = \tau_h/\tau_r$. When the Damkohler number is much less than one the influence of

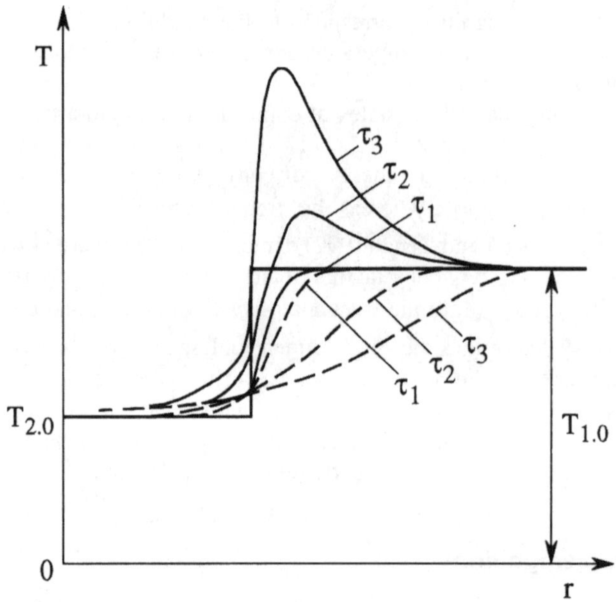

Fig. 3.11 Temperature distribution within the droplet and the surrounding gaseous medium at the moments $\tau_1 < \tau_2 < \tau_3$. The bold line corresponds to the initial state $(\tau = 0)$. The solid line corresponds to the reactive system, the dashed lines to the non-reactive system

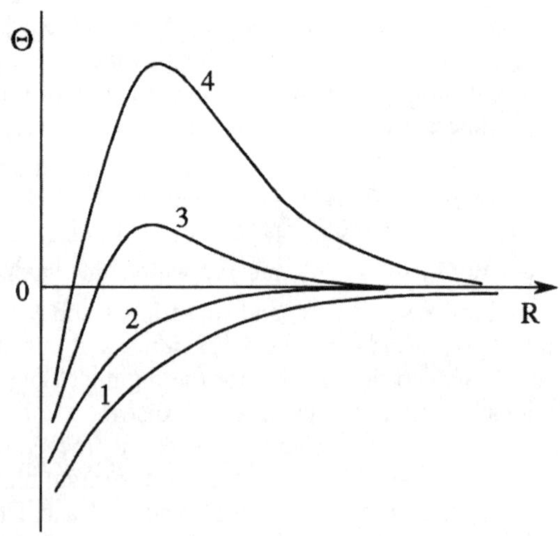

Fig. 3.12 Profiles $\Theta(R)$ for different values of the Damkohler numbers. Curve 1, $Da=Da_1$; curve 2, $Da=Da_2$; curve 3, $Da=Da_3$; curve 4, $Da=Da_4$ ($Da_1<Da_2<Da_{cr}$; $Da_{cr}<Da_3<Da_4$)

the chemical reaction on the temperature (concentration) field is negligible. At large Da the effect of the heat release is dominant. In this case the temperature profile has a characteristic maximum which is located within the region where the ratio of the mass fluxes of the reactants is close to the stoichiometric one.

In the quasi-steady state approximation it is possible to assume that evolution of the temperature field in the gaseous phase is determined solely by the Damkohler number. In this case peculiarities of the heat transfer for different stages of the process are clearly visible on the $\Theta - R$ diagram which is presented in Fig. 3.12 ($R = (r/r_s) - 1$, r_s is the droplet radius). As can be seen in Fig. 3.12, there are two kinds of the temperature profiles corresponding to small and large the Damkohler numbers, respectively. The transition from the monotonic profiles to the profiles with a characteristic maximum occurs at some critical value of the Damkohler number Da_{cr}.

When $Da < Da_{cr}$ the excess temperature Θ is negative for any values of R, whereas at $Da > Da_{cr}$ it is negative or positive for $R < R_m$ and $R > R_m$, respectively (R_m is the value of R corresponding to Θ_{max}). This means that the direction of the heat flux changes at $R = R_m$. For small values of the Damkohler number the heat flux is directed towards the droplet surface. It is determined by the difference between the ambient and the droplet temperature $T_{1\infty} - T_2$, whereas for large Da it is proportional to the difference between the maximum and the droplet temperature $T_{1.m} - T_2$. An increase in the local temperature in the mixing layer leads to an increase of the heat flux on the droplet surface, as well as of its vaporization rate.

The curves $\Theta(R)$ related to the sub- and supercritical values of the Damkohler number correspond, respectively, to low temperature oxidation of the vapor–oxidizer mixture and to its burning. At high activation energy the transition from the low to the high temperature state has an "explosive" character: small perturbations of the parameters near the critical states lead to a sharp change in the temperature and concentrations. That allows one to consider Da_{cr} as a criterion determining the droplet ignition.

Theoretical approach. Heating, ignition and burning of a liquid droplet are themselves unsteady processes proceeding under the conditions of continuous change of parameters of both phases. When the gas relaxes much faster than the liquid, the uncoupling between the liquid unsteady and gas quasi-steady states becomes plausible (Bellan and Summerfield 1976). Such a situation is typical for the gas–droplet systems under subcritical conditions when the ratio of the gas relaxation time to that of the liquid is small enough. In this case it is reasonable to use the quasi-steady approximation to describe the ignition of liquid fuel droplets.

From the physical point of view the ignition and combustion of vapor–oxidizer mixture in the vicinity of a burning droplet are similar to the ignition and combustion in flows of unpremixed gases. In both cases an intensive chemical reaction proceeds in a thin flame front to which the gaseous fuel and oxidizer are supplied separately. The transport of the reactants and of the combustion product occurs by means of convection and diffusion, whereas the heat transfer occurs by means of convection and conduction. Thus, in the frame of the quasi-steady approximation both problems reduce to solving the system of the mass, momentum, energy and species conservation equations containing distributed sources accounting for the heat release due to chemical reaction and conversion of the initial reagents into final products.

$$\nabla \cdot (\rho \mathbf{v}) = 0 \tag{3.18}$$

$$\rho(\mathbf{v} \cdot \nabla)\mathbf{v} = -\nabla P + \mu \nabla^2 \mathbf{v} \tag{3.19}$$

$$\rho(\mathbf{v} \cdot \nabla)C_k - \nabla \cdot (\rho D \nabla C_k) = -W_k \tag{3.20}$$

$$\rho(\mathbf{v} \cdot \nabla)(c_p T) - \nabla(\lambda \nabla T) = qW_k \tag{3.21}$$

Here $\mathbf{v}$ is the velocity, T is the temperature, ρ is the density, P is the pressure, $C_k = \rho_k/\rho$ and ρ_k are the concentration and the density of the k-th species, $\rho = \sum_k \rho_k$, W_k is the chemical reaction rate, and q is the heat of reaction.

The system of Eqs. (3.18) – (3.21) should be supplemented by the equation of state of the gas

$$P = \rho R_g T \tag{3.22}$$

where R_g is the gas constant.

The boundary conditions express that (i) the phases are at equilibrium at the droplet surface, (ii) mass and thermal fluxes are in balance on the droplet surface, and (iii) the parameters in the unperturbed flows in the vicinity of the droplet are known.

The general approach to study burning of unpremixed gases at infinite rate of reaction was developed by Burke and Schuman (1928), Schvab (1948) and Zel'dovich (1948). Later on Vulis et al. (1968) and Vulis and Yarin (1978) used this approach to study a wide class of diffusion flames. Burning of unpremixed gases at finite rate of reaction were studied by Zel'dovich (1948), Spalding (1951), Yarin (1961), Spalding and Jain (1962), Fendell (1965), Kassoy and Williams (1968), Kassoy et al. (1969), and Vulis and Yarin (1970).

The mechanism of burning of unpremixed gases with a finite rate of reaction was considered by Zel'dovich (1948). He studied the limit of combustion and found that the maximal decrease in temperature corresponding to extinction of the diffusion flame is of the order of

$$\Delta T_{max} \sim \frac{3RT_{ad}^2}{E} \tag{3.23}$$

where T_{ad} is the temperature of the adiabatic diffusion flame.

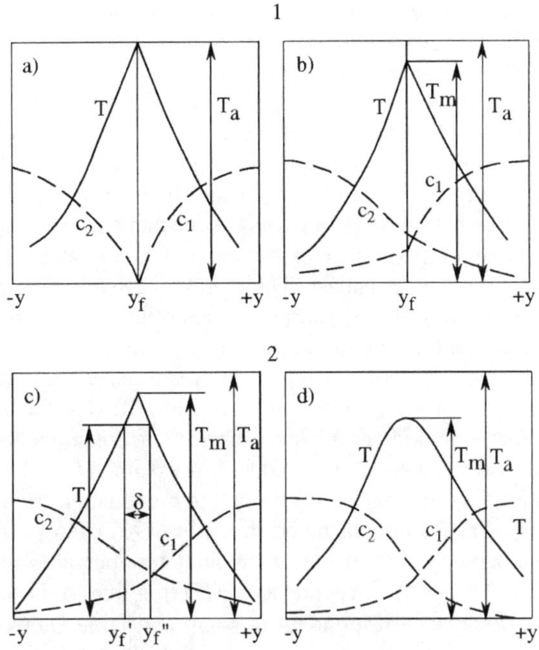

Fig. 3.13 Temperature and concentration distribution in diffusion flame. 1. Quasi-heterogeneous model. a) Infinity large rate of chemical reaction. b) Finite rate of chemical reaction. 2. Distributed reaction model. c) Finite thickness of the reaction zone. d) Actual structure of the diffusion flame. The adiabatic temperature of the flame is denoted by T_a, the maximum flame temperature T_m, T, C_1 and C_2 are the temperature and reagents concentrations, and δ is the thickness of the flame front. Subscript f corresponds to the locus of an infinitely thin front; f_f' and f_f'' are boundaries of a reaction zone of finite thickness

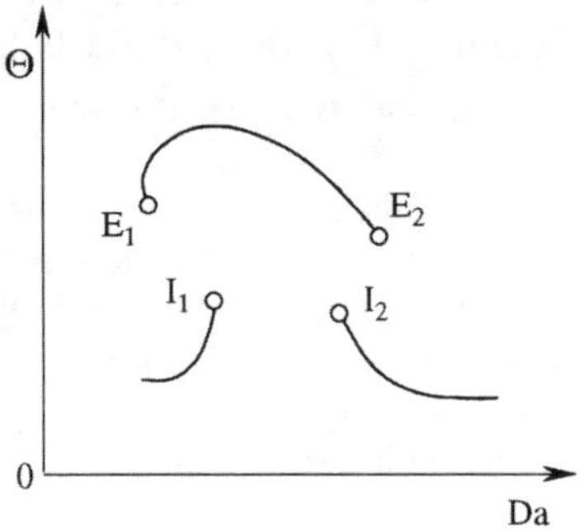

Fig. 3.14 Dependence of the flame temperature on the Damkohler number. I_1 and E_1 correspond to adiabatic ignition and extinction, while I_2 and E_2 correspond to thermal ignition and extinction affected by heat removal

From the physical point of view extinction of the diffusion flame results from the inequality of the rates of supply of reactants to the flame front and the kinetic rate of their conversion into combustion products. When the rate of supply of reagents to the flame front is less than the rate of their consumption, the flame temperature is close to the adiabatic one and the thickness of the reaction zone is relatively small. In this case the Burke–Schuman approximation is valid (Vulis et al. 1968). In contrast with that, when the rate of the chemical reaction is less than the rate of supply of reactants, there is an excess of reactants in the reaction zone. The excessive reactants do not participate in the reaction and, as a matter of fact, play a role of an inert component, a ballast. Their presence leads to a decrease of the flame temperature, of the order of Eq. (3.23), resulting in a dramatic decrease in the reaction rate and in the extinction of the diffusion flame.

The thermal regime of combustion of unmixed gases was studied by Yarin (1961). In this analysis of the possible states of the diffusion flame in the mixing layer of gaseous fuel and oxidizer the quasi-heterogeneous model of reactive zone was employed. In accordance with this model it is assumed that the chemical reaction proceeds with a final rate on an infinitely thin combustion front (Fig. 3.13a).

The solution of problem considered in Yarin (1961) shows that the dependence of the flame temperature on the Damkohler number presents an Ω shape (Fig. 3.14). Such a form of the dependence $\Theta(\mathrm{Da})$ is due to the existence of the two pairs of critical states corresponding to small and large Damkohler numbers. The first of them is due to the intensification of the transport of reactants towards the flame front, whereas the second is due to the heat losses (radiation) at low rates of heat release in the reaction zone.

The analysis of the thermal regimes of combustion in the mixing layer can be based on a more general model of the reaction zone. The latter is assumed to have a small but finite thickness (Vulis and Yarin 1970). This model assumes that combustion is localized in the regions of maximum temperature (Fig. 3.13b). In the general case the ratio δ/L varies in the range $0 \le (\delta/L) \le 1$ (where δ and L are the characteristic thickness of the reaction zone and the boundary layer thickness). The limiting values of δ/L equal to 0 and 1 correspond to the heterogeneous model of the flame front and to a fully distributed reaction within the mixing layer.

The characteristic size of the reaction zone in the case of laminar combustion of non premixed gases is of the order of the thickness of the flame front in a homogenous mixture. The latter reflects the physical similarity of the processes of homogeneous gas combustion and combustion within the reaction zone of the non premixed gas flame.

According to the dimensional arguments

$$\frac{k\delta^2}{D} \sim \mathrm{const} \tag{3.24}$$

where D is the diffusivity, and $k = k_0 \exp(-E/(RT))$ has the dimensionality $[s^{-1}]$.
Therefore, for the relative value of $\overline{\delta} = \delta/L$ we obtain

$$\overline{\delta} \sim \sqrt{\frac{D}{L^2 k}} = \sqrt{\frac{\tau_k}{\tau_d}} \cdot \exp\left(\frac{E}{2RT}\right) \sim Da^{-1} \tag{3.25}$$

where $\tau_k = 1/k_o$ is the characteristic kinetic time, and $\tau_d \propto L^2/D$ is the characteristic diffusion time.

The critical temperature of the flame of non premixed gases is expressed as

$$\theta_{cr} = 0.5\left\{\frac{2\theta_\infty + \vartheta}{1+2\vartheta}\left[1 \pm \sqrt{1 - 4\frac{\theta_\infty(\theta_\infty - \vartheta)}{(2\theta_\infty + \vartheta)^2}(1+2\vartheta)}\right]\right\} \tag{3.26}$$

where $\theta = (RT)/E$, and $\vartheta = [(0.5qC_{*\infty}R)/(c_p E)]\left[1 + \operatorname{erf}(\varphi_f \sqrt{Pr})\right]$, $C_{*\infty}$ is the fuel concentration far from the flame, φ_f is the dimensionless flame coordinate ($\varphi = \overline{\eta}/\sqrt{\overline{\xi}}$, $\overline{\eta} = \int_o^{\overline{y}} \overline{\rho} d\overline{y}$, $\overline{\xi} = \overline{x}$, $\overline{y} = y/L$, $\overline{x} = x/L$, $\overline{\rho} = \rho/\rho_\infty$), Pr is the Prandtl number, subscript ∞ refers to unperturbed flow. Plus and minus signs before the radical in Eq. (3.26) correspond to extinction and ignition, respectively.

At small θ_∞, when $4[(\theta_\infty(\theta_\infty + \vartheta)(1+\vartheta))/(2\theta_\infty + \vartheta)^2] < 1$ the ignition temperature is found from Eq. (3.26) as

$$\theta_I = \frac{\theta_\infty(\theta_\infty + \vartheta)}{(\vartheta + 2\theta_\infty)} + \frac{\theta_\infty^2(\theta_\infty + \vartheta)^2}{(2\theta_\infty + \vartheta)^3}(1+2\vartheta) \tag{3.27}$$

For high calorific fuels

$$\theta_I \simeq \theta_\infty + 2\theta_\infty^2 \tag{3.28}$$

Therefore, the relative heating of the system corresponding to ignition of the non premixed gases is small. According to Eq. (3.28) it is of order of $2((RT_\infty)/E)$.

A number of important results related to the combustion of non premixed gases were obtained by Fendell (1965). Using the asymptotic analysis (for small and large Damkohler numbers) and numerical calculations (for the intermediate Da) he studied the characteristics of the diffusion flame in impinging opposite

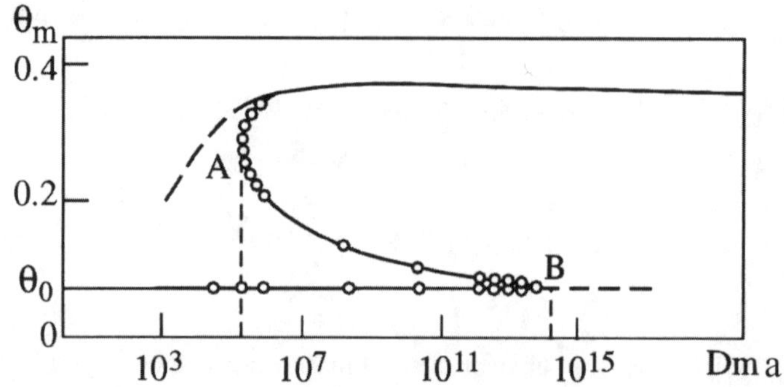

Fig. 3.15 Dependence of the maximum temperature on the Damkohler number. Reprinted from Fendell (1965), with permission. Circles correspond to the results of the numerical calculation, and the dashed line to the asymptotic analytical solution

flows of fuel and oxidizer. In particular, it was shown that the curve $\theta_m(\mathrm{Da})$ for the maximum temperature has a characteristic S-shaped configuration (Fig. 3.15). The upper and lower branches of this curve correspond to stable $(d\theta_m/d\mathrm{Da} > 0)$ high and low temperature regimes, whereas the intermediate branch corresponds to the unstable states $(d\theta_m/d\mathrm{Da} < 0)$. The sharp transition from the lower to the upper state occurs at point B in Fig. 3.15, whereas the reverse transition occurs at point A of the upper branch. Therefore, hysteresis phenomena are characteristic of the ignition and extinction of the non premixed gases. It should be noted the states at the points A and B correspond to the adiabatic extinction and ignition. The absence of the second pair of critical states on the curve $\theta_m(\mathrm{Da})$ is a result of neglecting the effect of the heat losses.

The ignition and extinction of a droplet of liquid fuel in an oxidizing atmosphere were studied by Peskin and Wise (1966), Peskin et al. (1967), Polymeropoulos and Peskin (1969), Rosser and Rajopakse (1969), Law (1975, 1978), Chiu et al. (1998), Pindera and Brzkustowski (1984), Shaygan and Prakash (1995) and by a number of the other researchers (see Williams 1973, Annamalai and Ryan 1992, Siriganano 1999, Chiu 2000). In these studies various models of the diffusion vapor–oxidizer flame enveloping a single droplet were employed. For instance, Peskin and Wise (1966) proposed the modified flame surface model which assumes that chemical conversion of the vapor–oxidizer mixture into combustion products occurs at the surface of an infinitely thin flame front. In accordance with this model, the source term in the species conservation equations was written as $W(C,T)\ \delta(R_* - R_f)$, where $W(C,T)$ is some function of the reactants, concentrations and temperature, $\delta(R_* - R_f)$ is the Dirac delta function, $R_* = r/r_s$, $R_f = r_f/r_s$, r_s and r_f are the droplet and the diffusion flame radii. The ignition temperature T_I is defined as the temperature at which a sharp transition from the kinetic-controlled to the diffusion-controlled regime occurs. The following expression for T_I was obtained:

$$T_I = \frac{E}{R}\frac{C_{o.\infty}+1}{\ln\frac{b_1}{b_*}} \tag{3.29}$$

where $C_{o.\infty}$ is the ambient oxygen concentration, $b_1=[(r_s^2 k_o)/D]\, k_o$ is the pre-exponential factor in the Arrhenius law, $b = b_1 \exp(-E/(RT))$, D is the diffusivity of the fuel vapor, and b_* characterizes the parameters at the flame front at the instant of ignition.

The predictions of the Peskin–Wise (1966) theory qualitatively agree with the experimental data of Wise and Agoston (1958). In particular, the theory predicts an increase in the ignition temperature with decreasing droplet diameter (Fig. 3.16), which agrees with the data. The effect of the ambient temperature and of the oxygen concentration on the critical states of a burning droplet was examined by Peskin et al. (1967). It was shown that there is a lower limit for the oxidizer concentration below which the ignition cannot occur at any ambient temperature. There also is some lower temperature limit at which droplet ignition is impossible at any ambient oxidizer concentration.

A more complex theory of droplet ignition and combustion was proposed by Polymeropoulos and Peskin (1969). The theory accounts for the existence of distributed chemical sources in the vapor–oxidizer mixture. It was used for the analysis of the behavior of fuel droplets at different ambient temperatures, oxidizer concentrations, reaction orders and stoichiometric ratios. The numerical solution of Polymeropoulos and Peskin (1969) revealed the character of the transition of the reactive system from the kinetic-controlled to the diffusion-controlled regime under different conditions. The dependences of the mass burning rate on the ambient temperature for different oxidizer concentrations are shown in Fig. 3.17. It is seen that the curves of dimensionless mass burning rate $\bar{u}_s = (u_s r_s)/\alpha$ (u_s is the radial velocity of fuel vapor at the surface, r_s is the radius of the fuel droplet, and α is the thermal diffusivity of the gaseous phase) on ambient temperature T_∞ are typically S-shaped with the lower and upper branches related to the kinetic-controlled and diffusion-controlled regimes. As the ambient concentration decreases, the S-shaped curves tend to flatten. As a result the extinction and ignition temperatures increase with the decrease in the ambient oxidizer concentration, whereas according to the linear theory of Peskin and Wise (1966) it increases with an increase in $C_{o.\infty}$. This shows that the model of an infinitely thin flame front is insufficient.

Law (1975) also developed a theory of droplet ignition and extinction and formulated the appropriate criteria. His theory is based on the method of matched asymptotic expansions, which was used earlier by Linan (1974) for studying the structure of the counterflow diffusion flame. The theory allows for calculation of the critical Damkohler numbers corresponding to the droplet ignition and extinction. It should be noted that in the general case of an unsteady ignition process the parameters can be found by using the dynamic ignition criteria (Chiu 2000).

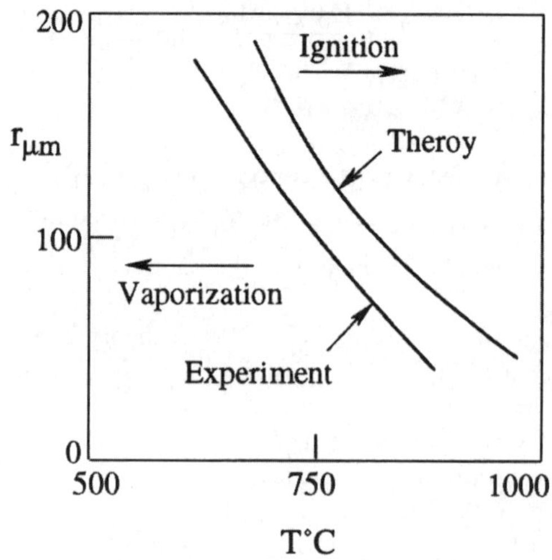

Fig. 3.16 The ignition temperature versus droplet radius. Reprinted from Peskin and Wise (1966), with permission. Experimental data by Wise and Agoston (1958)

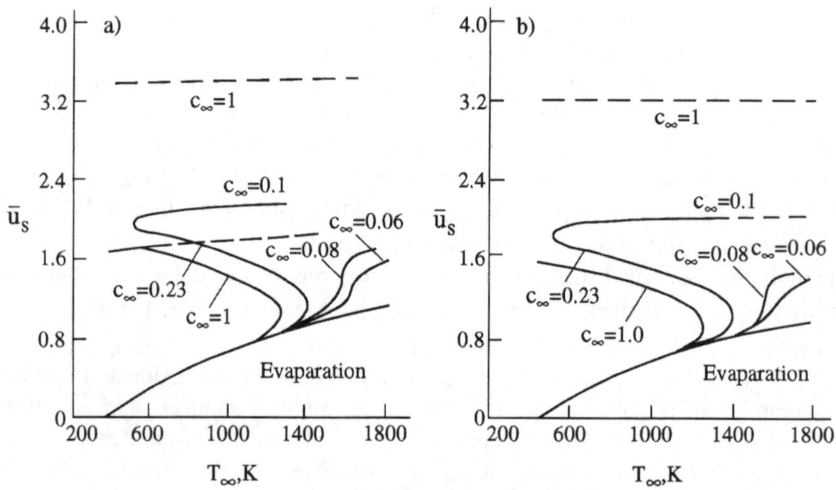

Fig. 3.17 Mass burning rate versus the ambient temperature. Reprinted from Polymeropoulos and Peskin (1969), with permission. a) $b_1 = 0.6 \cdot 10^{10}$, b) $b_1 = 10^4$, $b_1 = r_s^2 k_o / D$, k_o is the preexponent in the Arrhenius law, D is the diffusivity of fuel vapor

The ignition of a single component liquid droplet which is suddenly introduced into an oxidizing atmosphere was considered by Shaygan and Prakash (1995). Their numerical solution of the coupled liquid–gas phase problem showed that droplet ignition is governed by the state of the vapor–oxidizer mixture in the vicinity of the droplet, and not by its temperature. For the droplet with an initial temperature T_0 lower than the wet-bulb temperature $T_{w\text{-}b}$, the ignition occurs at $T < T_{w\text{-}b}$ and only close to the end of the droplet lifetime does its temperature approach $T_{w\text{-}b}$.

Another approach to the study of the critical states of a burning droplet was proposed by Rosser and Rajopakse (1969). They used the simpliest "quasi-solid" model which assumes that the droplet is enveloped by a reactive spherical shell within which the conductive heat transfer and heat release occur. As a matter of fact, the problem on droplet ignition reduces to the well-known problem on thermal explosion (Frank-Kamenetskii 1969). In both cases it consists of the calculation of a certain value of the parameter δ accounting for the geometric, thermal and kinetic characteristics. For values of δ much greater than the critical one, a steady-state solution becomes impossible (disappears). This value of the parameter δ is considered as the critical one corresponding to droplet ignition.

An attempt to study droplet ignition and extinction by application of the stability theory and specifically of Liapunov's first method, was made by Pindera and Brzkustowski (1984). In their approach, the equations governing regimes of droplet burning are transformed to the following nonlinear ordinary equations;

$$\frac{d\Theta}{dt} = -\mu\pi^2\Theta Le + (\frac{Da}{P_1})C_o C_f \exp\left(\frac{\Theta}{1+\beta\Theta}\right) \tag{3.30}$$

$$\frac{dC_o}{dt} = -\mu\pi^2(C_o - C_{o.\infty}) - m_o C_o C_f \exp\left(\frac{\Theta}{1+\beta\Theta}\right) \tag{3.31}$$

where Le is the Lewis number, $\mu = \left(T_s/T_\infty\right)^{4/3}$, $P_1 = (c_p R T_\infty^2)/(qE)$, $\beta = (RT_\infty)/E$, $\Theta = ((T - T_\infty)E)/(RT_\infty^2)$, Da is the Damkohler number, C is the concentration, m is the mass of the species involved in the reaction, and subscripts o, f and ∞ correspond to oxidizer, fuel and the ambient gas.

Equations (3.30) and (3.31) are dimensionless. They are similar to the equations used to study the stability of the lumped-parameter autonomous systems by Willems (1970), Hyseyin (1978).

The solution of Eqs. (3.30) and (3.31) determines the familiar S-shaped curves $\Theta(Da)$, and yields the detailed data on the transition from the pure evaporation state to the combustion state. The ignition (and extinction) is considered as the transition from the limiting low (high) temperature state to the high (low) temperature state at the given parameters of the system.

The equilibrium curve $\Theta(\mathrm{Da})$ is found from Eqs. (3.30) and (3.31) requiring the conditions of the steady state $d\Theta/dt = dC_o/dt = 0$. It is given by

$$z = \frac{\Theta}{1-P_2\Theta}\exp\left(-\frac{\beta\Theta}{1+\beta\Theta}\right) \tag{3.32}$$

where $z = (\mathrm{Da}C_f C_{fo\infty})/(\mu\cdot\pi^2 P_1 \mathrm{Le})$ and $P_2 = (m_o P_1 \mathrm{Le})/C_{o.\infty}$.

In accordance with the methods of the stability theory, the regions of stability of the steady states corresponding to Eqs. (3.30) and (3.31) are determined by the matrix **A:**

$$\mathbf{A} = \begin{vmatrix} \dfrac{\partial f_1}{\partial\Theta} & \dfrac{\partial f_1}{\partial C_o} \\ \dfrac{\partial f_2}{\partial\Theta} & \dfrac{\partial f_2}{\partial C_o} \end{vmatrix} \tag{3.33}$$

where f_1 and f_2 are right-hand sides of Eqs. (3.30) and (3.31).

The system is stable when det $\mathbf{A} > 0$ (the first static criterion). This leads to the following relation for the critical states which coincide with the extrema of the dependence $z(\Theta)$, and are given by

$$\Theta_{1.2} = \frac{(1-2\beta)\pm(1-4\beta-4P_2)^{\frac{1}{2}}}{2(\beta^2+P_2)} \tag{3.34}$$

The unstable states correspond to the intermediate branch of the curve $z(\Theta)$ which is confined by the points Θ_1 and Θ_2.

The second (dynamic) stability criterion is tr$\mathbf{A} < 0$. It implies that

$$\mathrm{Le}\left(-1+\frac{\Theta}{(1+\beta\Theta)^2}\right)-\frac{1}{1-P_2\Theta} < 0 \tag{3.35}$$

The limits of the dynamic stability correspond to tr$\mathbf{A} = 0$. This leads to the following expression:

$$\Theta-(1+\Theta\beta)^2)(1-P_2\theta)-(1+\Theta\beta)^2/\mathrm{Le} = 0 \tag{3.36}$$

Specific topics. The state of liquid droplets, in particular, their ignition and extinction depends on a number of different factors, such as the composition of the surrounding medium, its initial pressure and temperature, as well as the variation of the ambient parameters. In relation to these, we consider briefly two particular problems on droplet ignition: (i) ignition in a reactive environment, (ii) extinction under the conditions of depressurization. The first of these is of interest for simulations of spray ignition, the second one for studying operating regimes of diesel engines.

Botros et al. (1980) studied numerically droplet ignition and combustion in a reactive fuel–oxidizer environment. They revealed characteristics of the processes developing under the conditions of fuel-lean and fuel-rich environments. In the first case droplet ignition takes place close to the droplet surface where the composition of the vapor–oxidizer mixture is close to stoichiometric. Subsequently, the liquid vaporization leads to an increase in vapor concentration in the inner region of the diffusion flame. As a result, the flame propagates outward. In contrast with that, in the case of a fuel-rich environment the ignition occurs in a premixed reactive mixture. A similar pattern of the development of combustion in the vicinity of a burning droplet was revealed by Shaygan and Prakash (1995). Their numerical solution of the transient coupled liquid–gas phase problem on droplet combustion showed that the depending on the fuel vapor concentration in the environment, various scenarios of droplet burning may occur. They include the regimes of diffusion and premixed burning (corresponding to fuel-lean and fuel-rich mixtures), as well as a pure vaporization in a fuel-rich environment under conditions when all the oxidizer has been consumed.

The influence of depressurization on droplet extinction was investigated by Buchholz and Tapper (1978), and Bellan and Summerfield (1978). Two models of droplets (nonregressive, with a constant radius) and regressive (with decreasing radius) were considered. Some results of their numerical calculations of behavior of a nonregressive and regressive droplet are presented in Fig. 3.18 in the form of the dependence $\delta(t)$ for different values of the increment b. Here $\delta = Da_0/Da_E$, Da_0 and Da_E are the value of the Damkohler number corresponding to the zeroth order expansion and the extinction Damkohler number; $\delta = 1$ corresponds to extinction, b is the pressure increment in the expression $P/P_0 = \exp(b\tau)$, P_0 is the initial pressure. As can be seen in Fig. 3.18, the shapes of curves $\delta(\tau)$ corresponding to the nonregressive and regressive droplets at constant pressure $(b = 0)$ are very different. In the first case δ varies slightly with τ and exceeds one at all values of τ. This means that the extinction of the nonregressive droplet does not occur at the given pressure. In contrast with that, δ for the regressive droplet decreases significantly with τ and tends to unity when the droplet radius reduces to a critical size.

The effect of depressurization on droplet extinction is illustrated by the curves corresponding to b = 0.5, b = 1 and b = 2.0. An increase in the increment b is accompanied by a sharp decrease in δ. Therefore, an increase in the rate of depressurization leads to an early extinction of the droplet.

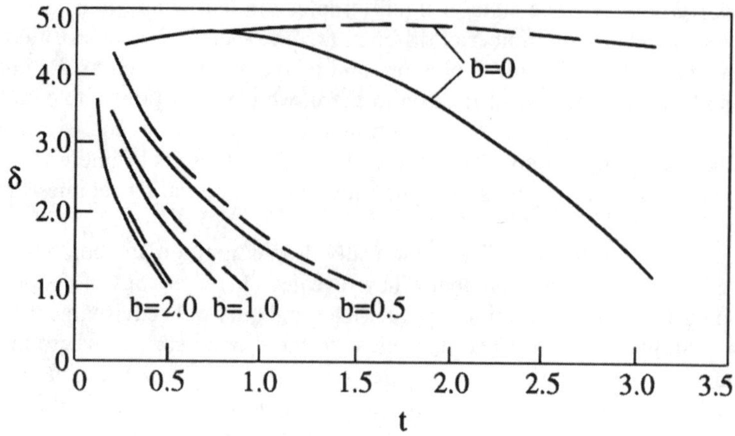

Fig. 3.18 Variation of the parameter δ with τ for both nonregression and regression droplets (Bellan and Summerfield 1978)

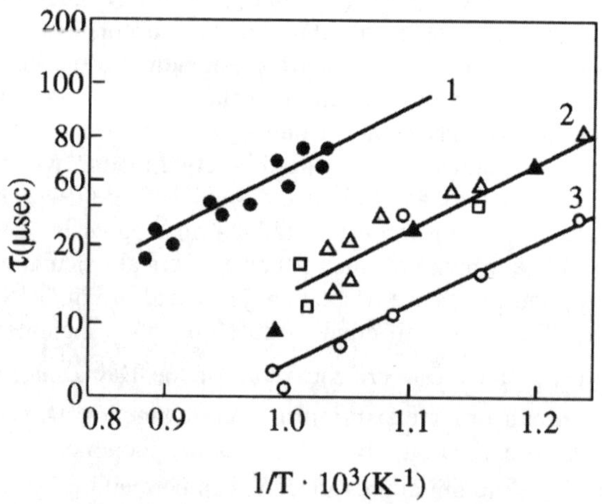

Fig. 3.19 The ignition lag for hexadecane droplets as a function of the oxidizer temperature and the droplet–droplet proximity. Reprinted from Wood and Rosser (1969), with permission. Curve 1, an isolated single drop; curve 2, a group of 3–4 drops, curve 3, a steady stream of drops generated at the rate 30 s^{-1}

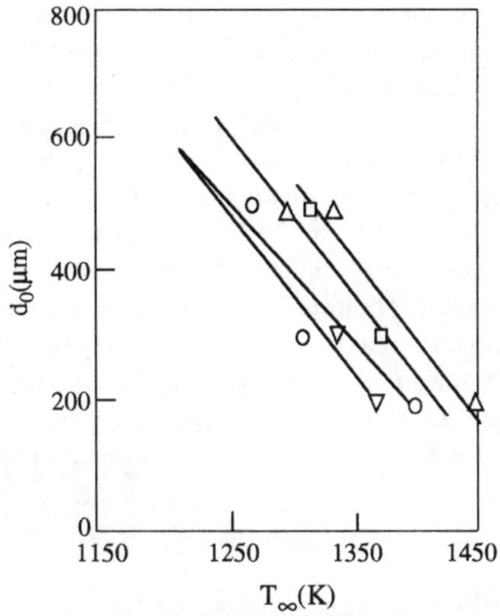

Fig. 3.20 The effect of the particle diameter on the ignition temperature. Reprinted from LeMott et al. (1971), with permission. ∇, hexadecane (molecular weight $\mu = 290$); O, JP-4 aircraft fuel ($\mu = 150$); □, N6 fuel oil ($\mu = 350$); Δ, N2 heating oil ($\mu = 180$).

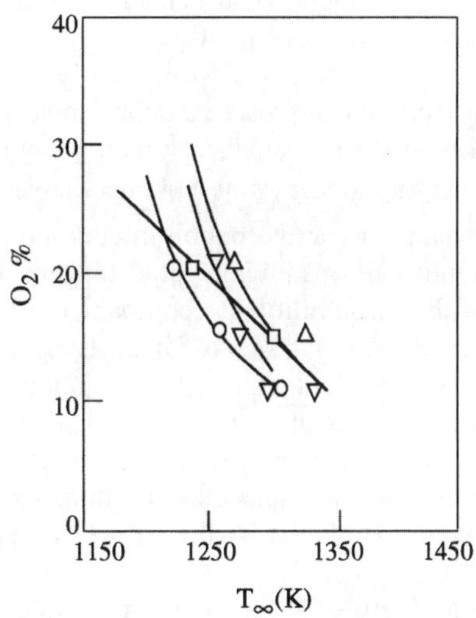

Fig. 3.21 The effect of the oxygen concentration on the ignition temperature. Reprinted from LeMott et al. (1971), with permission. ∇, hexadecane (molecular weight $\mu = 290$); O, JP-4 aircraft oil, ($\mu = 150$); □, N6 fuel oil ($\mu = 350$); Δ, N2 heating oil ($\mu = 180$)

Experiments. A number of experimental investigations of droplet ignition have been carried out during the past decades (Nishiwaki 1955, Tarifa et al. 1962, Wood and Rosser 1969, LeMott et al. 1971, Kauffman and Nicholls 1971, Saitoh et al. 1982, Takei et al. 1993). Most of the ignition measurements were directed to studying the effect of different parameters (such as the droplet size, the nature of liquid fuel, the oxidizer temperature and concentration, etc) on the ignition lag and temperature (cf. Figs. 3.19, 3.20, and 3.21). The experiments showed that the ignition lag depends significantly on the environmental temperature and only slightly on the oxidizer concentration. The ignition temperature is a steep function of the droplet size and it increases when the droplet diameter decreases. It was also shown that the ignition is controlled by the kinetic factors: by the pre-exponential in the Arrhenius law, by the energy activation and by the order of the reaction. The molecular weight of the fuel is not a discernible controlling parameter of droplet ignition.

1.3.3 Bubble ignition

Mechanism of ignition. Combustion of bubbly media is of significant interest in relation to many applications in chemical technology (Vol'fkovich et al. 1952), namely to production of organic and inorganic combinations (Dankverts 1973, Merzhanov 1973), and to the reduction metallurgy of titanium, zirconium, hafnium and other rare-earth elements (Garmata and Gulnitskii 1968, Barishnikov et al. 1979). A knowledge of bubble ignition is needed for understanding the mechanism of combustion wave propagation in bubbly media and improving numerous technological processes using reactive bubbly media.

Bubbly media with relatively small volumetric content of gaseous phase ($\xi_1 \sim 0.1$–0.2) are liquids saturated by fine bubbles diameter whose is much less than the interparticle distance. Reactive bubbly media are prepared, as a rule, by supplying oxidizer jets into still or moving liquid reagent. Disintegration of such jets as a consequence of their instability is accompanied by formation of fine bubbles that are filled primarily with gaseous oxidizer. Evaporation of liquid reagent and gas dissolving in the carrier fluid lead to the formation of a gas–vapor mixture inside the bubbles, as well as to the appearance of a continuous molecular solution of gas phase. Since the rate of chemical reaction in the gas–vapor mixture is, as a rule, much higher than that in the molecular solution, the dominant role in the combustion of bubbly media is played by the gas-phase reactions inside of bubbles.

At moderate pressure $P \sim 10^5$–10^6 Pa and typical values of volumetric content of gaseous phase $\xi_1 \sim 0.1$–0.2, the mass content of oxidizer in the bubbly media is much less than that needed for full burning-out of liquid reagent. The consumption of oxidizer in the chemical reaction and the corresponding heating of the gaseous phase lead to a sharp variation of temperature and pressure inside the bubble, whereas the temperature of the surrounding liquid varies only slightly. Under these conditions the development of combustion in a bubble depends on a number

of factors reflecting the initial composition of the reactive mixture, the rate of evaporation of liquid reagent and diffusion of its vapor, the rate of chemical reaction and the heat transfer between the bubble and the surrounding liquid. Depending on the physical properties of gaseous and liquid reagents, intensity of heat and mass transfer, as well as on the conditions of bubbly medium formation, the initial composition of the gaseous phase can range from pure oxidizer to an equilibrium mixture of oxidizer and saturated vapor of liquid reagent. Accordingly, different scenarios of bubble ignition and combustion are possible. The development of these processes in the bubbles primarily filled with pure oxidizer or with an equilibrium gas–vapor mixture was explored by Fridman et al. (1981). For a numerical solution of the problem on thermal and dynamic interaction of a gas bubble with the surrounding liquid reagent they used the system of equations of energy, species and momentum balance, the equation of state of the gas, as well as the Rayleigh equation for radial expansion/compression of the bubble in an infinite incompressible fluid: $rr'' + (3/2)r'^2 = (1/p)[P(r,t) - P_\infty(t)]$, were r is the bubble radius, $r' = dr/dt$, $r'' = d^2r/dt^2$, P(r,t) is the pressure at the bubble surface in the liquid, $P_\infty(t)$ is the ambient pressure (Leal 1992). The analysis showed that ignition and combustion of a combustible bubble is mostly determined by the intensity of
evaporation of liquid reagent and the characteristic times of heat transfer and chemical reaction. These factors are taken into account via the dimensionless parameters

$$St \sim \frac{q_e E}{R c_{p_0} T_0^2} \tag{3.37}$$

$$\delta \sim \frac{q k_o \rho_0 E}{\lambda_0 R T_0^2} r_0^2 \exp\left(-\frac{E}{RT_0}\right) \tag{3.38}$$

where q and q_e are the heats of combustion and evaporation, k_o is the pre-exponential factor in the Arrhenius law; c_p and λ are the gaseous phase specific heat and thermal conductivity, respectively; r is the bubble radius; subscript 0 refers to the initial state.

The ignition of a combustible gas bubble occurs when δ exceeds its critical value δ_{cr}, which depends on the initial composition of the reactive mixture. The critical value δ_{cr} increases with St. At relatively small values of St (St<150), the temperature profiles in the bubbles, which are primarily filled with an equilibrium gas–vapor mixture, are monotonic with the maximum at the bubble center. At St>150 the temperature profiles have a characteristic break at the interface where the dimensionless excessive temperature Θ is negative (Fig. 3.22). It is seen that a phase transition at the bubble surface occurs due to the heat transfer from the hot bubble, as well as from the surrounding liquid. As a result of the phase transition, the interfacial temperature is reduced.

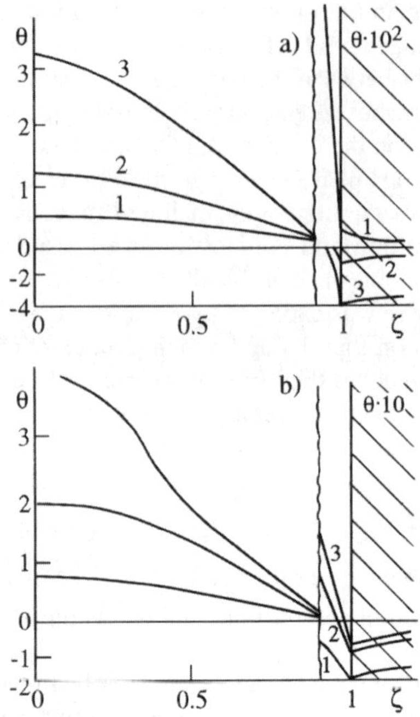

Fig. 3.22 Temperature distribution inside a bubble and near the interface. Reprinted from Fridman et al. (1981), with permission. a) Bubble is filled by an equilibrium gas–vapor mixture (curve 1, $\tau = 0.084$; curve 2, $\tau = 0.3$; curve 3, $\tau = 1.24$). b) The bubble contains pure oxidizer at the moment initial $\tau = 0$ (curve1, $\tau = 0.86$; curve2, $\tau = 1.8$; curve3, $\tau = 2.26$) τ and ζ are dimensionless time and current radius

The combustion of a bubble primarily filled with oxidizer is multi-stage. It involves the evaporation of liquid reagent, mixing of its vapor with oxidizer, ignition and combustion of the reactive mixture. At the values of δ close to δ_{cr}, the formation of the gas–vapor mixture inside the bubble terminates before the beginning of the intensive chemical reaction. Therefore, the subsequent process proceeds similarly to that described above.

At large values of δ ($\delta > \delta_{cr}$) chemical reaction overtakes the diffusion of the vapor of liquid reagent. Accordingly, the process develops under the conditions of a lack of the vapor in the central region of the bubble. In this case a combustion wave is formed in the reactive mixture. It moves from the outside towards the bubble center and stabilizes at some distance from the interface. This distance decreases with δ.

Bubble oscillations. It is well known that perturbation of a bubbly medium about its equilibrium parameters can initiate bubble oscillations. The results concerning the dynamics of gas and vapor bubbles are summarized in the monographs by Leal (1992) and Nigmatulin (1991). Here we confine ourselves to a brief consideration of the oscillations of combustible bubbles. These oscillations emerge due to compression of the gas–liquid system, its inertia, as well as due to the intensive heat release inside the bubble and the heat transfer from the bubble to the surrounding fluid.

The burning of a gas-vapor mixture is accompanied by an increase of the temperature of the gaseous phase inside the bubble. This leads to bubble expansion, as well as to the acceleration of the surrounding fluid and its motion in a radial direction (only special, symmetric oscillations are discussed). The intensity of such motion depends on the rate of combustion and decreases with the burn-out of the oxidizer contained in the bubble. In contrast with combustion, the heat transfer from a hot bubble to a cold liquid reagent promotes reduction of the temperature and pressure of the gaseous phase, as well as bubble compression. The competition of the above-mentioned factors determines the dynamics of combustible bubbles.

The behavior of a single combustible gas bubble suspended in an infinite inviscid liquid was considered by Gol'dshtein et al. (1998). They used a number of plausible assumptions. In particular, they assumed a large activation energy, an ideal gas behavior and a homobaric approximation, an invariable mass content of the gaseous phase in the absence of phase transformation. As a result, they reduced the problem to the following system of three ordinary nonlinear equations for the change of temperature, concentration of combustible component and bubble size

$$\gamma\frac{d\Theta}{d\tau}=\xi C\ \ R_b^3\exp\left(\frac{\Theta}{1+\beta\Theta}\right)-\delta R_b^2\Theta-\frac{\chi}{R_b}(1+\beta\Theta)\frac{dR_b}{d\tau} \tag{3.39}$$

$$\frac{dC}{d\tau}=-3\frac{C}{R_b}\frac{dR_b}{d\tau}-C\exp\left(\frac{\Theta}{1+\beta\Theta}\right) \tag{3.40}$$

$$R_b\frac{d^2R_b}{d\tau^2}+\frac{3}{2}\left(\frac{dR_b}{d\tau}\right)^2=\Omega\left[\left(1+\beta\Theta\right)R_b^{-3}-1\right] \tag{3.41}$$

Here $\delta=[(3h\xi\beta T_{1.0})/(k_o C_{f0}\mu_f q r_0)]\exp(1/\beta)$, $\gamma=(c_{pl}\rho_{1.0}T_{1.0})/(C_{f0}\cdot q\cdot\mu_{lf})$, $\xi=c_p/c_v$, $\beta=(RT_0)/E$, $\chi=(3\xi\rho_{1.0}RT_{1.0})/(q\mu_f^2C_{f0})$, $\Omega=(t_R/t_*)^2[(4\pi^2)/(3\xi)]$, $\Theta=(T_l-T_{1.0})[E/(RT_{1.0}^2)]$, $C=C_f/C_{f0}$ and $R_b=r_b/r_0$; Θ, C and R_b are the dimensionless temperature, the concentration of combustible component and the bubble radius. Also the dimensionless time is $\tau=t/t_R$ with $t_R=k_o^{-1}\exp\left(E/RT_{1.0}\right)$ being the characteristic time of chemical reaction, k_o is the pre-exponential factor in the Arrhenius law; h is the heat transfer coefficient, μ

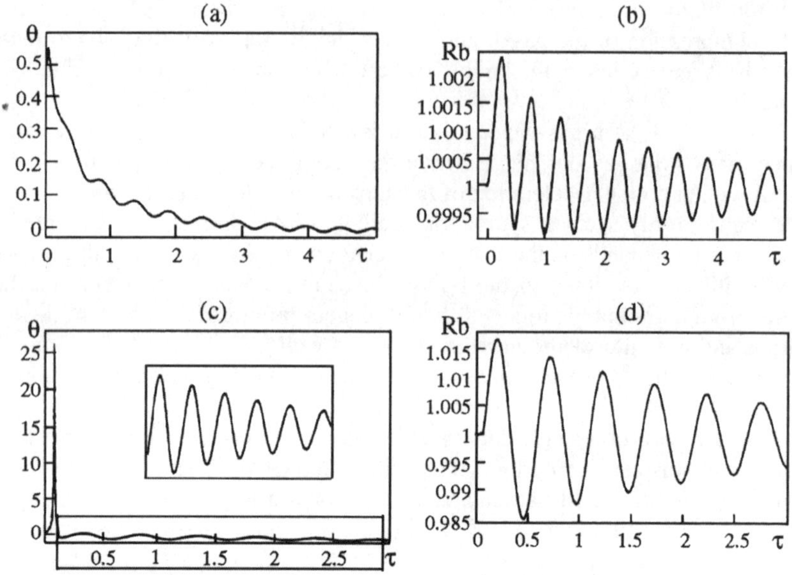

Fig. 3.23 Bubble oscillations. Reprinted from Goldshtein et al. (1998), with permission. 1. The slow reaction regime: $\gamma = 0.0333$, $\beta = 0.008$, $c = 1.4$, $\delta = 4$, $\chi = 5$, and $\Omega = 50$ a) and b) depict the temperature and radius time series, respectively). 2. The explosive regime: $\gamma = 0.0333$, $\beta = 0.008$, $c = 1.4$, $\delta = 2.5$, $\chi = 5$, and $\Omega = 50$ (c) and d) depict the temperature and radius history, respectively). The region of the temperature oscilations is marked in the inset

is the molecular weight, $t_* = 2\pi\left((\rho_2 r_0^2)/(3\xi P_0)\right)^{1/2}$ is the characteristic time of harmonic, oscillations, subscripts 1 and 2 refer to the gaseous and liquid phases, respectively; f and 0 refer to the combustible component fuel and to the initial state of the system, respectively.

The analysis of the problem showed that there is a critical value of the parameter δ which subdivides the thermal states of the bubble in two groups: (i) one corresponding to a slow (low temperature) reaction and (2) the other corresponding to explosion (high temperature). The critical value of the parameter δ and the temperature corresponding to the Semenov's critical conditions (the conditions of tangency of the heat release and heat losses curves, Eqs. (3.7) and (3.8)) are given by

$$\delta_{cr} = \xi e(1-\beta); \quad \Theta_{cr} = 1 + 2\beta \tag{3.42}$$

where e=2.718… .

The change of the temperature of the gaseous phase and of the bubble radius at ignition and combustion of the reactive mixture inside a single bubble suspended in an infinite bulk liquid is shown in Fig. 3.23 for both $\delta > \delta_{cr}$ and $\delta < \delta_{cr}$. In both cases the ignition of the reactive mixture is accompanied by a sharp increase of the temperature from its initial value $\Theta = 0$ up to a maximum $\Theta_m \simeq 0.55$ and $\Theta_m \simeq 26$, respectively, for the slow ($\delta = 4$) and explosion ($\delta = 2.5$) regimes. Then the bubble temperature decreases (oscillating) to a steady state value. The dimensionless frequency of the temperature oscillations is different from the resonance Minnaert frequency corresponding to the adiabatic conditions (Nigmatulin 1990, Nakoryakov et al. 1993). It is noteworthy that as a consequence of inertia of a system, significant oscillations of the gas temperature lead to relatively small oscillations the bubble radius and volume V. For example, at $\delta = 2.5$ (the explosive regime) $T/T_0 \sim 1.3$, $r_b/r_0 \sim 1.017$ and $v/v_0 \sim 1.05$

1.3.4 Ignition of metal particles

Background. The systematic study of metal particle combustion dates back to the second part of the 20th century (Cassel and Liebman 1959, Fridman and Macek 1962, Gurevich et al. 1970 a,b, Khaikin et al. 1970). Significant experimental and theoretical results concerning metal particle combustion were obtained by Fridman and Macek (1962, 1963), Bartlett et al. (1963), Macek (1967), Gurevich et al. (1970 a,b), and Merzhanov et al. (1977). These works and the subsequent researches revealed the mechanism of metal particle combustion. The peculiarities of the ignition and combustion of magnesium (Cassel and Liebman 1959, Gurevich et al. 1979 a,b, Shafirovich and Goldshleger 1992, Dreizin et al. (2000)), aluminium (Fridman and Macek 1962, 1963, Bartlett et al. 1963, Macek 1967, Gurevich et al. 1970 a,b, Merzhanov et al. 1977, Dreizin 1996), beryllium (Macek 1967, Macek and Semple 1969), boron (Macek and Semple 1971, Macek 1973, Glassman et al. 1984, Yeh and Kuo 1996), and titanum (Molodetsky et al. 1998) particles were studied. Also ignition and combustion of particles of copper, tungsten, steel, tantalum and molybdenum were considered (Dreizin et al. 1993). At the same time, theoretical models allowing for estimates of the critical parameters corresponding to the ignition of metal particles were developed. An effective approach to the solution of the problem was proposed by Khaikin et al. (1990). It is based on the application of the methods of the classical thermal explosion theory for studying the ignition of metal particles. Below we discuss briefly the main results obtained in the framework of this approach.

Governing equations. Consider the thermal balance equation for a metal particle introduced instantaneously into a gaseous oxidizer. We assume that the radiative heat transfer is negligible.

In the frame of the lumped-capacitance heat transfer model the heat balance equation for a spherical metal particle is

$$mc_p \frac{dT}{dt} = \rho Q \cdot S \frac{d\delta}{dt} - hS(T - T_\infty) \tag{3.43}$$

where m and S are the mass and the surface area of the particle; ρ and c_p are the density and specific heat of particle material, h is the heat transfer coefficient, δ is the thickness of the oxide film, at the surface, Q is the heat release per gram oxide multiplied by the oxide/metal density ratio, and T_∞ is the ambient gas temperature.

Bearing in mind that $m = \rho(4/3)\pi r_o^3$, $S = 4\pi r_o^2$ and $h = (Nu\lambda)/d$ (Nu and λ are the Nusselt number and the thermal conductivity of the gaseous phase, r_o is the particle radius), we write Eq. (3.43) as follows:

$$\frac{1}{3}\rho c_p r_o \frac{dT}{dt} = Q\rho \frac{d\delta}{dt} - \frac{\lambda}{2r_o} Nu(T - T_\infty) \tag{3.44}$$

Eq. (3.44) should be supplemented by a relation determining the rate of metal oxidation. The data on the dependence of δ on t, as a rule, can be approximated by

$$\delta^{n+1} = K_* t + C \tag{3.45}$$

where $n = 0;\ 1;\ 2\ldots;$ K_* is the rate of oxidation constant, and C is constant.

Depending on the value of the parameter n, it is possible to distinguish between the linear $(n = 0)$, parabolic $(n = 1)$ and cubic $(n = 2)$ laws of metal oxidation. In the case n=0 the rate of growth of the oxide film is constant. The case n=0 corresponds to a reaction which leads to the formation of a crumbly easily permeable oxide film. In this case the rate of transport of gaseous reagent is controlled by an absorption mechanism. The case of n = 1 corresponds to $d\delta/dt$ decreasing inversely proportionally to δ. Such a behavior is accompanied by the formation of a layer of a dense product, whereas the rate of oxidation is determined by the diffusion transport of the gaseous reagent through it. In the cases n = 2 and $n \geq 3$ the reaction that is characterized by an intensive hindering of the process by oxide layer.

It should be noted that the real laws of metal oxidation are more complicated than the law (3.45). During the initial and the later stages of the process, different processes play the dominant role in the oxide film formation. Therefore, in order to describe the oxide film formation a number of relations similar to (3.45) should be tested.

Differentiating Eq. (3.45) we obtain

$$\frac{d\delta}{dt} = \frac{K_*}{(n+1)}\frac{1}{\delta^n} \tag{3.46}$$

The dependence $K_*(T)$ may be approximated by the Arrhenius law

$$K_* = A\exp\left(-\frac{E}{RT}\right) \tag{3.47}$$

where the pre-exponent A is a weak function of pressure P.

With the help of Eqs. (3.46) and (3.47) the rate of growth of the oxide film is given by

$$\frac{d\delta}{dt} = \frac{k_n C_o^m}{\delta^n}\exp\left(-\frac{E}{RT}\right) \tag{3.48}$$

where $k_n = (k_o A)/(n+1)$, k_o is a constant, $A = C_o^m$, C_o is the oxidizer concentration at the particle surface, m is the order of reaction.

The initial conditions for the system of Eqs. (3.44) and (3.48) are given by

$$t = 0, \quad T = T_i, \quad \delta = \delta_i \tag{3.49}$$

where subscript i refers to the initial state of the particle.

We introduce the dimensionless variables

$$\eta = \frac{\delta - \delta_i}{\delta_i},\ \Theta = \frac{E}{RT_\infty^2}(T - T_\infty),\ \tau = t\frac{3Q}{c_p}\frac{E}{RT_\infty^2}\frac{k_n C_o^m}{r_o \delta_i^n}\exp\left(-\frac{E}{RT_\infty}\right);$$

$$\chi = \frac{Q\rho}{\delta_i^n}\frac{E}{RT_\infty}\frac{2r_o}{\lambda}k_n C_o^m \exp\left(-\frac{E}{RT_\infty}\right);\ \beta = \frac{RT_\infty}{E},\ \Theta_i = \frac{E}{RT_\infty^2}(T_\infty - T_i);\ \gamma = \frac{r_o}{3\delta_i}\frac{c_p RT_\infty^2}{EQ}$$

Then Eqs. (3.44) and (3.48) reduce to the following form

$$\frac{d\Theta}{d\tau} = \varphi(\eta)\exp\left(\frac{\Theta}{1+\beta\Theta}\right) - \frac{\Theta}{\chi} \tag{3.50}$$

$$\frac{d\eta}{d\tau} = \gamma\varphi(\eta)\exp\left(\frac{\Theta}{1+\beta\Theta}\right) \tag{3.51}$$

where $\varphi(\eta) = (1+\eta)^n$. The initial conditions for Eqs. (3.50) and (3.51) are

$$\tau = 0, \quad \eta = 0, \quad \Theta = -\Theta_i \tag{3.52}$$

Ignition conditions. Introducing the new variables $\eta = \gamma z - 1$, $\tau' = \tau/\gamma^n$ we arrive at the equations

$$\frac{d\Theta}{d\tau'} = \frac{1}{z^n}\exp\left(\frac{\Theta}{1+\beta\theta}\right) - \frac{\Theta}{\Omega} \tag{3.53}$$

$$\frac{dz}{d\tau'} = \frac{1}{z^n}\exp\left(\frac{\Theta}{1+\beta\Theta}\right) \tag{3.54}$$

with the initial condition

$$\tau' = 0, \quad z = \frac{1}{\gamma}, \quad \Theta = -\Theta_i \tag{3.55}$$

where $z = \frac{\delta}{r_o}\frac{3QE}{c_p RT_\infty^2}$; and $\Omega = \frac{\chi}{\gamma^n}$

In the case when the dimensionless particle radius $\gamma \gg 1$, the initial condition (3.55) takes the form

$$\tau' = 0, \quad z = 0, \quad \Theta = -\Theta_i \tag{3.56}$$

The results of the numerical solution of Eqs. (3.53) and (3.54) subject to the conditions (3.56) are presented in Fig. 3.24. It is seen that a small variation of the parameter Ω leads to a sharp change in the particle temperature. The value of Ω corresponding to an abrupt increase of the temperature can be considered as critical. The thickness of the oxide film changes significantly during the pre-ignition period.

Khaikin et al. (1970) showed that in the case $\gamma \gg 1$ the critical condition can be presented as

$$\Omega_{cr} = \left(\frac{3QE}{c_p RT_{\infty.cr}}\right)^{n+1} \frac{c_p\rho}{3\lambda}\frac{k_n C_o^m}{r_o^{n-1}}\exp\left(-\frac{E}{RT_{\infty.cr}}\right) = c^* \tag{3.57}$$

where $T_{\infty.cr}$ is the ignition temperature, i.e. the ambient-gas temperature, corresponding to particle ignition; c^* is a constant for which (i) in the case of n=1:

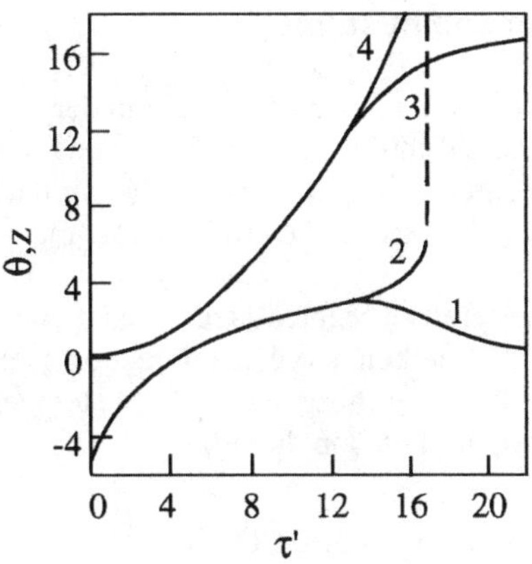

Fig. 3.24 Dimensionless particle temperature Θ and film thickness z as function of time τ' for the case $\gamma \gg 1$. Reprinted from Khaikin et al. (1979). Curve 1, $\Theta(\tau')$ for $\Omega = 2.33$; curve 2, $\Theta(\tau')$ for $\Omega = 2.34$; curve 3, $z(\tau')$ for $\Omega = 2.33$; curve 4, $z(\tau')$ for $\Omega = 2.34$

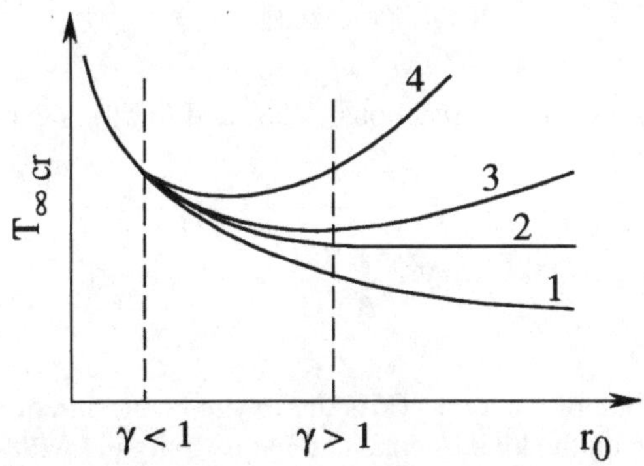

Fig. 3.25 Ignition temperature as a function of particle radius for various oxidation laws. Reprinted from Khaikin et al. (1979). Curve 1, n=0; curve 2, n=1; curve 3, n=2; curve 4, $n \gg 1$

c^*=1.57 for $\Theta_i = 0$, c^*=2.33 for $\Theta_i \gg 1$; (ii) in the case n= 2: c^*= 7.45, $\Theta_I = 0$ and $c^* = 18$ for $\Theta_I \gg 1$.

The condition (3.57) allows the effect of particle size on the ignition temperature (Fig. 3.25) to be studied. In the case n = 0 the ignition temperature decreases as r_0 increases, whereas in the case n = 1 it saturates and becomes constant depending on the particle size. At higher n the ignition temperature increases with r_0.

1.3.5 Coke particle combustion

Carbon–oxygen interaction. A coal particle undergoing heating and devolatilization is finally converted into a coke particle. The combustible component of such a particle is solid carbon. Accordingly, in the combustion of the coke particles the carbon–oxygen interaction determines the main characteristics of the process.

At present, it is generally accepted (see Predvoditelev 1949, Kantarovich 1958, Golovina 1983) that carbon–oxygen interaction is caused by a number of reactions. The most important of those are: (i) the heterogeneous reactions of carbon oxidation and reduction of carbon dioxide:

$$C + O_2 = CO_2 \tag{3.58}$$

$$CO_2 + C = 2\,CO \tag{3.59}$$

and (ii) the gas phase combustion reaction of carbon monoxide:

$$2CO + O_2 = 2CO_2 \tag{3.60}$$

The rate of the heterogeneous reactions (3.58) and (3.59) is expressed as (Khitrin 1957 a)

$$W = \frac{K'C}{A + C} \tag{3.61}$$

where W is the rate of reaction, C is the oxygen concentration, and A is a parameter depending on the kind of coke and varying slightly with temperature.

At a high rate of carbon–oxygen interaction the oxygen concentration at the reactive surface becomes very small. Then Eq. (3.61) takes the simplest form

$$W = KC \tag{3.62}$$

where $K = K'/A$ is defined by the Arrhenius law. Accordingly, the rate of the heterogeneous reaction is written as

$$W_j = K_j C \exp\left(-\frac{E}{RT}\right) \tag{3.63}$$

where subscript $j = O_2$ or CO_2 refers to the oxidation or reduction reactions respectively.

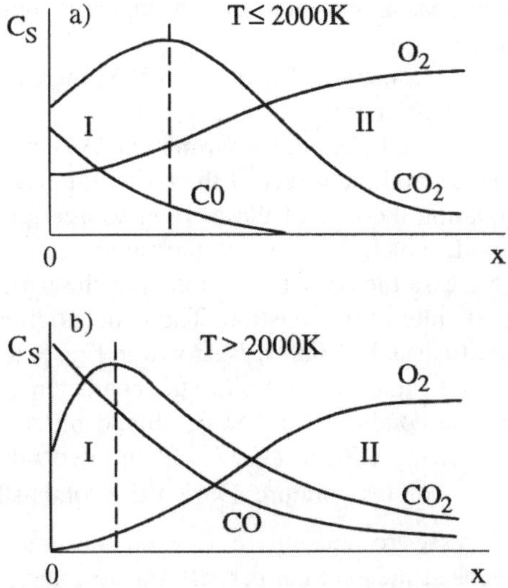

Fig. 3.26 Distribution of oxygen, and carbon mono- and dioxide in the vicinity of a coke particle. a) $T \leq 2{,}000$ K ; b) $T > 2{,}000$ K; I, the combustion region; II, the region of the diffusion mixing

The kinetic constants in Eq. (3.63) have the order: (i) $K_{O_2} \sim (1.15-10.6)\cdot 10^4$ [m/s], $E \sim (73-120)\cdot 10^3$ [J/mol], (ii) $K_{CO_2} \sim (3-3.64)\cdot 10^4$ [m/s], $E \sim (167-209)\cdot 10^3$ [J/mol], (the data by Vulis and Vitman—see Kantorovich 1958 and Tsukhanova—see Kantorovich 1958). The wide scatter in the values of the kinetic constants, especially of the pre-exponential factor, is due to the influence of the organic and mineral admixtures still contained in coke particles.

The contributions of the oxidation and reduction reactions (3.58) and (3.59) to the combustion of coke particles differ. They depend mostly on the temperature of the process. Under the conditions of a relatively low temperature (T<2,000 K) the rates of the heterogeneous reactions (3.58) and (3.59) are low enough so that the oxygen and carbon dioxide partially react at the particle surface and partially penetrate through the cracks and pores into the particle. The carbon monoxide that is formed as a result of the reduction reaction (3.59) partially burns with oxygen that is contained in the gaseous phase and partially penetrates inside the coke particle. The existence of CO_2, CO and O_2 in the porous material results in the reaction inside the particle. This process disappears when the particle temperature increases. Under these conditions the combustion reaction is located in a relatively thin gaseous layer adjacent to the carbon surface. Since the oxidation and the reduction reactions proceed directly at the particle surface, the combustion of carbon monoxide occurs in the volume adjoining the particle surface. The typical distributions of oxygen, and carbon mono- and dioxide in the vicinity of a coke particle are shown in Fig. 3.26a.

The rate of the reduction reaction (3.59) increases significantly at T>2,000 K. This leads to an increase of the amount of carbon monoxide reacting with oxygen which is supplied from the environment. As a result, oxygen does not reach the particle surface, and the effect of the oxidation reaction (3.58) becomes negligible. In this situation the role of the oxygen source for the coke particle is played by carbon dioxide which diffuses into the external domain, as well as to the particle surface. In this case the reduction reaction at the particle surface is dominant and determines the rate of combustion. The concentration distribution in the vicinity of the coke particle at T>2,000 K is shown in Fig. 3.26b.

Combined processes. In order to explore the characteristics of combustion of a coke particle under the conditions of the combined internal and external reactions, we consider following Khitrin (1957 a,b) the dependence of the specific oxygen consumption G_{O_2} on temperature. During the combustion of a porous particle, there are two processes responsible for oxygen consumption: (i) the chemical reaction which proceeds at the external particle surface, and (ii) the internal reaction that occurs within the porous material.

$$G_{O_2} = W_s + D_* \left(\frac{\partial C_*}{\partial n} \right)_s \tag{3.64}$$

Here G_{O_2} is the oxygen consumption per unit particle surface, W_s is the rate of the heterogeneous reaction, C is the oxygen concentration, $\left(\partial C_* / \partial n\right)_s$ is the oxygen gradient at particle surface, D_* is the coefficient of inner diffusion through the porous particle, asterisk $*$ and subscript s correspond to the particle bulk and its surface.

The equation for the overall process can be formally presented in a form similar to the law of a heterogeneous chemical reaction

$$G_{O_2} = \alpha \cdot f(C) \tag{3.65}$$

where α is the so-called reaction coefficient of mass exchange which can be considered as an effective kinetic rate constant, and $f(C)$ is a function of oxygen concentration.

To calculate the reaction coefficient of mass exchange, use is made of the equation describing oxygen diffusion in a porous spherical particle with an internal reaction

$$D_* \nabla^2 C_* = q \tag{3.66}$$

where $q = K_* C_*$ is the reaction rate, $K_* = K s_*$, K and K_* are the constants of heterogeneous and interior reactions, and s_* is the specific area of internal surface. The solution of Eq. (3.66) is

$$C_* = \sum_n A_n I_{n+1/2}(\lambda, r) \frac{Y_n(\theta, \varphi)}{\sqrt{r}} \tag{3.67}$$

where $\lambda = \sqrt{K s_* / D_*}$, $I_{n+1/2}$ (λ, r) is the modified Bessel function, $Y_n(\theta, \varphi)$ is the spherical function, r, θ and φ are the spherical coordinates, and A_n are constants which are found from a given oxygen concentration $C = C_R(\theta, \varphi)$, over the particle surface $r = r_o$.

$$C_R = \sum_n A_n I_{(n+1/2)}(\lambda, R_p) \frac{Y_n(\theta, \varphi)}{\sqrt{r_o}} \tag{3.68}$$

Here r_o is the particle radius.

Bearing in mind that for the first order reaction

$$G_{O_2} = \alpha_R = K C_R + D_* \left(\frac{\partial C_*}{\partial n} \right)_s \tag{3.69}$$

we determine the reaction coefficient of mass exchange as follows

$$\alpha = K + \lambda D_* \sum_n A_n I_{(n+1/2)}(\lambda_1, r_o) \frac{Y_n(\theta, \varphi)}{\sqrt{r_o}} \frac{\left[\frac{I'_{(n+1/2)}(\lambda, r_o)}{I_{(n+1/2)}(\lambda, r_o)} - \frac{1}{2\lambda r_o} \right]}{\sum_n A_n I_{(n+1/2)}(\lambda, r_o) \frac{Y_n(\theta, \varphi)}{\sqrt{r_o}}} \tag{3.70}$$

where $I' = (dI/dr)_s$.

For a spherically symmetric process C_R = const, and Eq. (3.70) takes the following form

$$\alpha = K + \frac{\lambda}{D_*} (\coth \lambda r_o - \frac{1}{\lambda r_o}) \tag{3.71}$$

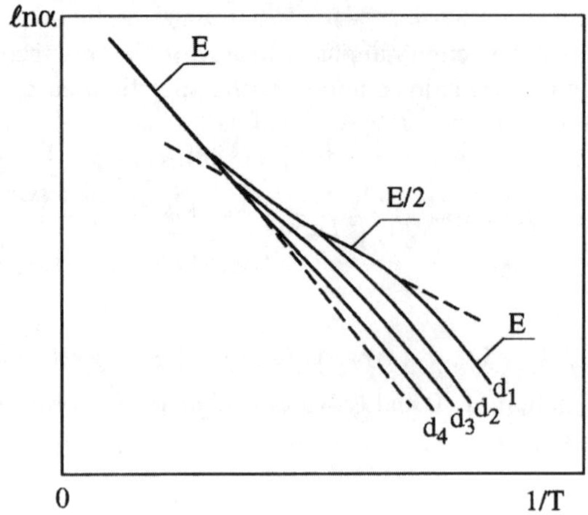

Fig. 3.27 Dependences of $\ln\alpha = \varphi(1/T)$ corresponding to several particles of different diameters $d_1 - d_4$. Reprinted from Khitrin (1957 a)

The dependences $\ln\alpha = \varphi(1/T)$ for several particles of different sizes are presented in Fig. 3.27. Depending on the value of the product λr_o it is possible to distinguish three sections on the curves $\alpha(1/T)$ which correspond to different temperature ranges. The first of these corresponds to small λr_o (low temperature and small particles).

At small enough values of λr_o $(\lambda r_o < 0.55)$

$$\coth\lambda r_o \approx \frac{1}{\lambda r_o} + \frac{\lambda r_o}{3} \tag{3.72}$$

Then the expression (3.71) takes the form

$$\alpha = K(1 + \frac{S_* r_o}{3}) \tag{3.73}$$

In this case the oxygen consumption equals

$$Q_{O_2} = 4\pi r_o^2 G_{O_2} = (S_p + s_* V) K C_R \tag{3.74}$$

where S_p is the external particle surface, and V is the particle volume.

At large s_* we obtain

$$Q_{O_2} \sim s_* V \tag{3.75}$$

Therefore, the oxygen consumption is proportional to the particle volume but not to its surface. This shows that the reaction occurs throughout the particle volume. The reaction coefficient of mass exchange α is directly proportional to the rate constant of chemical K which obeys the Arrhenius law. Accordingly the slope of the curves $\alpha(1/T)$ determines the values of the activation energy of the chemical reaction. The absolute value of the activation energy depends on the particle diameter and increases with d, as is seen in Fig. 3.27. In the intermediate temperature range the slope of the curve $\alpha(1/T)$ first decreases and then increases again up to a value corresponding to the low temperature region. Accordingly at some temperature $T = T_*$ the activation energy has a minimum $E_m = E/2$ (Zel'dovich 1939). At high enough temperature the curves $\alpha(1/T)$ merge into a single curve with the same slope as for the curves $\alpha(1/T)$ in the low temperature range.

At large values of λr_o $(\lambda r_o > 40)$ (high temperature and coarse particles) $\coth \lambda r_o \sim 1$. Then Eq. (3.71) takes the form

$$\alpha \approx K + \sqrt{s_* D_* K} \tag{3.76}$$

In the limiting case, corresponding to the high temperature regime, it yields

$$\alpha \simeq K \tag{3.77}$$

Thus, in this case oxygen is fully consumed at the particle surface.

Models of coke particle combustion. There are two limiting models of coke particle combustion: (i) the progressive conversion model, and (ii) the shrinking core model (Yang 1993). The former assumes that chemical reaction proceeds throughout the particle volume, whereas the latter assumes that the reaction occurs only at the particle surface (cf. Fig. (3.28) and (3.29), respectively). The rate of combustion corresponding to the volumetric reaction can be estimated as follows. In the frame of the quasi-steady state approximation the overall rate of combustion is determined by oxygen consumption through the particle surface.

$$W_{O_2} = 4\pi r_s^2 D_e \left(\frac{dC_{O_2}}{dr} \right)_{r=r_s} \tag{3.78}$$

where D_e is the effective diffusivity

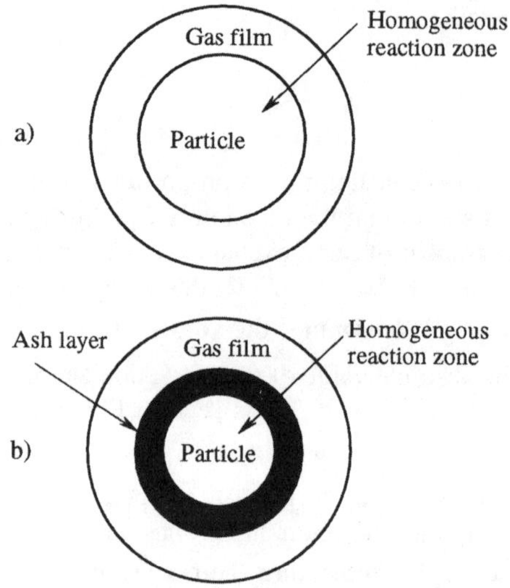

Fig. 3.28 Progressive conversion model. Reprinted from Yang (1993), with permission. a) Without an ash layer; b) with an ash layer

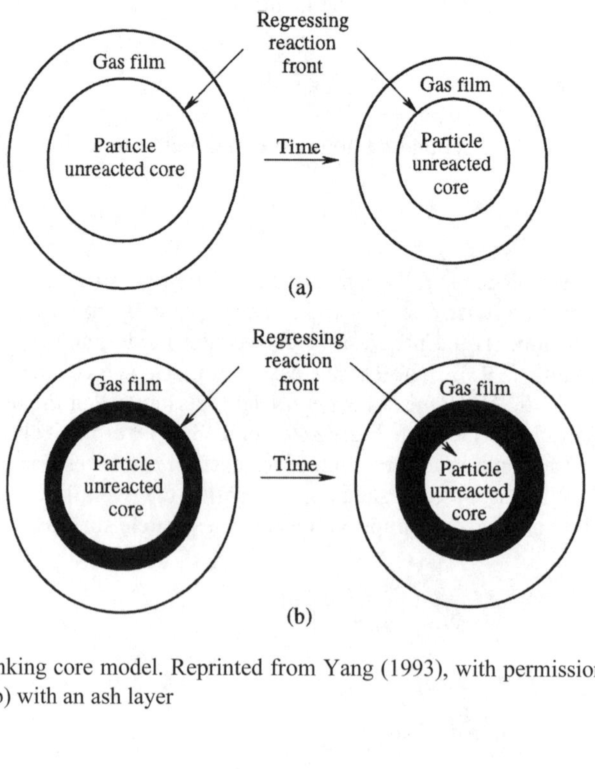

Fig. 3.29 Shrinking core model. Reprinted from Yang (1993), with permission. a) Without an ash layer; b) with an ash layer

The oxygen distribution within the particle is defined by Eq. (3.66) with the boundary conditions

$$r = 0 \qquad \frac{dC_{O_2}}{dr} = 0 \tag{3.79}$$

$$r = r_s \qquad D_e\left(\frac{dC_{O_2}}{dr}\right) = h_m(C_{O_2\infty} - C_{O_2}) \tag{3.80}$$

Note that the condition (3.80) implies that the diffusion resistance of the gas film surrounding the particle is negligible.

The solution of Eq. (3.66) has the following form

$$\frac{C_{O_2}}{C_{O_2\infty}} = \frac{r_s}{r}\frac{\sinh\left(\sqrt{\frac{Kr}{D_e}}\right)}{\sinh\left(\sqrt{\frac{Kr_s}{D_e}}\right)} \tag{3.81}$$

Using Eqs. (3.78) and (3.81) we arrive at

$$W_{O_2} = 4\pi r_s D_e C_{O_2\infty}(\phi \coth\phi - 1) \tag{3.82}$$

where $\phi = \sqrt{Kr_s/D_e}$.

A similar approach can be used to estimate the rate of combustion of a particle with an ash layer. In this case the problem reduces to solving the system of equations determining the oxygen diffusion within the ash layer and the reactive material.

The shrinking core models assume that conversion of the reactive material occurs at the reaction front that moves to the particle center. In this case the solution of the problem yields a law of particle shrinkage similar the d^2 law for evaporation. In the cases when the rate of the process is controlled by the gas film diffusion, or reagent or by the surface chemical reaction, the corresponding to d^2 law is expressed as

$$d_0^2 - d^2 = \frac{8D_e C_{O_2\infty} \nu_c}{\rho_c \nu_o} t \tag{3.83}$$

and

$$t_b = \frac{d_0^2 \rho_c \nu_o}{8 D_e C_{O_2\infty} \nu_c} \tag{3.84}$$

where d_0 and d are the initial and current particle radii, ρ_c is the particle density, ν_c and ν_o are the stoichiometric coefficients of the overall first order reaction $\nu_c[C] + \nu_o[O] \rightarrow \text{ash} + \text{products}$, and t_b is the time required for complete oxidation of the particle.

1.3.6 Droplet combustion

Model of droplet combustion. Burning of a single liquid fuel droplet in an oxidizing medium attracted the attention of scientists and engineers during the second part of the 20th century. Following the pioneering works by Varshavskii (1945), Spalding (1953), Godsave (1953), Wise et al. (1955), Hottel et al. (1955), numerous theoretical and experimental investigations were carried out. They considered a wide class of problems related to burning of droplets of one- and multicomponent fuels in a stagnant atmosphere and a gas flow. These results were summarized in a number of comprehensive surveys (Williams 1973, Law 1982, Annamalai and Ryan 1992, Yang 1993) and in several monographs (Khitrin 1957 a, Spalding 1955, Williams 1985, Kuo 1985, Glassman 1989, and Sirignano 1999).

Below we discuss briefly some of the most important results related to droplet combustion. First of all, we consider burning of a droplet of a one-component fuel in an oxidizing environment at rest and in a gas flow.

From the physical point of view droplet burning takes place by combustion of a reactive gaseous mixture formed due to mixing of the liquid fuel vapor with the oxidizer from the surrounding medium. The problem of droplet burning in a stagnant atmosphere can be considered as spherically symmetric when the buoyancy effect is negligible. The distributions of temperature and concentrations in the vicinity of a droplet in this case are depicted in Fig. 3.30. The vapor and oxidizer concentrations change monotonically with r from their values at the droplet surface $C_{v.s}$, and $C_{o,s.}$ The temperature profile has a maximum located at some distance $\Delta r = r_f - r_s$ from the droplet surface. At high activation energy an intensive chemical reaction proceeds within the region of high temperature which may be considered as the reaction zone. Outside of this zone the rate of chemical reaction decreases sharply. This region is shown shaded in Fig. 3.30.

In the cases where natural or forced convection is significant, the parameter distributions about a droplet are asymmetric. The reaction zone in these cases still represents a thin sheet, which envelops the burning droplet.

One-component droplet combustion. As was noted above, the process of droplet burning involves a number of successive stages such as liquid vaporization, mixing of gaseous reagents and combustion of vapor–oxidizer mixture. This process is accompanied by heat and mass transfer, as well as by the regression of

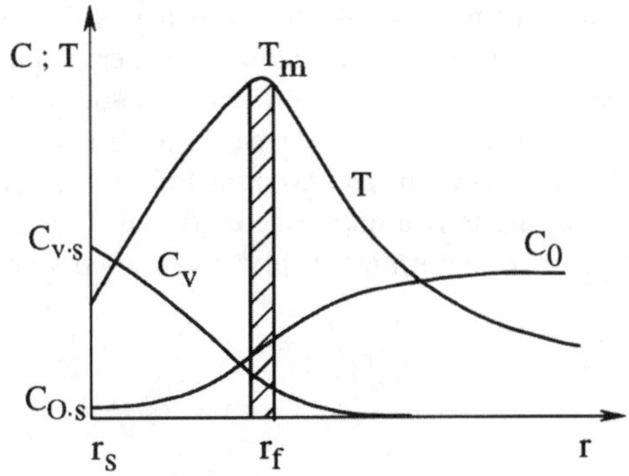

Fig. 3.30 Distribution of vapor and oxygen concentrations C_V and C_O, and temperature in the gaseous mixture in the vicinity of a droplet

the droplet surface and displacement of the reaction zone. Accordingly, a theoretical description of droplet combustion in a gas flow implies solving the non steady coupled liquid–gas phase equations governing the mass, momentum, energy, and species transfer. The non linear terms accounting for heat release and species consumption present in the equations make theoretical analysis of the problem extremely difficult. Therefore, as a rule, droplet burning is studied using models based on a number of simplifying assumptions. The most important of these are the following:

(i) The gas density is much smaller than the liquid density (at moderate pressure).
(ii) The chemical reaction rate is infinitely large.
(iii) The droplet temperature is uniform (the lumped-capacitance approximation).
(iv) The effects of buoyancy and radiative heat transfer are negligible.
(v) The Lewis number equals one.
(vi) The physical properties of the liquid and gaseous phases are constant and invariable.

A assumption (i) implies that the characteristic time in the gas phase is much shorter than the characteristic droplet surface regression time. This allows the application of the quasi-steady state approximation and reduces the problem to solving the system of the ordinary differential equations describing the processes in a reactive gaseous mixture.

A assumption (ii) allows for consideration of the reaction zone as an infinitely thin flame front which separates the flow field into two regions: the inner one (near the drop surface) filled with a mixture of vapor and combustion products, and the outer one filled with a mixture of oxidizer and combustion products. The vapor and oxidizer concentrations at the flame front equal zero in this case, whereas their flux ratio is stoichiometric.

The third and the fourth assumptions make it possible to use the lumped-capacitance heat transfer model, as well as to consider a spherically symmetric flame (for burning in a stagnant atmosphere). The flame temperature equals the adiabatic combustion temperature in this case. The droplet surface temperature changes only slightly during the droplet combustion process and can be taken as constant equal to the boiling temperature of liquid fuel. In this case the boundary conditions for the problem of a burning droplet of a one-component fuel are as follows:

$$r \to \infty: \quad C_o \to C_{o.\infty}, \quad C_f = 0, \quad T \to T_\infty \tag{3.85}$$

$$r = r_s: \rho v_s C_o - \rho D \frac{dC_o}{dr} = 0, \ \rho v_s C_f - \rho D \frac{\partial C_f}{dr} = \rho v_s, \ T = T_s, \ \lambda \frac{dT}{dr} = \rho v_s q_e \tag{3.86}$$

Here C_o and C_f are the oxidizer and fuel concentrations, respectively, v_s is the vapor velocity at the droplet surface, ρ is the gaseous phase density, q_e is the latent heat of vaporization, subscripts o, f, refer to oxidizer and fuel; s and ∞ refer to the droplet surface and the environment, respectively .

A significant simplification of the problem can be achieved by using assumption (v) implying that the Lewis number equals one. This allows the Schvab–Zel'dovich transformation (Schvab 1948, Zel'dovich 1948) to be used and to be removed the non-linear terms from the balance equations. Then the governing equations of the problem take the following form:

$$\nabla \cdot (\rho \mathbf{v}) = 0 \tag{3.87}$$

$$\nabla \cdot (\rho \mathbf{v} \beta_i - \rho D \nabla \beta_i) = 0 \tag{3.88}$$

where $\mathbf{v}$ is the velocity vector, $\beta = (C_i/\sigma_i) + (c_p T/q)$ is the so-called coupling function, σ_i are the stoichiometric oxidizer-to-fuel mass ratio ($\sigma_i = 1$ for fuel by definition), and q is the specific heat release due to the combustion reaction.

The integral of Eq. (3.87) is

$$\rho v r^2 = \rho v_s r_s^2 = M = \text{const} \tag{3.89}$$

Thus the total mass flux through any spherical surface enveloping the burning droplet $m = 4\pi\rho v r^2$ is constant. It equals the mass rate of droplet vaporization.

Integrating Eq. (3.88) taken into account Eq.(3.89), we obtain

$$-\frac{1}{r}=\frac{\rho D}{M}\ln\frac{b_i-a_i}{b_i} \tag{3.90}$$

where a_i and b_i are the integration constants.

Bearing in mind that at

$$r\to\infty:\ \beta_i\to\beta_{i\infty},\ (\beta_{o\infty}=\frac{C_{o\infty}}{\sigma_o}+\frac{c_pT_\infty}{q},\ \beta_f=\frac{c_pT_\infty}{q}) \tag{3.91}$$

and at

$$r=r_s:\ \beta_i=\beta_{is}=\frac{C_{is}}{\sigma_i}+\frac{c_pT_c}{q},\ \left(\frac{d\beta_o}{dr}\right)_s=\frac{c_p}{\lambda}\frac{M}{r_s^2}\left(\frac{C_{os}}{\sigma_o}+\frac{q_e}{q}\right)$$
$$\left(\frac{d\beta_f}{dr}\right)_s=\frac{c_p}{\lambda}\frac{M}{r_s^2}\left\{\frac{q_e}{q}-(1-C_{fs})\right\} \tag{3.92}$$

we find the constants a_i and b_i as

$$a_o=\frac{c_pT_s}{q}-\frac{q_e}{q},\quad a_f=\frac{c_pT_s}{q}-\frac{q_e}{q}$$
$$b_o=\frac{C_{o\infty}}{\sigma_o}+\frac{c_p(T_\infty-T_s)}{q}+\frac{q_e}{q},\quad b_f=\frac{c_p(T_\infty-T_s)}{q}+\frac{q_e}{q} \tag{3.93}$$

then Eq. (3.90) takes the form

$$-\frac{1}{r}=\frac{\rho D}{M}\cdot\ln\left\{\frac{(C_o/\sigma_o)q+c_p(T-T_s)+q_e}{(C_{o\infty}/\sigma_o)q+c_p(T_\infty-T_s)+q_e}\right\} \tag{3.94}$$

$$-\frac{1}{r}=\frac{\rho D}{M}\ln\left\{\frac{qC_f+c_p(T-T_s)+q_e-q}{c_p(T_\infty-T_s)+q_e-q}\right\} \tag{3.95}$$

The parameter distributions within the inner and outer regions of the diffusion flame, as well as the temperature values at the flame front and droplet surface correspond to:

$$r<r_f:\ C_o=0,\ r>r_f:\ C_f=0,\ r=r_f:\ C_o=C_f=0,\ r=r_s:\ T=T_s,\ r=r_f,:\ T=T_f \tag{3.96}$$

Using Eqs. (3.94) – (3.96) we find the temperature distributions within the inner and outer regions $r_s<r<r_f$ and $r_s<r<\infty$, as

$$-\frac{1}{r}=\frac{\rho D}{M}\ln\left\{\frac{c_p(T-T_s)+q_e}{(C_{o\infty}/\sigma_o)q+c_p(T_\infty-T_s)+q_e}\right\} \quad (3.97)$$

$$-\frac{1}{r}=\frac{\rho D}{M}\ln\left\{\frac{c_p(T-T_s)+q_e-q}{c_p(T_\infty-T_s)+q_e-q}\right\} \quad (3.98)$$

In order to determine the position of the flame front we use Eqs. (3.97) and (3.98). Assuming in these equations $r = r_f$ and $T = T_f$, we obtain two equations containing r_f and T_f. Eliminating the flame temperature T_f from these equations we find the coordinate of the flame front

$$r_f=\frac{M}{\rho D\ln(1+C_{0.\infty}/\sigma_o)} \quad (3.99)$$

The expression for the burning rate is found from Eq. (3.97) with $r=r_s$ and $T=T_s$, which yields

$$m_s=4\pi M=4\pi\rho D r_s\ln(1+B) \quad (3.100)$$

where $B=(1/q_e)\left[C_p(T_\infty-T_s)+(C_{o\infty}q/\sigma_o)\right]$ is the Spalding transfer number.

Using the above relations, the flame temperature and standoff ratio r_f/r_s are given as

$$T_f=\frac{(q-q_e)+c_pT_\infty(C_{o\infty}/\sigma_o+T_s/T_\infty)}{c_p(1+C_{o\infty}/\sigma_o)} \quad (3.101)$$

$$\frac{r_f}{r_s}=\frac{\ln(1+B)}{\ln(1+C_{o\infty}/\sigma_o)} \quad (3.102)$$

The fuel vaporization during liquid droplet burning is the sole cause of change in its mass. Accordingly, the mass balance equation for the burning droplet takes the following form

$$-\frac{d}{dt}\left(\frac{4}{3}\pi r_s^3\rho_2\right)=m_s=4\pi M \quad (3.103)$$

where ρ_2 is the liquid fuel density.

The integral of Eq. (3.103) taken into account Eq. (3.100) yields the dependence of the droplet diameter on time:

$$d_s^2 = d_{s0}^2 - Kt \tag{3.104}$$

where d_{s0} is the initial droplet diameter, $K=[(8\rho D)/\rho_2]\ln(1+B)$ is the burning rate constant.

Experimental verification of the d^2 law for one-component fuel. The dependence of d^2 on time for burning droplets of different fuels was studied experimentally by Isoda and Kumagai (1959), Faeth and Lazar (1971), Kumagai et al. (1971), Okajima and Kumagai (1974), Gokalp et al. (1988), and Hara and Kumagai (1990). The measurements that were carried out under the conditions of normal ($g = 9.81\ [m/s^2]$) and reduced gravity showed that the d^2 law (3.104) is a good approximation to the experimental data. A linear regression of the square of the droplet diameter with time was observed in all the experiments with freely falling and suspended droplets after a short initial time interval. The duration of this initial interval does not exceed 15–20% of the droplet lifetime. The deviation of the linear dependence $d^2(t)$ from the experimental data at small t is due to the process of droplet heating from its initial temperature up to a temperature corresponding to steady-state combustion (Law 1976, Law and Sirignano 1977). It is noteworthy that at large times the linear dependence $d^2(t)$ is also valid for suspended droplets of elliptical shape, if the volume equivalent diameter is used as the characteristic droplet size.

The simplest quasi-steady theory allows one to find not only the dependence $d^2(t)$ but also the other characteristics of a burning droplet, in particular, the flame front standoff ratio. This theory predicts that the flame front standoff is determined by the Spalding number, the oxidizer concentration in the environment, and by the stoichiometric oxidizer-to-fuel ratio. Moreover, the standoff ratio is independent of time. The latter contradicts the known experimental data and demonstrates on the insufficiency of the quasi-steady approach. Law et al. (1980) suggest that the cause of the variation of the standoff the ratio r_f/r_s is in the accumulation of fuel vapor in the inner region as a consequence of the difference between the vaporization and consumption rates. This implies actually an unsteady character of droplet burning.

A consistent theory of droplet combustion is based on the solution of the unsteady problem, which was developed by Waldman (1979), Crespo and Linan (1975), and Matalon and Law (1973). To solve this problem in the frame of the infinitely thin flame approximation, the method of matched asymptotic expansions was used. The comparison of the predictions of unsteady theory with the experimental data is presented in Fig. 3.31.

The *d^2 law for multicomponent droplet combustion.* Vaporization and combustion of multicomponent droplet are determined by a number of factors accounting for the difference in the physico-chemical properties of its components (i.e. volatility, boiling point, stoichiometric oxidizer-to-fuel ratio), for the difference in

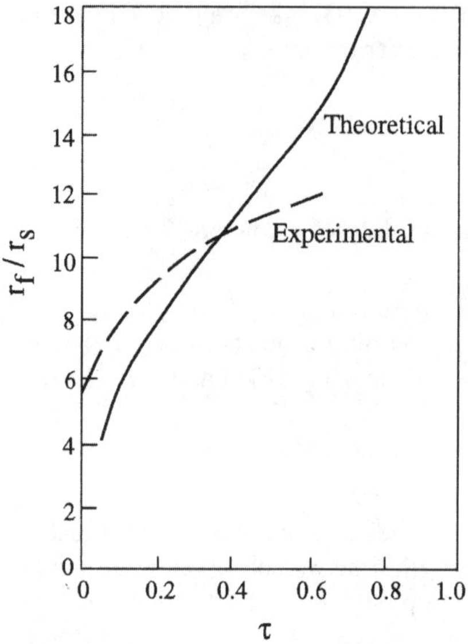

Fig. 3.31 Comparison of the predicted standoff ratio with the experimental data of Waldman (1979)

the rates of mass and heat diffusion in liquid fuel, as well as for the internal structure of liquid mixture (which may be a solution or an emulsion).

A number of works have attempted to express the combustion rate of multi-component droplets in terms of the d^2 law. Wood et al. (1960) suggested a pure phenomenological relation for the burning rate constant in the d^2 law for multi-component droplet

$$d^2 = d_0^2 - K_* t \tag{3.105}$$

Here K_* is the modified burning rate constant which depends on the components' concentration, their densities and latent heats, the ambient oxidizer's concentration, the droplet and ambient temperature, the specific heat capacity and on the heat of combustion of the mixture, as well as on its stoichiometric oxidizer-to-fuel mass ratio.

Wood's relation does not account completely for the individual properties of the fuel components, in particular, for their particular stoichiometric oxidizer-to-fuel mass ratios and heats of combustion (see Eq. (2.21), Chapter 1.2). An elaborate expression for the burning rate constant in the d^2 law for multi-component droplets was obtained by Law and Law (1982) on the basis of the model assuming the existence of a steady-state concentration boundary layer at the droplet surface and invariable liquid concentration distribution within the inner core of the droplet

$$K_* = \frac{8\lambda}{c_p(\sum C_{i\ell 0}/\chi_i)/(\sum C_{i\ell 0}/\chi_i\rho_i)} \times \\ \times \ln\left\{\frac{1 + c_p(T_\infty - T_s) + (C_{o\infty})/(C_{i\ell 0}\sigma_i)\left(\sum C_{i\ell 0}q_{\ell i}\right)}{\sum C_{i\ell 0}q_{ei}}\right\} \tag{3.106}$$

Here $q_{\ell i}$, q_{ei} and σ_i are the heat of combustion, the latent heat of vaporization and stoichiometric oxidizer-to-fuel mass ratio of the i-th component; $C_{i\ell 0}$ and $C_{o\infty}$ are the initial concentration of the i-th component of liquid fuel and the oxidizer concentration in the environment, respectively, $\chi_i = \frac{P_n}{P_\infty}\exp\left\{\frac{q_i}{R_i}\left(\frac{1}{T_{ib}} - \frac{1}{T_s}\right)\right\}$, P_n and P_∞ are the normal atmosphere pressure and the ambient pressure, T_{ib} is the normal boiling point; $T_s \simeq \sum X_{i\ell 0}T_{ib}$, $X_{i\ell 0}$ is the initial liquid-phase mole fraction, R_i is the gas constant of the i-th component, ρ_i is the density of the i-th liquid component, and λ is the gas-phase conductivity.

The comparison of the d^2 law (3.106) with experimental data on a burning droplet of the binary fuel mixture (50% undecane and 50% octanal) is shown in Fig. 3.32. It is seen that the data for d^2 change linearly with time at $t > t_{in} \approx 0.07$ s where t_{in} is the duration of the droplet heating which makes up about 10–15% of the droplet lifetime. The straight line corresponding to Eq. (3.106) agrees pretty well with the experimental data, shown by symbols in Fig. 3.32.

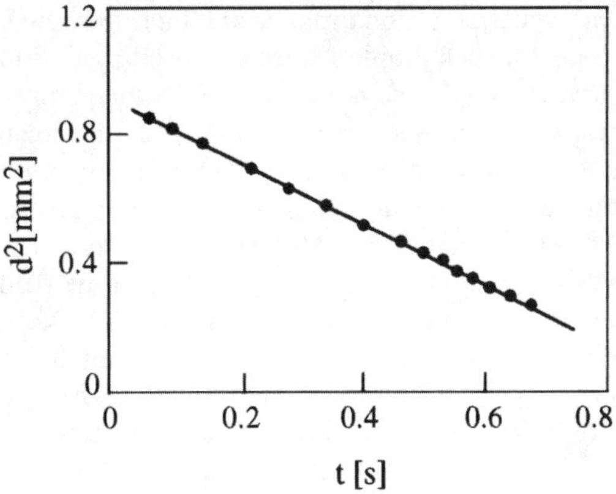

Fig. 3.32 The d^2 law for a binary fuel mixture droplet. Reprinted from Law and Law (1982), with permission

Microexplosion phenomenon. The disruptive combustion associated with burning multi-component droplets was observed in the experiments by Lasheras et al. (1979, 1980, 1981 a, 1981 b, 1984), Wang et al. (1984), Yap et al. (1984), Law and Law (1982), Nioka and Sato (1986), and Yang et al. (1990). This phenomenon result from homogeneous nucleation of the mixture vapor in the superheated droplet interior and from a subsequent growth and bursting of vapor bubbles leading to droplet destruction and to secondary fuel atomization.

Disruptive combustion is characteristic of droplets of solutions, as well as of emulsion droplets. However, the mechanism of this process is different according to whether the droplet is a solution or an emulsion. In the first case disruptive combustion consists in the following. During the combustion of a multicomponent droplet, vaporization of more volatile components of the mixture is accompanied by an increase in the concentration of the less volatile components in a thin layer near the droplet surface. This leads to an increase in the droplet temperature. As a result, the more volatile components that remain trapped in the droplet core due to the diffusion resistance become superheated. Then their homogeneous nucleation in the droplet core triggers the droplet bursting. Such a process can occur when the local temperature at some point located in the droplet core exceeds the superheat limit of the mixture

$$T_j > T_{\ell im} \tag{3.107}$$

Here T_j is the local temperature at point j; and $T_{\ell im}$ is the superheat limit of the mixture depending on all the species, mole concentrations and pressures.

The disruptive combustion of emulsion droplets is associated with dissolution of the dispersed phase in the continuous phase and the destruction of the internal emulsion structure during the transient heating of the drop. The model of combustion of oil–water emulsion droplets and thermodynamic criteria for the microexplosive events for such droplets were proposed by Law (1977).

Effect of natural and forced convection. The interaction of the external flow with the flow caused by droplet vaporization leads to violation of the spherical symmetry in resulting flow, changes of the flame front curvature and, accordingly, to changes of the burning rate, as well as of the drag coefficient, etc. A consistent theory of droplet combustion affected by natural and forced convection should include solutions of the hydrodynamic problem on the flow field around the droplet, as well as of the problem on heat and mass transfer due to the chemical reaction proceeding in the reactive vapor–oxidizer mixture. Fendell et al. (1966) elaborated the so-called thin-flame theory for burning droplets moving with small velocities corresponding to Stokes or Oseen flows. The calculation of the burning droplet characteristics at large Reynolds number presents a very complicated problem. Therefore, as a rule, the effect of natural or forced convection on the droplet combustion is accounted for using some empirical relations for the burning rate constant (Law and Willliams 1972).

$$K_i = K_o(1 + f_i) \tag{3.108}$$

Here K_o is the burning rate constant under the quiescent conditions, f is the correction factor, and subscript $i = n$ or f refers to natural or forced convection.

The combined effect of the natural and forced convection is accounted by

$$K_c = K_o(1 + f_n)(1 + f_f) \tag{3.109}$$

The correction factors f_n and f_t are expressed as

$$f_f = 0.276\,\mathrm{Re}_\infty^{\frac{1}{2}}\,\mathrm{Sc}_\infty^{\frac{1}{3}}$$

$$f_n = 0.533\,\mathrm{Gr}^{0.52}$$

where Re_∞ and Sc_∞ are the Reynolds and the Schmidt numbers determined by the ambient conditions; Gr is the Grashof number evaluated at the average (the flame and the ambient) temperature.

Effects of variable properties. The effect of variable properties in droplet combustion was considered by Kassoy and Williams (1968), Chung and Law (1984), Puri and Libby (1991), and Madoogly and Karagozian (1994). Their calculations show that variation of the gas properties with temperature and the gas composition significantly affects the burning rate, the flame standoff and the combustion temperature. Disregard of the temperature dependences of the thermal conductivity and diffusivity leads to errors which can sometime exceed one order of magnitude.

The influence of variation in transport properties manifests itself not only in changes in the burning rate, flame temperature and standoff distance, but also in the local values of the gas temperature and species concentrations. In particular, changes in the Lewis number result in significant changes in the parameter distributions in the space that is confined between the flame sheet and the droplet surface. Neglect of variation of the specific heat, the Prandtl and the Schmidt numbers in the vicinity of the flame front also leads to a noticeable deviation of the calculation result from the actual values of the parameters. It is noteworthy that by choosing some average values of the gas properties it is not possible to predict simultaneously all the characteristics of the burning droplet.

Effect of radiation. The smallest solid particles present in the reaction zone of burning droplets (e.g. soot particles, etc.) are the sources of flame radiation. Heat losses due to such radiation affect the thermal regime of the diffusion flame enveloping the droplet, in particular, its temperature, burning rate, etc.

Varshavskii (1945) estimated the effect of radiation on some of the characteristics of the burning droplets. To calculate the radius of the diffusion flame, its temperature and the rate of burning he used the thermal balance equation, which takes in the given case the following form

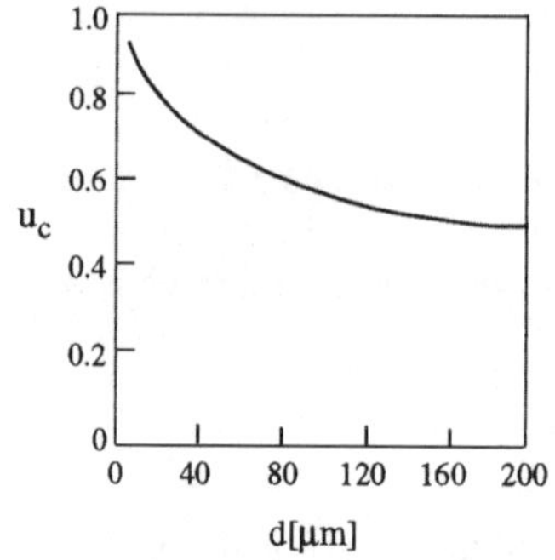

Fig. 3.33 The effect of the particle diameter on the combustion rate in the case of radiation cooling of the reaction zone. Reprinted from Varshavskii (1945). The rate of combustion, rendered dimensionless by the combustion rate unaffected by radiation, is denoted u_c

$$\int_{\xi_\infty}^{\xi_f} \varphi(\xi)d\xi + R_f \xi_f^4 = A\overline{f}_1(\xi_f,\ \xi_\infty) \tag{3.110}$$

where $R_f = r_f(\tilde{\sigma}T_0^3)/\lambda_0$, $A = C_\infty D_0 q/\lambda_0 T_0$, $\xi_i = T_i/T_0$, $\overline{f}_1(\xi_f, \xi_\infty) = \int_{\xi_\infty}^{\xi_f} \varphi(\xi)d\xi \Big/ \int_{\xi_\infty}^{\xi_f} \frac{\varphi(\xi)}{\psi(\varphi)} \xi d\xi$, $\varphi(\xi) = \lambda/\lambda_0$, $\psi(\xi) = D/D_0$; $\tilde{\sigma}$ is the Stefan–Bolzmann constant, r_f is the standoff radius of the diffusion flame, C is the oxygen concentration, λ and D are the thermal conductivity and diffusivity, and subscripts f, ∞, and 0 correspond to flame, the ambient and the initial state.

The system of the thermal and mass balance equations allows for calculation of the characteristics of the burning droplet. The results show that radiation from reaction zone leads to a significant decrease of the burning rate. This effect becomes mere severe for burning coarse droplets (Fig. 3.33).

References

Annamalai K, Durbetaki P (1977) A theory on transition of ignition phase of coal particles. Combust. Flame 29: 193–208

Annamalai K, Ryan W (1992) Interactive processes in gasification and combustion. Part I. Liquid drop arrays and clouds. Prog. Energy Combust. 18: 221–295

Annamalai K, Ryan W (1993) Interactive processes in gasification and combustion - II. isolated carbon, coal and porous char particles. Prog. Energy Combust. 19: 383–446

Babii BI, Kuvaev Ja F (1986) Combustion of coal dust and coal dust flame calculation. (in Russian) Energoatomizdat, Moscow

Barishnikov NN, Gerer VE, Denisova ND (1979) Metallurgy of zirconium and hafnium. (in Russian) Metallurgy, Moscow

Bartlett RW, Ong JNJr, Fassele WNJr, Papp CA (1963) Estimating aluminum particle combustion kinetics. Combust. Flame 7: 227–234

Bellan J, Summerfield M (1976) Quasi-steady gas phase assumption for a burning droplet. AIAA J. 14: 973–975

Bellan J, Summerfield M (1978) A preliminary theoretical study of droplet extinction by depressurization. Combust. Flame 32: 257–270

Botros P, Law CK, Sirignano WA (1980) Droplet combustion in a aeactive environment. Combust. Sci. Technol. 21: 123–130

Buchholz EK, Tapper ML (1978) Time to extinction of liquid hydrocarbon fuel droplets burning in a transient Diesel-like environment. Combust. Flame 31: 161–171

Burke SP, Schumann TE (1928) Diffusion flames. Ind. Eng. Chem. 20: 998–1004

Cassel HM, Liebman I (1959) The cooperative mechanism in the ignition of dust dispersions. Combust. Flame 3: 467–475

Chiu HH (2000) Advances and challenges in droplet and spray combustion. I. Toward a unified theory of droplet aerothermochemistry. Prog. Energy Combust. Sci. 26: 381–416

Chiu HH, Hu LH (1998) Dynamics of ignition transience and gasification partition of a droplet. In:Burges AR, Dryer FL (eds). The Twenty-seventh Symposium (International) on Combustion. The Combustion Institute, Pittsburgh, Pa., pp. 1889–1896

Chung SH, Law CK (1984) Importance of dissociation equilibrium and variable transport properties on estimation of flame temperature and droplet burning rate. Combust. Flame 55: 225–235

Crespo A, Linan A (1975) Unsteady effects in droplet evaporation and combustion. Combust. Sci. Technol. 11: 9–18

Dankverts PV (1973) Gas–liquid reaction. (in Russian) Chemistry, Moscow

Dreizin EL (1996) Experimental study of stages in aluminium particle combustion in air. Combust. Flame 105: 541–556

Dreizin EL, Suslov AV, Trunov MA (1993) General trends in metal particle heterogeneous combustion. Combust. Sci. Technol. 90: 79–99

Dreizin EL, Berman CH, Vicenzi EP (2000) Condensed-phase modifications in magnesium particle combustion in air. Combustion Flame 122: 30–42

Du X, Annamalai K (1994) The transition ignition of isolated coal particle. Combust. Flame 97: 339–354

Essenhigh RH, Misra MK, Shaw DW (1989) Ignition of cool particles: A review. Combust. Flame 77: 3–30

Faeth GM, Lazar RS (1971) Fuel droplet burning rates in a combustion gas environment. AIAA J. 9: 2165–2171

Fendell FE (1965). Ignition and extinction in combustion of initially unmixed reactants. J. Fluid Mech. 21: 281–303

Fendell FE, Sprankle ML, Dodson DS (1966) Thin-flame theory for a fuel droplet in slow viscous flow. J. Fluid Mech. 26: 267–280

Frank-Kamenetskii DA (1969) Diffusion and heat transfer in chemical kinetics. 2nd edn. Plenum Press, New York

Fridman R, Macek A (1962) Ignition and combustion of aluminium particles in hot ambient gases. Combust. Flame 6: 9–19

Fridman R, Macek A (1963) Combustion studies of single aluminium particles. The ninth symposium (international) on combustion. The Combustion Institute. Academic Press, New York, pp. 703–712

Fridman NB, Kitain MM, Shteinberg AS, Merzhanov AG (1981) On the mechanism of bubble ignition. Sov. Phys. Dokl. 258: 961–965

Fu WB, Zeng TF (1992) A general method for determining chemical kinetic parameters during ignition of coal char particles. Combust. Flame 88: 413–424

Fu WB, Zhang EZ (1992) A universal correlation between the heterogeneous ignition temperatures of coal char particles and coals. Combust. Flame 90: 103–113

Garmata BA, Gulnitskii BS (1968) Metallurgy of titanium. (in Russian) Metallurgy, Moscow

Glassman I (1989) Combustion. 2nd edn. Academic Press, New York

Glassman I, Williams FA, Antaki PJ (1984) A physical and chemical interpretation of boron particle combustion. The Twentieth Symposium (International) on Combustion. The Combustion Institute, Pittsburgh Pa., pp. 2057–2064

Godsave GA (1953) Studies of the combustion of drops on a fuel spray - the burning of single drops of fuel. The Fourth Symposium (International) on Combustion. The Williams and Wilkings, Baltimore, pp. 818–830

Gokalp I, Chauveau C, Richard JR, Kramer M, Leuckel W (1988) Observation of the low temperature vaporization and envelope or wake flame burning of n-heptane droplets at reduced gravity during parabolic flights. The Twenty-second Symposium (International) on Combustion. The Combustion Institute, Pittsburgh Pa., pp. 2027–2035

Gol'dshtein V, Goldfarb BI, Shreiber I, Zinoviev A (1998) Oscillations in a combustible gas bubble. Combust. Theory Modeling 2: 1–17

Golovina EC (1983) High temperature combustion and gasification of carbon. Energoatomizdat, Moscow (in Russian)

Gurevich MA, Lapkina KI, Ozerov ES (1970) Ignition limits of aluminum particles. Combust. Explos. Shock Waves 6: 154–157

Gurevich MA, Lydkin VM, Stepanov AM (1970a). Ignition and combustion of a gas suspension of magnesium particles. Combust. Explos. Shock Waves 6: 298–303

Gurevich MA, Ozerova GE, Stepanov AM (1970b) Heterogeneous ignition of an aluminum particle in oxygen and water vapor. Combust. Explos. Shock Waves 6: 291–297

Gurevich MA, Ozerova GE, Stepanov AM (1971) Critical avtoignion conditions for a polydisperse gas suspension of solid-fuel particles. Combust. Explos. Shock Waves 7: 7–14

Gururajan VS, Wall TF, Gupta RP, Truelove JS (1990) Mechanisms for the ignition of pulverized coal particles. Combust. Flame 81: 119–132

Hara H, Kumagai S (1990). Experimental investigation of free droplet combustion under microgravity. The Twenty-third Symposium (International) on Combustion. The Combustion Institute, Pittsburgh Pa., pp. 1605–1610

Hottel HC, Williams GS, Simpson HC (1955). Combustion of droplets of heavy liquid fuels. The Fifth Symposium (International) on Combustion. The Combustion Institute, Reinhold, New York, pp. 101–129

Hyseyin K (1978) Vibrations and stability of multiple parameter systems. Nordhoff

Isoda H, Kumagai S (1959) New aspects of droplet combustion. The seventh symposium (international) on combustion. The Combustion Institute. Academic Press, New York. pp. 523–531

Kantorovich BV (1958). Foundation of the theory of solid fuel combustion and gasification. (in Russian) AN SSSR, Moscow

Kassoy DR, Williams FA (1968) Effects of chemical kinetics on near equlibrium combustion in non premixed systems. Phys. Fluids 11: 1343–1351

Kassoy DR, Williams FA (1968) Variable property effects on liquid droplet combustion. AIAA J. 6: 1961–1965

Kassoy DR, Liu MK, Williams FA (1969) Comments on effects of chemical kinetics on near equilibrium combustion in non premixed systems. Phys. Fluids 12: 265–267

Kauffman CW, Nicholls JA (1971). Shock-wave ignition of liquid fuel drops. AIAA J. 9: 880–885

Khaikin BI, Bloshenko VN, Merzhanov AG (1970) On the ignition of metal particles. Combust. Explos. Shock Waves 6: 412–422

Khitrin LN (1957a) Fundamental principles of carbon combustion and factors intensifying the burning of solid fuels. The Sixth Symposium (International) on Combustion. The Combustion Institute. Reinhold New York, pp. 565–573

Khitrin LN (1957b) Physics of combustion and explosion. Moscow University, Moscow (in Russian)

Klyachko LA, Kuntsev GM (1978) A Theory of ignition for solid-fuel particles in a gaseous oxidizer. Combust. Explos. Shock Waves 14: 16–22

Kumagai S, Sakai T, Okajima S (1971) Combustion of free fuel droplets in a freely falling chamber. The Thirteenth Symposium (International) on Combustion. The Combustion Institute, Pittsburgh, Pa., pp. 779–785

Kuo KR (1986) Principles of combustion. Wiley, New York

Lasheras JC, Fernandez-Pello AC, Dryer FL (1979) Initial observations on the free droplet combustion characteristics of water- in-fuel emulsions. Combust. Sci. Technol. 21: 1–14

Lasheras JC, Fernandez-Pello AC, Dryer FL (1980) Experimental observations on the disruptive combustion of free droplets of multicomponent fuels. Combust. Sci. Technol. 22: 195–209

Lasheras JC, Fernandez-Pello AC, Dryer FL (1981a) On the disruptive burning of free droplets of ALCOHOL/n - Paraffin solutions and emulsions. The Eighteenth Symposium (International) on Combustion. The Combustion Institute, Pittsburgh, Pa., pp. 293–305

Lasheras JC, Kennedy IM, Dryer FL (1981b). Burning of distillate fuel droplets containing alcohol or water. Effect of additive concentration. Combust. Sci. Technol. 26: 161–169

Lasheras JC, Yap LT, Dryer FL (1984). Effect of the ambient pressure on the explosive burning of emulsified and multicomponent fuel droplets. The Twentieth Symposium (International) on Combustion. The Combustion Institute, Pittsburgh, Pa., pp. 1761–1772

Law CK (1975) Asymptotic theory for ignition and extinction in droplet burning, Combust. Flame 24: 89–98

Law C K (1976) Unsteady droplet combustion with droplet heating. Combust. Flame 26: 17–22

Law CK (1977) A model for the combustion of oil/water emulsion droplets. Combust. Sci. Technol. 17: 29–38

Law CK (1978) Theory of thermal ignition in fuel droplet burning. Combust. Flame. 31: 285–296

Law CK (1982) Recent advances in droplet vaporization and combustion. Prog. Energy Combust. Sci. 8: 171–201

Law CK, Law HK (1982) A d^2-Law for multicomponent droplet vaporization and combustion. AIAA J. 20: 522–527

Law CK, Sirignano WA (1977) Unsteady droplet combustion with droplet heating II: conduction limit. Combust. Flame 28: 175–186

Law CK, Williams FA (1972) Kinetics and convection in the combustion of Alkane droplets. Combust. Flame. 19: 393–405

Law CK, Chung SH, Srinivasan N (1980) Gas-phase quasi-steadiness and fuel vapor accumulation effects in droplet burning. Combust. Flame 38: 173–198

Leal LG (1992) Laminar flow and convective transport process. Scaling principles and asymptotic analysis. Butterworth-Heinemann, Mass. Boston

LeMott SR, Peskin RL, Levine DG (1971) Effect of fuel molecular weight on particle ignition. Combust. Flame 16: 17–27

Linan A (1974). The asymptotic structure of counterflow diffusion flames for large activation energies. Acta Astronaut 1: 1007–1039

Lisitsyn VI, Rumanov EN, Khaikin BI (1971) Induction period in the ignition of a particle system. Combust. Explos. Shock Waves 7: 1–6

Macek A (1967) Fundamentals of combustion of single aluminum and beryllium particles. The Eleventh Symposium (International) on Combustion. The Combustion Institute, Pittsburgh, Pa., pp. 203–217

Macek A (1973) Combustion of boron particles: experiment and theory. The Fourteenth Symposium (International) on Combustion. The Combustion Institute. Pittsburgh, Pa., pp. 1401–1411

Macek A, Semple JM (1969) Experimental burning rates and combustion mechanisms of single beryllium particles. The Twelfth Symposium (International) on Combustion. The Combustion Institute, Pittsburgh, Pa., pp. 71–81

Macek A, Semple JM (1971) Combustion of boron particles at elevated pressures. The Thirteenth Symposium (International) on Combustion. The Combustion Institute, Pittsburgh, Pa., pp. 859–868

Madooglu K, Karagozian AR (1994) Burning of a spherical fuel droplet in a uniform flowfield with exact property variation. Combust. Flame 94: 321–329

Matalon M, Law CK (1983) Gas-phase transition diffusion in droplet vaporization and combustion. Combust. Flame 50: 219–229

Merzhanov AG (1973) The problem of technological combustion. In: Merzhanov AG (ed) Combustion processes in chemical technology and metallurgy. AN SSSR, Chernogolovka, pp. 5–28 (in Russian)

Merzhanov AG, Grigorjev YuM, Gal'chenko YuA (1977) Aluminium ignition. Combustion Flame 29: 1–14

Molodetsky IE, Vicenzi EP, Dreizin EL, Law CK (1998) Phase of titaniuim combustion in air. Combust. Flame 112: 522–532

Nakoryakov VE, Pokusaev BG, Shreiber IR (1993) Wave propogation in gas-liquid media. 2nd. edn. CRC Press, Boca Raton Fla

Nigmatulin RI (1991). Dynamics of multiphase media v. 1 and 2. Hemisphere, London

Nioka T, Sato J (1986) Combustion and microexplosion behavior of miscible fuel droplets under high pressure. The Twenty-first Symposium (International) on Combustion. The Combustion Institute, Pittsburgh, Pa., pp. 625–631

Nishiwaki N (1955) Kinetics of liquid combustion processes: evaporation and ignition lag of fuel droplets. The Fifth Symposium (International) on Combustion. The combustion Institute. Reinhold, New York, pp. 148–158

Okajima S, Kumagai S (1974) Further investigations of combustion of free droplets in a freely falling chamber including moving droplets. The Fifteenth Symposium (International) on Combustion. The Combustion Institute, Pittsburgh, Pa., pp. 401–407

Peskin RL, Wise H (1966) A theory for ignition and deflagration of fuel drops. AIAA J. 4: 1646–1650

Peskin RL, Polymeropoulos CE, Yeh PS (1967) Results from a theoretical study of fuel drop ignition and extinction. AIAA J. 5: 2173–2178

Pindera MZ, Brzustowski TA (1984) A model for the ignition and extinction of liquid fuel droplets. Combust. Flame 55: 79–91

Polymeropoulos CE, Peskin RL (1969) Ignition and extinction of liquid fuel drops - numerical computations. Combust. Flame 13: 166–172

Predvoditelev AS, Khitrin LN, Tsukhanova OA, Kolodtsev KI, Grodzovsky MK (1949) Carbon combustion. (in Russian) AN SSSR, Moscow

Puri JK, Libby PA (1991) The influence of transport properties on droplet burning. Combust. Sci. Technol. 76: 67–80

Rosser WAJr, Rajapakse Y (1969) Thermal stability of a reactive spherical shell. Combust. Flame 13: 311–317

Saitoh T, Ishiguro S, Niioka T (1982) An experimental study of droplet ignition characteristics near the ignitable limit. Combus. Flame. 48: 27–32

Schvab BA (1948) Connection between temperature and velocity fields in gaseous flame. In: Investigation of processes of fossil fuel combustion. Gosenergoizdat, Moscow Leningrad, pp. 231–248 (in Russian).

Semenov NN (1935) Chemical kinetics and chain reactions. Oxford University Press, London

Shafirovich EYa, Goldshleger VI (1992) The superheat phenomenon in the combustion of magnesium particles. Combust. Flame 88: 425–42

Shaygan N, Prakash S (1995) Droplet ignition and combustion including liquid-phase heating. Combust. Flame 102: 1–10

Sirignano WA (1999) Fluid dynamics and transport of droplets and sprays. Cambridge. University Press, Cambridge

Solomon PR, Chien PL, Carangelo RM Serio MA, Markham JR, (1990) New ignition phenomenon in coal combustion. Combust. Flame 79: 214–215

Spalding DB (1953) The Combustion of liquid fuels. The Fourth Symposium (International) on Combustion. Williams and Wilkings, Baltimore, Md., pp. 847–864

Spalding DB (1955) Some fundamentals of combustion. Butterworth, London

Spalding DB (1951) A theory of the extinction of diffusion flames. Fuel 33: 255–273

Spalding DB, Jain VK (1962) A theoretical study of the effects of chemical kinetics on one-dimensional diffusion. Combust. Flame 6: 265–273

Takei M, Tsukamoto T, Niioka T (1993). Ignition of blended-fuel droplet in high temperature atmosphere. Combust. Flame 93:149–156

Tarifa CS, Notario PP, Moreno FG (1962) Comustion of liquid monopropellants and bipropellants in droplets. The Eighth Symposium (International) on Combustion. The Combustion Institute. Williams and Wilkins, Baltimore, Md., pp. 1035–1056

Varshavskii GA (1945) Burning of the droplet of liquid fuel. Diffusion theory. BNT MAP. Moscow (in Russian) (see also Khitrin LN (1957) The physics of combustion and explosion. pp. 347–349. Moscow University, Israel Program for Scientific Translations. Jerusalem, 1962)

Vol'fkovich CI, Egorov AP, Epshtein DA (1952) General chemical technology. (in Russian) Chemistry, Moscow Leningrad

Vulis LA (1961) Thermal regimes of combustion. McGraw-Hill, New York

Vulis LA, Yarin LP (1970). Combustion of unmixed gases at a finite reaction rate. Combust. Explos. Shock Waves 6: 423–428

Vulis LA, Yarin LP (1978). Aerodynamics of a torch. (in Russian) Energia, Leningrad

Vulis LA, Ershin ShA, Yarin LP (1968) Foundations of the theory of gas torch. (in Russian) Energia, Leningrad

Waldman CH (1974a) Theory of non-steady state droplet combustion. The Fifteenth Symposium (International) on Combustion. The Combustion Institute, Pittsburgh, Pa., pp. 429–442

Wang CH, Liu XQ, Law CK (1984) Combustion and microexplosion of freely falling multicomponent droplets. Combust. Flame 56: 175–197

Willems JL (1970) Stability theory of dynamical systems. Thomas Nelson

Williams A (1973) Combustion of droplets of liquid fuels: a review. Combust. Flame 21: 1–31

Williams FA (1985) Combustion theory. 2nd edn. Benjamin/Cummings, Menlo Park, Calif.

Wise H, Agoston GA (1958) Burning of a ignition droplet. In: Advances in chemistry, Ser. 20 American Chemical Society, New York, pp. 116–135

Wise H, Lorell J, Wood BJ (1955) The effects of chemical and physical parameters on the burning rate of a liquid droplet. The Fifth Symposium (International) on Combustion. The Combustion Institute. Reinhold, New York, pp. 132–141

Wood BJ, Rosser WAJr (1969) An experimental study of fuel droplet ignition. AIAA J. 7: 2288–2292

Wood BJ, Wise H, Inami SH (1960) Heterogeneous combustion of multicomponent fuels. Combust. Flame 4: 235–242

Yang JC (1993) Heterogeneous combustion. In: Puri JK (ed) Environmental Implications of Combustion Processes. CRC Press, London, pp. 97–137

Yang JC, Jackson GS, Avedisian CT (1990) Combustion of unsupported Methanol/Dedecanol mixture droplets at low gravity. The Twenty-third Symposium (International) on Combustion. The Combustion Institute, Pittsburgh, Pa., pp. 1619–1625

Yap LT, Kennedy IM, Dryer FL (1984) Disruptive and micro-explosive combustion of free droplets in highly convective environment. Combust. Sci. Technol. 41: 291–313

Yarin LP (1961) Thermal regime of combustion of unpremixed gases. In: Acad. Sci. of Kazakhstan. Energetics. Alma-Ata, 19: 47–55 (in Russian)

Yeh CL, Kuo KK (1996) Ignition and combustion of boron particles. Prog. Energy Combust. Sci. 22: 511–541

Zel'dovich YaB (1939). Towards the theory of reaction on porous or powder-like material. J. Phys. Chem. 13:163-168. (see Zel'dovich YaB (1984) Selected papers chemical physics and gas dynamics, (in Russian) Nauka, Moscow)

Zel'dovich YaB (1948). Toward the theory of unmixed gases combustion. J. Tech. Phys. 19: 1199–1210 (in Russian). (see also Zel'dovich YaB, Barenblatt GI, Librovich VB, and Makhviladze GM (1985) Mathematical theory of combustion and explosion. Plenum, New York)

1.4 Collective effects

1.4.1 Introduction

The results of the previous chapters are related to particle behavior in two-phase reactive flows with small volumetric content of dispersed reactant. In such flows the effects which could arise due to the particle-to-particle interaction are insignificant. Under these condition the particle drag, the intensity of heat and mass transfer between the continuous and dispersed phases, as well as the rate of particle burning practically do not depend on particle concentration. If the volumetric content of the condensed phase n_v is high enough (about $n_v \sim 10^{-2}$; Soo 1998), the interaction between the particles becomes significant. It manifests itself in changes in the particle velocities, drag, distribution of particles in the flow, etc.

There are two limiting types of particle-to-particle interaction: (i) direct interaction due to particle collisions; and (ii) interaction without particle collisions when particles affect each other via disturbances of the flow field of the carrier fluid. Direct particle collisions are characteristic only of very dense mixtures. They are accompanied by a drastic change of the properties of the dispersed phase because the particles crush one another, agglomerate and coagulate (Soo 1998, Boothroyd 1974, Crowe et al. 1997, and Sirignano 1999). Below we shall not consider complex phenomena associated with particle collisions. We only estimate the conditions under which such interactions become significant. The effect of direct particle-to-particle collisions on the behavior of a dispersed admixture in a two-phase flow depends on the collision frequency. The latter is determined by a number of factors, such as the numerical concentration of particles, their velocities, direction of particles motion, flow regime, etc. The problem is significantly simplified for several types of particle motion, in particular, for translational motion (Yarin and Sukhov 1987). In this case the collision frequency is directly proportional to the difference of the particle velocities, their numerical concentrations and the interception cross-section:

$$f \sim \Delta u n_{**} S \tag{4.1}$$

where f is the collision frequency, $\Delta u = u_* - u_{**}$, u is the particle velocity, n_{**} is the particles' numerical concentration, $S = [\pi(\delta_* + \delta_{**})^2]/4$ is the interception cross-section, and single and double asterisks in the subscripts correspond to particles of the diameters δ_* and δ_{**}, respectively $(\delta_* \le \delta_{**})$.

Expressing the numerical concentration n_{**} via of the mass concentration n_m, we obtain

$$n_{**} = 6\frac{n_m}{\pi\delta_{**}^3\rho_2} \tag{4.2}$$

where ρ_2 is the admixture density

Equations (4.1) and (4.2) yield the following estimates for the collision frequency of particles and the mean free path between two collisions

$$f \sim \left(\frac{3}{2}\right)\frac{n_m}{\delta_{**}\rho_2}\Delta u(1+\frac{\delta_*}{\delta_{**}})^2 \tag{4.3}$$

$$\ell \sim \left(\frac{2}{3}\right)\frac{\rho_2\delta_{**}u_{**}}{\Delta u(1+\frac{\delta_*}{\delta_{**}})^2 n_m} \tag{4.4}$$

Bearing in mind that the particle dynamic relaxation length in a Stokes flow is given by

$$\ell_r = \frac{u_{**}\delta_{**}^2\rho_{2.1}}{18\nu_1} \tag{4.5}$$

we obtain an estimate of the ratio of the particle mean free path to the particle relaxation length

$$\frac{\ell}{\ell_r} \sim \frac{12\nu_1\rho_1}{\left[\delta_{**}n_m(1+\frac{\delta_*}{\delta_{**}})^2\right]\Delta u} \tag{4.6}$$

where ν_1 is the kinematic viscosity of the carrier fluid.

Using Eqs. (4.4) and (4.6) we estimate the order of the collision frequency and the ratio ℓ/ℓ_r for the following conditions: $n_m \sim 0.1\,[\mathrm{kg/m^3}]$, $\rho_2 \sim 10^{-3}\ [\mathrm{kg/m^3}]$, $\delta_* = 10^{-4}\,\mathrm{m}$, $\delta_{**} = 1.1\cdot 10^{-4}\,\mathrm{m}$, $u_{**} \sim 10\,[\mathrm{m/s}]$, $\Delta u = 0.1\ [\mathrm{m/s}]$, $\nu_1 \sim 10^{-5}\ [\mathrm{m^2/s}]$. Then, we obtain $f = 0.55\,[\mathrm{s^{-1}}]$, and $\ell/\ell_r \sim 30$. These estimates show that the effect of direct particle-to-particle interaction is still insignificant under the conditions characteristic of industrial furnaces.

1.4.2 Hydrodynamic interaction

The method of reflections. The hydrodynamic interactions of particles moving in a viscous liquid without direct collisions were the subject of numerous theoretical investigations (see Happel and Brenner 1983, Soo 1998, Crowe et al. 1997, Sadhal et al. 1997, Sirignano 1999, Kim and Karrila 1991 and Subramanian and Balasubramanian 2001). For calculation of the flow field in particle-laden flows various methods were used, in particular, the method of reflections (Happel and Brenner 1983, Hetsroni and Haber 1971, 1978). In accordance with this method, the flow field in the vicinity of an arbitrary particle belonging to the moving ensemble of particles consists of its own field created by a given particle, and perturbations due to its interaction with the surrounding particles. At small Reynolds number corresponding to the Stokes flow, the velocity distribution is described by the Stokes equations

$$\mu\nabla^2\mathbf{u} = \nabla P \tag{4.7}$$

$$\nabla\cdot\mathbf{u} = 0 \tag{4.8}$$

Solutions are subject to the following conditions

$$\mathbf{u} \to \mathrm{U}_k \text{ at the particle surface} \tag{4.9}$$

$$\mathbf{u} \to 0 \text{ in the undisturbed flow} \tag{4.10}$$

where $\mathbf{u}$ is the fluid velocity, P is the pressure, μ is the viscosity, and $\mathbf{U}_k$ is the velocity of the k-th particle.

The velocity and pressure distributions may be presented as follows

$$\mathbf{u} = \sum_{j=1}^{\infty} \mathbf{u}^{(j)}; \qquad P = \sum_{j=1}^{\infty} P^{(j)} \tag{4.11}$$

where $\mathbf{u}^{(j)}$ and $P^{(j)}$ are the solutions of Eqs. (4.7) and (4.8) satisfying conditions (4.9) and (4.10).

Using Eqs. (4.7) and (4.8) and the conditions (4.9) and (4.10), we determine the flow field of an arbitrary particle $\mathbf{u}^{(1)}$ (denoted as the first one) and the reflection of this field from the other N–1 particles $\mathbf{u}^{(2)}$:

$$\mathbf{u}^{(2)} = \sum_{k=2}^{N} \mathbf{u}_k^{(2)} \tag{4.12}$$

where $\mathbf{u}_k^{(2)} = \mathbf{U}_k - \mathbf{u}^{(1)}$.

Then determining the reflection $\mathbf{u}^{(3)}$ of the field $\mathbf{u}^{(2)}$, we can find the velocity distribution in the vicinity of the first particle:

$$\mathbf{u} = \mathbf{u}^{(1)} + \mathbf{u}^{(3)} \tag{4.13}$$

The expression (4.13) yields the first approximation for the velocity field resulting from pure hydrodynamic particle interaction. Successive repetition of the procedure described above yields the subsequent approximations $\mathbf{u}^{(5)}$, $\mathbf{u}^{(7)}$, etc. which allow for an increase of the calculation accuracy.

The expression for the drag force acting on the k-th particle results from the summation of the forces due to the fields $\mathbf{u}^{(1)}$, $\mathbf{u}^{(3)}$, etc.

$$\mathbf{F}_k = \mathbf{F}_k^{(1)} + \mathbf{F}_k^{(3)} + \ldots \tag{4.14}$$

In the general case corresponding to motion of particles in an arbitrary velocity field, the system of Eqs. (4.7) and (4.8) should be solved taken into account the actual velocity distribution. Such a problem was considered by Hetsroni and Haber (1978). These results were corrected later by Zinchenko (1981) (see also Subramanian and Balasubramanian 2001).

Interaction of two particles. As an example illustrating the characteristics of the hydrodynamic interactions we consider the problem of creeping flow due to the motion of two spherical particles in an unbounded viscous fluid. In the works by Jeffery (1915), Stimson and Jeffery (1926), Cooley and O'Neill (1969), Happel and Brenner (1983), Davis (1969), Hetsroni and Haber (1971, 1978), Wacholder and Weihs (1972), Haber et al. (1973), Reed and Morrison (1974), and Rushton and Davies (1978), and Zinchenko (1979, 1981, 1982, 1984) several approximate and exact solutions of this problem were obtained. They are related to hydrodynamic interactions of two solid particles of equal or different sizes, as well as of two liquid droplets or bubbles moving in a uniform or in an arbitrary velocity field.

Below, following the work by Haber et al. (1973), we consider the results related to slow motion of two liquid droplets moving along their line of centers in an unbounded quiescent fluid with constant velocities $\mathbf{U}_a$ and $\mathbf{U}_b$ (Fig. 4.1). This work contains the most general solution of the problem which incorporates a number of particular cases which were considered by Bard (1968), Wacholder and Weihs (1972), Stimson and Jeffery (1926), Goldman et al. (1967), as well as the classical Hadamard–Rybczynski solution.

Assuming the creeping flow approximation we should find that the flow and pressure fields due to droplets a and b exterior to the droplets **u,** P and interior of them $\mathbf{u}_j$, P_j (j=a,b) satisfy Eqs. (4.7) and (4.8). Particle radii are denoted a and b and their velocities are $\mathbf{U}_j$. The viscosities inside the droplets are denoted μ_j, and those outside the droplets are denoted μ_e.

The boundary conditions are based on the following assumptions: (i) the velocities inside and outside the droplets are continuous at the interface; (ii) the shear stresses are continuous at the interface; (iii) the normal stresses at the interface jump due to the capillary pressure related to the surface tension σ_{tj} (j=a,b), (iv) the droplets are assumed to be undeformable and spherical, which means that surface tension effect is strong enough and $(\mu_e u)/\sigma_{tj} \ll 1$, where μ_e is the viscosity of the continuous phase (this question is discussed in Hetsroni and Haber 1970 and Hetsroni et al. 1970), (v) far from the droplets the fluid is unperturbed, so that the velocity vector vanishes.

For non evaporating droplets in the gas or the liquid systems the boundary conditions are

$$\mathbf{u}_j = \mathbf{u} \tag{4.15}$$

$$\mathbf{U_j} \cdot \mathbf{e_{n.j}} = \mathbf{u} \cdot \mathbf{e_{n.j}} \tag{4.16}$$

$$\mathbf{\Pi}_{nn.j} = \boldsymbol{\tau}_{nn.j} + \sigma_{tj}\left(\frac{1}{r_{1j}} + \frac{1}{r_{2j}}\right)\mathbf{e}_{n.j} \tag{4.17}$$

at the droplets interface, and

$$\mathbf{u} = 0 \tag{4.18}$$

far from the droplets.

Here $\mathbf{e}_n$ is the unit outer normal vector at the droplet, $\boldsymbol{\tau}_{nn.j}$ is the normal stress in the dispersed phase, $\mathbf{\Pi}_{nn.j}$ is the normal stress in the continuous phase, σ_{tj} is the surface tension, and r_1 and r_2 are the principal radii of curvature.

The problem reduces to solving the linear equation for the stream function ψ (Happel and Brenner 1983)

$$E^4\psi = 0 \tag{4.19}$$

where E^2 is the Stokes stream function operator in the bipolar coordinate system (ξ, η, φ), associated with the cylindrical coordinates (ρ, z, φ) by the relations

$$\rho = c\frac{\sin\eta}{\cosh\xi - \cos\eta}, \qquad z = c\frac{\sinh\xi}{\cosh\xi - \cos\eta}, \qquad \varphi = \varphi \tag{4.20}$$

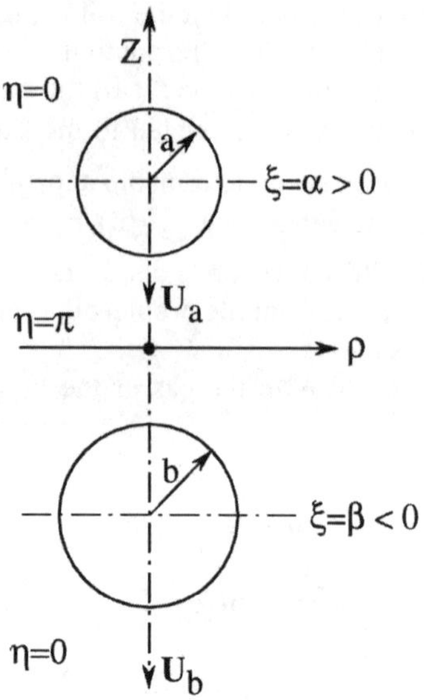

Fig. 4.1 The bipolar coordinate system

The interfaces of the droplets are defined by $\xi = \alpha > 0$ and $\xi = \beta < 0$, while their radii and the distance between centers ℓ are given by $a = c\,\mathrm{cosech}(\alpha)$, $b = c\,\mathrm{cosech}(\beta)$ and $\ell = c[\coth(\alpha) - \coth(\beta)]$, which also defines the value of the constant c.

The general solution of Eq. (4.19) that was given by Stimson and Jeffery (1926)

$$\psi = (\cosh\xi - \cos\eta)^{-\frac{3}{2}} \sum_{n=1}^{\infty} W_n(\xi) C_{n+1}^{-\frac{1}{2}}(\theta) \tag{4.21}$$

where
$W_n(\xi) = a_n \cosh(n - 1/2)\xi + b_n \sinh(n - 1/2)\xi + c_n \cosh(n + 3/2)\xi + d_n \sinh(n + 3/2)\xi$
$C_{n+1}^{-1/2}(\theta)$ is the Gegenbouer polynomials of order (n + 1) and degree $-1/2$ which is related to the Legendre plynomials via the relation $C_n^{-1/2}(\theta) = (P_{h-2}(\theta) - P_n(\theta))/(2n-1)$, $\theta = \cos\eta$, and the constants a_n, v_n, c_n and d_n are determined from the boundary conditions.

The drag forces acting on the trailing and leading droplets (cf. Fig.4.1) are expressed as

$$F_a = \frac{2\sqrt{2}}{c}\pi\mu_e \sum_{n=1}^{\infty} (a_n + b_n + c_n + d_n) \tag{4.22}$$

$$F_b = \frac{2\sqrt{2}}{c}\pi\mu_e \sum_{n=1}^{\infty} (a_n - b_n + c_n - d_n) \tag{4.23}$$

The equations (4.26) and (4.27) can be transformed to a form similar to the Stokes law, where $\Lambda_{11}, \Lambda_{12}, \Lambda_{21}$ and Λ_{22} are the correction factors dependent on the fluid viscosities in the droplet, which are determined numerically. Λ_{11} and Λ_{12} correspond to the trailing droplet, whereas Λ_{21} and Λ_{22} correspond to the leading one.

$$\mathbf{F}_a = -6\mu_e a[\Lambda_{11}\mathbf{U}_a + \Lambda_{12}\mathbf{U}_b] \tag{4.24}$$

$$\mathbf{F}_b = -6\mu_e b[\Lambda_{21}\mathbf{U}_a + \Lambda_{22}\mathbf{U}_b] \tag{4.25}$$

The dependencies of the correction factors Λ_{11} and Λ_{12} on the ratio between the radii of trailing and leading droplets $p = a/b$ are presented in Figs. 4.2. They correspond to cases when: (i) the droplets' viscosity is equal and very small compared to that of the carrier fluid (air bubbles in a viscous liquid); (ii) the droplets' viscosity is equal to that of the surrounding fluid; (iii) the droplets' viscosity is taken to be 67 times larger than that of the surrounding liquid (water droplets in the atmosphere). In all the cases the correction factors Λ_{11} and Λ_{12} increase as the radius of leading droplet increases (at a and ℓ/a constant). The correction factor of the trailing droplet Λ_{11} is less affected by the leading one than the correction factor Λ_{12}. Also, both correction factors increase without bound when the two droplets approach each other. The presence of the leading droplet significantly affects the drag of the trailing one. An increase of the radius of the leading droplet (when the radius of the trailing droplet a=const, $\ell/a = \text{const}$, and $\mathbf{U}_a = \mathbf{U}_b$) is accompanied by a decrease in the drag of the trailing droplet. In fact the drag acting on the trailing droplet becomes less than the drag of an isolated droplet.

Two equal droplets in contact. In this case the coefficients b_n and d_n in Eqs. (4.22) and (4.23) vanish and the forces $\mathbf{F}_a$ and $\mathbf{F}_b$ are equal. They are given by

$$\mathbf{F} = -6\pi\mu_e a\mathbf{U}\Lambda^* \tag{4.26}$$

where $\mathbf{F} = \mathbf{F}_a = \mathbf{F}_b$ and Λ^* is the correction factor for the two droplets in contact.

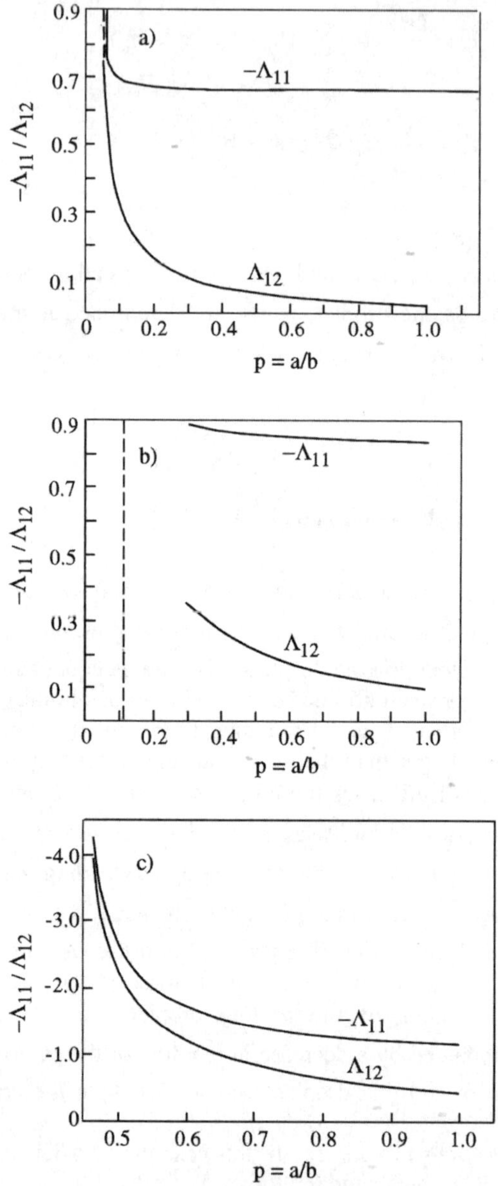

Fig. 4.2 The correction factors Λ_{11} and Λ_{12} in the Eqs. (4.28) and (4.29) for the drag force versus the ratio between the droplets' radii. Reprinted from Haber et al. (1973), with permission. a) $\bar{\mu}_a = \bar{\mu}_b = 0$; $a/\ell = 0.05$. b) $\bar{\mu}_a = \bar{\mu}_b = 1$' $a/\ell = 0.1$. c) $\mu_a = \mu_b = 67$; $a/\ell = 0.3$, $\bar{\mu} = \mu/\mu_e$. ℓ is the distance between the droplet centers

Equation (4.26) corresponds to the drag of two droplets with the same viscosity in contact. The correction factor Λ^* was calculated by Faxen (1927). In the general case of two equal-sized droplets possessing viscosities $\mu_a = \mu_b$ different from the viscosity of surrounding fluid μ_e the drag force is given by

$$F = -6\pi\mu_e aU \frac{2/3 + \bar{\mu}_a}{1 + \bar{\mu}_a}\Lambda \tag{4.27}$$

when the droplet are in contact. Here $\bar{\mu} = \mu/\mu_e$,

$$\begin{aligned}
&\Lambda = \frac{1+\bar{\mu}_a}{2/3+\bar{\mu}_a}\frac{1}{2}\int_0^\infty \frac{C_0 + \bar{\mu}_a C_1 + \bar{\mu}_b C_2 + \bar{\mu}_a\bar{\mu}_b C_3}{A_o + (\bar{\mu}_a + \bar{\mu}_b)A_1 + \bar{\mu}_a\bar{\mu}_b A_2}dx,\\
&C_o = 64e^{-x}\sinh^2 x(\cosh x + xe^x),\\
&C_1 = 8e^{2x}(2x^2 + 2x + 1) - 8e^{-2x}(2x - 1) - 8e^{-4x} - 48x^2 - 32x - 8,\\
&C_2 = 8\bar{e}^x(2x^2 - 2x + 1) + 8e^{2x}(2x + 1) - 8e^{4x} - 48x^2 - 32x - 8,\\
&C_3 = (16 + 3x^2)\sinh 8x + 32x\cosh 2x + 8e^{-4x} - 64x^3 - 64x^2 - 32x - 8,\\
&A_o = 4\sinh^2 2x; \quad A_1 = 2\sinh 4x - 8x; \quad A_2 = 4\sinh^2 2x - 16x^2.
\end{aligned} \tag{4.28}$$

The calculations show that the correction factor Λ varies slightly with $\bar{\mu}$. For example, $\Lambda = 0.69309765$ at $\bar{\mu} = 0$ (bubble) and $\Lambda = 0.64514946$ at $\bar{\mu} = 1000$ (virtually a solid particle). Reed and Morrison (1974) obtained the correction factor to the Hadamard–Rybczynski law (equal sized sphere with identical viscosities μ_1 and μ_2, $\mu_1 = \mu_2 = \mu$) in the form

$$\Lambda = \frac{\bar{\mu}+1}{3\bar{\mu}+2}\int_0^\infty \frac{e^{-x}\left\{\left[\bar{\mu}(x^2 + x + 1) + x\right]\sinh x + (x+1)(\bar{\mu}x + 1)\cosh x\right\}}{\bar{\mu}(x + \cosh x \cdot \sinh x) + \cosh^2 x}dx \tag{4.29}$$

Equation (4.29) reduces to the result of Faxen for infinite viscosity ratio of the dispersed phase $\mu_1 \to \infty$ (solid particles)

$$\Lambda = \frac{1}{3}\int_0^\infty \left(1 + \frac{2\sinh^2 x - 2x^2}{\sinh 2x + 2x}\right)dx \tag{4.30}$$

For the second limited case corresponding to bubble motion in liquid (zero viscosities of the dispersed phase $\mu_1 = 0$) the correction factor is expressed as

$$\Lambda = \frac{1}{2}\int_0^\infty \frac{e^{-x}[x\sinh x - (x+1)\cosh x}{\cosh^2 x}dx \tag{4.31}$$

It should be noted that in the work by Morrison and Reed (1978) the integral in Eq. (4.31) was evaluated in a closed form. It was found that $\Lambda = \ln 2$.

Interaction of three particles. This problem was studied by Kim (1987) and Ramachandran et al. (1991). In the first of these works a uniform creeping flow past three spherical particles forming a triangular cluster was considered. For the drag force acting on the leading particle in a three-sphere symmetrical cluster settling with the apex directed downwards or horizontally the following relation was obtained

$$\overline{F} = \sum_{j=0}^{5} C_j \overline{r}^j \tag{4.32}$$

where $\overline{F} = F/F_{St}$, F and F_{St} are the drag force acting on the leading particle and the Stokes drag force of an isolated particle, $\overline{r} = r/R$, r and R are the particle radius and the center-to-center separation; $C_0 = 1$; $C_1 = -21/8$; $C_2 = 45/8$; $C_3 = 6{,}191/512$; $C_4 = 135{,}326/4{,}096$; $C_5 = 689{,}923/8{,}192$ for the cluster settling with apex directed downwards, and $C_0 = 1$; $C_1 = -15/8$; $C_2 = 153/32$; $C_3 = 5{,}447/512$; $C_4 = 102{,}885/4{,}096$; $C_5 = -1085{,}373/16{,}384$ for the cluster settling with the apex directed horizontally.

Ramachandran et al. (1991) explored the laminar flow past three closely spaced spherical solid particles and non evaporating droplets that are located one after another in the range of intermediate Reynolds numbers ($1 < Re < 200$). The dimensional particle spacing was $2 < d_{ij} < 6$ $(d_{ij} = d_i/d_j)$, where d_i and d_j are the diameters of the neighboring particles The calculations show that the separation angles and the effective drag coefficient of the interacting spheres differs significantly from that for an isolated particle at all the Reynolds numbers studied. The effect of the internal circulation on the drag of the non evaporating droplets is insignificant. The drag coefficient remains affected by the first sphere when $\alpha_{23} \to \infty$. In all the cases particle interaction leads to a decrease in the drag coefficient compared to the drag coefficient of an individual particle.

For the first , second and third particles the following correction factors were obtained

$$\alpha_1 = 1 - 0.096\,Re^{0.2475}\,d_{12}^{-0.965}\,e^{0.4764/d_{23}} \tag{4.33}$$

$$\alpha_2 = 1 - \mathrm{Re}^{0.1593}\left(0.2932/d_{12}^{0.4876} + 0.1341/d_{23}^{0.4242}\right) \tag{4.34}$$

$$\alpha_3 = 1 - 0.325(\ln \mathrm{Re} + 1)^{0.603} e^{-0.282/d_{12}} d_{23}^{-0.385} \tag{4.35}$$

Here $\alpha_i = C_{d.i}/C_{d.s}$, where $C_{d.i}$ is the drag coefficient of a particle in the cluster, whereas $C_{d.s}$ is the drag coefficient of the particle if it were alone, $C_{d.s} = \dfrac{24}{\mathrm{Re}} + \dfrac{6}{(1+\sqrt{\mathrm{Re}})} + 0.4$, for $1 < \mathrm{Re} < 2 \cdot 10^5$. It is seen that, when the distance between the first and the second particles tends to infinity without limit $(d_{12} \to \infty)$, the drag coefficient of the leading particle approaches that of an individual particle. An increase of the distance between the second and the third particles leads to an increase of the drag coefficient of the third particle up to the value corresponding to an individual particle.

Periodic arrays of spheres. The drag of spherical particles in periodic arrays was investigated analytically and numerically by Hasimoto (1959) and Zick and Homsy (1982). It was shown that the force acting on any of the particles in the lattices considered (simple cubic, body-centered cubic, face-centered cubic) is larger than that given by the Stokes law. Therefore, a correction factor k, which is defined as the ratio of the drag force of a particle in an array to the Stokes force, is larger than one (Fig. 4.3).

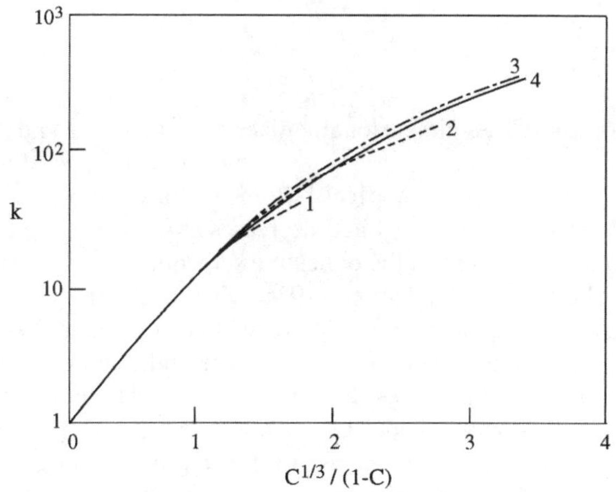

Fig. 4.3 Correction factor as a function of volume concentration of particles in an array. Reprinted from Zick and Homsy (1982) with permission. Curve 1, simple cubic; curve 2, body-centered cubic; curve 3, face-centered cubic; curve 4, Carman empirical correlation $k = (10C)/(1-c)^3$, where $C = (4\pi r^3)/(3\tau_o)$ is volume concentration, r is particle radius, τ_o is the unit cell volume: $\tau_o = d^3$, $(4/3\sqrt{3})d^3$, $2^{-1/2}d^3$ for simple cubic, body-centered cubic and face-centered cubic packing, respectively, andd is the lattice characteristic size

Finite ensembles of particles. Yarin (1987) used the renormalization group technique for examination of the hydrodynamic drag force acting on a spherical particle surrounded by a large but finite number of similar particles in an unrestricted flow field. He estimated the drag force for the creeping and inertial (with moderate Reynolds numbers) flow regimes for various particle arrangement in space: on lines, with equal distances between the neighboring particles, as well as on two- and three- dimensional (square and cubic) lattices. For creeping flow normal to a linear arrangement of spheres the correction to the Stokes drag force is given by

$$h_1 = 1 - \frac{3}{2}\frac{\ln N}{y} \tag{4.36}$$

where N is the number of the particles, $y = \ell/a$, ℓ is the distance between the particle centers, and a is the particle radius.

Equation (4.36) is valid in the case $(\ln N)/y \ll 1$. For two- and three- dimensional finite lattices facing the normal creeping flow the correction to the Stokes force is given by

$$h_{2.3} = \exp\left\{ -\frac{1}{1-\frac{1}{d}} \frac{k_1}{y} \left(N^{1-\frac{1}{d}} - 1\right) \right\} \tag{4.37}$$

Here for the two- and three-dimensional lattices $k_1 = 2.03$, d=2 and $k_1 = 5.7$, d=3, respectively.

It is seen that the collective effect becomes stronger in spaces of higher dimension. In all the cases the drag force decreases sharply in systems with a large number of particles as a result of the collective screening.

Comparison of the results of Hasimoto (1956), Zick and Homsy (1982) and Yarin (1987) shows that there is a principal difference between the hydrodynamic particle-to-particle interactions in periodic arrays and clouds, and in ensembles with a large but finite number of particles. It manifests itself via the dependences of the drag force on particle concentration. In a periodic infinite array an increase in the particle concentration leads to an increase of the drag force acting on an individual particle. On the other hand, in finite ensembles an increase in the number of particles is accompanied by a decrease of the drag force acting on individual particle. The difference is due to the difference between the flow conditions in the infinite and finite systems of particles. In the case of an infinite array the mean velocity in each of the lattice cross-sections does not vary from one opening to another (Fig. 4.4a). In contrast with that, when a finite ensemble of particles in a cloud is subject to a flow, the latter easily circumvents the particles (Fig. 4.4b).

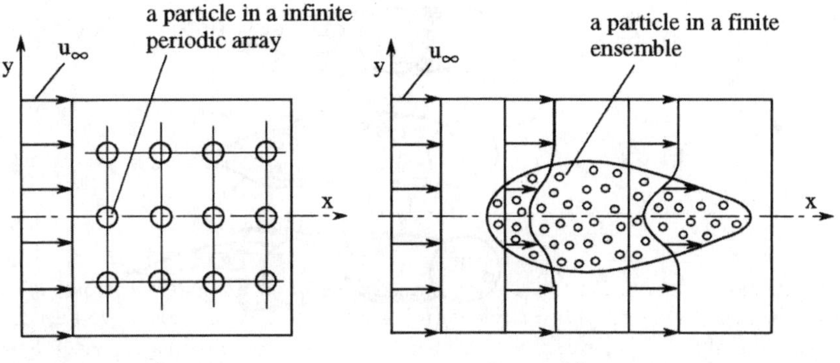

Fig. 4.4 Periodic arrays and finite ensembles (clouds). a) Periodic array. b) Cloud

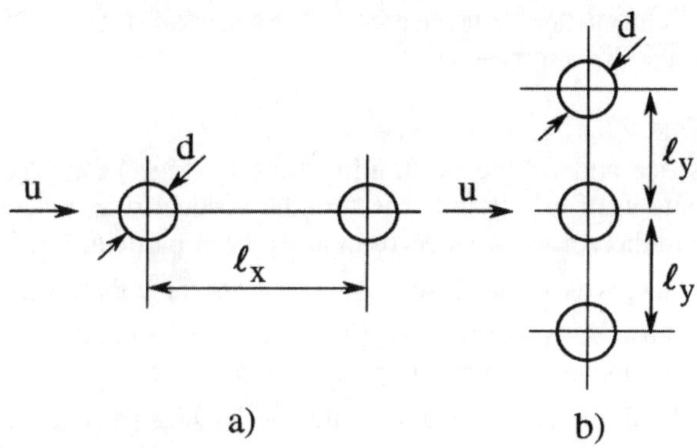

Fig. 4.5 Particle arrangements. a) Longitudinal orientation of particles. b)Transverse orientation of particles

Under these conditions the particles in the frontal part of the cloud screen the particles located in its core and rear parts. That is the cause of the reduction of the drag force in clouds of particles.

Experiments. Hydrodynamic interactions between particles arranged in different configurations were investigated experimentally by Lee (1979), Tsuji et al. (1982, 1985), Mulholland et al. (1988), Zhu et al. (1994) and Liang et al. (1996). The major efforts were focused on measuring of the flow structure and particle drag versus the Reynolds number and the interparticle distance.

The vortical patterns of the flow about spherical particles arranged in the longitudinal and transverse directions (Fig. 4.5) were studied by Tsuji et al. (1982). Experiments showed that the intensity of the hydrodynamic interactions is mostly determined by the dimensionless distance between the particles $\overline{\ell}_x = \ell_x / d$ (ℓ_x is the interparticle distance and d is the particle diameter). When $\overline{\ell}_x$ is of the order of one (two spheres are attached to each other), two separated vortices are formed in the flow field. They are located, respectively, in the

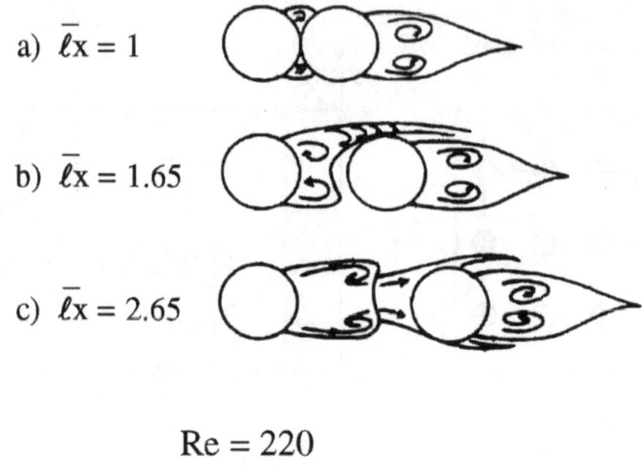

Fig. 4.6 Vortex configurations in the longitudinal arrangement at Re = 220. Reprinted from Tsuji et al.(1982), with permission

interparticle space and behind the trailing particle (Fig. 4.6). The first of these vortices presents itself as a symmetric toroidal vortical ring, whereas the second one is similar to the vortex shedded from an isolated particle. The vortical pattern significantly changes as $\overline{\ell}_x$ increases. At $\overline{\ell}_x \sim 1.56$ the vortex ring is transformed into an asymmetric vortical sheet which envelops the frontal part of the trailing particle, whereas the rear vortex remains practically the same as in Fig. 4.6a. When $\overline{\ell}_x \sim 2.5$, the vortex shedding from the leading particle becomes regular (the Strouhal number is about 0.2). At large enough $\overline{\ell}_x$ the hydrodynamic interactions between the leading and trailing particles vanish and the vortical structure of each of the individual particles becomes independent and identical to that of an isolated particle.

The distance between the neighboring particles significantly affects the intensity of the interaction in the transverse direction. In this case the vortical structure is more complicated because of the three-dimensional character of the flow. At small $\overline{\ell}_y$ $(1 \le \overline{\ell}_y < 1.48)$, $\overline{\ell}_y$ is the transverse distance between the particles is rendered dimensionless by d. The vortices behind the particles are similar to the Karman vortex. When $\overline{\ell}_y$ exceeds 2.2 the hydrodynamic interaction is negligible. A typical flow pattern about two equal-sized particles in a uniform flow normal to the line connecting their centers is shown in Fig. 4.7 where the results of the numerical calculations by Kim et al. (1993) are presented. The measurements by Tsuji et al. (1982, 1985) and Zhu et al. (1994) show that hydrodynamic particle-to-particle interactions significantly affect the particle drag. The magnitude of the effect differs significantly between the leading and trailing particles. The disturbances generated by the trailing particle only slightly affect the flow field near the leading particle. On the other hand the disturbances generated by the leading

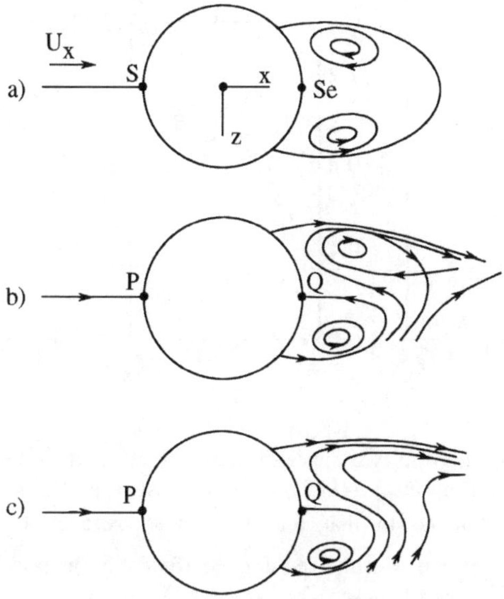

Fig. 4.7 Stream lines patterns near one of the two solid particles at Re = 100. Reprinted from Kim et al. (1993) with permission. a) Axisymmetric flow pattern (single sphere, Re=100). b) Asymmetric flow pattern, dimensionless separation equals 2 (one or two sphere, Re=100). c) Asymmetric flow pattern, dimensionless separation equals 1.5 (one or two sphere, Re=100)

particle markedly affect the flow structure around the trailing particle. Accordingly, the total drag of the leading particle is close to that for an isolated particle, whereas the total drag of the trailing particle changes sharply with interparticle distance (Fig. 4.8). It is seen that the hydrodynamic interactions can be neglected for $\overline{\ell}_x > 8$ and $\overline{\ell}_y > 3$.

For calculation of the drag force acting on the trailing particle Zhu et al. (1994) proposed the following empirical correlation

$$\frac{f_d}{f_{do}} = 1-(1-\alpha)\exp(\beta\overline{\ell}_*) \tag{4.38}$$

where f_d and f_{do} are the drag forces of the trailing and an equivalent isolated particle, $\overline{\ell}_* = \ell_*/d$, ℓ_* is the gap between the leading and trailing particles, and $\alpha = \alpha(\mathrm{Re})$ and $\beta = \beta(\mathrm{Re})$ are functions of the Reynolds number:

$\alpha = 1-\exp(-0.483+3.45\cdot 10^{-3}\,\mathrm{Re}-1.07\cdot 10^{-5}\,\mathrm{Re}^2)$ and

$\beta = -0.115-8.75\cdot 10^{-4}\,\mathrm{Re}+5.61\cdot 10^{-7}\,\mathrm{Re}$.

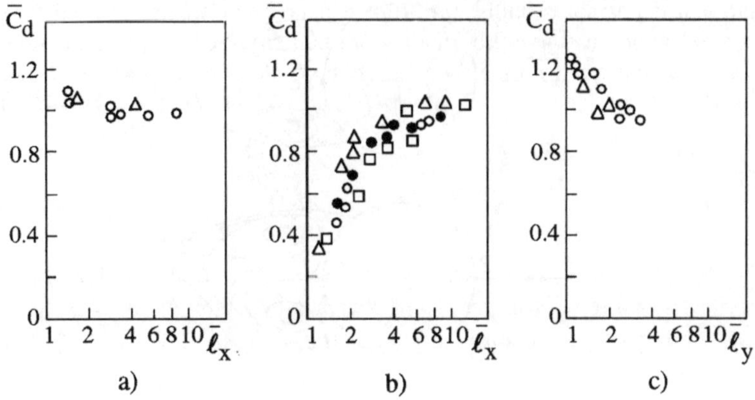

Fig. 4.8 Drag coefficients foe two spherical particles affected by hydrodynamic interactions. Reprinted from Tsuji et al. (1982), with permission. a) The drag coefficient of the leading particle, normalized to that for an isolated particle: $\bar{C}_d(\bar{\ell}_x)$; ○, Re = 540; Δ, Re = 245; □, Re = 860. b) Normalized drag coefficient of the trailing particle: $\bar{C}_d(\bar{\ell}_x)$; Δ, Re ~ 138; ○, Re = 312 – 330; ●, Re = 556 – 580; ▲, Re = 715 – 800. c) Normalized drag coefficient of the leading particle: $\bar{C}^*_d(\ell_y)$; ○, Re = 360 ~ 370; Δ, Re = 650 – 680; ∇, Re = 810 – 870 . $\bar{C}_d = C_d / C_{do}$, $\bar{C}^*_d = C^*_d / C_{do}$, C_{do} is the drag coefficient of a single particle; $\bar{\ell}_x = \ell_x / d$, $\bar{\ell}_y = \ell_y / d$, ℓ_x and ℓ_y are the distances between particle centers in the x and y directions

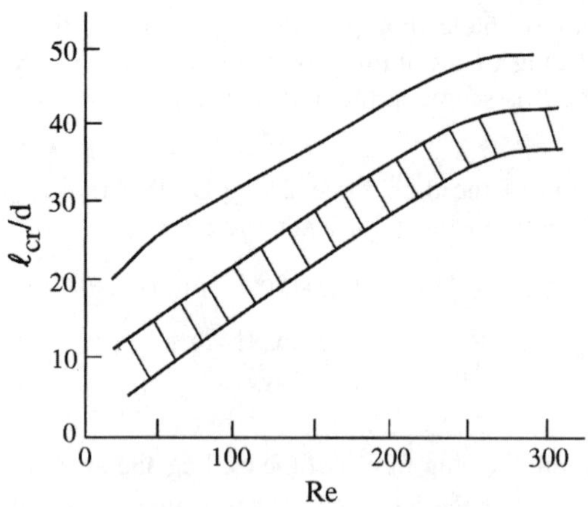

Fig. 4.9 Dependence of the critical interparticle distance on the Reynolds number. Reprinted from Mulholland et al. (1988) with permission. The shaded region corresponds to wake length behind a circular cylinder

The critical droplet spacing above which the interparticle distance does not affect the particle drag was determined by Mulholland et al. (1988). The measurements of the trajectories of a non evaporating droplet were performed for several values of the initial Reynolds number in the range 90–200 and for the dimensionless interparticle spacing from 1 to 1700. The experiments show that the critical droplet spacing ℓ_{cr} increases almost linearly with Re at Re<250 (Fig. 4.9). It was also noticed that the dependence $\ell_{cr}(\mathrm{Re})$ is similar to the dependence of the wake length behind a particle on Re. This allows one to assume that the value of the critical distance between the interacting particles is determined by the vortex flow structure in the interparticle space.

1.4.3 Mass transfer

Aminzadeh et al. (1974) calculated the mass transfer rates for the equal-sized particles in a Stokes flow directed parallel to their line of centers for Peclet numbers in the range of 0–50 ($\mathrm{Pe} = \mathrm{ud/D}$, D is the diffusivity). They showed that the effect of the particle-to-particle interactions on the mass transfer is significant. In particular, it was found that the overall Sherwood number for either particle is always less than that of a single isolated particle. At low Peclet numbers, the overall Sherwood number based on the particle diameter is less than 2.

The mass transfer of a chain of solid spherical particles in a translational creeping flow was considered by Gupalo et al. (1985). For the total diffusion flux to the surface of the k-th particle, as well as for the whole chain of k particles, the following relations were obtained:

$$I_k = I_1[k^{\frac{2}{3}} - (k-1)^{\frac{2}{3}}] \tag{4.39}$$

$$I_\Sigma^{(k)} = I_1 k^{\frac{2}{3}} \tag{4.40}$$

where I_1 and I_k are the total diffusion fluxes to the first and k-th particles of the chain; $I_\Sigma^{(k)}$ is the total mass flux to the chain of k particles.

Equation (4.39) shows that the total diffusion flux to the k-th particle decreases when k increases:

$$I_k \to \frac{2}{3} I_1 k^{-\frac{2}{3}}, \quad \text{at} \quad k \to \infty \tag{4.41}$$

This effect is a result of screening of the subsequent particles by the previous ones. In accordance with that $I_1 > I_2 ...... > I_k > I_{k+1}$. The total diffusion flux to all k particles of the chain is proportional to $k^{2/3}$, which is significantly smaller than the total diffusion flux to k non interacting particles proportional to k ($k \gg 1$). The mean Sherwood number for the whole chain of spherical particles of equal diameters is given by

$$\mathrm{Sh} = \mathrm{Sh}_1\, k^{-\frac{2}{3}} \tag{4.42}$$

where Sh_1 is the Sherwood number of an isolated particle. Thus, the Sherwood number of the particle chain decreases sharply with k.

1.4.4 Interaction of burning particles

This problem was considered by Labowski (1978, 1980), Brzustiwski et al. (1979), Umemura et al. (1981(1), 1981(2)), Miyasaka and Law (1981), Sangiovanni and Labowski (1982), and Marberry et al. (1984). In the present section a summary of the most significant results of these works is given.

From the physical point of view the mechanism of burning of an individual droplet and of a system of droplets is similar. In both cases a chemical reaction proceeds in the gaseous reactive mixture that is formed due to mixing of the vapor of a liquid fuel with an oxidizer contained in the environment. The existence of a number of interacting droplets in a cloud affects the conditions of liquid fuel evaporation and mixing of its vapor with the oxidizer. Accordingly, the interactions affect that location of the reaction zone, its configuration, as well as the rate of combustion.

As in the case of an isolated particle the assumption of the quasi-steady character of the process is employed. Also it is assumed that: (i) the state of the ambient gas far from the droplet is uniform and the pressure is below the critical point, (ii) the temperature is uniform in each droplet and the phases are in thermodynamic equilibrium at the surface droplet, (iii) all the chemical reactions occur under stoichiometric conditions and combine into a single-step reaction $F + O \rightarrow P$ (F, O and P are fuel, oxidizer and combustion product), (iv) the physical properties of the reactants and combustion products are constant, (v) the effects of natural convection and radiative heat transfer are negligible. As has already been mentioned, it is advantageous to use the Schvab–Zel'dovich transformation under these conditions. This allows for removal of the highly non linear source terms from the species and energy balance equations and reduces them to the linear equations for coupling functions β_j

$$\nabla \cdot (\rho \mathbf{v} \beta_j - \rho D \nabla \beta_j) = 0, \quad j = 1;2 \tag{4.43}$$

where $\beta_1 = [C_F/(M_F\sigma_F)] - [C_O/(M_O\sigma_O)]$; $\beta_2 = [(c_p(T - T_m))/q] - [C_O/(M_O\sigma_O)]$ are the coupling functions; C is the mass fraction , M is the molecular weight, σ

is the stoichiometric coefficient, subscripts F and O correspond to fuel and oxidizer, q is the heat of reaction, and T_m is the flame temperature.

The boundary conditions corresponding to the problem of burning of a system of N droplets are written as

$$C_F = C_{F.S}; \quad T = T_S; \quad \rho \mathbf{v} C_{O.S} - \rho D \nabla C_{O.S} = 0 \tag{4.44}$$

at the droplet surface, and

$$C_F \to 0, \quad C_O \to C_{O.\infty}, \quad T \to T_\infty \tag{4.45}$$

far from the droplet surface. Here subscripts S and ∞ refer to the droplet surface and environment.

Assuming that the flow in the vicinity of the droplet is a potential, we express the velocity in terms of the potential as

$$\rho \mathbf{v} = \nabla \phi \tag{4.46}$$

where ϕ is the potential.

As was noted by Marberry et al. (1984), Eq. (4.46) is not valid in the space occupied by the droplets. Thus, a number of discontinuities exists in the concentration, temperature and flow fields. In order to circumvent such difficulties the modified Laplace equation is used

$$\nabla^2 \phi + \sum_{i=1}^{N} g_i \delta(x - x_i)\delta(y - y_i)\delta(z - z_i) = 0 \tag{4.47}$$

where g_i is a point source located at the center (x_i, y_i, z_i) of the i-th droplet. Its strength is determined by the conditions $m_i = \int_{S_i} (\rho \mathbf{v}) \cdot \mathbf{n} dS_i$ and $m_i q_e = \lambda \int_{S_i} (\nabla T) \cdot \mathbf{n} dS_i$, S_i is the surface of the i-th droplet, $\mathbf{n}$ is the outer unit normal vector at the droplet surface, q_e is the latent heat of evaporation, δ is the Dirac delta function.

The solution of Eqs. (4.43) and (4.46) reads

$$\beta_j = A_j + B_j \exp\left(\frac{\psi}{\rho D}\right) \tag{4.48}$$

$$\phi = \phi_\infty + \sum_{i=1}^{N} \frac{g_i}{r_i} \tag{4.49}$$

where $r_i = [(x - x_i)^2 + (y - y_i)^2 + (z - z_i)^2]^{1/2}$ is the radial position from the I-th droplet, and A_j, B_j and ϕ_∞ are constants.

The corresponding burning rate of each droplet in a system of two droplets is expressed by the following relations

$$m_1 = m_{iso}(a_1)\left\{\frac{C_1}{a_1} + \frac{1}{6}\left(\frac{C_2}{a_1}\right)\varphi(P_1)\right\} \tag{4.50}$$

$$m_2 = m_{iso}(a_2)\left\{\frac{C_2}{a_2} + \frac{1}{6}\left(\frac{C_1}{a_2}\right)\varphi(P_2)\right\} \tag{4.51}$$

where $m_{iso}(a_i)$ is the burning rate of an isolated droplet of radius a_i; $Pi = d/a_i$, $d = [(x_i - x_j)^2 + (y_i - y_j)^2 + (z_i - z_j)^2]^{1/2}$ is the center-to-center distance between droplets, $C_1 = a_1 \dfrac{P_1 - a_2/a_1}{P_1 - 1/P_2}$, $C_2 = a_2 \dfrac{P_2 - a_1/a_2}{P_2 - 1/P_1}$, and $\varphi(P_i) = \left[\dfrac{1}{(P_i - 1)^2} + \dfrac{1}{(P_i + 1)^2} + \dfrac{4}{(P_i^2 + 1)^{3/2}}\right]$.

The burning rate of droplets corresponding to several particular types of finite arrays (Fig. 4.10) of droplets of the same radius is expressed as

$$m = m_{iso}(a)\frac{g \cdot a}{3(\phi_s - \phi_\infty)}\sum_{i=1}^{N}\frac{\ell_{1.i}}{r_{1.i}} \tag{4.52}$$

where $\ell_{1,i} = |x_1 - x_i| + |y_1 - y_i| + |z_1 - z_i|$. The expressions for the strength of the point source g for droplets arranged in various arrays are: $g = (a\Delta\phi)/\psi(a/b)$, $\Delta\phi = \phi_s - \phi_\infty$, $\psi^{-1}(a/b) = 1 + 2(a/b)$ for triangular $\psi^{-1}(a/b) = 1 + 3(a/b)$ for tetrahedral, $\psi^{-1}(a/b) = \left[1 + \left(2 + 1/\sqrt{2}\right)(a/b)\right]$ for square and $\psi^{-1}(a/b) = \left[1 + \left(3 + (3/\sqrt{2}) + (1/\sqrt{3})\right)(a/b)\right]$ for cubical arrays.

The effect of the interparticle spacing on the burning rate of two droplets of different size is illustrated in Fig. 4.11, where the dependences of the correction factor $\eta = m/m_{iso}$ on the dimensionless distance between the droplets are presented. The curve corresponding to $a_{21} = 1$ presents the results for the interaction of two equal droplets and the correction factors η_1 and η_2 related to the larger and the smaller droplet. It is seen that the burning rate of both droplets decreases significantly when the droplets approach each other. The effect is more evident for the small droplet. The data on the effect of droplet interactions on the burning rate

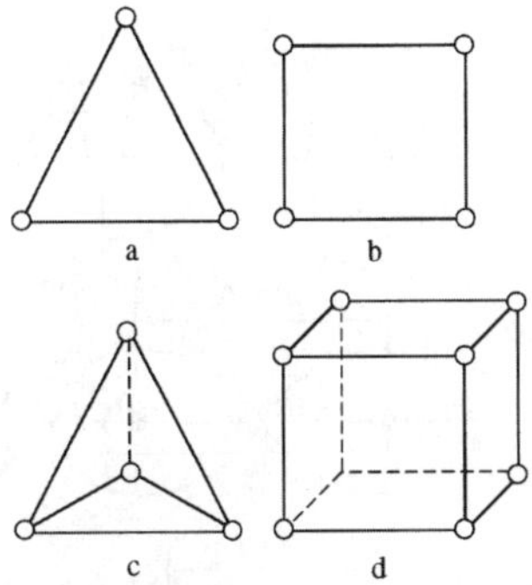

Fig. 4.10 Finite arrays of different types. a) Triangular array. b) Square array. c) Tetrahedral array. d) Cubical array

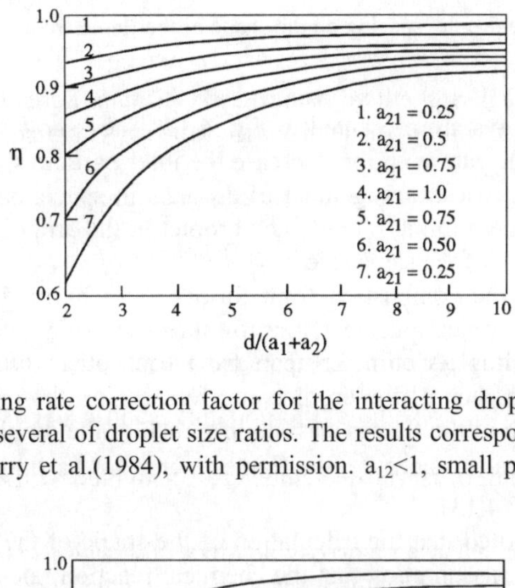

Fig. 4.11 The burning rate correction factor for the interacting droplets versus the inter droplet spacing for several of droplet size ratios. The results correspond to Eq. (4.52) Reprinted from Marberry et al.(1984), with permission. $a_{12}<1$, small particles; $a_{12}>1$, large particles

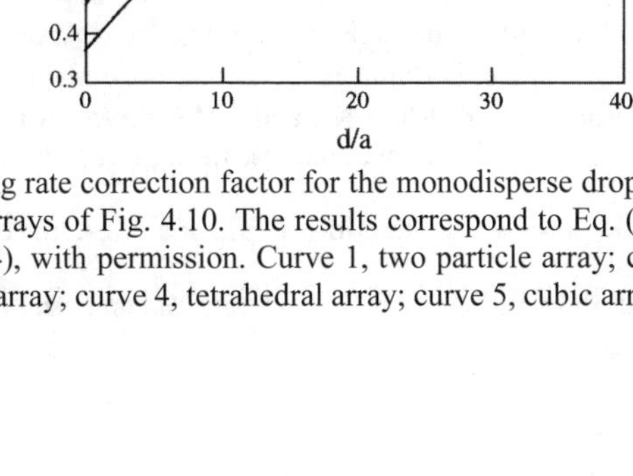

Fig. 4.12 The burning rate correction factor for the monodisperse droplets located in the vicinity of the finite arrays of Fig. 4.10. The results correspond to Eq. (4.52) Reprinted from Marberry et al.(1984), with permission. Curve 1, two particle array; curve 2, triangular array; curve 3, square array; curve 4, tetrahedral array; curve 5, cubic array

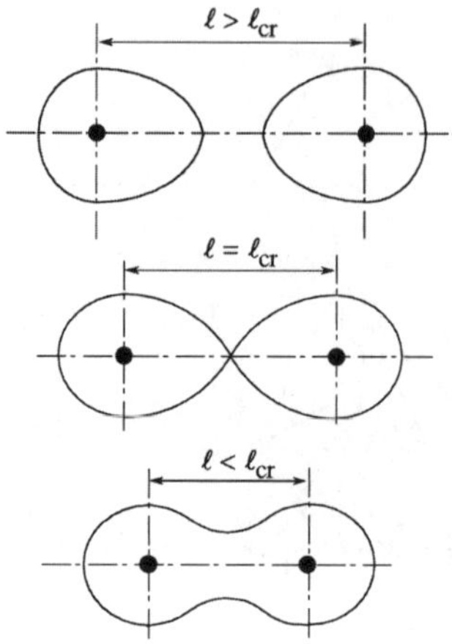

Fig. 4.13 Effect of the droplet spacing on the shape of the flame front

of several finite arrays are presented in Fig. 4.12. They show that the fastest decrease in the burning rate with the decrease the inter particle spacing corresponds to arrays with the densest arrangement of droplets in space, namely to the cubic array. In all cases the burning rate of each droplet in the array is smaller than that of an isolated droplet of the same size.

The shape of the combustion front formed by a burning droplet array depends on the interdroplet spacing (Brzustowski et al. 1979 and Umemura et al. 1981 a,b). If two droplets burn far apart from each other, two nearly spherical flames that envelop each of the droplets are formed. As the interdroplet spacing decreases, the shape of the flames surrounding each of the droplets changes markedly. At some critical distance ℓ_{cr} the flames merge into a single front that envelopes the array (Fig. 4.13).

It should be noted that the calculation of the shape of the flame front in arrays is based on the assumption that the chemical reaction rates exceed the transport rates. In this case combustion takes place within a thin reaction zone which can be presented as an infinitely thin flame front where fuel vapor and oxidizer fluxes meet in stoichiometric ratio, whereas the reagent concentrations equal zero.

References

Aminzadeh K, Taha TRAl, Cornish ARH, Kolansky MS, Pfeffer R (1974) Mass transport around two spheres at low Reynolds numbers. Int. J. Heat Mass Transfer 17: 1425–1436

Bard E (1968) The slow unsteady settling of a fluid sphere towards a flat fluid interface. Chem Eng. Sci. 23: 193–210

Boothroyd RG (1974) Following gas-solid suspensions. Chapman and Hall, London

Brzustowski TA, Twardus EM, Wojcicki S, Sobiesiak A (1979) Interaction of two burning fuel droplets of arbitrary size. AIAA J. 17: 1234–1242

Cooley MDA, O'Neill ME (1969) On the slow motion generated in a viscous fluid by the approach of a sphere to a plane wall or stationary sphere. Mathematika 16: 37–49

Crowe CT, Sommerfeld M, Tsuji Y (1997) Multiphase flows with droplets and particles. CPC Press, New York

Davis MH (1969) The slow translation and rotation of two unequal spheres in a viscous fluid. Chem. Eng. Sci. 24: 1769–1776

Faxen H (1927) Die geschwindigkeit zwen kugeln, die unter ein wirkung der schwere in einen zahen flussigkeit fallen. ZAMM 7: 79–80

Goldman AJ, Cox RG, Brenner H (1967) Slow viscous motion of a sphere parallel to a plane wall. Part I. Motion through a quiescent fluid Chem. Eng. Sci. 22: 637–651 Part II 22: 653

Gupalo YuP, Polyanin AD, Ryazantsev YuS (1985) Mass transfer reactive particles with flow. (in Russian) Nauka, Moscow

Haber S, Hetsroni G, Solan A (1973) On the low Reynolds number motion of two droplets. Int. J. Multiphase Flow 1: 57–71

Happel J, Brenner H (1983) Low reynolds number hydrodynamics. Martinus Nijhoff, The Hague

Hasimoto H (1959) On the periodic fundamental solutions of the Stokes equations and their application to viscous flow past a cubic array of spheres. J. Fluid Mech. 5: 317–328

Hetsroni G, Haber S (1970) The flow fields in and around a droplet or bubble submerged in an unbounded arbitrary velocity field. Rheol. Acta 9: 488–496

Hetsroni G, Haber S (1971) Low Reynolds number motion of raindrops in the atmosphere. Department of Mechanical Engineering, Technion, Haifa

Hetsroni G, Haber S (1978) Low Reynolds number motion of two drops submerged in an unbounded arbitrary velocity field. Int. J. Multiphase Flow 4: 1–17

Hetsroni G, Haber S, Wacholder E (1970) The flow fields in an around a droplet moving axially within a tube. J. Fluid Mech. 41: 689–705

Jeffery G B (1915) On the steady rotation of a solid of revolution in a viscous field. Proc. London Math. Soc. (Ser 2) 14: 327–338

Kim I, Elghobashi S, Siriganano WA (1993) Three-dimensional flow over two spheres placed side by side. J Fluid Mech. 246: 465–488

Kim S (1987) Stokes flow past three spheres: an analytic solution. Phys. Fluids 30: 2309–2314

Kim S, Karrila SJ (1991) Microhydrodynamics: principles and selected applications. Butterworth – Heinemann, Boston Mass

Labowsky M (1978) A formalism for calculating the evaporation rates of rapidly evaporating interacting particles. Combust. Sci. Technol. 18: 145–151

Labowsky M (1980) Calculation of burning rates of interacting fuel droplets. Combust. Sci. Technol. 22: 217–226

Lee KC (1979) Aerodynamic interaction between two spheres at Reynolds numbers about 10^4. Aero. Q. 30: 371–385

Liang S-C, Hong T, Fan L-S (1996) Effects of particle arrangements on the drag force of a particle in intermediate flow regime. Int. J. Multiphase Flow 22: 285–306

Marberry M, Ray AK, Leung K (1984) Effect of multiple particle interactions on burning droplets. Combust. Flame 57: 237–245

Miyasaka K, Law CK, (1981) Combustion of strongly – interacting linear droplet arrays. The Eighteenth Symposium (International) on Combustion. The Combustion Institute, Pittsburgh, Pa., pp. 283–292

Morrison FAJr, Reed LD (1978) The slow motion of two touching fluid spheres along their line of centers: addendum. Int. J. Multiphase Flow 4: 433–434
Mulholland JA, Srivastava RK, Wendt JOL (1988) Influence of droplet spacing on drag coefficient in nonevaporating, monodisperse streams. AIAA J. 26: 1231–1237
Ramachandran RS, Wang T-Y, Kleinstreuer C, Chiang H (1991) Laminar flow past three closely spaced monodisperce spheres or nonvaporating droplets. AIAA. J. 29: 43–51
Reed L, Morrison FAJr (1974) The slow motion of two touching fluid spheres along their line of centers. Int. J. Multiphase Flow 1: 573–584
Rushton E, Davies G A (1978) The slow motion of two spherical particles along their line of centers. Int. J. Multiphase Flow 4: 357–381
Sadhal SS, Ayygawamy PS, Chung JN (1997) Transport phenomena with drops and bubbles. Springer, New York
Sangiovanni JJ, Labowsky M (1982) Burning times of linear fuel droplet arrays: a comparison of experiment and theory. Combust. Flame 47: 15–30
Siriganano WA (1999) Fluid dynamics and transport of droplets and sprays. Cambridge University Press, Cambridge
Soo SL (1998) Multiphase fluid dynamics. Science Press and Gower Technical, Bejing
Stimson M, Jeffery GB (1926) The motion of two spheres in viscous fluid. Proc. R. Soc. London, Ser. A 111: 110–116
Subramanian RS, Balasubramaniam R (2001) The motion of bubbles and drops in reduced gravity. Cambridge University Press, Cambridge
Tsuji Y, Morikawa Y, Terashima K (1982) Fluid - dynamic interaction between two spheres. Int. J. Multiphase Flow 8: 71–82
Tsuji Y, Morikawa Y, Fujiwara Y (1985) Pipe flow with solid particles fixed in space. Int. J. Multiphase Flow 11: 177–188
Umemura A, Ogawa S, Oshima N (1981a) Analysis of the interaction between two burning fuel droplets with different sizes. Combust. Flame 43: 111–119
Umemura A, Ogawa S, Oshma N (1981b) Analysis of the interaction between two burning droplets. Combust. Flame 41: 45–55
Wacholder E, Weihs D (1972) Slow motion of a fluid sphere in the vicinity of another sphere or a plane boundary. Chem. Eng. Sci. 27: 1817–1828
Yarin AL (1987). Collective effect in disperse systems. J. Exp. Theor. Phys. 93: 1256–1259 (in Russian)
Yarin LP, Sukhov GS (1987) Fundamentals of combustion theory of two-phase media. (in Russian) Energoatomizdat, Leningrad
Zhu C, Liang S-C, Fan L-S (1994) Particle wake effects on the drag force of an interactive particle. Int. J. Multiphase Flow 20: 117–129
Zick AA, Homsy GM (1982) Stokes flow through periodic arrays of spheres. J. Fluid Mech. 115: 13–26
Zinchenko AZ (1979) Calculation of hydrodynamic interaction between drops at low Reynolds number. Appl. Math. Mech. 42: 1046–1051
Zinchenko AZ (1981) The slow asymmetric motion of two drops in a viscous medium. Appl. Math. Mech. 44: 30–37
Zinchenko AZ (1982) Calculation of close interaction between drops, with internal circulation and slip effect taken into account. Appl. Math. Mech. 45: 564–567
Zinchenko AZ (1984) Hydrodynamic interaction of two identical spheres in linear flow field. Appl. Math. Mech. 47: 37–43

1.5 Particle–turbulence interaction

1.5.1 Models of interaction

Particle–turbulence interaction is of major importance in understanding the mechanism of turbulence generation in two-phase flows. This problem is also relevant in the development of calculation methods for equipment using multiphase mixtures, such as industrial furnaces, combustion chambers of gas turbines and jet engines, etc. During the last decade this problem motivated a number of investigations covering various aspects of the phenomenon (Owen 1969, Hinze 1972, Hetsroni 1989, Gore and Crowe 1989, Rashidi et al. 1990a, 1990b, Yuan and Michaelides 1992). The following situations have been studied experimentally: particle-laden turbulent jets (Hetsroni and Sokolov 1971, Laats and Frishman 1973, Levy and Lockwood 1981, Shuen et al. 1985, Berlow and Morrison 1990, Sheen et al. 1994, and Parthasarathy and Faeth 1987); two-phase pipe flows (Zisselmar and Molerus 1979, Maeda et al. 1980, Kulick et al. 1993, Tsuji and Morikawa 1982, Tsuji et al. 1984, Alajbegovich et al. 1994, Nouri et al. 1987, Savolainen and Karvinen 1998, and Varaksin et al. 1998); two-phase boundary layers: hydrodynamics and heat transfer (Rashidi et al. 1990b, Hetsroni and Rozenblit 1994, and Hetsroni et al. 1995, 1997), and two-phase homogeneous turbulence (Parthasarathy and Faeth 1990, and Mizukami et al. 1992).

In the early investigations on particle-laden turbulent jets, it was discovered that the presence of solid particles leads to modulation of the turbulence of the carrier fluid (Hetsroni and Sokolov 1971, and Laats and Frishman 1973). This effect is manifested not only in the turbulence intensity, but also in the energy spectra—mostly in reduction of the high-frequency component and a corresponding change in the distribution of the fluctuation energy. Subsequent measurements (Shuen et al. 1985, Parthasarathy and Faeth 1987, 1990, Tsuji et al. 1988, and Sheen et al. 1994), carried out under wide variation of the particle size, phase density ratio, total mass content etc., corroborated these results. In parallel, new effects reflecting specific features of the particle-turbulence interaction have been discovered. It was shown (Tsuji and Morikawa 1982, Tsuji et al. 1984, and Lee and Durst 1982) that turbulence intensity of the carrier fluid in particle-laden flows with coarse particles does not always decrease: it may also increase. Such a phenomenon is of general interest and may be observed in various types of particle-laden flows: jets, pipe flows, flows in boundary layers and in homogeneous turbulence. It is significant that the turbulence intensity of the carrier fluid in a two-phase mixture may be several times higher than in a single-phase fluid.

The above characteristics of particle-laden flows indicates that the mechanism of turbulence modulation in two-phase turbulent flows is highly complicated. It depends on numerous factors, associated with generation and dissipation of the turbulent energy by the particles (via particle acceleration, vortex shedding, transfer from mean to fluctuating flow, etc.). Depending on the flow velocity, particle sizes and concentrations, and phase density ratios different flow regimes may be realized. When dissipation of the kinetic energy is predominant, the particle–turbulence interaction leads to suppression of the turbulence, but when generation is predominant, the turbulence augmentation takes place. In both cases a heavy admixture affects the mean and fluctuative flow structures, as well as the rates of the mass, momentum and heat transfer.

A number of investigations dealt with the theoretical description of the particle–turbulence interaction (Abramovich 1970, Danon et al. 1974, Elghobashi and Abou-Arab 1983, Shuen et al. 1985, Abou-Arab and Roco 1988, and Shraiber et al. 1990). Their approach was based on the application of the extended single-phase models of turbulence, of the mixing length theory of Prandtl (1925), and of the $\kappa - \varepsilon$ model (Crowe et al. 1996). Several theoretical works contain the results of direct numerical simulations of the phenomenon (Squires and Eaton 1990, Elghobashi and Truesdell 1992, 1993, Pan and Banerjee 1997, Boivin et al. 1998, and Sundaram and Collins 1999). In parallel, simplified models of particle/turbulence interaction were proposed (Owen 1969, Hetsroni, 1989, Yuan and Michaelides 1992, Sato and Hishida 1996, Kenning and Crowe 1977, and Crowe 2000).

One of the first attempts to describe particle-turbulence interaction is due to Abramovich (1970), who used Prandtl's mixing length theory and derived a simple expression for the dependence of the velocity fluctuation intensity on the carrier fluid, the particles' physical properties, and the flow regime parameters. This theory is based on the notion of mixing length as the transverse distance which an element of fluid has to cover at its original mean velocity, so that the change in the latter would equal the mean transverse fluctuation. Although this idea is quite old, it shows remarkable insight into the physics of turbulent flows. It seems that Prandtl's mixing-length theory modeled coherent structures long before they were actually discovered.

Over the last 20 years the above approach was successfully used for solving turbulent two-phase flow problems, preferrably from "simple" flows to more complicated ones, including the effects of gravity (Abramovich et al. 1984, Girshovich and Leonov 1979), the average velocity difference of particles and carrier fluid (Frishman 1979) and particle diffusion (Abramovich and Girshovich 1973, Yarin and Hetsroni 1994 a, 1994 b) used the modified mixing length theory for calculating the effect of the particle size distribution on the turbulence of the carrier fluid and particles, and estimating the level of temperature fluctuations in two-phase monodisperse and polydisperse mixtures. It should be noted that Abramovich's model does not hold for enchanced turbulence and, naturally, is valid only for flows laden with fine particles. In this case, fairly good agreement was achieved between the theoretical and experimental data.

Several theoretical models were developed for turbulent flows laden with coarse particles. The turbulence modulation was investigated by Al-Tawell and Landau (1977), Parthasarathy and Faeth (1990), and Yuan and Michaelides (1992); the last researchers were using the turbulent kinetic energy balance for a simple solution. Gore and Crowe (1989) sought to generalize experimental data on particle-turbulence interaction and suggested the ratio of particle size to the characteristic turbulence scale as a basic criterion. Analysis of the data for different types of particle-laden flows showed that there exists a fairly distinct particle size/turbulence scale ratio below which particles suppress the turbulence and above which they enchance it (Gore and Crowe 1989, Crowe 2000). The boundary between these regimes corresponds to a ratio of approximately 0.1. Hetsroni (1989) suggested that turbulence in coarse-particle laden two-phase flows is due to the vortex-shedding phenomenon which manifests itself at $Re_p > 110$ (Achenbach 1974). Such a model assumes that vortices are indeed the cause for intensification of the turbulence through energy transfer from the mean to the fluctuating motion by the particle wakes. With the particle Reynolds number as the determining criterion, the effects due to viscosity of the carrier fluid and to particle inertia can be accounted for. Yarin and Hetsroni (1994 c) developed a theory of particle-turbulence interaction in a dilute two-phase flow with arbitrary particle sizes. This theory takes into account two sources of turbulence in particle-laden flows: (i) the carrier fluid velocity gradients, (ii) the turbulent wakes behind the coarse particles. The theoretical description of the process is based on the modified mixing-length theory and the turbulent kinetic energy balance method. The Yarin–Hetsroni theory permits prediction of the fluctuation intensity in various types of turbulent particle-laden flows. It agrees fairly well with known experimental data on turbulence modulation in two-phase flows laden with coarse particles (see Section 1.5.3).

1.5.2 The effect of particle-size distribution on the turbulence of the carrier fluid

Governing equations. The effect of polydisperse particles on the turbulence of the carrier fluid is naturally of practical importance, since certain processes can be controlled by varying the particle sizes and proportions. The analysis given below is based on the modified mixing-length theory (Yarin and Hetsroni 1994 a).

Consider a steady, incompressible turbulent flow laden with polydisperse particles. The particle diameters $d_1, d_2, \ldots d_n$, are relatively small, such that the effect of turbulent wakes can be neglected, i.e. the particle Reynolds number $Re_p <$ 110. The particles are contained in a fluid element (we use the terminology adopted in Schlichting (1979): fluid particle or fluid element) as an independent structure, i.e. the phases coexist together during its lifetime but move at different velocities depending on their mass, on the fluid velocity etc. The number of particles, of each diameter, in that fluid element are $k_1, k_2 \ldots, k_n$, respectively.

Further we shall consider flows where transverse pressure gradients are absent (boundary layers, pipe flows). We also restrict that analysis to flows with slightly changing longitudinal velocity, where the relative average velocity of the carrier fluid and the particles is negligibly small (developed turbulent flow in a jet etc.).

Let the mass of the particles in the i-th fraction be $M_{pi} = m_i k_i$, where m_i is the mass of a single particle. The mass of all the particles in the fluid element is $M_p = \sum_{i=1}^{n} m_i k_i$, and it is assumed to be constant over the fluid element lifetime, i.e. the particles do not leave the fluid element as long as it persists as an entity. This assumption is fairly reasonable for a number of real two-phase flows (Yarin and Hetsroni 1994 a).

In accordance with Prandtl's fundamental idea of mixing length, we assume that the fluid element acquires at time t = 0, momentum J_o under the influence of a hydrodynamic field disturbance (pressure fluctuation). Further, it moves by inertia and interacts with the particles inside it. The carrier fluid and particles are then described by the following equations of conservation of momentum

$$d\left[v' + \sum_{i=1}^{n} \gamma_i v_p'^{(i)}\right] = 0 \tag{5.1}$$

and

$$m_i \frac{dv_p'^{(i)}}{dt} = \frac{1}{2} C_d^{(i)} f_i \rho \tilde{v}_i \left|\tilde{v}_i\right| \tag{5.2}$$

where v' is the fluctuating velocity of the carrier fluid and $v_p'^{(i)}$ are those of the particles in fraction i (namely, those with diameter d_i) M is the mass of the fluid in the element; $\gamma = M_p/M$ is the overall mass content of the particles in the element and, $\gamma_i = M_{pi}/M$ are those of the particles of fraction i, such that $\gamma = \sum_{\varepsilon=1}^{n} \gamma_i$; $C_d^{(i)}$ is the drag coefficient of the particles in fraction i, f_i is the cross-section area of the particles of fraction i, $\tilde{v}^{(i)} = v' - v_p'^{(i)}$ are the relative velocities between the carrier fluid and the particles of fraction i, and ρ is the density of the fluid.

Note that on the left-hand side of Eq. (5.1) we omitted the term representing the drag force exerted by the surrounding fluid on the given element, which means that we assume the latter to move in an inviscid fluid. It is known that this approximation is fairly satisfactory for relatively large eddies, where the drag/inertia force ratio is small and the element is much larger than a particle. Note also that on the left-hand side of Eq. (5.2) we omitted the terms accounting for the unsteady character of the particle motion in the element (the inertial force due to virtual mass, the Basset force, etc.; these effects are negligibly small in particle-laden gas flows—Boothroyd (1971), and Nigmatulin (1991)). For example, in a submerged particle-laden jet with the parameters $\gamma = 0.1;\ d = 5\cdot 10^{-5}\,\text{m};\ \bar{\rho} = 8\cdot 10^{2}$ ($\bar{\rho} = \rho_p/\rho$ being the phase density ratio, ρ_p the particle density), $\nu = 10^{-5}\ [\text{m}^2/\text{s}]$ and $v_0' = 1\,[\text{m/s}]$ (v_0' being the initial fluctuating velocity of carrier fluid), the ratio of the pressure to the drag force, the force due to the virtual mass to the drag force, the Basset force to the drag force and the gravity force to the drag force are 10^{-2}, 10^{-2}, 10^{-1} and 10^{-1}, respectively. Thus, Eq. (5.2) is merely a statement that the change of particle momentum is due to the drag force of the fluid. It is emphasized that the approach used to describe the fluctuating velocities of the particles and fluid does not omit pressure effects: they are accounted for by the initial momentum of the fluid element, $J_0 = M \cdot v'_0$, which determines the characteristic scales of the fluid and particle velocities.

The initial conditions for Eqs. (5.1) and (5.2) are

$$t = 0 \quad v' = v'_0, \quad v'^{(i)}_p = 0 \tag{5.3}$$

They correspond to the physical mechanism of the fluctuating motion in turbulent two-phase flows. First, the carrier fluid fluctuation begins, then the momentum is transmitted to the particles by the drag. Thus, the cause of the fluctuating motion of the particles is that of the fluid. Therefore, under the conditions of finite propagation speed of perturbations and finite relaxation times (which is the case), the fluctuating particle velocity at $t = 0$ (the moment of eddy onset) should be zero. It is emphasized that this assumption leads to physically reasonable results, as well as to fairly good agreement with experimental data (Abramovich et al. 1984).

Integrating Eq. (5.1) we obtain, in dimensionless form, the following expression

$$\overline{v}' + \sum_{i=1}^{n} \gamma_i \overline{v}'^{(i)}_p = 1 \tag{5.4}$$

where $\overline{v}' = v'/v'_0$ and $\overline{v}'^{(i)}_p = v'^{(i)}_p / v'_0$, are the fluctuating velocities normalized against the initial velocity v'_0.

Equation (5.4) can also be formulated as

$$\overline{v}' + \sum_{i=1}^{n} [\gamma_i (\overline{v}'^{(i)}_p - \overline{v}') + \gamma_i \overline{v}'] = 1 \tag{5.5}$$

or

$$\overline{v}' = (1+\gamma)^{-1} \left[1 + \sum_{i=1}^{n} \tilde{v}'^{(i)}_p \right] \tag{5.6}$$

Eq. (5.6) states that the fluctuating velocity of the carrier fluid (normalized against its initial value) is determined by the total mass content of the particles γ, and by the relative velocities of the particles of various fractions $\tilde{v}_p^{(i)}$, the former represent the inertial characteristics of the admixture, and the latter the viscous interaction (drag) between the particles and the carrier fluid.

One can examine an extreme case—say, the case when the particles are very fine ($d \to 0$) and the relative velocities negligible, $\tilde{v}_p^{(i)} = 0$, which leads to the minimal fluctuating velocities of the carrier fluid

$$\overline{v}' = (1+\gamma)^{-1} \tag{5.7}$$

Assume some form of a drag force, say the Stokes law, $C_d^{(i)} = 24/Re_i$, where the Reynolds number is defined based on the relative velocity

$$Re_i = \frac{\tilde{v}_p^{(i)} d_i}{\nu} \tag{5.8}$$

We rewrite Eq. (5.2) as follows

$$\frac{dv_p'^{(i)}}{dt} = \frac{1}{2}\rho \frac{24}{\tilde{v}_p^{(i)} d_i} \nu \tilde{v}_p^{(i)} \left|\tilde{v}_p^{(i)}\right| \frac{f_i}{m_i}$$

or simply

$$\frac{dv_p'^{(i)}}{dt} = \frac{18\mu\tilde{v}^{(i)}}{d_i^2 \rho_i} \tag{5.9}$$

where ρ_i is the particle density of the i-th fraction.

Combining Eq. (5.4) and Eq. (5.9), we obtain

$$\frac{d\overline{v}_p'^{(i)}}{dt} = \tau_i^{-1}\left[1 - \sum_{i=1}^{n} \gamma_i \overline{v}_p'^{(i)} - \overline{v}_p'^{(i)}\right] \tag{5.10}$$

where the relaxation time is defined as

$$\tau_i = \frac{d_i^2 \rho_i}{18\mu} \tag{5.11}$$

The set of equations (5.4) and (5.10) with the initial condition (5.3) describes the turbulent fluctuations of the particles of various fractions, and of the carrier fluid.

Bidisperse admixture. Consider a bidisperse mixture, i.e. a carrier fluid with particles of two sizes in it. Then Eqs. (5.4) and (5.10) read

$$\overline{v}' + \gamma_1 \overline{v}_1' + \gamma_2 \overline{v}_2' = 1 \tag{5.12}$$

$$\frac{d\overline{v}_1'}{dt} = \tau_1^{-1}\left[1 - (1+\gamma_1)\overline{v}_1' - \gamma_2 \overline{v}_2'\right] \tag{5.13}$$

and

$$\frac{d\overline{v}_2'}{dt} = \tau_2^{-1}\left[1 - \gamma_1 \overline{v}_1' - (1-\gamma_2)\overline{v}_2'\right] \tag{5.14}$$

where $\overline{v}_1' = \overline{v}_p'^{(1)}$, $\overline{v}_2' = \overline{v}_p'^{(2)}$ and superscripts 1 and 2 refer to the large and small particles, respectively.

The initial condition (5.3) reduces to

$$t = 0 \quad \overline{v}' = 1, \quad \overline{v}_1' = \overline{v}_2' = 0 \tag{5.15}$$

One can examine some general properties of the solutions to the set of Eqs. (5.12)–(5.14) with conditions (5.15). Eliminating time from Eqs. (5.13) and (5.14), we obtain

$$\frac{d\overline{v}_1'}{d\overline{v}_2'} = \frac{a\overline{v}_2' + b\overline{v}_1' + c_*}{\alpha\overline{v}_2' + \beta\overline{v}_1' + c_{**}} \tag{5.16}$$

where $a = -\gamma_2\omega$, $b = -(1+\gamma_1)\omega$, $c_* = \omega$, $\alpha = -(1+\gamma_2)$, $\beta = -\gamma_1$, $c_{**} = 1$, $\omega = d_2^2/d_1^2$ $(0 < \omega \le 1)$, and $\omega = 1$ corresponds to flow with monodisperse particles.

Let $\overline{v}_2' = x + \varepsilon_2$, $\overline{v}_1' = y + \varepsilon_1$; ε_1 and ε_2 be arbitrary parameters. Then the numerator and denominator on the right-hand side of Eq. (5.16) read

$$a\overline{v}_2' + b\overline{v}_1' + c_* = ax + by + (a\varepsilon_2 + b\varepsilon_1 + c_*) \tag{5.17}$$

and

$$\alpha \overline{v}_2' + \beta \overline{v}_1' + c_{**} = \alpha x + \beta y + (\alpha\varepsilon_2 + \beta\varepsilon_1 + c_{**}) \tag{5.18}$$

Since the values of ε_1 and ε_2 are arbitrary, one can choose them to satisfy the equations $(a\varepsilon_2 + b\varepsilon_1 + c_*) = 0$ and $(\alpha\varepsilon_2 + \beta\varepsilon_1 + c_*) = 0$. Thus,

$$\varepsilon_1 = \frac{a - \alpha\omega}{\alpha b - a\beta} = \frac{1}{1+\gamma}, \qquad \varepsilon_2 = \frac{b - \beta\omega}{a\beta - \alpha b} = \frac{1}{1+\gamma} \tag{5.19}$$

Taking into account Eqs. (5.17) and (5.18), we write Eq. (5.16) in the following form

$$\frac{dy}{dx} = \frac{ax + by}{\alpha x + \beta y} \tag{5.20}$$

Defining a new variable $z = \frac{y}{x}$ $\left(\frac{dy}{dx} = x\frac{dz}{dx} + z\right)$, we obtain

$$x\frac{dz}{dx} + z = \frac{a + bz}{\alpha + \beta z} \tag{5.21}$$

Separating the variables in Eq. (5.21), we have

$$\frac{\alpha + \beta z}{a + (b-\alpha)z - \beta z^2}dz = \frac{dx}{x} \tag{5.22}$$

Since the expression in the folllowing equation is negative

$$\Delta = -4\beta a - (b-\alpha)^2 = -4\gamma_1\gamma_2\omega - \{(1+\gamma_1)\omega + (1+\gamma_2)\}^2 < 0 \tag{5.23}$$

integration of Eq. (5.22) yields

$$\frac{\alpha}{\sqrt{-\Delta}}\ln\frac{-2\beta z + (b-\alpha) - \sqrt{-\Delta}}{-2\beta z + (b-\alpha) + \sqrt{-\Delta}} + \beta\left\{-\frac{1}{2\beta}\ln[a + (b-\alpha)z - \beta z^2] + \right. \tag{5.24}$$

$$\left. + \frac{b-\alpha}{2\beta}\frac{1}{\sqrt{-\Delta}}\ln\frac{-2\beta z + (b-\alpha) - \sqrt{-\Delta}}{2\beta z + (b-\alpha) + \sqrt{-\Delta}}\right\} = \ln x + \ln c$$

or in the rearranged form

$$\frac{b+\alpha}{2\sqrt{-\Delta}}\ln\frac{-2\beta z+(b-\alpha)-\sqrt{-\Delta}}{-2\beta z+(b-\alpha)+\sqrt{-\Delta}}-\frac{1}{2}\ln[a+(b-\alpha)z-\beta z^2]=\ln x+\ln c \tag{5.25}$$

Equation (5.25) in turn yields

$$x[a+(b-\alpha)z-\beta z^2]^{\frac{1}{2}}\left[\frac{-2\beta z+(b-\alpha)-\sqrt{\xi}}{-2\beta z+(b-\alpha)+\sqrt{\xi}}\right]^{-\frac{(b+\alpha)}{2\sqrt{\xi}}}=c \tag{5.26}$$

where $\xi=-\Delta$, $x=\overline{v}_2'-(1+\gamma)^{-1}$, $y=\overline{v}_1'-\overline{v}(1+\gamma)^{-1}$, $z=y/x$ $(1\le z<\infty)$, $z=1$ (corresponds to flow of monodisperse admixture), and c is a constant.

Using the initial conditions $t=0$, $v_1'=0$ and $v_2'=0$, we find

$$c=-\sqrt{\frac{1-\omega}{1+\gamma}}\left\{\frac{2\gamma_1-(1+\gamma_1)\omega+(1+\gamma_2)-\sqrt{\xi}}{2\gamma_1-(1+\gamma_1)\omega+(1+\gamma_2)+\sqrt{\xi}}\right\}^{-\frac{(b+\alpha)}{2\sqrt{\xi}}} \tag{5.27}$$

There are interesting cases which can be considered, resulting in significant simplifications of Eq. (5.26). It can be solved numerically for any case, but here we shall confine ourselves to an example, demonstrating the plausibility of the solution. Consider an admixture with bidisperse particles, such that d_1>>d_2 (i.e. $\omega\ll 1$, $z\gg 1$); then Eq. (5.26) reduces to

$$\overline{v}_1'=(1+\gamma)^{-1}+c\gamma_1^{\frac{1}{2}} \tag{5.28}$$

and

$$\overline{v}_2'=(1+\gamma)^{-1} \tag{5.29}$$

These expressions, together with Eq. (5.12), enable us to calculate the fluctuating velocity of the carrier fluid in this particular case

$$\overline{v}'=1-\gamma(1+\gamma)^{-1}-c\gamma_1^{\frac{1}{2}} \tag{5.30}$$

where c, defined above, can also be simplified if $\gamma_1 < \gamma_2$. In this case we have

$$c \approx -\gamma_1^{\frac{1}{2}}(1+\gamma)^{-1} \tag{5.31}$$

Therefore,

$$\overline{v}_1' \approx 0, \quad \overline{v}_2' = (1+\gamma), \quad \overline{v}' = (1+\gamma)^{-1}(1+\gamma_1) \tag{5.32}$$

This result implies that the fluctuation intensity of the small particles (subscript 2) is independent of the mass content of the large particles (subscript 1). It also shows that, for a fixed total mass content of the admixture γ, the fluctuation intensity of the carrier fluid is proportional to the mass content of the large particles γ_1, as was shown experimentally to be qualitatively correct.

Equation (5.26) interrelates the intensity of the velocity fluctuations of particles of various sizes, but does not actually yield the absolute values of these fluctuations. In order to obtain them, Eqs. (5.13) and (5.14) have to be integrated. They may be presented in the form of the following vector equation

$$\frac{d\mathbf{Y}}{d\tau} = A\mathbf{Y} + \mathbf{F} \tag{5.33}$$

where $\mathbf{Y} = \begin{pmatrix} y_1 \\ y_2 \end{pmatrix}; \quad \mathbf{F} = \begin{pmatrix} f_1 \\ f_2 \end{pmatrix}; \quad A = \begin{pmatrix} a_{11} & a_{12} \\ a_{21} & a_{22} \end{pmatrix};$

$$y_1 = \overline{v}_1', \quad y_2 = \overline{v}_2', \quad a_{11} = -(1+\gamma_1)\tau_1^{-1}, \quad a_{12} = -\gamma_2\tau_1^{-1},$$
$$a_{21} = -\gamma_1\tau_2^{-1}, \quad a_{22} = -(1+\gamma_2)\tau_2^{-1}, \quad f_1 = \tau_1^{-1}, \quad f_2 = \tau_2^{-1},$$
$$\tau_1 = \frac{d_1^2\rho_2}{18\mu} \quad \text{and} \quad \tau_2 = \frac{d_2^2\rho_2}{18\mu}.$$

First of all let us find a particular solution of Eq. (5.33). Let us assume that the solution $\mathbf{Y} = \mathbf{Y_p}$ does not depend on time τ. Thus we obtain

$$A\mathbf{Y_p} + \mathbf{F} = 0 \tag{5.34}$$

or

$$\begin{pmatrix} a_{11} & a_{12} \\ a_{21} & a_{22} \end{pmatrix} \begin{pmatrix} y_{p1} \\ y_{p2} \end{pmatrix} + \begin{pmatrix} f_1 \\ f_2 \end{pmatrix} = 0 \tag{5.35}$$

Equation (5.35) may be rearranged in the form

$$a_{11} y_{p1} + a_{12} y_{p2} = -f_1, \qquad a_{21} y_{p1} + a_{22} y_{p2} = -f_2 \tag{5.36}$$

which yields

$$y_{p1} = \frac{-f_1 a_{22} + f_2 a_{12}}{a_{11} a_{22} - a_{21} a_{11}} \tag{5.37}$$

and

$$y_{p2} = \frac{f_2 a_{11} + f_1 a_{21}}{a_{11} a_{22} - a_{21} a_{12}} \tag{5.38}$$

We see that the denominator

$$a_{11} a_{22} - a_{21} a_{12} = \frac{1+\gamma}{\tau_1 \tau_2} \neq 0 \tag{5.39}$$

Now let us seek the solution $\mathbf{Y} = \mathbf{Y}_h$ of the homogeneous equation

$$\frac{d\mathbf{Y}}{d\tau} = a\mathbf{Y} \tag{5.40}$$

Representing the solution of Eq. (5.40) in the form

$$\mathbf{Y}_h = \mathbf{c} e^{p\tau}, \quad \mathbf{c} = \begin{pmatrix} c_1 \\ c_2 \end{pmatrix} \tag{5.41}$$

we obtain

$$p\mathbf{Y_h} = A\mathbf{Y_h} \tag{5.42}$$

which is equivalent to

$$py_{h1} = a_{11}y_{h1} + a_{12}y_{h2}, \quad py_{h1} = a_{21}y_{h1} + a_{22}y_{h2}, \tag{5.43}$$

and

$$(a_{11} - p)y_{h1} + a_{12}y_{h2} = 0, \quad a_{11}y_{h1} + (a_{22} - p)y_{h2} = 0 \tag{5.44}$$

To obtain a nontrivial solution, we need to satisfy the following condition

$$\begin{vmatrix} a_{11} - p & a_{12} \\ a_{21} & a_{22} - p \end{vmatrix} = 0 \tag{5.45}$$

or

$$a_{11}a_{22} - p(a_{22} + a_{11}) + p^2 - a_{12}a_{21} = 0 \tag{5.46}$$

We find p from Eq. (5.46) in the following form

$$p_{12} = \frac{a_{11} + a_{22}}{2} \pm \sqrt{\frac{(a_{11} + a_{22})^2}{4} + a_{11}a_{21} - a_{11}a_{22}} \tag{5.47}$$

Then

$$\begin{aligned} y_{h1}^{(1)} &= c_1^{(1)} \exp(p_1\tau), \quad y_{h2}^{(1)} = c_2^{(1)} \exp(p_1\tau) \\ y_{h1}^{(2)} &= c_1^{(2)} \exp(p_2\tau), \quad y_{h2}^{(2)} = c_2^{(2)} \exp(p_2\tau) \end{aligned} \tag{5.48}$$

Substituting Eqs. (5.48) in Eqs. (5.43) we obtain

$$c_1^{(1)} = -\frac{a_{22} - p_1}{a_{21}} c_2^{(1)}, \quad c_1^{(2)} = -\frac{a_{22} - p_2}{a_{21}} c_2^{(2)} \tag{5.49}$$

Therefore, the solution of Eq. (5.40) is as follows:

$$\mathbf{Y}_{h_1} = \begin{bmatrix} -\dfrac{(a_{22}-p_1)}{a_{21}} \cdot c_2^{(1)} \\ c_2^{(1)} \end{bmatrix} \exp(p_1\tau); \quad \mathbf{Y}_{h_2} = \begin{bmatrix} -\dfrac{(a_{22}-p_2)}{c_2^{(2)}} c_2^{(2)} \end{bmatrix} \exp(p_2\tau) \tag{5.50}$$

Hence

$$Y_h = A\begin{bmatrix} \dfrac{-(a_{22}-p_1)}{a_{21}}\exp(p_1\tau) \\ \exp(p_1\tau) \end{bmatrix} + B\begin{bmatrix} \dfrac{-(a_{22}-p_2)}{a_{21}}\exp(\rho_2\tau) \\ \exp(p_2\tau) \end{bmatrix} \tag{5.52}$$

where $A = c_2^{(1)}$ and $B = c_2^{(2)}$.

Thus, the solution of Eq. (5.33) may be obtained by using Eqs. (5.37), (5.38) and (5.51) in the form

$$y_1 = -A\frac{(a_{22}-p_1)}{a_{21}}\exp(\rho_1\tau) - B\frac{(a_{22}-p_2)}{a_{21}}\exp(p_2\tau) + \frac{-f_1a_{22}+f_2a_{12}}{a_{11}a_{22}-a_{21}a_{12}} \tag{5.52}$$

and

$$y_2 = A\exp(p_1\tau) + B\exp(p_2\tau) + \frac{-a_{11}f_2 + a_{21}f_1}{a_{11}a_{22}-a_{21}a_{12}} \tag{5.53}$$

Using the initial condition $\tau = 0$ and $y_1 = y_2 = 0$, we obtain the equations for the constants A and B

$$-A\frac{a_{22}-p_1}{a_{21}} - B\frac{a_{22}-p_2}{a_{21}} + \frac{-f_1a_{22}+f_2a_{12}}{a_{11}a_{22}-a_{21}a_{12}} = 0 \tag{5.54}$$

$$A + B + \frac{-a_{11}f_2 + a_{22}f_1}{a_{11}a_{22}-a_{21}a_{12}} = 0 \tag{5.55}$$

whence

$$A = \varepsilon\left[\frac{a_{22}-p_1}{a_{21}} + 1\right]\frac{a_{21}}{p_2-p_1} - \varepsilon, \quad B = -\varepsilon\left[\frac{a_{22}-p_1}{a_{21}} + 1\right]\frac{a_{21}}{p_2-p_1} \tag{5.56}$$

where $\varepsilon = (1+\gamma)^{-1}$.

Now we are able to write the expressions for $\overline{v}_1'$ and $\overline{v}_2'$

$$\frac{\overline{v}_1'}{\varepsilon}=\left[\left(\frac{a_{22}-p_1}{a_{21}}+1\right)\frac{a_{21}}{p_2-p_1}-1\right]\frac{a_2-p_1}{a_{21}}\exp(p_1\tau)+ \\ +\left(\frac{a_{22}-p_1}{a_{21}}+1\right)\frac{a_{21}}{p_2-p_1}\cdot\frac{a_{22}-p_2}{a_{21}}\exp(p_2\tau)+1 \tag{5.57}$$

and

$$\frac{\overline{v}_2'}{\varepsilon}=\left[\left(\frac{a_{22}-p_1}{a_{21}}+1\right)\frac{a_{21}}{p_2-p_1}-1\right]\exp(p_1\tau)-\left(\frac{a_{22}-p_1}{a_{21}}+1\right)\frac{a_{21}}{p_2-p_1}\exp(p_2\tau)+1 \tag{5.58}$$

These expressions may be rearranged in the following form:

For the large particles

$$\overline{v}_1' =(1+\gamma)^{-1}\left\{-(\varphi_2-\varphi_1)^{-1}\left[\gamma_1^{-1}(1+\gamma)-\varphi_2\right]\left[\gamma_1^{-1}(1+\gamma_2)-\varphi_1\right]\exp(-\gamma_1\varphi_1\overline{\tau}^*)+ \\ +(\varphi_2-\varphi_1)^{-1}\left[\gamma_1^{-1}(1+\gamma)-\varphi_1\right]\left[\gamma_1^{-1}(1+\gamma_2)-\varphi_2\right]\exp(-\gamma_1\varphi_2\overline{\tau}^*)+1\right\} \tag{5.59}$$

and for the small particles

$$\overline{v}_2' =(1+\gamma)^{-1}\left\{(\varphi_2-\varphi_1)^{-1}\left[\gamma_1^{-1}(1+\gamma)-\varphi_2\right]\exp(-\gamma_1\varphi_1\overline{\tau}^*)- \\ -(\varphi_2-\varphi_1)\left[\gamma_1^{-1}(1+\gamma)-\varphi_1\right]\exp(-\gamma_1\varphi_2\overline{\tau}^*)+1\right\} \tag{5.60}$$

where

$$\varphi_{1,2}=0.5\left\{\gamma_1^{-1}\left[(1+\gamma_1)\omega+(1+\gamma_2)\right]\pm\sqrt{\gamma_1^2\left[(1+\gamma_1)\omega\right]^2+4\omega\gamma_1^{-1}\gamma_2}\right\}$$

and $\overline{\tau}^*=t^*/\tau_2$, t^* being the time of interaction of the particle and carrier fluid.

The velocity fluctuations of the carrier fluid $\overline{v}'$ must be computed from Eq. (5.12). In two-phase flows a fluid element, carrying a polydisperse admixture persists as an entity until the largest particle leaves. The distance which this particle transverses during the interaction time may be estimated as $\ell\geq\ell_1\geq\ell-(d_f/2)$.

Here ℓ and ℓ_1 are the fluid element and particle mixing lengths, respectively; d_f is the characteristic size of the fluid element. Therefore, we can write the following expression for the interaction time: $t^* = \ell_1/v'_{1.m} = \ell/v'_{1.m}(1-d_f/2\ell)$, where v'_{1m} is the mean velocity of a large particle during the interaction time. The latter may be estimated (linear approximation) as $v'_{1.m} = (1/2)\left(v'_{1.0}+v'_1\right)$. Taking into account the initial condition $t=0$, $v'_{1.0}=0$ we obtain $v'_{1,m} = (1/2)v'_1$. The dimensionless time $\overline{\tau}^*$, i.e. the ratio of the iteraction time t^* to the relaxation time τ_2 is

$$\overline{\tau}^* = 36\frac{\overline{\ell}}{\overline{v}'_1\,\overline{\rho}\,\mathrm{Re}_p} = \frac{\tau^*}{\overline{v}'_1} \tag{5.61}$$

where $\overline{\rho} = \rho_p/\rho$, ρ_p and ρ are density of the particles and the fluid, respectively, $\overline{\ell} = \ell/d_2$ and $\tau^* = 36(\overline{\ell}/(\overline{\rho}\,\mathrm{Re}_p))$, $\mathrm{Re}_p = \left(d_2|\ v'_0\ |\right)/\nu$

If the interaction time is significantly longer than the relaxation time, $\overline{\tau}^* \gg 1$, i.e. when the particles follow the fluids fluctuations closely, their fluctuational velocities tend to the limit $(1+\gamma)^{-1}$. For finite values of $\overline{\tau}^*$, the level of the fluctuations is a function of the ratio of particle diameters, the total mass content of the admixture, the mass content of the fine and coarse particles, the ratio of particles and fluid densities, the particle Reynolds number, the physical properties of the fluid and the particles and finally the ratio of the characteristic scale of turbulence and the fine particle size. It should be noted that the mixing length in two-phase flow depends on the admixture inertia. However, the effect of this dependence on the velocity fluctuations is small for $\overline{\tau}^* > 1$, so that this effect can be neglected in estimating the fluctuation intensity.

Turbulence intensity of polydisperse mixture. In Figs. 5.1–5.3 the dependences of the dimensionless fluctuation intensities of the carrier fluid $\overline{v}'$ and the particles, $\overline{v}'_1$ and $\overline{v}'_2$, are plotted as per Eqs. (5.12) and (5.26). Figure 5.1 shows the fluctuation intensity of the carrier fluid, $\overline{v}'$ (solid lines), and of the large particles, $\overline{v}'_1$, (dashed line) as function of that of the small particles, $\overline{v}'_2$, with the total mass content of the admixture γ as parameter. It can be seen that an increase of $\overline{v}'_2$ is accompanied by a decrease of $\overline{v}'$ and an increase of $\overline{v}'_1$. Also, at a certain value of $\overline{v}'_2$ (depending on the parameter γ) the curves of $\overline{v}'$ and $\overline{v}'_1$ meet. At this point the fluctuation intensities of the particles of both sizes and of the fluid are equal and a function of only the total load, as $\overline{v}'_2 = \overline{v}'_1 = \overline{v}' = (1+\gamma)^{-1}$. This also corresponds to the case when $\overline{\tau}^* \gg 1$, and all

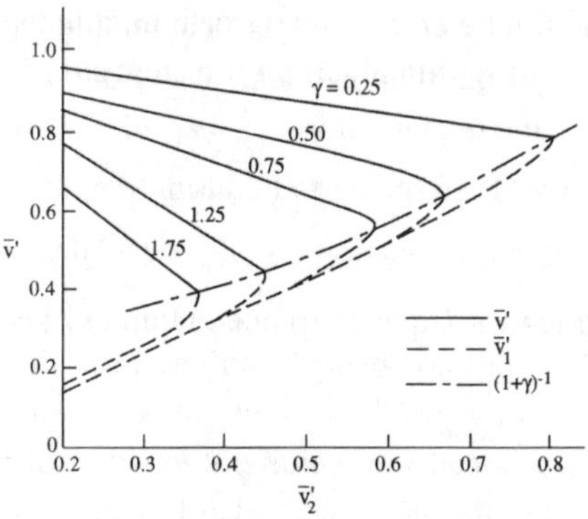

Fig. 5.1 Turbulence intensity of the carrier fluid and large particles as a function of the turbulence intensity of the small particles, for different mass contents of a polydisperse admixture $(\omega = 0.8,\ \gamma = \gamma_2 = \gamma/2)$. Reprinted from Yarin and Hetsroni (1994 a), with permission

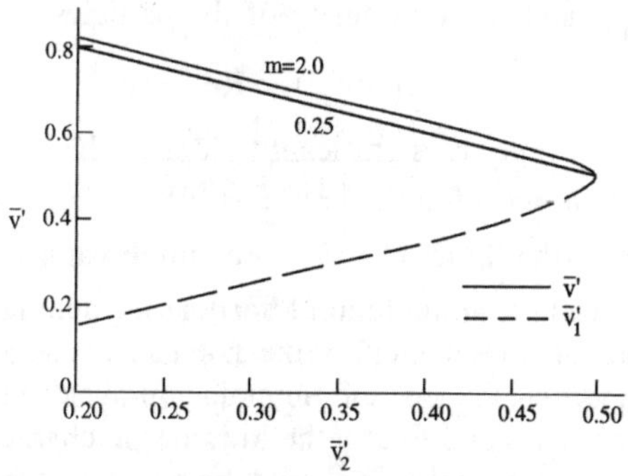

Fig. 5.2 Turbulence intensity of the carrier fluid and large particles as a function of the turbulence intensity of the small particles, for different mass content ratios of the large and small particles $(\gamma = 1,\ \omega = 0.8)$. Reprinted from Yarin and Hetsroni (1994 a), with permission

the particles follow the fluid very closely. Note that an increase of the fluctuational velocity of a particle due to a decrease of that of the fluid is related to momentum transfer from the fluid to the particle as a result of viscous interaction (the total momentum of the fluid element–particles system being conserved over a period equal to the integral turbulence time scale).

The dependence of $\overline{v}'$ and $\overline{v}_1'$ on $\overline{v}_2'$ for different ratios of the mass content of large to small particles, $m = m_1/m_2$ (with fixed total mass content) is plotted in Fig. 5.2. The velocity fluctuations of the carrier fluid increase only slightly for an increase of m from 0.25 to 2.0. This is due to a decrease of the mass content of small particles, with the corresponding decrease in energy dissipation due to their presence.

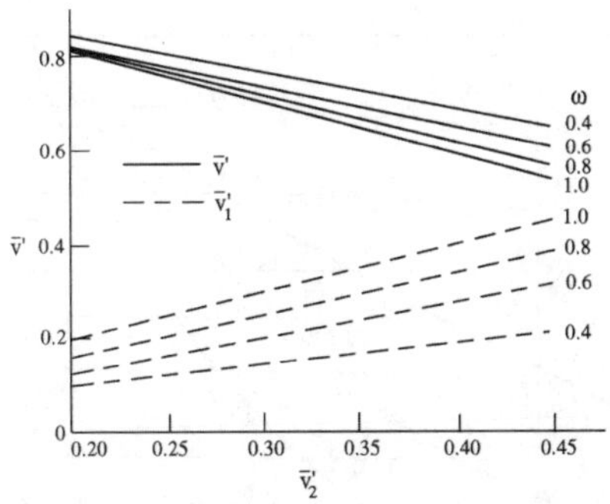

Fig. 5.3 Turbulence intensity of the carrier fluid and large particles as a function of the turbulence intensity of the small particles for different size ratios of the large and small particles $(\gamma_1 = \gamma_2 = \gamma/2 = 0.5)$. Reprinted from Yarin and Hetsroni (1994 a), with permission

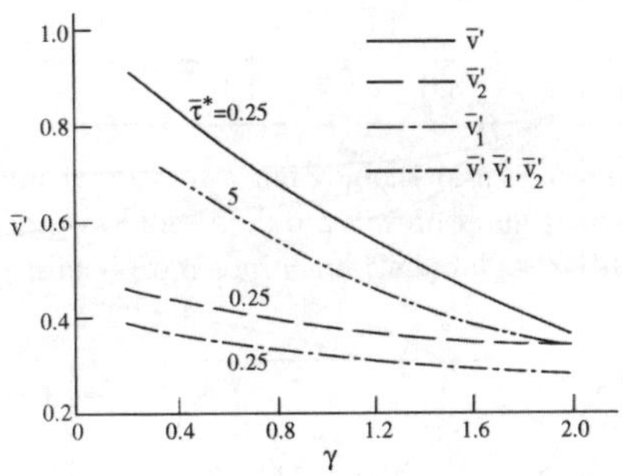

Fig. 5.4 Dependence of the velocity fluctuations of the polydisperse mixture on the total mass content ratio of the admixture $(\gamma_1 = \gamma_2 = \gamma/2,\ \omega = 0.8, \bar{\tau}^* = 0.25)$. Reprinted from Yarin and Hetsroni (1994 a), with permission

The dependences of $\bar{v}'$ and $\bar{v}_1'$ on $\bar{v}_2'$ for different ratios of particle size $\left(\omega = d_2^2/d_1^2\right)$ are plotted in Fig. 5.3. The variation is almost linear. Again it is seen that the fluctuation velocity of the carrier fluid decreases with an increase of that of the small particles, while that of the large particle increases. These effects are more pronounced as ω tends to unity. The dependences of $\bar{v}'$, $\bar{v}_2'$, and $\bar{v}_1'$, on the total mass content of the admixture γ are plotted in Fig. 5.4. An increase in γ is invariably accompanied by reduced fluctuation. As $\bar{\tau}^*$ increases, the curves for $\bar{v}'(\gamma)$, $\bar{v}_1'(\gamma)$ and $\bar{v}_2'(\gamma)$ converge, and at $\bar{\tau}^* = 5$ they practically merge. In Fig. 5.5 one can see that a change in the ratio $m = m_1/m_2$, for a constant total mass content of the admixture, does not affect the level of the fluctuations. An increase in the large-particle load leads to a small increase in the turbulence intensity of the carrier fluid, for $0.25 < \bar{\tau}^* < 5$.

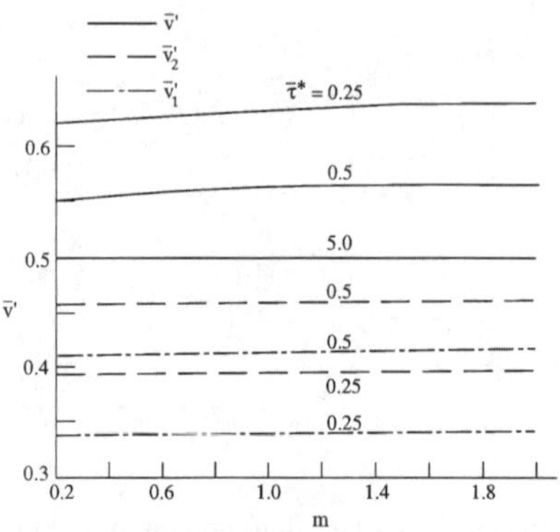

Fig. 5.5 Dependence of the velocity fluctuations of the polydisperse mixture on the mass content ratio of the large and small particles ($\omega = 0.8,\ \gamma = 1.0$). Reprinted from Yarin and Hetsroni (1994 a), with permission

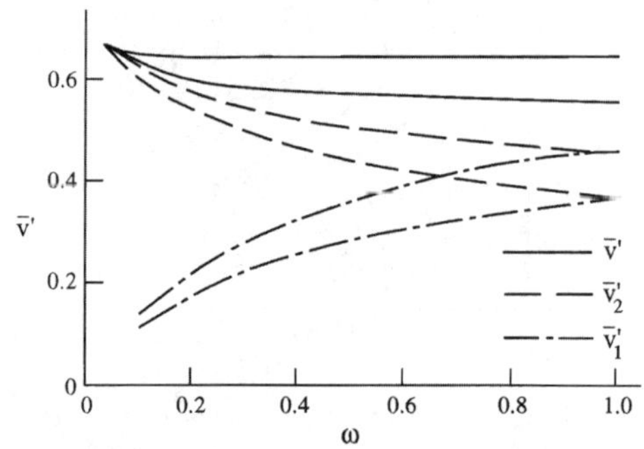

Fig. 5.6 Dependence of the velocity fluctuations of the polydisperse mixture on the size ratio of the large and small particles ($\gamma_1 = \gamma_2 = \gamma/2 = 0.5$). Reprinted from Yarin and Hetsroni (1994 a), with permission

The particle-size ratio $\omega = d_2^2/d_1^2$ affects the turbulence intensity of the carrier fluid and particles, as is seen from Fig. 5.6. It is shown that the effect of the parameter ω on the turbulent fluctuations of particles of different sizes is significant. When $\overline{\tau}^*$ is fixed, a decrease in the parameter ω leads to an increase in the turbulent fluctuations of both the fine and coarse particles.

The dependence of the carrier fluid turbulence on the particle size ratio is shown in Fig. 5.7. The results for the turbulence intensity in polydisperse and monodisperse systems with equal total mass content of the admixture, are shown in that figure as the ratio $[(v'_{pol} - v'_{mon})/v'_{mon}]\times 100\%$ versus ω (v'_{pol} and v'_{mon} being the carrier fluid fluctuations in a polydisperse and a monodisperse system, respectively). Curve 1 corresponds to the case when d_2 and $\overline{\tau}^*$ vary and d_1 is

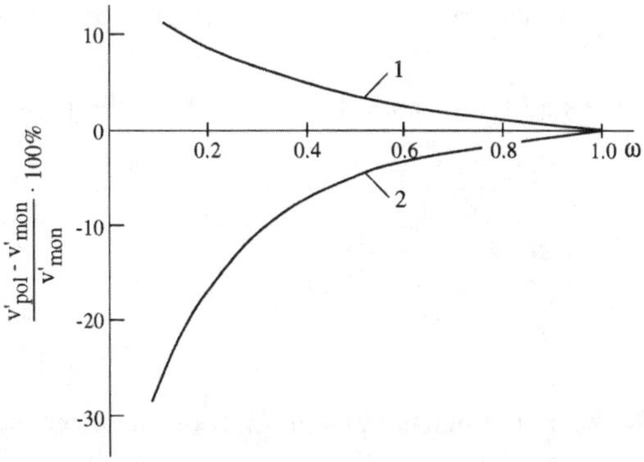

Fig. 5.7 Increase in the turbulence intensity of a polydisperse admixture relative to that in a monodisperse one versus the particle size ratio. Reprinted from Yarin and Hetsroni (1994 a), with permission. Curve 1, d_2=const., $\overline{\tau}^* = \text{const}$, $d_1 = \text{variable}$ $(\gamma_1 = \gamma_2 = \gamma/2 = 0.5$, $\omega = 0.5)$; curve 2, $d_1 = \text{const}$, $d_2 = \text{variable}$, $\overline{\tau}^* = \text{variable}$ $(\gamma_1 = \gamma_2 = \gamma/2 = 0.5)$

constant. The end point $\omega = 1$ of curves 1 and 2 corresponds to a monodisperse system. It is seen that the turbulence intensity in a polydisperse system differs significantly from that in a monodisperse one.

It is emphasized that the turbulence in a polydisperse system may be higher or lower than that in a monodisperse one. For example, if the size of the coarse particles is the same in the two systems, whereas that of the fine particles is smaller in the polydisperse than in the monodisperse system, the turbulence intensity in the former is lower than that in the latter. In this case a decrease in the fine particle size leads (at fixed γ, γ_1 and γ_2) to a larger deviation of v'_{pol} from v'_{mon}. The latter effect is related to an increase in the energy spent on acceleration of the admixture, since a decrease in size leads to an increase in the velocity fluctuations of the fine particles. An increase in the coarse particle size is accompanied by a decrease in the spent energy, which leads to an increase in the turbulence intensity of the carrier fluid.

Fluctuations in a submerged jet. Now we consider the effect of a polydisperse admixture on the turbulent properties of a submerged jet, using Eqs. (5.12) and (5.59)–(5 .61). Let us rearrange Eqs. (5.59) and (5.60) as follows

$$\frac{v'_1}{u_m} = \frac{v'_0}{u_m} F_1, \quad \frac{v'_2}{u_m} = \frac{v'_0}{u_m} F_2, \quad \frac{v'}{u_m} = 1 - \frac{v'_0}{u_m}(\gamma_1 F_1 + \gamma_2 F_2) \tag{5.62}$$

where $F_{1,2}$ and u_m are, respectively, the right-hand side of Eqs. (5.59) and (5.60) and the mean velocity at the jet axis. The multiplier v'_0/u_m is a function of the exit velocity, the diameter of the jet and the location. According to Prandtl's hypothesis $v_0' = \ell(du/dy)$, where ℓ the mixing length and du/dy is the mean velocity gradient.

Considering as usual the mean velocity profile in the jet as $u/u_m = f(\eta)$ we obtain

$$\frac{v_0'}{u_m} = \frac{\ell}{\delta} f(\eta) \tag{5.63}$$

where $\eta = y/\delta$ is the nondimensional lateral coordinate measured from the jet centerline, δ is the jet width and $f'(\eta)$ is a function of η with prime denoting differentiation with respect to η. These characteristics may be described as (Abramovich et al. 1984)

$$u_m = \left(\frac{K_0}{2\pi \cdot 0.067}\right)^{\frac{1}{2}} \frac{1}{\delta}, \quad \delta = 27.3\ \beta^2 x, \quad f(\eta) = (1-\eta^{\frac{3}{2}})^2 \tag{5.64}$$

Here $K_0 = I_0/\rho$, I_0 is the total jet momentum in the axial direction x, and $\beta = 0.09$ is an empirical constant.

The distribution of the turbulent viscosity and turbulent shear stress in the flow field of a submerged jet may be determined by using the known correlations of turbulence theory

$$\nu_T = \ell |\,\overline{v'}\,| \quad \text{and} \quad \tau_{tur} \approx \rho\overline{u'v'} = |\overline{u'}| \quad |\overline{v'}| \tag{5.65}$$

where ν_T is the turbulent viscosity, τ_{tur} is the turbulent shear stress, u' and v' are components of the fluctuation velocity, $|\overline{u'}|$ and $|\overline{v'}|$ are time-averages of the absolute values of these fluctuations.

The set (5.59)–(5.61) and (5.64) permits computation of the fluctuating velocity in a two-phase jet. In Fig. 5.8 such data are presented for jets with exit diameter $d = 2 \cdot 10^{-2}$ m, velocity $u_0 = 70$ [m/s] and air kinematic viscosity $\nu = 10^{-5}$ [m^2/s]. The particles in the jet have a diameter $d_2 = 5 \cdot 10^{-5}$m and a density ratio $\rho_p/\rho = 2 \cdot 10^3$. In the figure the turbulent shear stress is plotted versus the dimensionless distance from the jet axis, at a location downstream from the nozzle exit x/d =22.5, for a polydisperse admixture ($\omega = 0.5$) and for a monodisperse one ($\omega = 1.0$). It is clear that the turbulent shear stress in a two-phase jet with a polydisperse admixture is smaller than in that with a monodisperse one.

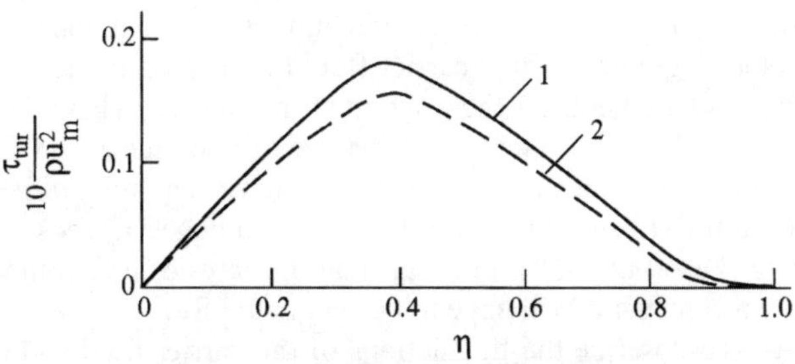

Fig. 5.8 Radial variation of the turbulent shear stress in monodisperse and polydisperse two-phase flow in a submerged jet. Reprinted from Yarin and Hetsroni (1994 a), with permission. Curve 1, jet with monodisperse admixture $\omega = 1$; curve 2, jet with polydisperse admixture $\varpi = 0.5$

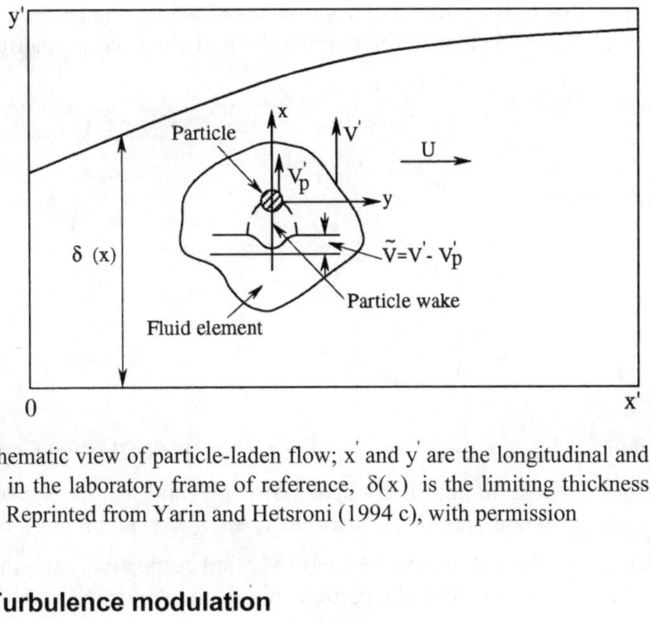

Fig. 5.9 Schematic view of particle-laden flow; x' and y' are the longitudinal and transverse coordinates in the laboratory frame of reference, $\delta(x)$ is the limiting thickness of the external flow. Reprinted from Yarin and Hetsroni (1994 c), with permission

1.5.3 Turbulence modulation

Physical model. Consider a particle-laden turbulent flow with a velocity gradient, as per Fig. 5.9. A fluid element with the admixture moves along the Oy' axis to a distance equal to the mixing length. In single-phase flow, such motion of a fluid element results in the average velocity gradient

$$v_0' = \ell \frac{\partial u}{\partial y'} \tag{5.66}$$

where ℓ is the mixing length and u the average velocity.

In two-phase flow with a velocity gradient, these fluctuations will depend not only on the velocity gradient of the carrier fluid but also on the mass of the particles inside the fluid element, on their size distribution etc. These, in turn, determine the viscous and inertial interactions between the particles and the fluid. The experimental evidence indicates that small particles suppress turbulence, most likely due to inertia effects; while large particles enhance it, most likely due to vortex shedding (Hetsroni 1989). In describing the physics of turbulence modulation, both these mechanisms will have to be accounted for.

Modeling. To describe the fluctuations of the carrier fluid and the particles, we use the momentum balance equations for a fluid element and a single particle, as well as those for the turbulent flow behind the coarse particles. The former equations govern the fluctuations due to motion of the fluid element in a field with velocity gradients, while the latter govern turbulence modulation in the particle wakes.

Assuming that only inertial and viscous forces act on a particle, the momentum equations for a fluid element (containing the particles) and a spherical particle are

$$\frac{d}{dt}(v' + \gamma v'_p) = 0 \qquad (5.67)$$

and

$$m_p \frac{dv'_p}{dt} = \frac{1}{2} C_d \rho \; \tilde{v} \mid \tilde{v} \mid f_p \qquad (5.68)$$

where v' and v'_p are the carrier fluid and particle velocity fluctuations due to the turbulent fluid element motion in gradient flow, m_p and f_p are respectively the mass and cross-sectional area of the particle, $\gamma = M_p/M$ is the mass content of the particles in the fluid element, M and M_p are respectively the mass of the fluid in the fluid element and of the particle, C_d is the drag coefficient of the particle, $\tilde{v} = v'_f - v'_p$ is the relative velocity between the carrier fluid and the particle and ρ is the fluid density.

The drag coefficient in Eq. (5.68) is given by (Schiller and Naumann 1933, Chapter 1.1)

$$C_d = \frac{24}{Re}(1 + 0.15\, Re^{0.687}) \qquad (5.69)$$

where $Re = (|v| d)/\nu$ is the Reynolds number and d is the diameter of the particle.

Equations (5.67) and (5.68) are to be integrated in the time interval $0 < t < t^*$, where the interaction time t^* is determined according to Abramovich (1970) as

$$t^* = \frac{\ell}{|v'_p|} \tag{5.70}$$

The initial conditions for Eqs. (5.67) and (5.68) are

$$t = 0 \quad v' = v'_0, \quad v'_p = 0 \tag{5.71}$$

In describing the turbulent motion in the wake of a coarse particle, we use the set of equations of the carrier fluid momentum balance and of continuity and the equation for the specific turbulent kinetic energy. In the boundary layer approximation, assuming a quasi-stationary character of the wake, these equations read

$$u\frac{\partial u}{\partial x} + v\frac{\partial u}{\partial y} = -\frac{1}{y}\frac{\partial}{\partial y}(y\overline{u'v'}) \tag{5.72}$$

$$\frac{\partial u}{\partial x} + \frac{\partial v}{\partial y} = 0 \tag{5.73}$$

and

$$u\frac{\partial \frac{q^2}{2}}{\partial x} + v\frac{\partial \frac{q^2}{2}}{\partial y} + \frac{1}{y}\frac{\partial}{\partial y}\left[yv'\left(\frac{P}{\rho} + \frac{q^2}{2}\right)\right] + \overline{u'v'}\frac{\partial u}{\partial y} + 15v\overline{\left(\frac{\partial u'}{\partial x}\right)^2} = 0 \tag{5.74}$$

Here we use the frame of reference associated with the particle, u and v are the mean velocity components, $q^2/2$ is the specific turbulent kinetic energy and P is the pressure.

The boundary conditions for the set (5.72)–(5.74) are

$$\left.\begin{aligned} y = 0 \quad & \frac{\partial u}{\partial y} = 0, \quad v = 0, \quad \frac{\partial \frac{q^2}{2}}{\partial y} = 0 \\ y = \delta \quad & u = u_\delta, \quad \frac{q^2}{2} = 0 \end{aligned}\right\} \tag{5.75}$$

where δ is the half-width of the wake, $u_\delta = \tilde{v}$.

Turbulence modulation. Consider the flow behind a coarse particle moving in the fluid element, and assume that the velocity difference in the wake is small. In this case the solution of the hydrodynamic problem for the mean velocity away from the particle may be found in the self-similar form (Swain 1929)

$$\frac{u_1}{u_{1m}} = f(\eta), \qquad u_{1m} = Ax^{\alpha} \tag{5.76}$$

where $u_1 = u_\delta - u$ is the velocity defect ($u_1 \ll u_\delta$ at a large distance from the particle), $u_{1m} = u_\delta - u_m$ (u_m being the velocity at the wake axis), $f(\eta)$ is some function of the variable $\eta = y/\delta$, $\delta = Bx^{\xi}$; A, B and ξ are constants. The function $f(\eta)$ is taken from Abramovich et al. (1984).

$$f(\eta) = \left(1 - \eta^{\frac{3}{2}}\right)^2, \quad u_{1m} = Ax^{-\frac{3}{2}} \tag{5.77}$$

where $A = u_\delta C_d^{\frac{1}{3}} d^{\frac{2}{3}} \left(\frac{0.138^3}{4\cdot\beta^4}\right)^{\frac{1}{3}}$, $B = C_d^{\frac{1}{3}} d^{\frac{2}{3}} \left(\frac{27\beta^2}{4\cdot 0.129}\right)^{\frac{1}{3}}$, $\xi = \frac{1}{3}$, β is an empirical constant $(\beta = \ell/\delta)$ and d is the particle diameter.

Neglecting the effect of the pressure gradient on the turbulent kinetic energy, we arrive at the integral representation of Eq. (5.74)

$$\frac{d}{dx}\int_0^\delta \frac{1}{2} uq^2 y dy + \int_0^\delta \overline{u'v'} \frac{\partial u}{\partial y} y dy = -\int_0^\delta \varepsilon y dy \tag{5.78}$$

where ε is the dissipation, given by Kolmogorov (1942)

$$\varepsilon = c\frac{(q^2)^{\frac{3}{2}}}{L} \tag{5.79}$$

Here c is a constant, L is the integral scale of turbulence, which is proportional to the size of the mixing zone, and $L = c_1\delta$; c_1 is a constant. The values of the constants c and c_1 in the expressions for the dissipation and the integral scale of turbulence may be taken as $c = \left(\sqrt{3}\right)^{-1}$ and $c_1 = 0.2$ (Townsend 1976).

Assume that the turbulent fluctuation energy distribution in the cross-sections of the wake also has a self-similar form

$$\frac{q^2}{q_a^2} = \varphi(\eta) \tag{5.80}$$

Here q_a^2 is the turbulence fluctuation energy at the wake axis and $\varphi(\eta)$ is a function of the variable η.

Using correlation (5.80) and the expression $\overline{u'v'} = \ell^2 \left| \partial u/\partial y \right| \cdot (\partial u/\partial y)$ we rearrange Eq. (5.78) by substituting (5.76), (5.79) and (5.80). Then we obtain

$$\frac{1}{2} u_\delta \frac{d}{dx}(q_a^2 \delta^2) I_1 + \frac{c}{c_1}(q_a^2)^{\frac{1}{2}} \delta I_2 - \beta^2 u_{1m}^2 \delta I_3 = 0 \tag{5.81}$$

where

$$I_1 = \int_0^1 \varphi(\eta)\eta d\eta \tag{5.82}$$

$$I_2 = \int_0^1 \varphi(\eta)^{\frac{3}{2}} \eta d\eta \tag{5.83}$$

$$I_3 = \int_0^1 [f'(\eta)]^2 \left| f'(\eta) \right| \eta d\eta \tag{5.84}$$

To evaluate the integrals I_1 and I_2, we use the experimental data of Uberoi and Freymuth (1970) on the turbulent kinetic energy distribution across the wake of a sphere. As a result, we obtain $I_1 = 0.132$ and $I_2 = 0.112$. To evaluate the integral I_3, we use expression (5.76) and obtain $I_3 = 0.599$. Using the experimental data on the mean velocity distribution along the wake axis, we find the value of the constant $\beta = 0.453$. This value is larger than that of a turbulent axisymmetric jet; a similar result was also obtained for planar wakes (Abramovich et al. 1984).

Assume that the turbulent kinetic energy distribution along the wake axis has the form

$$q_a^2 = N x^n \tag{5.85}$$

where N and n are constants.

Substituting expression (5.85) in Eq. (5.81) and requiring that the result be independent of x, we arrive at $n = -4/3$. We also obtain an algebraic equation for the constant N

$$e_1 N^{\frac{3}{2}} + e_2 N + e_3 = 0 \tag{5.86}$$

where

$$e_1 = \frac{c}{c_1} I_2, \quad e_2 = -\frac{5}{6} u_\delta B I_1, \quad e_3 = -\beta^2 A^3$$

Introducing the new variable Z=n $+(e_2/3e_1)$, (n= $N^{1/2}$), we rearrange Eq. (5.86) in the form

$$Z^3 + 3pZ + 2g = 0 \tag{5.87}$$

where

$$2g = \frac{2e_2^3}{27e_1^3} + \frac{e_3}{e_1}, \qquad 3p = -\frac{e_2^2}{3e_1^2}$$

The third-order polynomial in Eq. (5.87) can be solved, and the solution depends on the sign of $D = g^2 + p^3$. The latter may be written in the form

$$D = \frac{e_3}{e_1}\left(\frac{e_2^3}{27e_1^3} + 0.25\frac{e_3}{e_1}\right) \tag{5.88}$$

As $e_3/e_1 < 0$ and $(e_2^3/e_1^3) < 0$, it is seen that $D > 0$. Therefore, Eq. (5.87) has one real and two imaginary roots. The real root is

$$Z = Z' + Z'' \tag{5.89}$$

where

$$Z' = \sqrt[3]{-g + \sqrt{g^3 + p^3}}, \qquad Z'' = \sqrt[3]{-g - \sqrt{g^3 + p^3}}$$

Using the expressions for the coefficients e_1, e_2 and e_3, we obtain

$$Z' = u_\delta C_d^{\frac{1}{3}} d^{\frac{2}{3}} \sqrt[3]{M_1 + \sqrt{M_1^2 + M_2^3}} \tag{5.90}$$

and

$$Z'' = u_\delta C_d^{\frac{1}{3}} d^{\frac{3}{3}} \sqrt[3]{M_1 - \sqrt{M_1^2 + M_2^3}} \tag{5.91}$$

where

$$M_1 = \left(\frac{c_1}{c}\right)\beta^2 \left\{\frac{125}{216} \frac{1}{4\cdot 0.129}\left(\frac{I_1}{I_2}\right)^3 \left(\frac{c_1}{c}\right)^2 + 0.5\frac{0.138^3}{4\cdot\beta^4}\left(\frac{I_3}{I_2}\right)\right\}$$

and

$$M_2 = -\frac{1}{9}\cdot\frac{25}{36}\left(\frac{27\cdot\beta^2}{4\cdot 0.129}\right)^{\frac{2}{3}}\left(\frac{I_1}{I_2}\right)\left(\frac{c_1}{c}\right)^2$$

The real root of Eq. (5.87) is

$$Z = u_\delta C_d^{\frac{1}{3}} M_3 \tag{5.92}$$

where

$$M_3 = \sqrt[3]{M_1 + \sqrt{M_1^2 + M_2^3}} + \sqrt[3]{M_1 - \sqrt{M_1^2 + M_2^2}}$$

By using expression (5.92), we find the value of the constant N

$$N = u_\delta^2 C_d^{\frac{2}{3}} d^{\frac{4}{3}} M_5 \tag{5.93}$$

where

$$M_5 = (M_3 + M_4)^2, \quad M_4 = \frac{5}{3\cdot 6}\left(\frac{c_1}{c}\right)\left(\frac{I_1}{I_2}\right)\left(\frac{27\cdot\beta^2}{4\cdot 0.129}\right)^{\frac{1}{3}}$$

Note that the coefficients M_1, M_2, M_3 and M_4 depend only on the parameter β.

Now we obtain the expression for the turbulent energy on the axis of the wake

$$q_a^2 = \tilde{v}^2 C_d^{\frac{2}{3}} d^{\frac{4}{3}} M_5 x^{-\frac{4}{3}} \tag{5.94}$$

Equations (5.76), (5.88), (5.79) and (5.94) enable one to determine the local characteristics of the flow in the wake of a single particle. To estimate the effect of turbulence modulation on the fluctuation intensity in particle-laden flow, we have to account for an ensemble of particles. To do this, we first calculate the overall fluctuation energy in the wake of a single particle q_t^2, integrating the energy distribution

$$q_t^2 = 2\pi \int_0^{\ell_w} \left(\int_0^{\delta} q^2 y dy \right) dx \tag{5.95}$$

where ℓ_w is the length of the wake.

Note that the integral in expression (5.95) converges on account of the finite wake length. Substitution of expressions (5.80) and (5.85) in (5.95) yields

$$q_t^2 = 6\pi I_1 \; NB^2 \ell_w^{\frac{1}{3}} \tag{5.96}$$

The energy of the fluctuations generated by the particles in a volume of a two-phase mixture is

$$q_v^2 = \frac{q_t^2 n}{L_f^3} \tag{5.97}$$

where n is the number of particles in the volume L_f^3.

Using the equation of mass balance, one obtains

$$L_f^3 = \frac{n\pi d^3 \bar{\rho}}{6\gamma} \tag{5.98}$$

where $\bar{\rho} = \rho_p / \rho$ is the phase density ratio.

Substitution of fluctuation energy (5.76), (5.93), (5.96) and (5.98) in Eq. (5.97) yields the expression for the total energy of the fluctuations

$$q_v^2 = \frac{36}{\bar{\rho}} I_1 u_\delta^2 C_d^{\frac{4}{3}} \gamma M_5 \left(\frac{27 \cdot \beta^2}{4 \cdot 0.129} \right)^{\frac{2}{3}} \left(\frac{\ell_w}{d} \right)^{\frac{1}{3}} \tag{5.99}$$

Since it was assumed that the particles remain inside the fluid element throughout their lifetime, one may also assume that the maximum wake length is of the order of the diameter of the fluid element, i.e. $\ell_w \sim L_f$. This, with expression (5.98), yields the following value of the ratio $(\ell_w/d)^{1/3}$ for dilute $(n = 1)$ two-phase flows

$$\left(\frac{\ell_w}{d}\right)^{\frac{1}{3}} = \Omega\left(\frac{\bar{\rho}}{\gamma}\right)^{\frac{1}{9}} \tag{5.100}$$

where Ω is an empirical constant of the order of unity.

Dividing the left- and right-hand sides of Eq. (5.99) by v'_0 and substituting in the latter expression the values of the integral I_1, $u_\delta = \tilde{v}$ and constants β, c, c_1, we obtain

$$\bar{q}_v^2 = 52.55\Omega\bar{\tilde{v}}^2 \left(\frac{\gamma}{\bar{\rho}}C_d^{\frac{3}{2}}\right)^{\frac{8}{9}} \tag{5.101}$$

where $\bar{q}_v^2 = q_v^2/v_0'^2$ and $\bar{\tilde{v}} = \tilde{v}/v'_0 = [1-\bar{v}'_p(1+\gamma)]$.

With the fluctuation intensity determined by the turbulence modulation (the effect of the coarse particles), the turbulence intensity of the carrier fluid is

$$\frac{\sqrt{\overline{u'^2}}}{\tilde{v}} = \text{const}\left(\frac{\gamma}{\bar{\rho}}C_d^{\frac{3}{2}}\right)^{\frac{4}{9}} \tag{5.102}$$

Namely, in a flow laden with coarse particles, the logarithm of the turbulence intensity $\ln\left(\sqrt{\overline{u'^2}}/\tilde{v}\right)$ is a linear function of the ratio $(\gamma C_d^{3/2})/\bar{\rho}$, with the coefficient of proportionality equal to $4/9$. One may also refer to this result as the $4/9$-power law.

In particle-laden flows with extremely coarse particles and high turbulence intensity, the drag coefficient is a very weak function of the relative velocity ($Re_p > 10^3$). In this case the fluctuation energy depends only on the total mass content of the admixture and the phase density ratio (as in the $4/9$-power law). This effect may be substantial when the gravity forces are of the same order as or even higher than the aerodynamic drag. For example, $Re_p = 780$ for free-falling particles with $d = 2$ mm (Mizukami et al. (1992), see the section Effect of gravity).

Dimensional analysis. A better insight into the nature of the particle-turbulence interaction may be achieved by analyzing the problem in terms of dimensionless parameters. We use as the scales of time, velocity, fluctuation energy, and density the following quantities: τ_0 ($\tau_0 = \ell/v'_0$), ℓ, v'_0 and ρ. Using the dimensionless variables, $\bar{\tau} = t/\tau_0$, $\bar{d} = d/\ell$, $\bar{v}' = v'/v'_0$, $\bar{v}'_p = v'_p/v'_0$, $\bar{q}^2 = q^2/v_0'^2$ and $\bar{\rho} = \rho_p/\rho$, the equations which describe the system are:

$$\frac{d\overline{v}'_p}{d\tau} = \frac{3}{4}C_d(\overline{d}\cdot\overline{\rho})^{-1}\left[1-\overline{v}'_p(1+\gamma)\right]\left|1-\overline{v}'_p(1+\gamma)\right| \tag{5.103}$$

$$C_d = \frac{24}{Re_p}\left[1+0.15\,Re_p^{0.687}\right] \tag{5.104}$$

$$Re_p = Re_{po}\left[1-\overline{v}'_p(1+\gamma)\right], \qquad Re_{po} = \frac{dv'_0}{\nu} \tag{5.105}$$

$$\overline{q}_v^2 = 52.55\Omega\left(\frac{\gamma}{\overline{\rho}}C_d^{\frac{3}{2}}\right)^{\frac{8}{9}}\left[1-\overline{v}'_p(1+\gamma)\right]^2 \tag{5.106}$$

and

$$\overline{q}_\Sigma^2 = \overline{q}_f^2 + \overline{q}_v^2 \tag{5.107}$$

where $\overline{q}_\Sigma^2$ is the total turbulent energy, and $\overline{q}_f^2 = 3\overline{v}'^2$ and $\overline{q}_v^2$ are the turbulent kinetic energy due to the velocity gradient and the turbulent modulation in the wakes behind the particles, respectively.

It is seen that Eqs. (5.103)–(5.107) contain four dimensionless parameters: $\overline{d}$, Re_{p0}, γ and $\overline{\rho}$, Therefore, the experimental data for particle–turbulence interaction may be described with the former used as criteria of similarity

$$\overline{d} = \text{idem}, \quad Re_{p0} = \text{idem}, \quad \gamma = \text{idem}, \quad \overline{\rho} = \text{idem} \tag{5.108}$$

Clearly an attempt to describe the particle-turbulence interaction in terms of less than four dimensionless variables would fail. This is evident from Figs. 5.10a and 5.10b, where data of a number of typical particle-laden flows are shown (taken from Gore and Crowe 1989). Note that the particle-turbulence interaction may be characterized by the ratio of the turbulence change due to the presence of particles in to the turbulence intensity in the corresponding single-phase flow (Gore and Crowe 1989)

$$\sigma = \frac{\left(\sqrt{\overline{u_p'^2}} - \sqrt{\overline{u'^2}}\right)}{\sqrt{\overline{u'^2}}} \tag{5.109}$$

where $\sqrt{\overline{u_p'^2}}$ and $\sqrt{\overline{u'^2}}$ are the turbulence intensities in the particle-laden and pure flows, respectively. The coefficient σ may be referred to as the "enhancement coefficient". The data in Figs. 5.10a,b show that the particle diameter/turbulence scale ratio does not collapse the experimental data for different

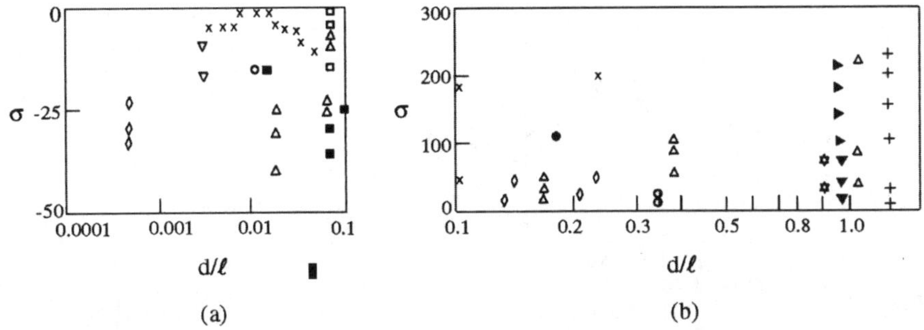

Fig. 5.10 a) Dependence of the enhancement coefficient σ on the particle size/ turbulence scale ratio (enlarged from Gore and Crowe 1989). Experimental data by: □, Levy and Lockwood (1981); ◇, Hetsroni and Sokolov (1971); Δ ,Tsuji et al. (1984); ∇ , Moderress et al. (1984a); O, Moderress et al. (1984b); × , Shuen et al. (1985), ■, Zisselmar and Molerus (1979); ▲, Maeda et al. (1980); Reprinted from Yarin and Hetsroni (1994 c), with permission. b) Dependence of the enhancement coefficient on the particle size/turbulence scale ratio (enlarged from Gore and Crowe 1989). Experimental data by: +, Tsuji and Morikawa (1982); Δ , Tsuji et al. (1984); ×, Shuen et al. (1985); ●, Lee and Durst (1982); ◇, Sun and Faeth (1986); ◑ , Parthasarathy and Faeth (1987); ✡, Wang al. (1987); ►, Theofanous and Sullivan (1982), ▼, Serizawa et al. (1975). Reprinted from Yarin and Hetsroni (1994 c), with permission

particles sizes onto a single curve; it cannot be used as a sole criterion of similarity, but only for estimating the domains of decrease and increase of the turbulence intensity.

In particular cases where the value of the criterion $\overline{d}$ is significantly smaller or larger than unity (fine or coarse particles), the analysis of the problem may be simplified drastically.

The first particular case corresponds to two-phase flows with relatively small $Re_p<400$, i.e. small relative velocities. Naturally, in that case modulation by the wake shedding mechanism and the particle-turbulence interaction process is determined by a single parameter: the total mass content. This result reflects, as a matter of fact, the leading role of particle inertia in particle-laden flows with fine particles. In the second particular case, that of coarse particles, their absolute velocities are relatively small and the fluctuation level is determined by the wake-shedding modulation effect. Correspondingly, the turbulent kinetic energy depends only on the ratio $(\gamma/\overline{\rho})C_d^{3/2}$. For very large particles ($Re_p \geq 10^3$), the drag coefficient is approximately constant. In that case, the turbulent fluctuation level is determined by the quotient of the total mass content of the admixture by the phase-density ratio $(\gamma/\overline{\rho})$. These results are illustrated in Figs. 5.11–5.14, where the enhanced turbulence intensity is plotted versus γ and $\gamma/\overline{\rho}$ for particle-laden turbulent jets and pipe flows.

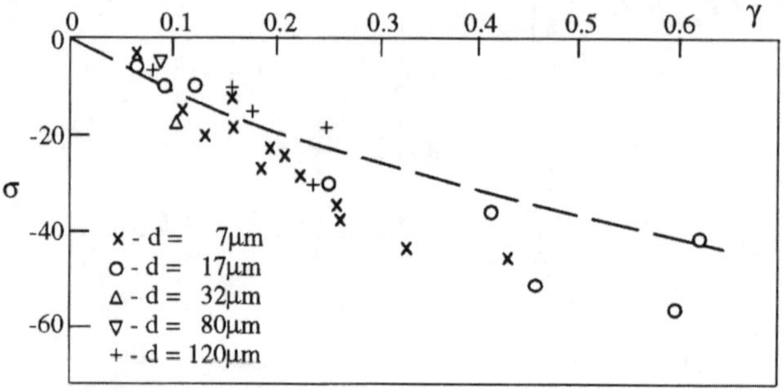

Fig. 5.11 Dependence of the enhancement coefficient on the total mass content of the admixture in a turbulent jet. Reprinted from Yarin and Hetsroni (1994 c), with permission. Data by Laats and Frishman (1973), are shown by symbols, the dashed curve is $\sigma = -\gamma(1+\gamma)^{-1}, \gamma$ is the mass content of the admixture

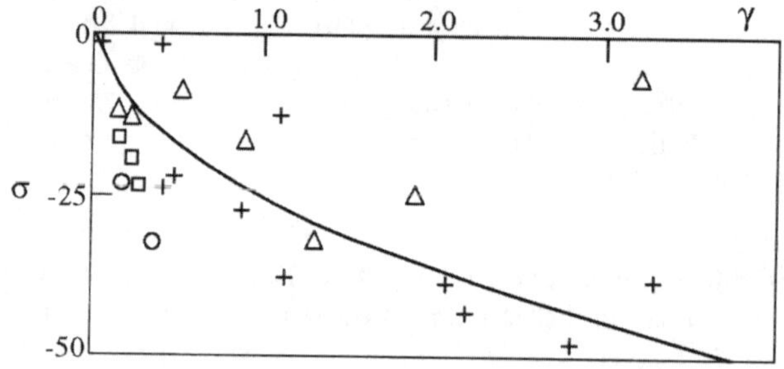

Fig. 5.12 Dependence of the enhancement coefficient on the total mass content of the admixture. Reprinted from Yarin and Hetsroni (1994 c), with permission. Experimental data by: +, Tsuji and Morikawa (1982); △, Tsuji et al. (1984); □, Zisselmar and Molerus (1979); ○, Maeda et al. (1980)

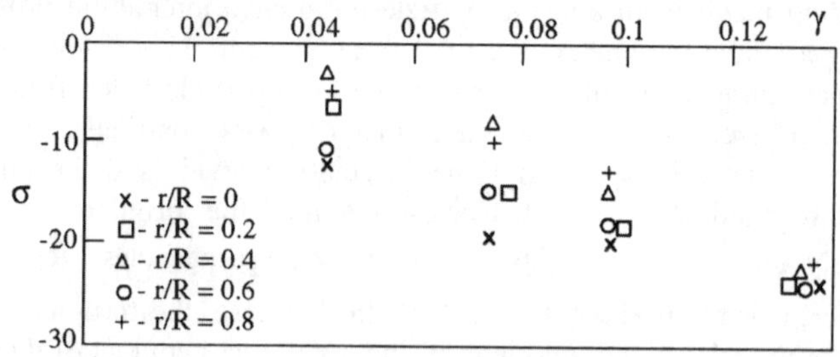

Fig. 5.13 Dependence of the enhancement coefficient on the total mass content at different cross-sections in pipe flow. Reprinted from Yarin and Hetsroni (1994 c), with permission. Experimental data by Zisselmar and Molerus (1979)

In Fig. 5.11, experimental data for the carrier fluid fluctuations at the axis of an axisymmetric jet (Laats and Frishman 1973) are presented. The particle sizes in these experiments varied in the range 7–120 μm and the total mass content within $0<\gamma<0.6$. The graph also contains the curve $\sigma=-\gamma(1+\gamma)^{-1}$, corresponding to the asymptotic case for the turbulence intensity at the limit $\overline{d}\to 0$. As is seen from the figure, in the $\sigma-\gamma$ plane all the points corresponding to the different particle sizes are grouped about a single curve $\sigma(\gamma)$. A similar behavior is seen in Fig. 5.12, where the experimental data for pipe flows are plotted in the form of the turbulence enhancement coefficient versus the total mass content of the admixture (the data of Tsuji and Morikawa 1982, Tsuji et al 1984, Zisselmar and Molerus 1979 and Maeda et al. 1980). It is seen that in all cases, an increase in the total mass content of the fine particles leads to a decrease in the fluctuation level.

The effect of the admixture mass content on the enhancement coefficient is shown in Fig. 5.13 for different zones of the pipe. It is seen that, in spite of the large scatter of the experimental data, the above-mentioned tendency is observed at each r/R position in the central part of the flow where the effect of the pipe walls is weaker.

In Fig. 5.14, experimental data for air with coarse particles suspended in it, flowing in horizontal and vertical pipes (Tsuji and Morikawa 1982, Tsuji et al. 1984) are plotted in dimensionless form. This representation supports the conclusion that there is a dependence of σ on the single parameter $\gamma/\overline{\rho}$ for coarse particles of different sizes. The figure also shows pronounced distinctness of the respective curves for particle-laden flows with fine (d=0.5 mm) and coarse (d=1 to 3 mm) particles. This means that the dependence of the enhancement coefficient on the ratio $\gamma/\overline{\rho}$ has been proved only for sufficiently large particles. In a mixture with particles of smaller sizes, the effect of the particle size on the drag coefficient is more important, hence the remoteness of the curve for the smaller particles.

Effect of the particle-size distribution. The velocity fluctuation in a polydisperse mixture will now be calculated. In this case of a bidisperse admixture (particles in two sizes), one has to solve a set of nonlinear equations which describe the carrier fluid and particle velocity fluctuations and the corresponding relative velocities $\tilde{v}_1=v'-v_1'$ and $\tilde{v}_2=v'-v_2'$ (subscripts 1 and 2 refer to the two groups of particles)

$$\overline{v}'+\gamma_1\overline{v}_1'+\gamma_2\overline{v}_2'=1 \qquad (5.110)$$

$$\frac{d\overline{v}_1'}{d\overline{\tau}}=\frac{3}{4}C_d^{(1)}(\overline{d}_1\overline{\rho})^{-1}\left\{1-\overline{v}_1(1+\gamma_1)-\gamma_2\overline{v}_2'\right\}\ \left|\,1-\overline{v}_1'\,(1+\gamma_1)-\gamma_2\overline{v}_2'\,\right| \qquad (5.111)$$

$$\frac{d\overline{v}_2'}{d\overline{\tau}}=\frac{3}{4}C_d^{(2)}\left(\overline{d}_1\overline{\rho}\right)^{-1}\left\{1-\gamma_1\overline{v}_1'-(1+\gamma_2\overline{v}_2')\right\}\ \left|\,1-\gamma_1\overline{v}_1'-(1+\gamma_2)\overline{v}_2'\,\right| \qquad (5.112)$$

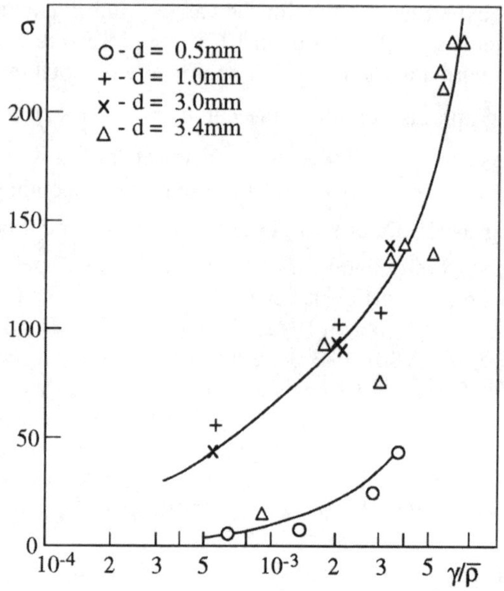

Fig. 5.14 Dependence of the enhancement coefficient on the total mass content of the admixture/phase density ratio for the pipe axis. Reprinted from Yarin and Hetsroni (1994 c), with permission. Experimental data by: Δ , Tsuji and Morikawa (1982); x, +, O, Tsuji et al. (1984)

$$C_d^{(i)} = \frac{24}{Re_p^{(i)}}\left[1 + 0.15\, Re_p^{(i)0.687}\right], \qquad i = 1,\ 2 \tag{5.113}$$

$$Re_p^{(1)} = Re_{po}^{(1)}\left[1 - \bar{v}_1'\,(1+\gamma_1) - \gamma_2 \bar{v}_2'\right], \qquad Re_{po}^{(1)} = \frac{d_1 \left|v_o'\right|}{\nu} \tag{5.114}$$

$$Re_p^{(2)} = Re_{po}^{(2)}\left[1 - \gamma_1 \bar{v}_1' - (1+\gamma_2)\bar{v}_2'\right], \qquad Re_{po}^{(2)} = \frac{d_2 \left|v_o'\right|}{\nu} \tag{5.115}$$

where $\bar{v}_1' = v_p'^{(1)}$; $\bar{v}_2' = \bar{v}_p'^{(2)}$, the superscripts (1) and (2) refer to the larger and smaller particle sizes, respectively.

It is emphasized that, where the Stokes formula $C_d^{(i)} = 24/Re_p^{(i)}$ is applicable, the set (5.110)–(5.115) has an analytical solution (Yarin and Hetsroni 1994 a).

The average energy of the fluctuations per unit volume of a polydisperse mixture with particles of different sizes will now be calculated. For particles of some fraction with diameter d, we can write the following equation

$$q_t^2 = 6\pi I N_i B_i \ell_w^{\frac{1}{3}} \tag{5.116}$$

Assume that the polydisperse mixture contains k_i particles with diameter d_i. Then,

$$\overline{q}_v^2 = 36 I_1 \frac{\gamma}{\overline{\rho}} \left(\frac{27 \cdot \beta^2}{4 \cdot 0.129} \right)^{\frac{2}{3}} \Omega M_5^{i-1} \frac{\sum_{i=1}^{n} k_i \tilde{\overline{v}}^2 C_d^{3/4} \left(\frac{\overline{\rho}}{\gamma} \right)^{\frac{1}{9}} \cdot d_i^3}{\sum_{i=1}^{n} k_i d_i^3} \tag{5.117}$$

Noting that

$$\frac{k_i d_i^3}{\sum_{i=1}^{n} k_i d_i^3} = \frac{\gamma_1}{\gamma}$$

and substituting in Eq. (5.117), we find

$$\overline{q}_v^2 = 52.55\Omega \sum_{i=1}^{n} \tilde{\overline{v}}_i^2 \left(\frac{\overline{\rho}}{\gamma_i} C_d^{\frac{3}{2}} \right)^{\frac{8}{9}} \tag{5.118}$$

Applying the general equations (5.110)–(5.117) to a bidisperse mixture, we see that the turbulence modulation in it depends on the particle Reynolds numbers $Re_{po}^{(1)}$ and $Re_{po}^{(2)}$, the sizes $\overline{d}_1$ and $\overline{d}_2$, the mass content of the fine and coarse particles γ_1 and γ_2 and the phase density ratio.

Effect of gravity. Above we considered the modulation of turbulence due to the inertial and aerodynamic forces, excluding gravity. We consider now an admixture with coarse particles in the gravitational field. Actually, the gravity forces may be of the same order as, or even higher than, the aerodynamic drag.

Girshovich and Leonov (1979) showed that the effect of gravity on the turbulent structure of a particle-laden jet is significant. It manifests itself in changes in the fluctuation intensity, in the turbulent shear stresses, etc. The effect of gravity depends on the ratio of the initial velocity fluctuations to the terminal (levitation) velocity, the particle sizes, the total mass content, etc.

Here we confine ourselves to the effect of gravity on turbulence modulation in vertical particle-laden flows. Other effects, such as the variation of the particle concentration in horizontal flows, are not considered.

Assume that the gravity force is in equilibrium with the aerodynamic drag force

$$\frac{1}{2}C_d\tilde{v}^2\rho_f f_p = mg \tag{5.119}$$

where g is the gravity acceleration, m is the particle mass , and $\tilde{v}$ is the relative velocity which is equal to the terminal (levitation) velocity in this case.

From Eq. (5.119) we obtain

$$\tilde{v}^2 = \frac{4}{3}\frac{\overline{\rho}}{C_d}gd \tag{5.120}$$

It shows that in the given case the relative velocity is determined by the phase density ratio, the drag coefficient, the particle size and the gravity acceleration.

Substituting Eq. (5.120) in the expression for the fluctuation energy (5.101), we arrive at

$$q_v^2 = 70.07\Omega\frac{\overline{\rho}}{C_d}dg\left(\frac{\gamma}{\overline{\rho}}C_d^{3/2}\right)^{\frac{8}{9}} \tag{5.121}$$

It is seen that the energy of the carrier fluid velocity fluctuations is proportional to the particle diameter. For a light admixture, it was found previously that v' depends only on γ when gravity is neglected, which is natural since the total mass content of the admixture determines only the inertial properties of the admixture.

Results. Consider now the results of the theoretical study of the fluctuation intensity in two-phase flows, with both turbulence generation due to the velocity gradient and turbulence modulation via wake shedding taking place. First, we show the effect of the particle size ratio on the characteristics of particle-laden flows. In Fig. 5.15 the effect of the parameter $\overline{d} = d/\ell$ on the carrier fluid and the particle fluctuations is shown (the fluctuations are due to the velocity gradient v' and the relative velocity is proportional to $\sqrt{\overline{q}_v^2}$). The dependence of the energy of the fluctuations $\overline{q}_v^2$ due to turbulence modulation in the particle wake is also presented in Fig. 5.15. It is seen that an increase in $\overline{d}$ is accompanied by an increase in the carrier fluid fluctuations due to the velocity gradient, and by a decrease in the particle fluctuation intensity. The latter in turn causes an increase in the relative velocity and a decrease in the drag coefficient. The energy of the fluctuations due to turbulence modulation in the particle wakes increases monotonically with the particle size/turbulence scale ratio.

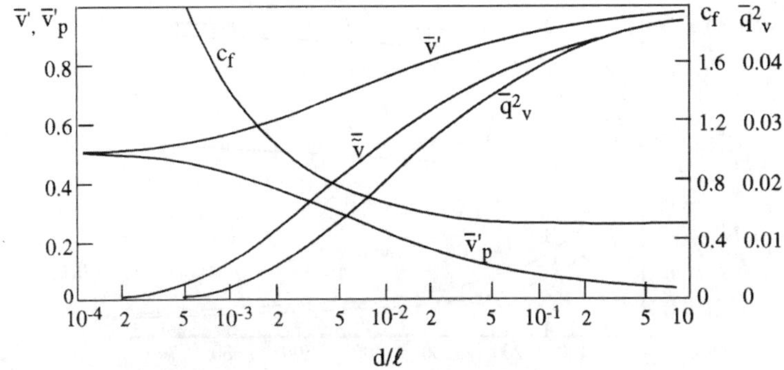

Fig. 5.15 Dependence of the particle and carrier fluid velocity fluctuation on the particle size/turbulence scale ratio: $\gamma = 1$, $\mathrm{Re}_{p0} = 500$. Reprinted from Yarin and Hetsroni (1994 c), with permission

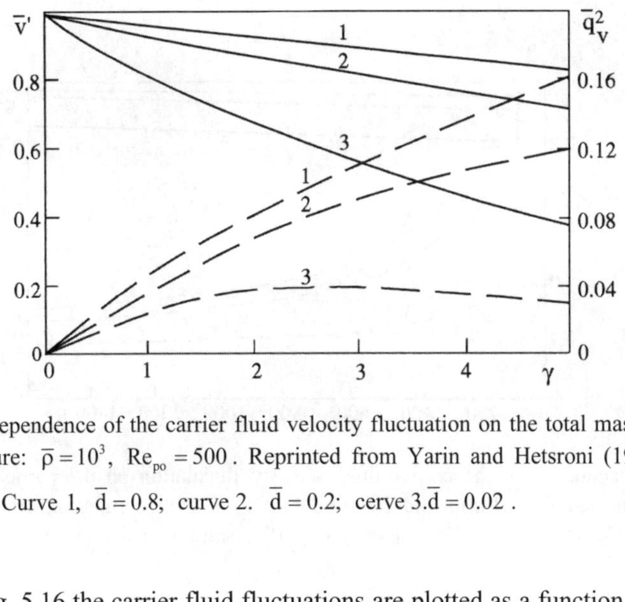

Fig. 5.16 Dependence of the carrier fluid velocity fluctuation on the total mass content of the admixture: $\bar{\rho} = 10^3$, $\mathrm{Re}_{po} = 500$. Reprinted from Yarin and Hetsroni (1994 c), with permission. Curve 1, $\bar{d} = 0.8$; curve 2. $\bar{d} = 0.2$; cerve 3.$\bar{d} = 0.02$.

In Fig. 5.16 the carrier fluid fluctuations are plotted as a function of the total mass content of the admixture. For a large $\bar{d}$ ratio, an increase in the parameter γ leads to a decrease in the level of the fluctuations due to the velocity gradient and to an increase in the energy of the fluctuations due to the turbulence in the particle wakes. For small values of $\bar{d}$, the character of the dependence $\bar{q}_v^2(\gamma)$ is different: here an increase in the total mass content of the admixture leads for relatively small γ to an increase in $\bar{q}_v^2$;, and for large γ to a decrease in $\bar{q}_v^2$; the latter effect is due to the decrease in the relative velocity at large γ .

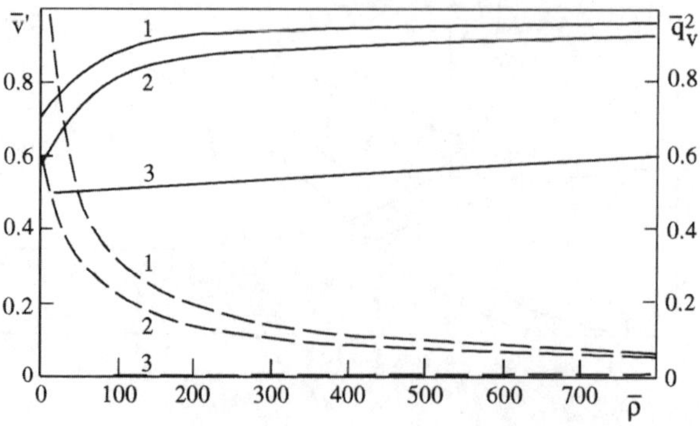

Fig. 5.17 Dependence of the carrier fluid velocity fluctuation on the phase density ratio. Reprinted from Yarin and Hetsroni (1994 c), with permission. $\gamma = 1$, $\mathrm{Re}_{p0} = 500$. Curve 1, $\bar{d} = 0.8$; curve 2, $\bar{d} = 0.2$; curve 3, $\bar{d} = 0.02$, solid lines depicts $\bar{v}'$, dashed lines depict $\bar{q}_v^2$

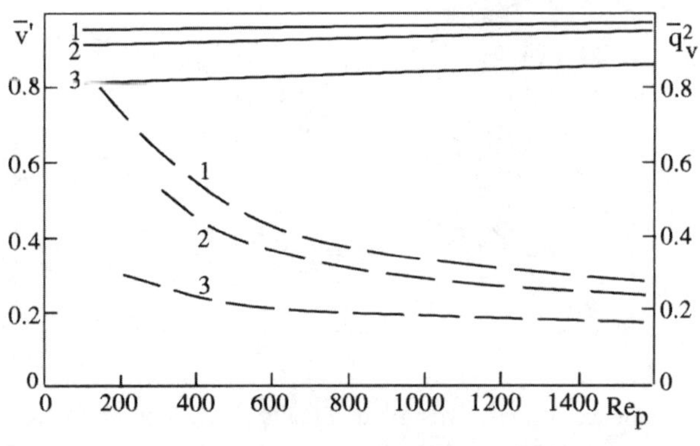

Fig. 5.18 Dependence of the carrier fluid velocity fluctuation on the particle Reynolds number. Reprinted from Yarin and Hetsroni (1994 c), with permission. $\bar{\rho} = 10^3$, $\gamma = 1$ Curve 1, $\bar{d} = 0.8$; curve 2, $\bar{d} = 0.2$, curve 3, $\bar{d} = 0.02$, solid line depicts $\bar{v}'$, dashed lines depict $\bar{q}_v^2$

The effect of the phase density ratio on the carrier fluid turbulence is illustrated in Fig. 5.17. It is seen that an increase in $\bar{\rho}$ leads to a weak increase in the fluctuations due to the velocity gradient, and a decrease in the energy of those due to turbulence modulation in the particle wakes.

The effect of Re_{p0} on the carrier fluid fluctuations and the turbulent kinetic energy is illustrated in Fig. 5.18. It is seen that an increase in Re_{p0} is accompanied by a decrease in the energy of the fluctuations due to the turbulence modulation and a weak increase of those due to the velocity gradient.

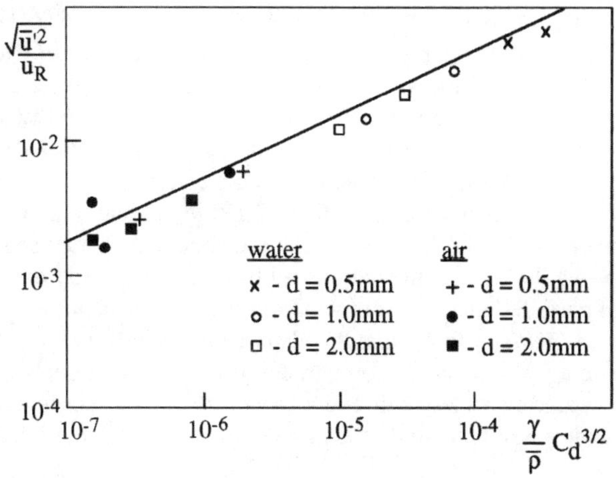

Fig. 5.19 Dependence of the carrier fluid velocity fluctuation on the complex $(\gamma/\overline{\rho})C_d^{3/2}$ (homogeneous turbulence: u_R is the relative velocity). Reprinted from Yarin and Hetsroni (1994 c), with permission. Experimental data by Parthasarathy and Faeth (1990) and Mizukami et al. (1992). The straight lines correspond to the 4/9 power law

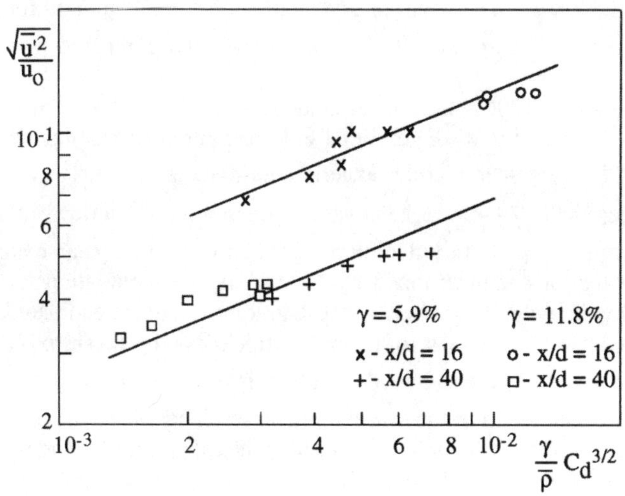

Fig. 5.20 Dependence of the carrier fluid velocity fluctuation on the complex $(\gamma/\overline{\rho})C_d^{3/2}$ (turbulent jet; u_o is the initial average velocity). Reprinted from Yarin and Hetsroni (1994 c), with permission. Experimental data by Parthasarathy and Faeth (1987). The straight lines correspond to the 4/9 power law

Note that for a wide range of the particle/carrier fluid density ratio $(\bar{\rho} > 100)$ the absolute values of the energy of the fluctuations are related to the flow around the particles. They are significantly smaller than their counterparts due to the velocity gradient. Probably, the modulation of turbulent energy in two-phase flows with coarse heavy particles is affected not only by the fluctuational motion of the particles and carrier fluid due to the velocity gradient, but also by gravity. The latter fact leads to a significant increase in the relative velocity and a corresponding increase in the energy of the fluctuations. These conclusions are used in the sequel comparing the results of the calculations with the experimental data on turbulent fluctuations in particle-laden flows in vertical pipes.

Comparison with experiments. A: *Homogeneous turbulence*. Homogeneous turbulence is, in a way, ideal for comparison of theoretical models of fluctuations generated by coarse particles with experimental data. The absence of a velocity gradient makes it possible to study the process of turbulence modulation in "pure" form.

Some experimental data can be found in Parthasarathy and Faeth (1990) and in Mizakami et al. (1992). In these works, the flow generated by uniform free fall of spherical particles in a space filled with water or air was studied. The range of parameters in these experiments was: $0.5 \le d < 2$ mm; $2.45 \le \bar{\rho} \le 2094$; $0.47 \le C_d \le 2.08$; and $38 \le Re_p \le 780$. The intensity of the fluctuations varied in the range $3.3 < \sqrt{\overline{u'^2}} < 6.1$ mm/s and $8.6 < \sqrt{\overline{u'^2}} < 29$ mm/s for water and air, respectively. The results of the two works are compared with theory in Fig. 5.19.

Expression (5.120) is used to calculate the relative velocity in the flow, and the velocity fluctuations of the carrier fluid are plotted versus. the dimensionless ratio $(\gamma/\bar{\rho})C_d^{3/2}$. It is seen that the experimental data are in fairly good agreement with the theoretical $4/9$-power law over a wide range of variation of the flow.

B. Turbulent jet. In turbulent particle-laden jets with a coarse admixture the level of fluctuations is determined by the turbulence modulation in the particle wakes and by the velocity gradients. These gradients may be estimated as the ratio of the velocity at the jet axis u_m to the jet thickness δ. In an axisymmetric jet $u_m \sim x^{-1}$, $\delta \sim x$, hence the velocity gradient is proportional to x^{-2} and decreases steeply as the distance from the nozzle x increases. Therefore, one may expect that in the far field of the jet the process of turbulence modulation in the particle wakes is the dominant mechanism. In this case the $4/9$-power law holds.

In Fig. 5.20 the results of the present theory are compared with experimental data on the turbulence intensity in a submerged turbulent water jet in still water (Parthasarathy and Faeth 1987). In these experiments glass particles of diameter $d = 0.5$ mm and density $\rho_p = 2450$ $[\mathrm{kg/m^3}]$ were used. The initial load ratio varied in the range 5.9–11.8%; and the initial velocity in the range 1.66–1.72 $[\mathrm{m/s}]$.

The results of the measurements of the turbulent fluctuations in the cross-sections of the jet are plotted in Fig. 5.20 as a function of the dimensionless parameter $(\gamma/\overline{\rho})C_d^{3/2}$, and the straight line corresponding in the logarithmic coordinates to the $4/9$ power law is also shown. It is seen that in the present case, the experimental data are in fairly good agreement with theory. The curves depicting the ratio $\sqrt{u'^2}/u_0$ versus $(\gamma/\overline{\rho})C_d^{3/2}$, corresponding to different cross-sections of the jet, vary in accordance with the difference in the particle and carrier-fluid relative velocities at different distances from the nozzle.

C. Pipe flow. Particle-laden pipe flow has a number of characteristic features due to the effect of the velocity gradient and particle-wake turbulence interplay. This interplay manifests itself in an essentially nonuniform distribution of the velocity fluctuations in the pipe cross-sections with the emergence of zones of enhancement and suppression of the turbulence, and the attendant discrepancy between the experimental data and the $4/9$ power law.

Some experimental studies are available on the velocity fluctuations in turbulent pipe flows with coarse particles. For comparing the theoretical predictions with experimental data we chose the work of Tsuji et al. (1982), containing detailed data on the $\sqrt{\overline{u}'^2}$ fluctuation distributions for different values of the total mass content. The other data of Tsuji et al. (1984) concerning the axisymmetric $\sqrt{\overline{u}'^2}$ profiles in vertical flow, were also used and are replotted in Fig. 5.21 in the form of the dependence of the enhancement coefficient σ on the dimensionless radius. It is seen that in the region with a small velocity gradient (near the pipe axis) there is a noticeable increase in the fluctuations. This is a result of turbulence modulation in the wakes of the particles. Near the wall the enhancement coefficient decreases. In particle-laden flow with a fine admixture $(d \leq 1 \text{ mm})$ there is a suppression of the turbulence, associated with the interplay of the inertial behavior of the admixture (suppression of the gradient-oriented turbulence and reduction of the turbulence intensity in the near-wall zone).

The present theory accounts for the above effects, as is seen in Fig. 5.21, where a comparison with the experimental data for the flow in a vertical pipe is shown. In Fig. 5.22 the dependence of $\left(\sqrt{\overline{u}'^2}/u_m\right)^{9/4}$ on $(\gamma/\overline{\rho})C_d^{3/2}$ for different distances from the pipe axis is plotted. The dashed line represents the $4/9$ power law. It is seen that the data for the flow axis are close to the line, whereas those for the off-axis zones are not. The latter manifests itself most clearly as the particle size and turbulence intensity modulation decrease (Fig. 5.23). The effect of the total mass content of the admixture differs for fine and coarse particles. For the former, an increase in γ leads to a decrease in the turbulence intensity as a result of the higher inertia of the admixture, and to an increase in the turbulence intensity, as a result of the larger number of particle wakes per unit volume of the mixture.

The quantitative agreement between the theoretical $4/9$ power law and the experimental results is of considerable theoretical and practical interest. What remains to be done is to determine the empirical constant Ω in expression (5.101) and the constant in Eq. (5.102).

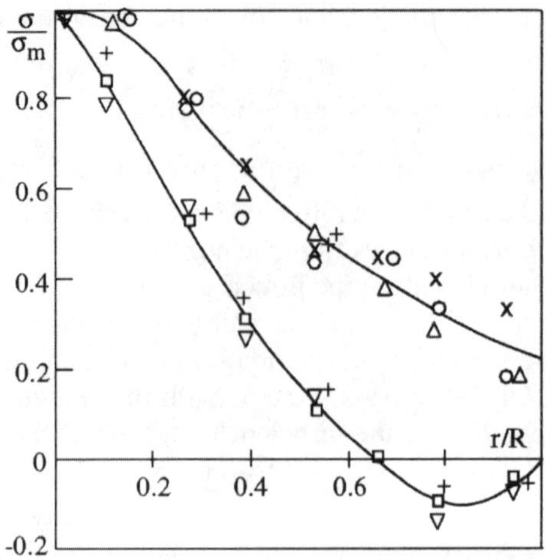

Fig. 5.21 Distribution of the enhancement coefficient in the cross section of the pipe, σ_m corresponds to the pipe axis. Reprinted from Yarin and Hetsroni (1994 c), with permission. Experimental data by Tsuji et al. (1984): d=1 mm ×, $\gamma/\overline{\rho} = 5.88 \cdot 10^{-4}$; Δ, $\gamma/\overline{\rho} = 3.3 \cdot 10^{-1}$; ○, $\gamma/\overline{\rho} = 2.2 \cdot 10^{-1}$; ∇, $\gamma/\overline{\rho} = 5.88 \cdot 10^{-4}$; □, $\gamma/\overline{\rho} = 1.99 \cdot 10^{-3}$; +, $\gamma/\overline{\rho} = 2.94 \cdot 10^{-3}$

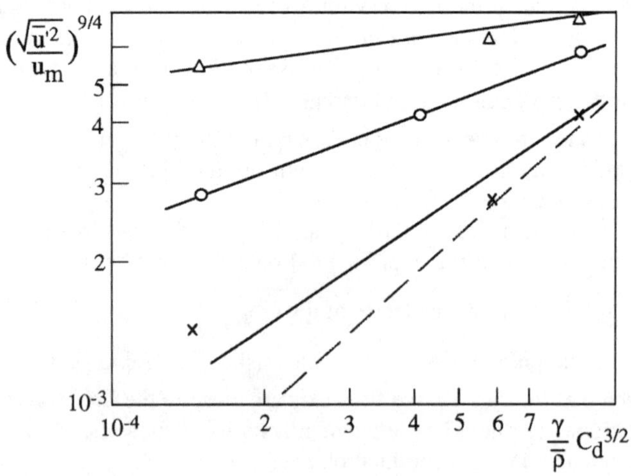

Fig. 5.22 Dependence of the carrier fluid velocity fluctuation on the complex $(\gamma/\overline{\rho})C_d^{3/2}$ for different distances from the axis (pipe flow): u_m is the velocity at the pipe axis. Reprinted from Yarin and Hetsroni (1994 c), with permission. Experimental data by Tsuji et al. (1984). x, $r/R = 0$; o, $r/R = 0.532$; Δ, $r/R = 0.937$ the dashed line shows the $4/9$ power law

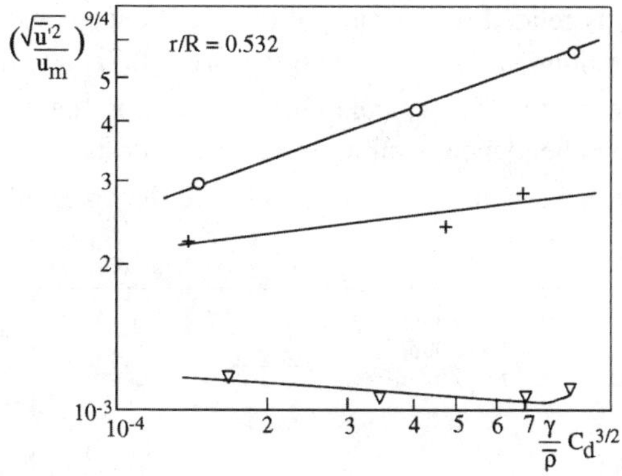

Fig. 5.23. Dependence of the carrier fluid velocity fluctuation on the complex $(\gamma/\bar{\rho})C_d^{3/2}$ for different particles sizes: u_m is velocity at the pipe axis. Reprinted from Yarin and Hetsroni (1994 c), with permission. Experimental data by Tsuji et al. (1984). O, d=3 mm; +, d=1mm; ∇, d=0.5 mm

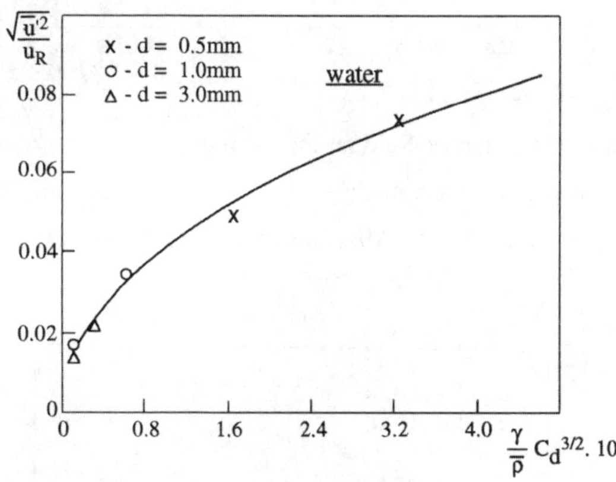

Fig. 5.24 Dependence of the carrier fluid fluctuation on the complex $(\gamma/\bar{\rho})C_d^{3/2}$ (homogeneous turbulence; $\Omega = 0.36$, u_R is the relative velocity). Reprinted from Yarin and Hetsroni (1994 c), with permission. Experimental data by Parthasarathy and Faith (1990). The curve shows the results of the calculations

In Figs. 5.24–5.27 the calculation results of the fluctuation intensity for pipe flow and homogeneous turbulence are presented, and compared with experimental data. It is seen that at the relevant value of Ω=0.36 for homogeneous turbulence, and $\Omega = 0.6$ for pipe flows, there is fairly good agreement between theory and experiment. This supports the supposition that in flows with coarse particles the physical mechanism of wake and vortex shedding is the main turbulence generation process.

Note that Ω is related to the ratio of the turbulence constants c and c_1. The above-mentioned values of Ω correspond to $c_1/c = 0.7$, and an increase in c_1/c leads to a decrease in Ω. The calculations show that in the vicinity of the above value of c_1/c, the dependence is rather weak. It should be emphasized that in the range of reasonable variation of c and c_1, Ω is of the order of unity.

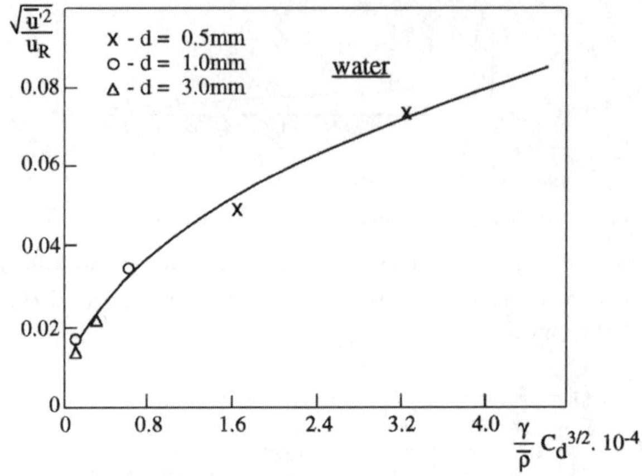

Fig. 5.25 Dependence of the carrier fluid on the complex $(\gamma/\overline{\rho})C_d^{3/2}$ (homogeneous turbulence; $\Omega = 0.36$, u_R is the relative velocity). Reprinted from Yarin and Hetsroni (1994 c), with permission. Experimental data by Mizukami et al. (1992). The curve shows the results of the calculations

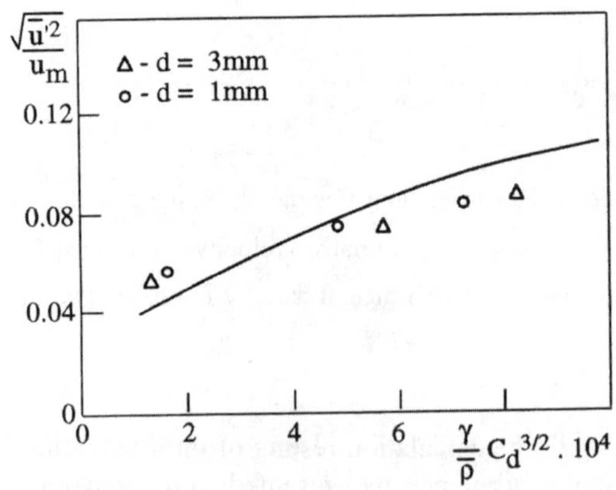

Fig. 5.26 Dependence of the carrier fluid fluctuation on the complex $(\gamma/\overline{\rho})C_d^{3/2}$ (pipe flow: $\Omega = 0.6$, $C_d = 0.4$, u_m is the velocity at the pipe axis). Reprinted from Yarin and Hetsroni (1994 c), with permission. Experimental data by Tsuji et al. (1984). The curve shows the results of the calculations (pipe axis)

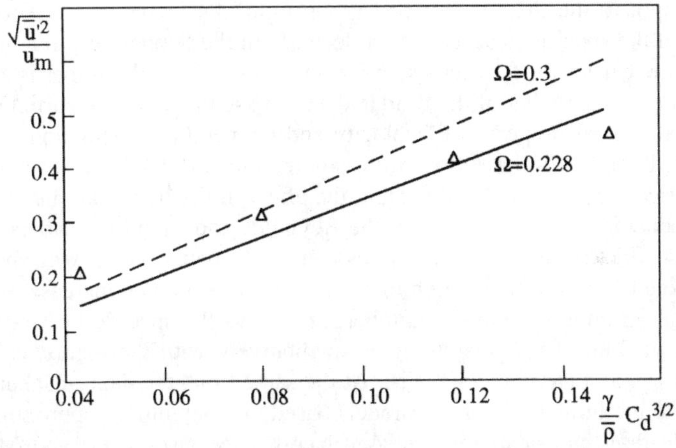

Fig. 5.27 Dependence of the carrier fluid velocity (axial) fluctuation on the complex $(\gamma/\bar{\rho})C_d^{3/2}$. Reprinted from Haam and Brodkey (2000), with permission

1.5.4 Temperature fluctuations in particle-laden flows

Background. In the preceding section we examined the effect of particle size and concentration on the intensity of turbulent fluctuations of the carrier fluid and particles in two-phase flows. Here we expand the problem to include the heat transfer and temperature distributions. In particular, we examine here the effect of temperature fluctuations corresponding to the same average temperature on the average rate of chemical reaction. The latter, besides being of scientific interest, is of extreme practical importance since average reaction rates depend on temperature fluctuations (Zel'dovich 1949, Vulis 1960). An example is the rate of the reaction of NO_x formation. Considerable variations in NO_x concentration corresponding to the some average temperature result from the temperature fluctuation about the average temperature (Kuznetsov and Sabel'nikov 1989).

There are only a few investigations on temperature fluctuation in two-phase flows. Soo (1990) formulated the problem and considered some statistical properties of the temperature fluctuation. Shraiber et al. (1990) also evaluated the level of temperature fluctuations and heat transfer. In particular, they showed that an increase in particle size and heat capacity leads to a decrease in energy transfer from the fluid to the particles. Lisin (1988) gives some data on the effect of the radiative heat transfer on the intensity of temperature fluctuations; it was shown that radiation leads to a decrease in turbulent energy transfer. Yarin and Hetsroni (1994 b) used the modified mixing-length theory to evaluate the intensity of temperature fluctuations in two-phase polydisperse flow. They showed that it is determined by the total mass content of the admixture, the mass contents of the particles of fine and coarse fractions, their diameter ratio, specific heats and density ratios of the particle and the carrier fluids, the Reynolds, Nusselt and Prandtl numbers and the mixing length to the diameter of one group ratio. An increase in the total mass content and the specific heat leads to a decrease in the temperature fluctuation intensity of the carrier fluid, whereas an increase in the Prandtl number is accompanied by an increase in that of the fluid and a decrease in that of the particles.

The statistical properties of velocity and temperature fluctuations in homogeneous turbulent two-phase flow were studied recently by Jaberi (1998) via direct numerical simulation. As the result, the effect of the basic parameters (the total mass content of solid admixture, the Reynolds and Prandtl numbers, etc.) on temperature fluctuation intensity was evaluated. In particular, it was shown that the particle temperature fluctuation intensity decreases with an increase of the particle relaxation time, the Prandtl number, as well as the specific heat ratio of the particles and fluid. These results agree qualitatively with the theoretical predictions by Yarin and Hetsroni (1994 b). At the same time, the data of Jaberi (1998) demonstrate the limitations of the models based on a simplified approximation of the particle drag and heat transfer coefficients (see also Jaberi and Mashayck 2000 and Derevich 2001).

Governing equations. We will consider temperature fluctuation in the steady flow of a clear incompressible fluid, laden with spherical particles with diameters $d_1, d_2 \ldots, d_n$ at number concentrations $k_1, k_2 \ldots. k_n$, respectively. To this end we resort to a modified Prandtl mixing-length theory. As above we assume that an element of the carrier fluid retains its identity over the traversal of one mixing length. If the particles are relatively small and their relaxation times shorter than the hydrodynamic characteristic time, one can assume that the average velocities of the particles are approximately equal to that of the fluid. However, due to radiation, the average temperature of the particles is not the same as that of the fluid; let T_p be the temperature of the particle and T_w that of the walls surrounding the flow, when $T_w > T_p$, the latter exceeds the temperature of the carrier fluid. Only in the case of intensive convective heat transfer (characteristic of fine particles) and moderate radiation level, can the average temperature of the particles be assumed to equal that of the carrier fluid.

In the sequel we confine the analysis to flows with weakly varying longitudinal temperature, where the relative average temperature of the carrier fluid and the particles is negligibly small (developed turbulent flow in jets, etc.).

The equations describing the conservation of momentum of the fluid element and particles in it were given in Sect. 1.5.1. In addition we have the energy balances as follows:

$$\frac{c}{\gamma}\left\{\sum_{i=1}^{n} m^{(i)}k_i\right\}\frac{dT'}{dt} = -4\sum_{i=1}^{n} h_i f_p^{(i)} k_i (T' - T_p'^{(i)}) \tag{5.122}$$

$$m^{(i)}c_p^{(i)}\frac{dT_p'^{(i)}}{dt} = 4h_i f_p^{(i)}(T' - T_p'^{(i)}) - 4q_R f_p^{(i)} \tag{5.123}$$

where T' is the temperature fluctuation of the fluid and T_p' that of the particles, t is time, $m_p^{(i)}$ the mass of a particle in size group i, $c_p^{(i)}$ is its specific heat, c the specific heat of the carrier fluid, h_i the heat transfer coefficient, $f_p^{(i)}$ the particle cross-section, γ the total mass content of the particles, and q_R the radiative heat exchange between the particles and the walls.

The initial conditions for the velocities will be the same as in Section 1.5.1, and those for the the temperature.

$$t = 0, \quad T' = T_0', \quad T_p' = T_{p0}' = 0 \tag{5.124}$$

Note that these conditions actually describe the physical mechanism of the fluctuating motion in turbulent two-phase flows. The carrier fluid fluctuation sets in first, and is followed by momentum and heat transfer to the particles via viscous and thermal interaction. Accordingly, the initial particle temperature fluctuations are zero since the characteristic time of thermal interaction is nonzero. It is emphasized that the initial conditions used in Sect. 1.5.1 and above lead to physically reasonable results, in fairly good agreement with experimental data (Abramovich et al. 1984).

In the particular case $q_R = 0$ in Eq. (5.123), we can multiply it by k_i and perform a summation

$$\sum_{i=1}^{n} m^{(i)} c_p^{(i)} k_i \frac{dT_p'^{(i)}}{dt} = 4 \sum_{i=1}^{n} h_i f_p^{(i)} k_i (T' - T_p'^{(i)}) \tag{5.125}$$

Supplementing the above with Eq. (5.122), we arrive at the energy balance equation in the following form:

$$\frac{d}{dt}\left\{ T' c \gamma^{-1} \sum_{i=1}^{n} m^{(i)} k_i + \sum_{i=1}^{n} m^{(i)} c_p^{(i)} k_i T_p'^{(i)} \right\} = 0 \tag{5.126}$$

Integrating Eq. (5.126) with the initial conditions (5.124), we obtain

$$\overline{T}' + \sum_{i=1}^{n} \gamma_i \overline{c}^{(i)} \overline{T}_p'^{(i)} = 1 \tag{5.127}$$

where $\overline{T}' = T'/T_0'$ and $\overline{T}_p' = T_p'^{(i)}/T_0'$ are the dimensionless temperatures and $\overline{c}^{(i)} = c_p^{(i)}/c$ is the dimensionless specific heat.

The velocity fluctuation is also given in Section 1.5.1. For the simple case $\overline{T}' = \overline{T}_p'^{(i)}$ and $\overline{c}^{(i)} = \overline{c}$, the temperature fluctuation is

$$\overline{T}' = (1+\gamma\overline{c})^{-1} \tag{5.128}$$

which is analogous to its velocity counterpart as per above, meaning that the temperature fluctuation of the carrier fluid depends on the total mass content of the admixture γ and the specific heat ratio. The latter implies that when the particle specific heat is smaller than that of the fluid, i.e. when $\overline{c} < 1$, the temperature fluctuation is larger than the velocity fluctuation, as shown in Fig. 5.28. For most metals in air we have $\overline{c} < 1$, and a relatively small heat flux to the particles causes a strong increase in their temperature.

Monodisperse mixture. Consider the temperature fluctuation in a two-phase flow, with particles of uniform size. Eliminating t from Eqs. (5.123) and (5.123), we arrive at

$$\frac{d\overline{T}_p'}{d\overline{v}_p'} = \frac{Nu}{3\,Pr\,\overline{c}}\frac{1-(1+\overline{c}\gamma)\overline{T}_p'}{1-(1+\gamma)\overline{v}_p'} \tag{5.129}$$

where Nu and Pr are the Nusselt and Prandtl numbers, respectively.

For small particles one can take $Nu = 2$, i.e. treat it as is constant. Neglecting also the variations in the physical properties due to the temperature fluctuation, Eq. (5.129) can be integrated. This yields

$$\overline{T}_p' = (1+\gamma\overline{c})^{-1}\{1-[1-(1+\gamma)\overline{v}_p']^{\Omega}\} \tag{5.130}$$

where

$$\overline{\Omega} = \frac{Nu}{3\,Pr\,\overline{c}}\frac{(1+\gamma\overline{c})}{(1+\gamma)}.$$

The temperature fluctuation for a monodisperse system is plotted in Fig. 5.29 as a function of the velocity fluctuation, with $\overline{\Omega}$ as parameter. Clearly, for $\overline{\Omega}=1$, $\overline{\Omega}>1$ and $\overline{\Omega}<1$ the temperature fluctuation is proportional to, higher than and is lower than the velocity fluctuation, respectively.

Now, using Eqs. (5.127), (5.130) and (5.4) we obtain

$$\overline{T}' = (1+\gamma\overline{c})^{-1}\cdot\left\{1+\gamma\overline{c}\left(\gamma^{-1}\left[\overline{v}'(1+\gamma)-1\right]\right)^{\overline{\Omega}}\right\} \tag{5.131}$$

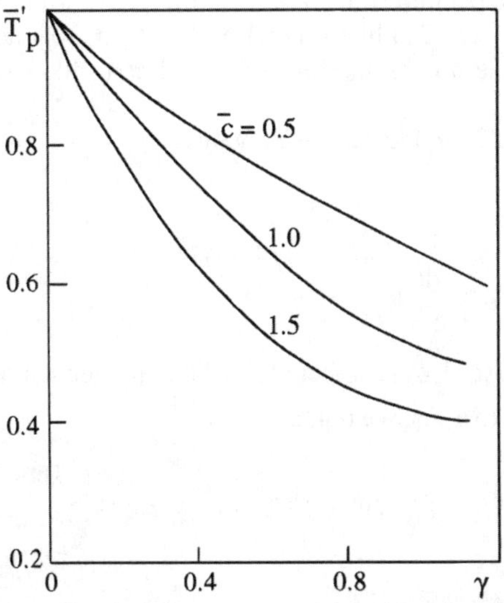

Fig. 5.28 Dependence of the temperature and velocity (curve $\overline{c} = 1$) fluctuations of the particles on the total mass content of the admixture for different specific heat ratios of the particles and carrier fluid. Reprinted from Yarin and Hetsroni (1984 b), with permission

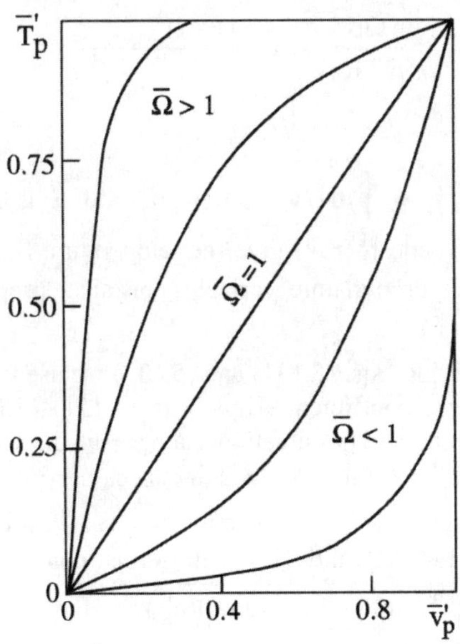

Fig. 5.29 Dependence of the particle temperature fluctuation on their velocity fluctuation for a monodisperse mixture. Reprinted from Yarin and Hetsroni (1984 b), with permission

which relates the temperature fluctuations in the carrier fluid to the velocity fluctuation. Note that Eq. (5.131) like Eq.(5.130), is completely universal for a monodisperse system, since no assumption was made as to the structure of the turbulence.

From Eqs. (5.123) and (5.127) we obtain

$$\frac{d\overline{T}_p'}{dt} = \omega^*\left[1-(1+\gamma\overline{c})\overline{T}_p'\right] \tag{3.132}$$

where $\omega^* = 6(\mathrm{Nu}\lambda)/(c_p d^2 \rho_p)$, λ being the thermal conductivity of the fluid.

Integrating Eq. (5.132), we obtain

$$\overline{T}_p' = (1+\gamma\overline{c})^{-1}(1-e^{-K}) \tag{5.133}$$

where $K = \omega^*(1+\gamma\overline{c})t^*$, t^* being the interaction time between the particles and fluid, as per Sect. 1.5.1.

Using Eq. (5.130), we find

$$K = 36\frac{\overline{\Omega}\beta}{\overline{\rho}\ \ \mathrm{Re}}\ \frac{(1+\gamma)^2}{1-\left\{1-(1+\gamma\overline{c})\overline{T}_p'\right\}^{\frac{1}{\Omega}}} \tag{5.134}$$

where $\beta = \ell/d$, $\mathrm{Re} = \left(\left|\ v_o'\ \right|\ d\right)/\nu$, $\overline{\rho} = \rho_p/\rho$, and ℓ is the mixing length.

Equations (5.133) and (5.134) give the temperature fluctuation of the particles in terms of the hydrodynamic variables, physical properties $(\overline{\Omega}, \overline{c}, \overline{\rho})$ and loading (γ).

Note, that in deriving Eqs. (5.131) and (5.133) we use the Stokes approximation for the particle drag coefficient. Thus, Eqs. (5.131)–(5.134) are valid at low particle Reynolds numbers. The situation changes in the case of non-Stokesian particle motion. At $\mathrm{Re}_p \gg 1$ and $\mathrm{Nu} \neq 2$ it is necessary to account for the actual character of the dependences $C_d(\mathrm{Re}_p)$ and $\mathrm{Nu}(\mathrm{Re}_p, \mathrm{Pr})$. That leads to a set of nonlinear equations for v' and T' with terms containing the expressions $\overline{(v'-v_p')\ (\overline{\mathrm{Re}}_p+\mathrm{Re}')^{2/3}}$, $\overline{(T'-T_p')(\overline{\mathrm{Re}}_p+\mathrm{Re}_p')^{1/2}}$, etc. (Varaksin 1998). The problem can be simplified for some types of turbulent flows (e.g. for the quasi-equilibrium flow with small or significant fluctuating slip, or for the non-equilibrium flows with small time-averaged slip, etc.) when the contribution of the above expressions is negligible. In this case the right-hand sides of the governing equations become explicit functions of v', T' and Re'_p. Thus for the quasi-equilibrium flow with significant fluctuating slip, the equations for v'_p and T'_p acquire the following form (Varaksin 1998)

$$\frac{dv'_p}{dt} = \frac{v' - v'_p}{\tau}\left(1 + \frac{1}{6} Re_p'^{\frac{2}{3}}\right) \tag{5.135}$$

$$\frac{dT'_p}{dt} = \frac{T' - T'_p}{\tau_T}(1 + 0.3 Re_p'^{\frac{1}{2}} Pr^{\frac{1}{3}}) \tag{5.136}$$

where τ and τ_T are the dynamic and thermal relaxation times.

It should be noted that the effects of non Stokesian dependences of the drag and heat transfer coefficient are very weak under the conditions corresponding to dust flows. The calculations carried out on the Reynolds and Prandtl numbers by Jaberi (1998) for two types of the dependences $C_d(Re_p)$ and $Nu(Re_p, Pr)$

$$C_d = 24/Re_p, \quad Nu=2 \tag{5.137}$$

$$C_d = 24/Re_p(1 + 0.15 Re_p^{0.687}), \quad Nu = 2 + 0.6 Re_p^{0.5} Pr^{0.33} \tag{5.138}$$

showed that the particle to carrier fluid ratios of the velocity and temperature fluctuations change by 10.4% and 16%, respectively, as the particle Reynolds number increases from 0 to 1.9. (Table 5.1). The theoretical predictions for the correlation between the particle temperature and velocity fluctuations (Eq. 5.130) are compared with the results of direct numerical simulation of the problem for different dependences $C_d(Re_p)$ and $Nu(Re_p, Pr)$ in the last column of Table 5.1. It is seen that the data differ by not more than 3% for $0<Re_p<1$ and by up to 11% in the range $0<Re_p<1.9$. The effect of non - Stokesian flow on drag and the nonlinearity relation of the heat transfer coefficients manifests itself more noticeably at high Reynolds and Prandtl numbers, as well as at high ratios of the specific heat of the particle to that of the fluid (Fig. 5.30). The figure shows that agreement between Eq. (5.130) and the results of the direct numerical simulation is reasonable at $\bar{c} Pr < 1$, while at $\bar{c} Pr > 1$ there is a significant difference. It should be noted that for a number of important gas–particle reactive mixtures the product $\bar{c} Pr$ is close to unity. For example $\bar{c} Pr = 0.917$ for coal dust–air mixtures.

Bidisperse mixture. Consider a two-phase mixture in the form of a carrier fluid in which particles of two sizes are suspended. The energy equations for the particles are

$$m^{(1)} c_p \frac{dT'^{(1)}}{dt} = 4 h_1 f_1 (T' - T_p'^{(1)}) \tag{5.139}$$

Table 5.1 Reprinted from Jaberi(1998),with permition

ρ_p [kg/m^3]	C_d	Nu	$\overline{Re_p}$	$\overline{v'^2}/\overline{v_p'^2}$	$\overline{T_p'^2}/\overline{T'^2}$	$\overline{(T'-T_p')^2}/\overline{T'^2}$	$(TR)_{theor}/(TR)_{DNS}$
250	$24(0.15Re_p^{0.687})/Re_p$	$2+0.6Re_p^{0.5}Pr^{0.33}$	1.9	0.566	0.536	o.685	1.113
500	$24(0.15Re_p^{0.687})/Re_p$	$2+0.6Re_p^{0.5}Pr^{0.33}$	1.37	0.556	0.525	0.694	1.098
1,000	$24(0.15Re_p^{0.687})/Re_p$	$2+0.6Re_p^{0.5}Pr^{0.33}$	0.97	0.553	0.509	0.704	1.082
4,000	$24(0.15Re_p^{0.687})/Re_p$	$2+0.6Re_p^{0.5}Pr^{0.33}$	0.50	0.533	0.497	0.718	1.063
16,000	$24(0.15Re_p^{0.687})/Re_p$	$2+0.6Re_p^{0.5}Pr^{0.33}$	0.25	0.523	0.48	0.727	1.046
1,000	$24/Re_p$	2	0.0	0.507	0.448	0.752	0.997
1,000	$24(0.15Re_p^{0.687})/Re_p$	$2+0.6Re_p^{0.5}Pr^{0.33}$	0.97	0.546	0.448	0.749	1.037
1,000	$24/Re_p$	2	0.97	0.507	0.513	0.706	1.050

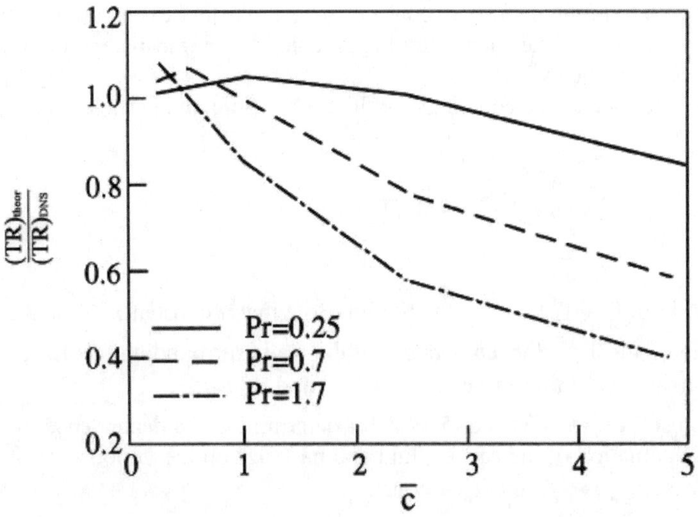

Fig. 5.30 Variation of the ratio of the Yarin and Hetsroni model (1984 b) (to the direct numerical simulation (DNS) values of $(TR)_{theor}/(TR)_{DNS}$ versus $\overline{c}$. Reprinted from Jaberi (1998), with permission

$$m^{(2)}c_p \frac{dT'^{(2)}}{dt} = 4h_2 f_2 (T' - T_p'^{(2)}) \tag{5.140}$$

where subscripts 1 and 2 correspond to coarse and fine particles, respectively.

Using Eqs. (5.127), (5.139) and (5.140), we find the absolute value of the temperature fluctuation of the coarse and fine particles

$$\overline{T}'^{(1)} = (1+\gamma\overline{c})^{-1}[(\varphi_2-\varphi_1)^{-1}[\gamma_1^{-1}\overline{c}^{-1}(1+\gamma\overline{c})-\varphi_2]\times[\gamma_1^{-1}\overline{c}^{-1}(1+\gamma_2\overline{c})-\varphi_1]\times \exp(-\gamma_1\overline{c}\varphi_1\tilde{\tau}^*)+[\gamma_1^{-1}\overline{c}^{-1}(1+\gamma\overline{c})-\varphi_1](\varphi_2-\varphi_1)^{-1}[\gamma_1^{-1}\overline{c}^{-1}(1+\gamma_2\overline{c})-\varphi_2]\times \exp(-\gamma_1\overline{c}_1\varphi_2\tilde{\tau}^*)+1] \tag{5.141}$$

$$\overline{T}_p'^{(2)} = (1+\gamma\overline{c})^{-1}\left\{ (\varphi_2-\varphi_1)^{-1}\left[\gamma_1^{-1}\overline{c}^{-1}(1+\gamma\overline{c})-\varphi_2 \right]\exp(-\gamma_1\overline{c}\varphi_1\tilde{\tau}^*) - (\varphi_1-\varphi_2)\left[\gamma_1^{-1}c^{-1}(1+\gamma\overline{c})-\varphi_1\right]\exp\left(-\gamma_1\overline{c}\varphi_2\tilde{\tau}^*\right)+1\right\} \tag{5.142}$$

where $\tilde{\tau}^* = t^*\omega_2^*$,

$$\varphi_{1.2} = 0.5\left\{\gamma_1^{-1}\overline{c}^{-1}\left[(1+\gamma_1\overline{c})\omega+(1+\gamma_2\overline{c})\right]\pm\sqrt{\gamma_1^{-2}\overline{c}^{-2}\left[(1+\gamma_1\overline{c})\omega-(1+\gamma_2\overline{c})\right]^2+4\omega\gamma_1^{-1}\gamma_2}\right\}$$

Effect of radiation. Let us consider now the effect of radiation on the temperature fluctuation. The carrier fluid is assumed to be transparent and laden with monodisperse particles.

The energy balance equations for the carrier fluid and particles are

$$mc_p \frac{dT_p'}{dt} = 4hf_p(T' - T_p') - q_R 4f_p \tag{5.144}$$

where $q_R = \tilde{\sigma}(T_p^4 - T_w^4)$, $\tilde{\sigma}$ is the Stefan–Boltzmann constant, T_p is the particle temperature, and T_w is the temperature of the walls surrounding the flow.

Consider two limiting cases: $T_w \gg T_p$, and $T_p \gg T_w$.

We use Eqs. (5.143) and (5.144) for determining the dependence of the temperature fluctuation of the carrier fluid and particles on the various parameters for the first case, $T_w \gg T_p$, and $q_R = \text{const}$.

$$\overline{T}' = 1 - \gamma\overline{c}(1+\gamma\overline{c})^{-1}\left[1 - \overline{q}_R(1+\gamma\overline{c})^{-1}\right]\left[1 - \exp(N\tilde{\tau}^*)\right] - \gamma\overline{c}(1+\gamma\overline{c})^{-2}\overline{q}_R N\tilde{\tau}^* \tag{5.145}$$

and

$$\overline{T}_p' = (1+\gamma c)\left\{\left[1 - \overline{q}_R(1+\gamma\overline{c})^{-1}\right]\left[1 - \exp(-N\tilde{\tau}^*)\right] - \gamma\overline{c}(1+\gamma\overline{c})^{-1}\overline{q}_R N\tilde{\tau}^*\right\} \tag{5.146}$$

where $N = \omega^*(1+\gamma\overline{c})$, and $\overline{q}_R = -(\tilde{\sigma}T_w^4)/(hT_o')$.

The second limiting case, $T_p \gg T_w$, is described by Eqs. (5.143) and (5.144), which can be integrated numerically.

Intensity of temperature fluctuation. The results obtained indicate that the intensity of the temperature fluctuation in a two-phase mixture depends on several parameters: the physical properties of the fluid and particles, the total mass contents of the particles, the fraction of particles of each size, etc. Mainly, however, it depends on the density and specific heat ratios of the particles to the fluid.

In Fig. 5.31 the temperature fluctuation is plotted as a function of the particle velocity fluctuation (monodisperse mixture, with the load as parameter). It is seen that an increase in the velocity fluctuation is accompanied by an increase in the temperature fluctuation of the particles and a decrease in that of the carrier fluid. The curves meet at certain points, which represent the state of equilibrium of the system, i.e. the temperature difference $\Delta T' = 0$. An increase in the total mass content γ leads to a decrease in the equilibrium temperature. Figure 5.31 corresponds to the specific heat ratio $\overline{c} = 0.13$, while Fig. 5.32 corresponds to $\overline{c} = 2.0$; a significant difference can be seen in the shape of the curves.

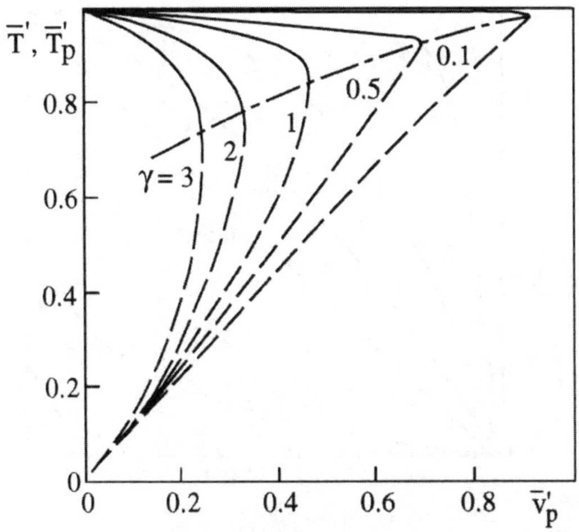

Fig. 5.31 Temperature fluctuations of the carrier fluid and the particles as a function of the particle velocity fluctuation ($\overline{c} = 0.13$, $Pr = 4$). Reprinted from Yarin and Hetsroni (1984 b), with permission. —— $\overline{T}'$, - - - $\overline{T}_p'$, -·- $\overline{T}' = (1+\gamma\overline{c})^{-1}$

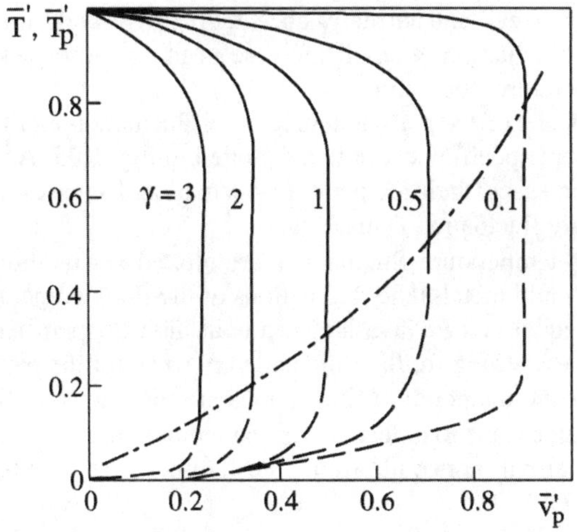

Fig. 5.32 Temperature fluctuations of the carrier fluid and the particles as a function of the particle velocity fluctuation ($\overline{c} = 0.13$, $Pr = 4$). Reprinted from Yarin and Hetsroni (1984 b), with permission. —— $\overline{T}'$, - - $\overline{T}_p'$, -·- $\overline{T}' = (1+\gamma\overline{c})^{-1}$

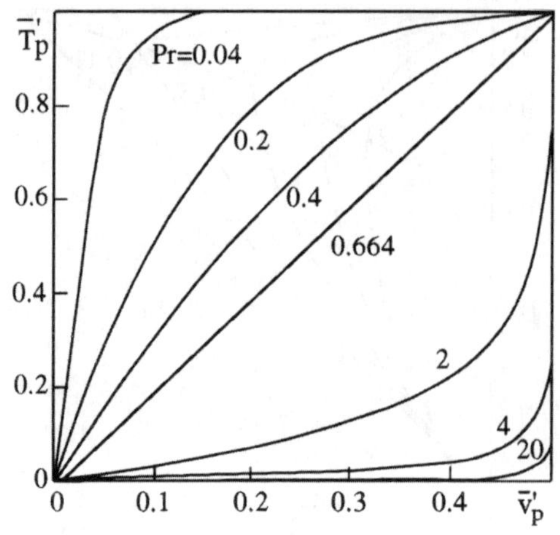

Fig. 5.33 The particle temperature fluctuation for different values of $Pr(\overline{c}=1, \gamma=1)$ Reprinted from Yarin and Hetsroni (1984 b), with permission

The contribution of the Prandtl number to the dependence of the temperature fluctuations of particles (Fig. 5.33) and fluid (Fig. 5.34), on the particle velocity fluctuations is also seen to be substantial. For high Pr the particle temperature fluctuation is very weak and virtually absent. At low Prandtl numbers and weak particle velocity fluctuation, a small increase in these causes a sharp increase in the particle temperature fluctuation.

The dependence of the particle temperature fluctuation on their velocity fluctuation for different specific heat ratios is plotted in Fig. 5.35. As the particle specific heat ratio increases, their temperature fluctuation decreases. At small vales of $\overline{c}$, the temperature fluctuation is maximal.

In Fig. 5.36 temperature fluctuations are plotted versus the Prandtl number. At very low Pr (liquid metals) the fluctuations of the fluid temperature and particle temperature are equal and Pr invariant, while at high Pr (petroleum) temperature fluctuations are very strong for the fluid and very weak for the particles.

In Fig. 5.37 the temperature fluctuations are plotted as a function of $\overline{c}$. An increase in the latter leads to a decrease in the temperature fluctuations, and an increase in the thermal nonequilibrium. This effect was mentioned earlier by Shraiber et al. (1990).

The effect of the total particle mass content on the temperature fluctuations is shown in Fig. 5.38. The maxima in the curves for the particles are due to the conflicting effect of the load: on the one hand, an increase of γ leads to a decrease of the particle velocity fluctuation, and consequently to an increase of their interaction time, with the attendant increase of the temperature fluctuation. On the other hand, an increase of γ leads to an increase of the total heat capacity of the admixture, and consequently to a decrease of the temperature fluctuation, as is seen in Fig. 5.38.

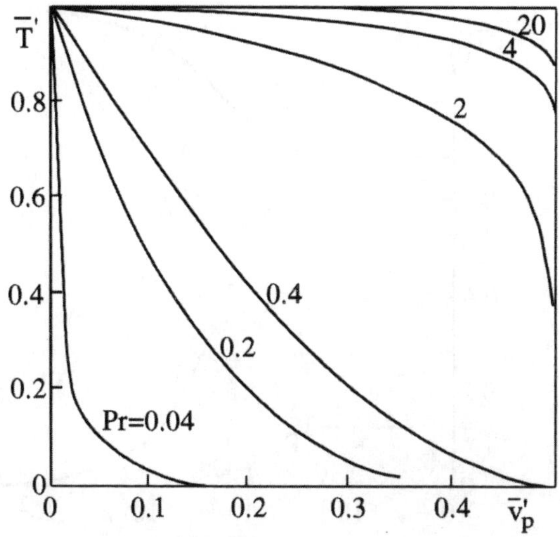

Fig. 5.34 The carrier fluid temperature fluctuation as a function of the particle velocity fluctuation for different values of $\Pr(\overline{c}=1, \gamma=1)$ Reprinted from Yarin and Hetsroni (1984 b), with permission

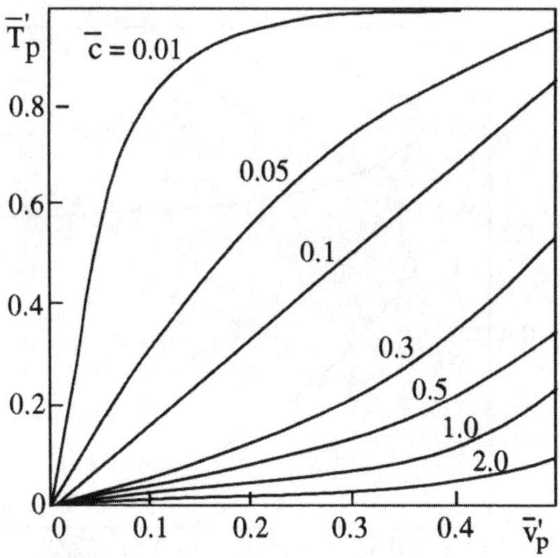

Fig. 5.35 The particle temperature fluctuation as a function of the particle velocity fluctuation for different values of the particle and carrier fluid specific heats $(\gamma=1, \quad \Pr=4)$. Reprinted from Yarin and Hetsroni (1984 b), with permission

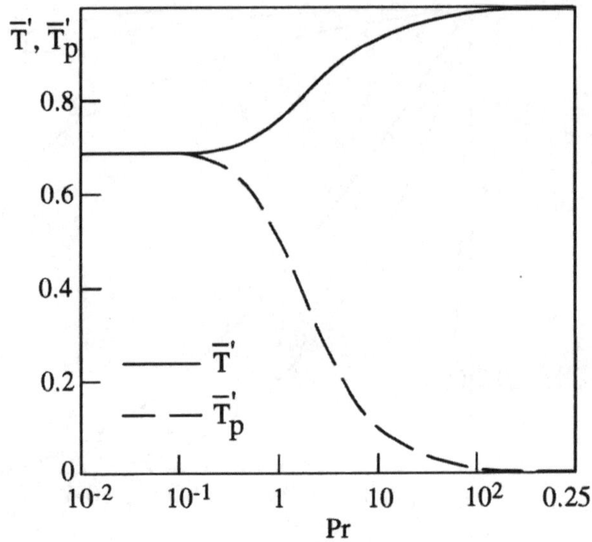

Fig. 5.36 Dependences of the carrier-fluid and particle temperature fluctuations on $Pr(\beta = 75,\ \ \overline{\rho} = 2 \cdot 10^3,\ \ Re = 1,\ \ \gamma = 1)$. Reprinted from Yarin and Hetsroni (1984 b), with permission

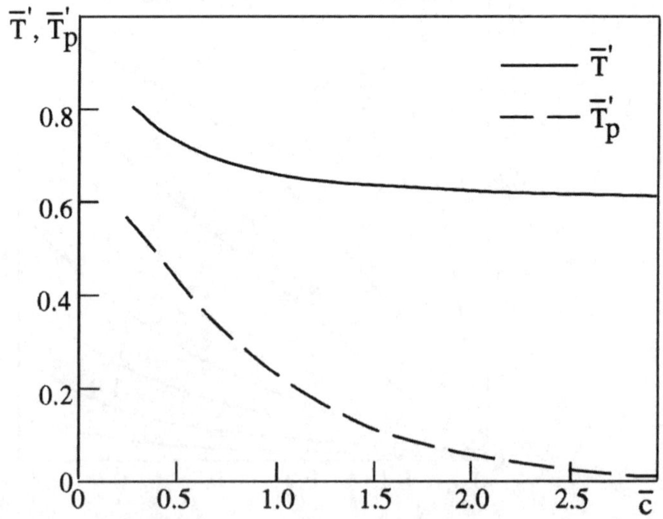

Fig. 5.37 The carrier fluid and particle temperature fluctuations as functions of the particle and fluid specific-heat ratio $(\rho = 75,\ \ \overline{\rho} = 2 \cdot 10^3,\ \ Re = 1,\ \ Pr = 4)$ Reprinted from Yarin and Hetsroni (1984 b), with permission

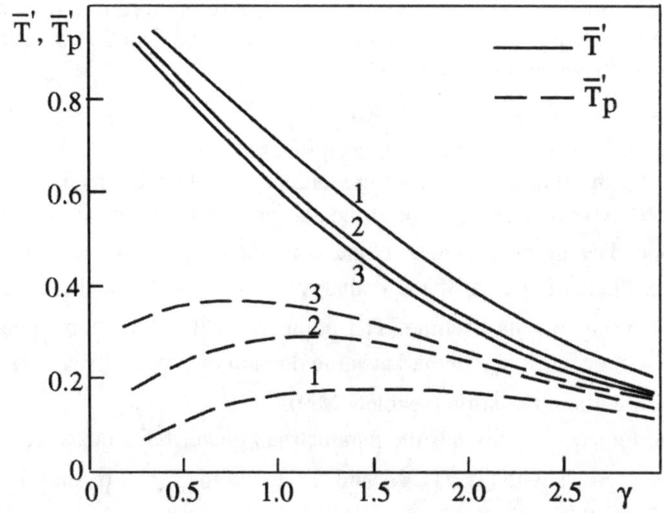

Fig. 5.38 The carrier fluid and the particle temperature fluctuations as functions of the total mass content of the admixture. Reprinted from Yarin and Hetsroni (1984 b), with permission $\bar{\rho} = 2 \cdot 10^3$, $\mathrm{Re} = 1$, $\bar{c} = 0.5$, $\mathrm{Pr} = 4$: 1 $\beta = 50$; 2 $\beta = 75$; 3 $\beta = 100$

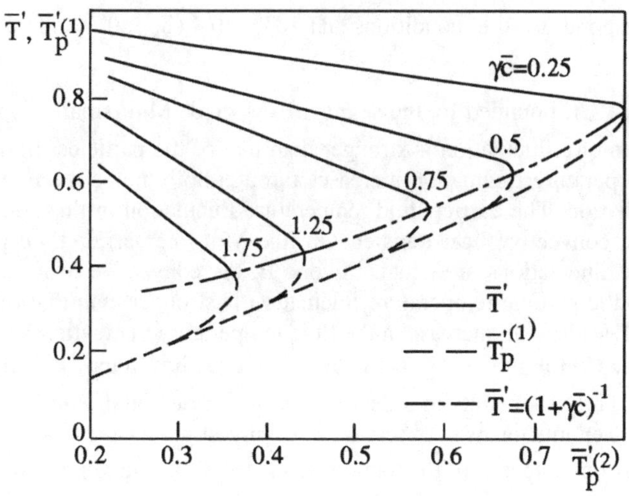

Fig. 5.39 The carrier fluid ($\bar{T}'$) and coarse particle ($\bar{T}'^{(1)}_p$) temperature fluctuations as a function of that of the fine particle ($T'^{(2)}_p$) for different values of the total mass content in a polydisperse mixture [$\bar{T}' = (1 + \gamma\bar{c})^{-1}$]. Reprinted from Yarin and Hetsroni (1984 b), with permission

The temperature fluctuation in a polydisperse mixture is shown in Fig. 5.39. Here the fluctuation of the carrier fluid is plotted as a function of that of the fine particles, with $\gamma\bar{c}$ as a parameter.

The effect of radiation is shown in Figs. 5.40–5.42 at $T_p \gg T_w$ and $T_w \gg T_p$ for a monodisperse admixture. It is seen that the effect is strong both for the particles and for the fluid. In Fig. 5.40 the temperature fluctuations of the carrier fluid are plotted as a function of those of the particles, with $\bar{q}_R$ (positive or negative) as parameter. The graph includes a line corresponding to flow without radiation, when the fluctuating part of the enthalpy $J_m' = \bar{T}' + \gamma\bar{c}\bar{T}_p'$ of the fluid element is constant throughout its lifetime. The point M on this line, corresponding to the limiting regime $\bar{T}' = \bar{T}_p'$, demarkates the domain of physically realizable states of the system without radiation (segment MN).

The line $\bar{q}_R = 0$ divides the parametrical plane $\bar{T}'\bar{T}_p'$ into zones corresponding to the regimes with $T_w > T_p$ (A and B) and with $T_w < T_p$ (C and D). The curve $\bar{T}'(\hat{T}_p')$ for both positive or negative $\bar{q}_R$ issues from the point N, corresponding to the initial state of the system ($t = 0$, $\bar{T}' = 1$, $\bar{T}_p' = 0$). Each point of these curves has its own value of $\bar{\tau}^*$, which increases the farther the point is from N.

Figure. 5.40 shows two lines joining the extrema of the $\bar{T}'(\bar{T}_p')$ curves. These lines correspond to the conditions $d\bar{T}'/d\bar{T}_p' = 0$ $(\bar{q}_R < 0)$ and $d\bar{T}_p'/d\bar{T}' = 0$ $(\bar{q}_R > 0)$.

In zone A, bounded by the segment NM and Mm (small $\tilde{\tau}^*$), the carrier fluid temperature fluctuation is stronger than that of the particles. In this case the particle temperature fluctuation increases through both the convective heat transfer and radiation. The carrier fluid temperature fluctuation in this zone decreases through the convective heat transfer. On line Mm, the carrier fluid and particle temperature fluctuations are equal. In zone B, the convective heat transfer is reversed, i.e. the particle temperature fluctuation is stronger than that of the carrier fluid, which leads to an increase in the fluid temperature fluctuation.

In zone C in Fig. 5.40 $(\bar{q}_R > 0)$, the convective heat removal from the carrier fluid to the particles results in a decrease in the carrier fluid temperature fluctuation. This phenomenon was observed when investigating the effect of radiative heat transfer on temperature fluctuation (Lisin 1988). For larger $\tilde{\tau}^*$, when the heat losses increase, both the carrier fluid and particle temperature fluctuations decrease (zone D), as a result of which the carrier fluid temperature fluctuation is stronger than that of the particles.

In Fig. 5.41, the carrier fluid and the particle temperature fluctuations are plotted as functions of the parameter $N\tilde{\tau}^*$. The particle temperature fluctuation gradually increases with the interaction time $\tilde{\tau}^*$. As for the dependence $\bar{T}(N\tilde{\tau}^*)$, it has an extremum at the point where the convective heat flux is reversed.

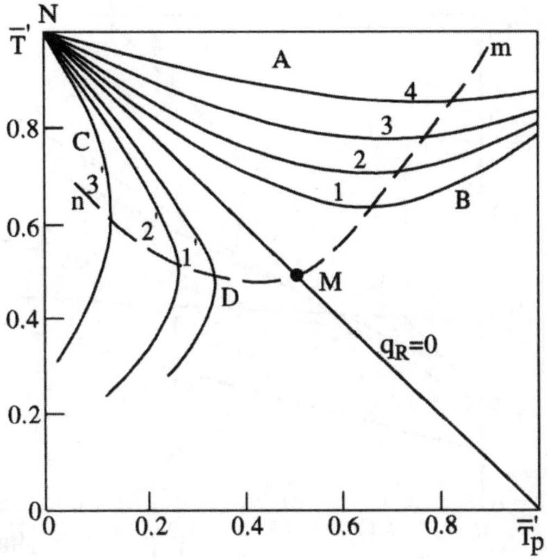

Fig. 5.40 Dependence of the carrier fluid temperature fluctuation on that of the particles. Reprinted from Yarin and Hetsroni (1984 b), with permission. Curve 1, $|\overline{q}_R| = 0.25$; curve 2, $|\overline{q}_R| = 0.5$; curve 3, $|\overline{q}_R| = 1$; curve 4, $|\overline{q}_R| = 2$ ($T_w >> T_p$); curve 1′, $|\overline{q}_R| = 0.125$; curve 2′, $|\overline{q}_R| = 0.25$; curve 3′, $|\overline{q}_R| = 0.5$ ($T_w << T_p$)

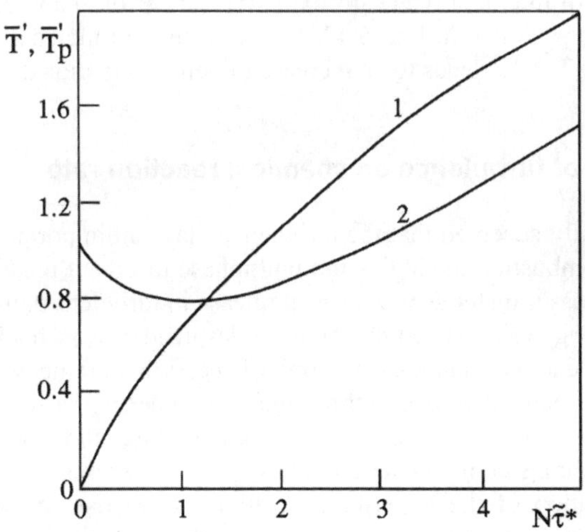

Fig. 5.41 Dependence of the carrier fluid (2) and particle (1) temperature fluctuations on the parameter $N\tilde{\tau}^*(\gamma\overline{c} = 1,\ |\overline{q}_R| = 1,\ T_w >> T_p)$. Reprinted from Yarin and Hetsroni (1984 b), with permission

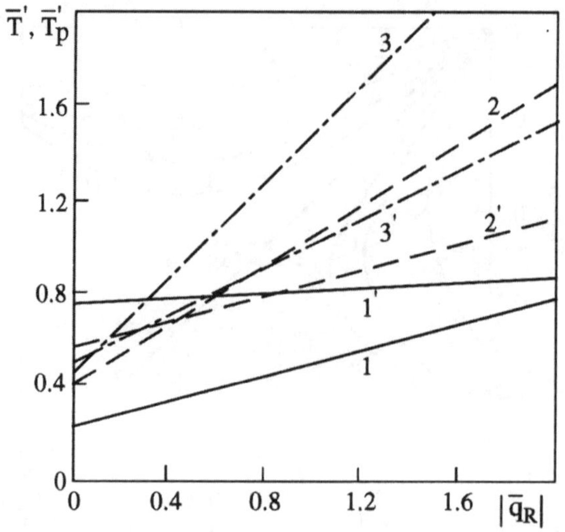

Fig. 5.42 Dependence of the carrier fluid and the particle temperature fluctuations on the parameter $|\overline{q}_R|(T_w >> T_p)$. Reprinted from Yarin and Hetsroni (1984 b), with permission. Curve 1, $\overline{T}_p'$; curve 1′, $\overline{T}'(N\tilde{\tau}^* = 0.6)$; curve 2, $\overline{T}_p'$; curve 2′, $\overline{T}'(N\tilde{\tau}^* = 1.8)$; curve 3, $\overline{T}_p'$; curve 3′, $\overline{T}'(N\tilde{\tau}^* = 30)$

The effect of the radiative heat flux on the carrier fluid and particle temperature fluctuations is shown in Fig. 5.42. It is seen that an increase in the radiative flux (in the case $T_w > T_p$) leads to an increase in both fluctuations.

1.5.5 Effect of turbulence on chemical reaction rate

It is generally accepted that the turbulence plays an important role in the development of combustion in single- and multiphase turbulent reactive flows. It affects heat and mass transfer as well as heat release in turbulent flames.

In describing turbulent combustion, widespread use is made of the mass, momentum and energy equations in terms of the flow parameters. These can be derived from the Navier–Stokes and the mass and energy conservation equations through resolution of the actual velocity, temperature and concentrations into mean and fluctuating components and subsequent averaging of these equations. Such transformation of the governing equations gives rise to problems of two kinds. The first type concerns the presence, in the averaged conservation equations, of additional terms accounting for the tensor components of the turbulent stresses and turbulent heat and mass fluxes, the second concerns the presence in the energy and mass conservation equations, of the source terms containing the averaged rate of chemical reaction.

The above may be illustrated as follows. Consider low velocity turbulent flow of a reactive mixture. Assuming that the specific heat and thermal conductivity of the medium are constant, we write energy and mass balance equations in the following form:

$$\rho c_p \frac{\partial T}{\partial t} + \rho c_p (\mathbf{v} \cdot \nabla) T = \lambda \nabla^2 T + q_h W(C,T) \tag{5.147}$$

$$\frac{\partial \rho}{\partial t} + \nabla \cdot (\rho \mathbf{v}) = 0 \tag{5.148}$$

where q_h and $W(C,T)$ are the heat and rate of the chemical reaction; ρ, v, T and C are the actual density, velocity, temperature and concentration, respectively.

Combination of Eqs. (5.147) and (5.148) yields

$$c_p \frac{\partial \rho T}{\partial t} + c_p \nabla \cdot (\rho \mathbf{v} T) = \lambda \nabla^2 T + q_h W(C,T) \tag{5.149}$$

Following the Reynolds procedure, we resolve the actual parameters into mean and fluctuating components

$$p = \overline{p} + p' \tag{5.150}$$

where p, $\overline{p}$ and p' are the actual, averaged and fluctuating values of the given parameters of interest: $p = \rho, v, T,$ and C.

Substituting Eq. (5.150) in Eq. (5.149) and averaging the result, we obtain the equation for the average parameters

$$\overline{\rho} \frac{\partial \overline{T}}{\partial t} + \nabla \cdot (\overline{\rho \mathbf{v}} \overline{T}) = \frac{\lambda}{c_p} \nabla^2 \overline{T} - \overline{T} \frac{\partial \overline{\rho}}{\partial t} - \frac{\partial \overline{\rho' T'}}{\partial t} - \nabla \cdot (\overline{\rho' T'} \overline{\mathbf{v}} + \overline{\rho' \mathbf{v}'} \overline{T} + \overline{T' \mathbf{v}'} \overline{\rho} + \overline{\rho' \mathbf{v}' T'}) + \tag{5.151}$$
$$+ \frac{q_h}{c_p} \overline{W(C,T)}$$

The correlations $\overline{T'\mathbf{v}'}$, $\overline{\rho' T'}$, $\overline{\rho' \mathbf{v}'}$ and $\overline{\rho' \mathbf{v}' T'}$ can be estimated with the aid of known models of turbulence (Schlichting 1979, Monin and Yaglom 1971). The average chemical reaction rate $W(C,T)$ depends not only on the average temperature and concentration, but also on their fluctuations. It differs substantially from the one corresponding to the average $\overline{T}$ and $\overline{C}$ (Zel'dovich 1949, Jones and Whitelaw 1982 and Lentini 1993). Furthermore, since the chemical source is a function, unknown in advance, of the average and fluctuating temperatures and concentrations, the set of governing equations is non closed. This problem was considered in detail in the context of turbulent combustion (Spalding 1976, Libby and Williams 1980, Bray 1978, Bray et al. 1984, 1988, Kuznetsov and Sabel'nikov 1989). The most convenient way to determine $\overline{W(T,C)}$ is via the probability density function. In this case the reaction average rate is expressed as

$$\overline{W(C,T)} = \int \int W(C,T)P(C,T)dCdT \tag{5.152}$$

where P(C,T) is the joint probability density function (pdf) of the temperature and concentration. In the general case $\overline{W(C_j,T)} = \int .. \int W(C_j,T)P(C_j,T)dC_j dT$, j=1,2,… .

Under this approach, the problem reduces to finding the pdf for the given type of turbulent flow. At present several models of turbulent combustion are available, in which various approximations for the pdf were proposed (Spalding 1971, Kent and Bilger 1977, Pope 1977 and Lockwood 1977). At the same time the effect of turbulence on the reaction rate may be estimated by direct calculation of the dependences $\overline{W(T,C)}$ and $W\left(\overline{T},\overline{C}\right)$ at known average and fluctuating temperatures and concentrations at a given point of turbulent flow and for a given form of the dependence W(T, C). This problem was studied by Tsuge and Sagara (1978) and Sagara (1980). Such an approach is fruitful in studying the effect of the fluctuations of T and C in two-phase reactive flows, and in particular, in the context of particle ignition.

The first attempt to estimate the effect of turbulent fluctuation on the chemical reaction rate was apparently made by Vulis (1960). Neglecting the change in the concentration, Vulis derived the following relation for $\overline{W(T)}$

$$\overline{W(T)} = W(\overline{T})\cos h\Omega \tag{5.153}$$

where $\overline{W(T)}$ and $W\left(\overline{T}\right)$ are the average reaction rate and that corresponding to the average temperature, $\Omega = \varepsilon_T/\overline{\theta}$, $\varepsilon_T = \sqrt{\overline{T'^2}}/\overline{T}$ is the fluctuation temperature intensity, and $\overline{\theta} = (R\overline{T})/E$.

It is seen that the ratio $\chi = \overline{W(T)}/W(\overline{T})$ is determined by two factors: (i) the intensity of temperature fluctuation, and (ii) the average temperature level. Increase of the turbulence intensity and decrease of the average temperature lead to substantial increase of χ. For example, under realistic physical conditions: $\varepsilon_T \sim 0.1$, $\theta \sim 0.05$, and $\chi \sim 3.76$.

The effect of concentration fluctuation on the average chemical reaction rate was considered by Kuznetsov (1969) and by Libby and Williams (1980). It was shown that it tends to reduce the reaction rate relative to that corresponding to the average temperature. Thus, the effects of temperature and concentration fluctuation are mutually opposite in this respect, and, in the general case the average reaction rate in turbulent flow may be higher or lower than that corresponding to the average temperature (Kuznetsov and Sabel'nikov 1989, Vulis 1972).

In considering the above effects on the average reaction rate, we will use the following simplified assumptions: (i) the characteristic reaction time is much shorter than the time scale of the turbulence, hence the actual reaction rate is determined by the actual temperature and concentration of the reactive mixture, (ii) the reactive mixture is an incompressible fluid with invariable physical properties, (iii) the mass and thermal diffusivities are equal, (iv) a simple one-step chemical reaction of the n-th order is involved, expressed as the product of two factors depending only on the actual temperature and concentration, respectively

$$W'(C,T) = \varphi'(C)\psi(T) \tag{1.154}$$

where $\varphi'(C)$ and $\psi(T)$ are the relevant dependences.

For mathematical convenience, the chemical reaction rate can be expediently expressed as a function of the actual temperature. To this end, we seek the relationship between the actual temperature and concentration, using the energy and fuel conservation equations

$$\frac{\partial h}{\partial t} + (\mathbf{v}\cdot\nabla)h = \alpha\nabla^2 h + q_h W'(C,T) \tag{5.155}$$

$$\frac{\partial c}{\partial t} + (\mathbf{v}\cdot\nabla)C = D\nabla^2 C - W'(C,T) \tag{1.156}$$

where α and D are thermal and mass diffusivity coefficients, h is the enthalpy ($h = c_p T$, c_p being the specific heat).

Eliminating the source terms from Eqs. (5.155) and (5.156), we obtain the equation for the total (physical and chemical) enthalpy

$$\frac{\partial H}{\partial t} + (\mathbf{v}\cdot\nabla)H = \alpha\nabla^2 H \tag{1.157}$$

where $H = h + q_h c$

For an insulated system $(\partial H/\partial n)_\Sigma = 0$, Σ being the system envelope, n the normal to it. In this case Eq. (5.157) is integrated as

$$H = H_* = \text{constant} \tag{1.158}$$

The above constant is found from the condition that the maximum temperature T_m corresponds to complete combustion, i.e. C = 0. Accordingly, we obtain the concentration as per

$$C = \frac{q_h}{c_p(T_m - T)} \tag{5.159}$$

With this in mind, we can present $\varphi'(C)$ as

$$\varphi'(C) = \varphi'\left[\frac{q_n}{c_p(T_m - T)}\right] = \varphi(T)$$

It should be noted that Eqs. (5.158) and (5.159) are valid only when the Lewis number $Le = \alpha/D$ equals unity. In the general case, with $Le \neq 1$ the distribution of the total enthalpy within the reactive system is more complicated (Fig. 5.43). For example, in heterogeneous combustion the total enthalpy has an extremum near the outer edge of the boundary layer. A change in the total enthalpy takes place also in combustion of a gaseous mixture (homogenous combustion), in which case H has extrema in the vicinity of the flame. It was shown by Zel'dovich et al. (1985) that this is the result of energy redistribution due to dissimilar rates of heat and mass transfer at Le $\neq$ 1.

Because of the unequal variation of the physical and chemical enthalpies their sum in some cross-section of the boundary layer may be larger (or smaller) than that far from that cross-section. In particular, at Le>1 (thick thermal layer at the solid surface) the rise of the temperature precedes the drop in the concentration, and the energy influx to the given zone from the neighboring ones leads to increase of the local total enthalpy (Zel'dovich et al. 1985 and Vulis 1961).

Bearing in mind correlation (5.159), we present the reaction rate as

$$W'(C,T) = W(T) = \varphi(T)\psi(T) \tag{6.160}$$

Assuming that $W'(C,T) = z'C^n \exp(-E/RT)$, we write the explicit expression for W(T)

$$W(T) = z(T_m - T)^n \exp\left(-\frac{E}{RT}\right) \tag{5.161}$$

where z' and z are the pre-exponential factor and the modified one, E the activation energy, R the universal gas constant, and n the order of the reaction.

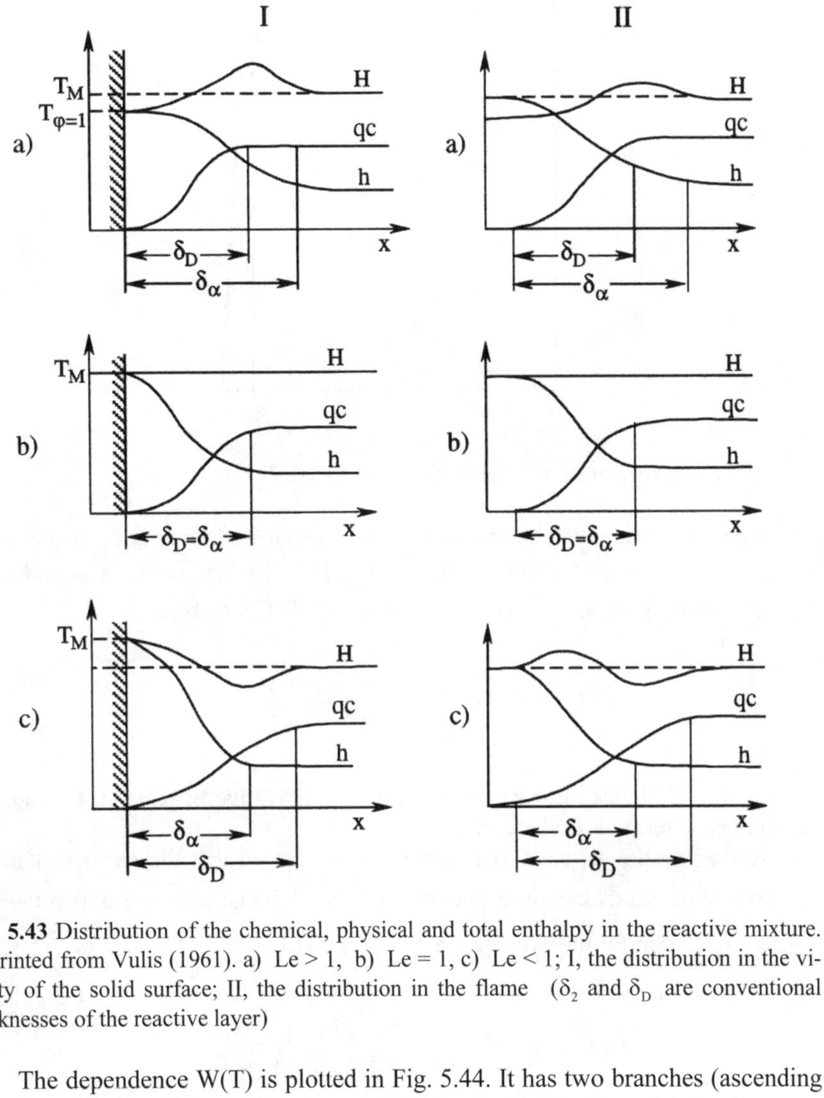

Fig. 5.43 Distribution of the chemical, physical and total enthalpy in the reactive mixture. Reprinted from Vulis (1961). a) $Le > 1$, b) $Le = 1$, c) $Le < 1$; I, the distribution in the vicinity of the solid surface; II, the distribution in the flame (δ_2 and δ_D are conventional thicknesses of the reactive layer)

The dependence W(T) is plotted in Fig. 5.44. It has two branches (ascending and decending) which represent, respectively, insignificant and intensive consumption of the reactive mixture. The maximum of W(T) corresponds to a certain temperature T_* close to T_m

$$\theta_* = \theta_m - 2n\theta_m^2 \approx \theta_m \tag{5.162}$$

where $\theta = (RT)/E$.

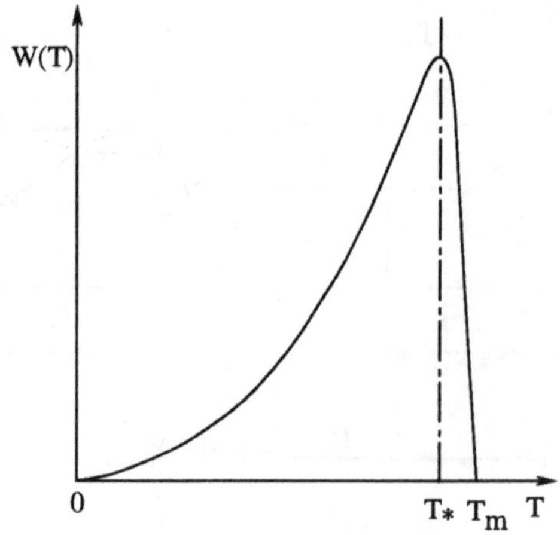

Fig. 5.44 Dependence of the reaction rate on temperature

Resolving the actual temperature into average and fluctuating components $T=\overline{T}+T'$, we expand $W(T) = W(\overline{T}+T')$ in a Taylor series. Assuming that W(T) is a continuous and differentiable function of T we obtain

$$W(T)=W(\overline{T})+\sum_{k=1}^{\infty}\frac{d^{\boldsymbol{K}}W(\overline{T})}{d\overline{T}^{\boldsymbol{K}}}\frac{(T')^{k}}{k!} \tag{5.163}$$

where the ***K*** italics indicate the order of the derivative, whereas the k indicate the number of term of Taylor series.

Averaging the series (5.163) and bearing in mind that the terms containing odd powers of the temperature fluctuation $\overline{(T')^{2k-1}}$ equal zero, we obtain the following expression for the average reaction rate $\overline{W(T)}$

$$\overline{W(T)}=W(\overline{T})+\sum_{k=1}^{\infty}\frac{dW^{2\boldsymbol{K}}(\overline{T})}{d\overline{T}^{2\boldsymbol{K}}}\cdot\frac{\overline{T'^{2k}}}{(2k)!} \tag{5.164}$$

Using the formula

$$\{\varphi(\overline{T})\psi(\overline{T})\}^{2\boldsymbol{K}}=\sum_{m}^{2K}C_{2k}^{m}\varphi^{\boldsymbol{m}}(\overline{T})\psi^{2\boldsymbol{K}-\boldsymbol{m}}(\overline{T}) \tag{5.165}$$

we rewrite Eq. (5.164) in the following form

$$\chi = 1 - \sum_{k=1}^{\infty} W(\bar{T})^{-1} \left\{ \sum_{m}^{2k} C_{2k}^{m} \varphi^{m}(\bar{T}) \psi^{2K-m}(\bar{T}) \right\} \frac{\overline{T'^{2k}}}{(2k)!} \tag{5.166}$$

where $\chi = \overline{W(T)}/W(\bar{T})$ is ratio of the average reaction rate to that corresponding to the average temperature, $C_{2k}^{m} = (2k)!/(m!(2k-m)!)$, $C_{2k}^{o} - C_{2k}^{2k} = 1$, $\varphi^{o}(\bar{T}) = \varphi(\bar{T})$, $\psi^{o}(\bar{T}) = \psi(\bar{T})$, for integer n, $\varphi^{m}(\bar{T}) = 0$ at $m > n$, $\varphi^{m}(\bar{T}) = (-1)^{m} n!$ at $m = n$.

To calculate the sum on the right hand side of Eq. (5.166) we use the Frank-Kamenetskii approximation (1969). Presenting the function $\psi(\bar{T})$ in the form

$$\psi(\bar{T}) = \exp\left(-\frac{E}{R\bar{T}}\right) \simeq \exp\left(-\frac{E}{RT_{av}}\right) \exp\left(\frac{E}{R_{av}}(\bar{T} - T_{av})\right) \tag{5.167}$$

we calculate the derivative $\psi^{2K}(\bar{T})$

$$\psi^{2K} = \psi(\bar{T}) \left(\frac{E}{RT_{av}^{2}}\right)^{2k} \tag{5.168}$$

where T_{av} is some average temperature close to $\bar{T}$. Substituting expression (5.168) in Eq. (5.166) we obtain

$$\chi = K_{1}(\Omega\omega) \cos h\Omega - K_{2}(\Omega\omega) \sin h\Omega \tag{5.169}$$

where

$$K_{1}(\Omega\omega) = 1 + \frac{n(n-1)}{2!}(\Omega\omega)^{2} + \frac{n(n-1)(n-2)(n-3)}{4!}(\Omega\omega)^{4} + \ldots$$

$$\ldots + \frac{n(n-1)(n-2)(n-3)\ldots(n-2k+1)}{(2k)!}(\Omega\omega)^{2k}$$

$$K_{2}(\Omega\omega) = \frac{n}{1!}(\Omega\omega) + \frac{n(n-1)(n-2)}{3!}(\Omega\omega)^{3} +$$

$$\ldots + \frac{n(n-1)(n-2)(n-3)\ldots(n-2K+2)}{(2k-1)!}(\Omega\omega)^{2K-1}$$

$\omega = \theta_{av}^{2}/(\bar{\theta}_{m} - \theta_{av})$, $\Omega = \varepsilon_{T}/\theta_{av}$, $\varepsilon_{T} = (\overline{\theta'^{2}})^{1/2}/\theta_{av}$ is intensity of the temperature fluctuation, and $\theta_{av} = RT_{av}/E$.

Since the Frank–Kamenetskii approximation is valid at $\overline{\theta}/\theta_{av} \sim 1$, the parameters ω and Ω can be presented as follows

$$\omega \simeq \frac{\overline{\theta}^2}{\theta_m - \overline{\theta}}; \qquad \Omega = \frac{\varepsilon_T}{\theta_{av}} \approx \frac{\varepsilon_T}{\overline{\theta}} \tag{5.170}$$

Bearing in mind that $T_m \geq \overline{T} + \sqrt{\overline{T'^2}}$ (Fig. 5.45), we can write the following inequality, which is satisfied at any point of the turbulent reactive flow

$$\varepsilon_T \leq \frac{\theta_m}{\overline{\theta} - 1} \tag{5.171}$$

In accordance with Eq. (5.171) we obtain the following estimate for the product $\Omega\omega$

$$\Omega\omega = \frac{\varepsilon_T}{\dfrac{\theta_m}{\overline{\theta} - 1}} \tag{5.172}$$

For chemical reactions of zero and first orders, Eq. (5.169) takes the following form

$$\chi = \cos h\Omega \tag{5.173}$$

$$\chi = \cos h\Omega - \omega\Omega \sin h\Omega \tag{5.174}$$

The first expression is identical to the Vulis formula (5.153) which does not incorporate the effect of concentration fluctuations due to consumption of the reactive mixture and its mixing with the environment. In this case an increase of the parameter Ω leads to a progressive increase of the average reaction rate. Note that such a dependence is typical of certain physical characteristics of high-temperature media (electrical conductivity of low-temperature plasmas, etc.) which are exponential with temperature. Thus, the effect of temperature fluctuation on the reaction constant can be modeled by measuring the electric conductivity in turbulent flow (Vulis et al. 1968).

In contrast to the above case, at $n \geq 1$ (reactions of first or higher order) the dependence $\chi(\Omega)$ has a complicated character (Fig. 5.46). The $\chi(\Omega)$ curves, where form depends on the value of the product $\Omega\omega$, have descending and ascending branches corresponding to the states at which the turbulence leads to decrease and increase of the $\overline{W(T)}/W(\overline{T})$ ratio respectively. The first case takes

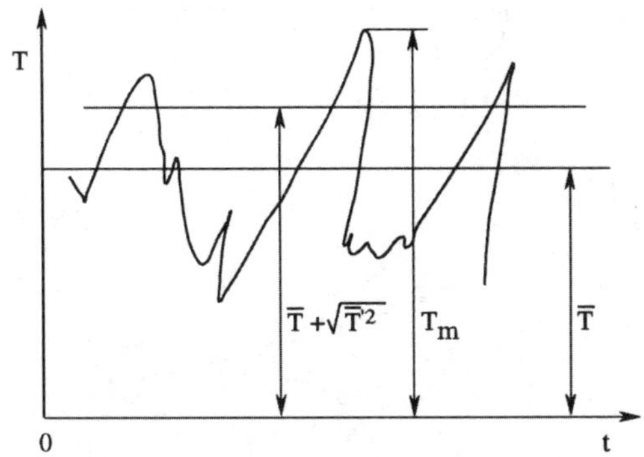

Fig. 5.45 The change of temperature in turbulent flow

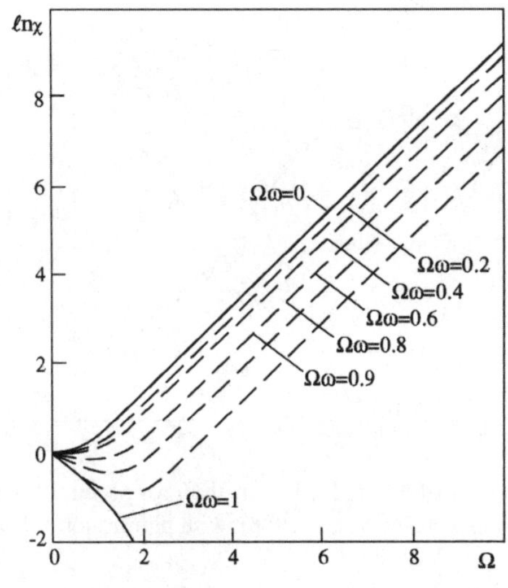

Fig. 5.46 The dependences $\ln\chi = f(\Omega)$

place when $\omega\Omega \sim 1$ and Ω is sufficiently small (high temperature), whereas the second one is realized at large Ω. The situations corresponding to a negligibly weak and a strong effect of the fluctuation of concentration on the reaction rate are represented by curves $\omega\Omega = 0$ and $\Omega\omega = 1$ respectively.

The dependence (5.174) is presented in Fig. 5.47 in the form of a family of $\omega(\Omega)$ curves for different values of χ. The curve $\omega(\Omega)_{\chi=1}$ corresponding to $\chi = 1$ subdivides the parametric plane $\omega - \Omega$ in two domains, corresponding to states at which the turbulence leads to an increase and a decrease of the average reaction rate (χ>1 and χ<1), respectively. This curve issues from the point $\Omega = 0,\ \omega = 0.5$ on the axis of ordinates and descends monotonically with Ω.

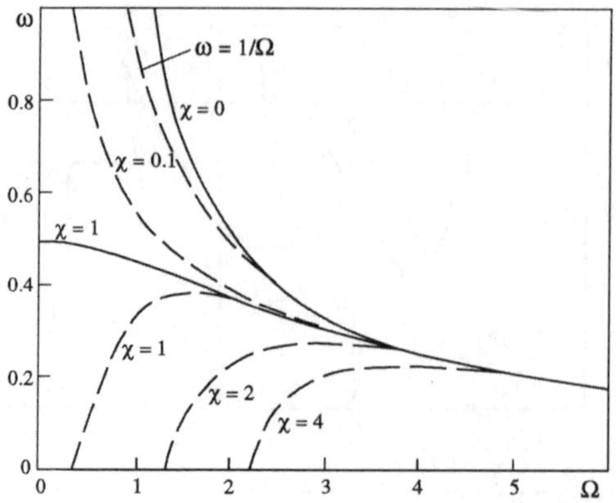

Fig. 4.47 The dependences $\omega(\Omega)$

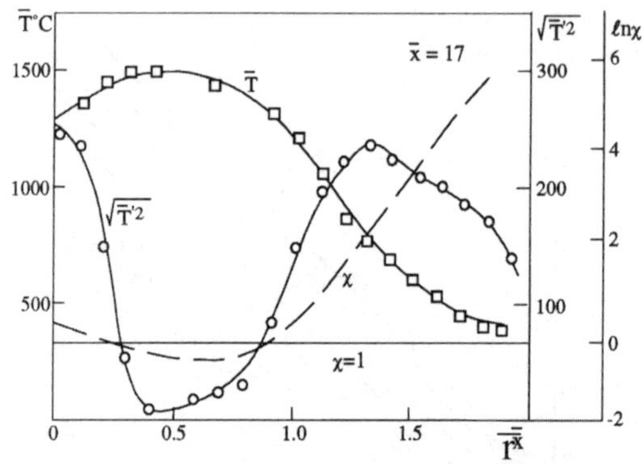

Fig. 5.48 The distributions of $\bar{T}$ and $\sqrt{\overline{T'^2}}$ in the cross-section of a turbulent premixed torch. Reprinted from Yoshida and Tsuji (1978), with permission. $---\chi$ (calculation)

Hence, the maximum value of the parameter ω at which the effect of the turbulence is positive $(\ln\chi > 0)$ equals 0.5. Accordingly, the lower temperature level corresponding to a regime with $\chi \le 1$ is

$$\theta_{\text{low}} = 0.25(\theta_{\text{m}} - 8\theta_{\text{m}}^2) \tag{5.175}$$

Thus, increase of the average reaction rate takes place at relatively low temperatures $(\theta < \theta_{\text{low}})$, when the effect of the concentration fluctuation is negligible and the dominant factor is that of the temperature. In situations where consumption of the reactive mixture is substantial (high-temperature regimes) the dominant factor is the concentration fluctuation. With the effect of temperature fluctuation sufficiently weak, the concentration fluctuation leads to decrease of the average reaction rate.

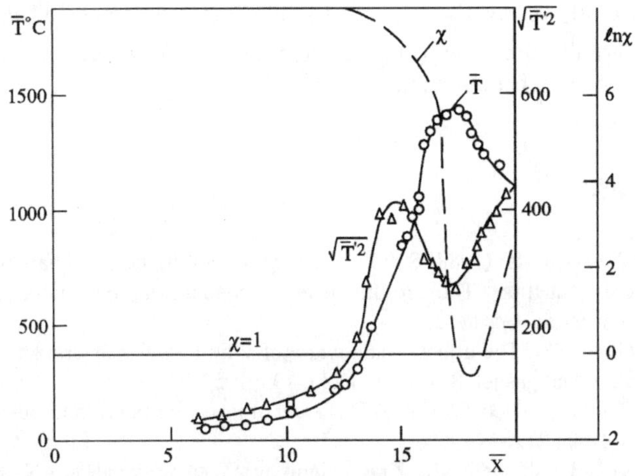

Fig. 5.49 The distributions of $\bar{T}$ and $\sqrt{\bar{T}'^2}$ along the axis of a turbulent premixed torch. Reprinted from Yoshida and Tsuji (1978), with permission. – – –χ (calculation)

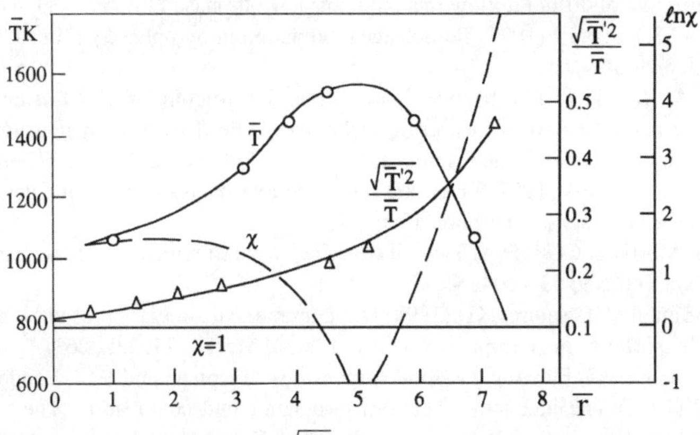

Fig. 5.50 The distributions of $\bar{T}$ and $\sqrt{\bar{T}'^2}/\bar{T}$ in the cross-section of a turbulent unpremixed torch Reprinted from Ballantyne and Bray (1976), with permission. – – –χ (calculation)

A number of experimental works contain data on the thermal structure of turbulent premixed and unpremixed torches, permitting estimation of the effect of fluctuation on the reaction rate (Kunugi and Jinno 1959, Lockwood and Odidi 1975, Ballantyne and Moss 1977, Ballantyne and Bray 1977, Yoshida and Tsuji 1979). Some calculation results of the ratio $\overline{W(T)}/W(\bar{T})$ in turbulent torches are presented in Figs. 5.48 - 5.50 (the values of χ were estimated for: m=1, $E/R = 2\cdot 10^4$ K^{-1}, $T_m = \bar{T} + \sqrt{\bar{T}'^2}$). They show that in the flow field of turbulent flames (torches) there are zones in which the effect of turbulent fluctuation is negative ($\ln \chi < 0$) or positive ($\ln \chi > 0$), the first zone being located in the vicinity of the combustion front (high temperature region) where the effect of concentration fluctuation is dominant.

References

Abou-Arab TW, Roco MC (1988) Solid phase contribution in the two-phase turbulence kinetic energy equation. The 3rd International Symposium on Liquid–Solid Flows. ASME, New York, pp. 13–28

Abramovich GN (1970) The effect of admixture of solid particles or droplets on the structure of a turbulent gas jet. Sov. Phys. Dokl. 190: 1052–1055

Abramovich GN, Girshovich TA (1973) Diffusion of heavy particles in turbulent gas flows. Sov. Phys. Dokl. 212: 573–576

Abramovich GN, Girshovich TA, Krasheninnikov S Yu, Sekundov AN, Smirnova IP (1984) Theory of turbulent jet. (in Russian) Science, Moscow

Achenbach E (1974) Vortex shedding from spheres. J. Fluid Mech. 62: 209–221

Alajbegovich A, Assad A, Bonetto F, Lahey RT Jr (1994) Phase distribution and turbulence structure for solid/fluid upflow in a pipe. Int. J. Multiphase Flow 20: 453–479

Al-Tawell AM, Landau J (1977) Turbulence modulation in two-phase jet. Int. J. Multiphase Flow 3: 341–351

Ballantyne A, Bray KNC (1976) Investigations into the structure of jet diffusion flames using time-resolved optical measuring techniques. The Sixteenth Symposium (International) on Combustion. The Combustion Institute, Pittsburgh, Pa., pp. 777–787

Ballantyne A, Moss JB (1977) Fine wire thermocouple measurements of fluctuating temperature. Combust. Sci. Technol. 17: 63–72

Berlow RS, Morrison CQ (1990) Two-phase velocity measurements in dense particle-laden jets. Exp. Fluids 9: 93-–104

Boivin M, Simonin O, Squires KD (1998) Direct numerical simulation of turbulence modulation by particles in isotropic turbulence. J. Fluid Mech. 375: 235–263

Boothroyd RG (1971). Flowing gas-solid suspensions. Chapman and Hall, London

Bray KNC (1978) The interaction between turbulence and combustion. The Seventeenth Symposium (International) on Combustion. The Combustion Institute, Pittsburgh, Pa., pp. 223–233

Bray KNC, Libby PA, Moss JB (1984) Flamelet crossing frequencies and mean reaction rates in premixed turbulent combustion. Combust. Sci. Technol. 41: 143–172

Bray KNC, Champion M, Libby PA (1988) Mean reaction rates in premixed turbulent flames. The Twenty-second Symposium (International) on Combustion. The Combustion Institute, Pittsburgh, Pa., pp. 763–769

Crowe CT (2000) On models for turbulence modulation in fluid-particle flows. Int. J. Multiphase Flow 26: 719–729

Crowe CT, Troutt TR, Chung JN (1996) Numerical models for two-phase turbulent flows. In: Lumley L, Van Dake M, Reed HL (eds) Annu. Rev. Fluid Mech. 28: 11–43

Danon H, Wolfshtein M, Hetsroni G (1974) Numerical calculations of two-phase turbulent jets. Int. J. Multiphase Flow 3: 223–234

Derevich IV (2001) Influence of internal turbulent structure on intensity of velocity and temperature fluctuations of particles. Int. J. Heat Mass Transfer 44: 4505–4521

Elghobashi SE, Abou-Arab TW (1983) A two-equation turbulent model for two-phase flows. Phys. Fluids 26: 931–938

Elghobashi S, Truesdell GC (1992) Direct simulation of particle dispersion in a decaying isotrorpic turbulence. J. Fluid Mech. 242: 655–700

Elghobashi S, Truesdell GC (1993) On the two-way interaction between homogeneous turbulence and dispersed solid particles. 1: Turbulence modulation. Phys. Fluids A, 5: 1790–1801

Frank-Kamenetskii DA (1969) Diffusion and heat transfer in chemical kinetics. 2nd edn., Plenum, New York

Frishman FM (1979) Effect of phase relative motion on turbulence intensity. In: Turbulent two-phase flows. (in Russian) A.N. ESSR, Tallinn, pp. 134–136

Girshovich TA, Leonov BA (1979) On the effect of admixture weight on the turbulent structure of vertical two-phase jet. In: Laats MK (ed) The III All-Union cConference in Theoretical and Applied Aspects of Turbulent Flows. Part II turbulent two-phase flows. (in Russian) AN ESSR, Talllinn, pp. 127–133

Gore RA, Crowe CT (1989) Effect of particle size on modulating turbulent intensity. Int. J. Multiphase Flow 15: 279–285

Haam SJ, Brodkey RS (2000) Laser Droplet anemometry measurements in an index of refraction column in the presence of dispersed beads. Part II. Motions of dispersed beads obtained by particle tracking velocimetry measurements. Int. J. Multiphase Flow 26: 1419–1438

Hetsroni G (1989) Particle-turbulence interaction. Int. J. Multiphase Flow 15: 735–746

Hetsroni G, Rozenblit R (1994) Heat transfer to a liquid-solid mixture in a flume. Int. J. Multiphase Flow 20: 671–689

Hetsroni G, Sokolov M (1971) Distribution of mass, velocity and intensity of turbulence in a two-phase turbulent jet. Trans. ASME. J. Appl. Mech. 38: 315–327

Hetsroni G, Rozenblit R, Lu DM (1995) Heat transfer enhancement by a particle on the bottom of a flume. Int. J. Multiphase Flow 21: 963–984

Hetsroni G, Rozenblit R, Yarin LP (1997) The effect of coarse particles on the heat transfer in a turbulent boundary layer. Int. J. Heat Mass Transfer 40: 2201–2217

Hinze JO (1972) Turbulent fluid and particle interaction. In: Hetsroni G, Sideman S, Hartnett JP (eds) Prog. Heat Mass Transfer 6: 433–452

Jaberi FA (1998) Turbulence fluctuations in particle-laden homogeneous turbulent flows. Int. J. Heat Mass Transfer 41: 4081–4090

Jaberi FA, Mashayek F (2000) Temperature decay in two-phase turbulent flows. Int. J. Heat Mass Transfer 43: 993–1005

Jones WP, Whitelaw JH (1982) Calculation methods for reacting turbulent flows. A review. Combust. Flame 48: 1–26

Kenning VM, Crowe CT (1977) On the effect of particles on carrier phase turbulence in gas-particle flows. Int. J. Multiphase Flow 23: 405–408

Kent JH, Bilger RW (1976) The prediction of turbulent diffusion flame fields and nitric oxide formation. The Sixteenth Symposium (International) on Combustion. The Combustion Institute, Pittsburgh, Pa., pp. 1643–1656

Kolmogorov AN (1942) Equations of turbulent motion of incompressible fluid. Jzv. Akad. Nauk SSSR, Phys. Ser. 1–2, 56–58 (in Russian)

Kulick JD, Fessler JR, Eaton JK (1993) On the interactions between particles and turbulence in fully-developed channel flow in air. Report No. MD-66. Standford University

Kunugi M, Jinno H (1959) Measurements of fluctuating flame temeprature. The Seventh Symposium (International) on Combustion. The Combustion Institute. Butterworths, pp. 942–948

Kuznetsov VR (1969) The effect of temperature and concentration fluctuations on average rate of rection in turbulent flow. The second all-union symposiuim on combustion and explosion. (in Russian) AN SSR, Chernogolovka 99–104

Kuznetsov VR, Sabel'nikov BA (1989) Turbulence and combustion. Hemisphere, New York

Laats MK, Frishman FM (1973) The development of the methodics and investigation of turbulence intensity at the axis of two-phase turbulent jet. Fluid Dynam. 8: 153–157

Lee SL, Durst F (1982) On the motion of particles in turbulent duct flow. Int. J. Multiphase Flow. 8: 125–146

Lentini D (1993) Models for Turbulent Combustion. In: Puri JK(ed). Environmental Implications of Combustion Processes. CRC Press, London, pp. 221–229

Levy Y, Lockwood FC (1981) Velocity measurements in a particle-laden turbulent free jet. Combust. Flame 40: 333–339

Libby PA, Williams FA (1980) In: Libby PA, Williams FA (eds) Turbulent reacting flows. Springer, Berlin, p.1

Lisin FN (1988) On temperature fluctuations in turbulent flow. J. Eng. Phys. 5: 731–735 (in Russian)

Lockwood FC (1977) The modelling of turbulent premixed and diffusion combustion in the computation of engineering flows. Combust. Flame 29: 111–122

Lockwood FC, Odidi AOO (1974) Measurement of mean and fluctuating temperature and of ion concentration in round free-jet turbulent diffusion and premixed flames. The Fifteenth Symposium (International) on Combustion. The Combustion Institute, Pittsburgh, Pa., pp. 561–571

Maeda M, Hishida K, Furutani T (1980) Optical measurements of local gas and particle velocity in an upward following dilute gas–solid suspension. In: polyphase flow and transport technology. Century 2 ETC, San Francisco Calif., pp. 216–221

Mizukami M, Parthasarathy RN, Faeth GM (1992) Particle-generated turbulence in homogenous dilute dispersed flows. Int. J. Multiphase Flow 18: 397–412

Moderress D, Tan H, Elghobashi S (1984a) Two-component LDA measurement in a two-phase turbulent jet. AIAA J. 22: 624–630

Moderress D, Wuerer J, Elghobashi SF (1984b) An expereimetnal study of a turbulent round two-phase jet. Chem. Eng. Commun 28: 341–354

Monin AS, Yaglom AM (1971) Statistical fluid mechanics. MIT Press, Cambridge, Mass

Nigmatulin RI (1991) Dynamics of multiphase media Hemisphere, London, vols. 1 and 2

Nouri JM, Whitelaw JH, Yianneskis M (1987) Particle motion and turbulence in dense two-phase flows. Int. J. Multiphase Flow 13: 729–739

Owcn PR (1969) Pneumatic transport. J. Fluid Mech. 39: 407–432

Pan Y, Banerjee J (1997) Numerical investigation of the effects of large particles on wall turbulence. Phys. Fluids 9: 3786–3807

Parthasarathy RN, Faeth GM (1987) Structure of particle-laden turbulent water jets in still water. Int. J. Mutliphase Flow 13: 699–716

Parthasarathy RN, Faeth GM (1990) Turbulence modulation in homogeneous dilute particle-laden flows. J. Fluid Mech. 220: 485–537

Pope SB (1977) The implications of the probability equations for turbulent combustion models. Combust. Flame 29: 235–246

Prandtl L (1925) Ueber die ausgebildete turbulenz. Z. Angew. Math., Mech. 5: 136–139
Rashidi M, Hetsroni G, Banerjee S (1990a) Particle–turbulence interaction in a boundary layer. Int. J. Multiphase Flow 16: 935–949
Rashidi M, Hetsroni G, Banerjee S (1990b) Mechanisms of heat and mass transport at gas-liquid interfaces. Int. J. Heat Mass Transfer 34: 1799–1810
Sagara K (1980) Exact turbulence correction to Arrhenius law in the asymptotic limit of high activation energy. Sci. Technol. 21: 191–197
Sato Y, Hishida K (1996) Transport process of turbulent energy in particle-laden turbulent flow. Int. J. Heat Fluid Flow 17: 202–210
Savolainen K, Karvinen R (1998) The effect of particles on gas turbulence in a vertical upward pipe flow. The Third International Conference on Multiphase Flow. ICMF-98, Lyon, France, June 8–12
Schiller L, Naumann A (1933) Ueber die grundlegenden berchnungen bei der schwerkraftaufbereitung. Ver. Deut. Ing. 77: 318–320
Schlichting H (1979) Boundary-layer theory. McGraw-Hill, New York
Serizawa A, Kataoka I, Michiyoshi I (1975) Turbulence structure of air-water bubbly flow. II Local properties. Int. J. Multiphase Flow 2: 235–246
Sheen HJ, Jou BH, Lee YT (1994) Effect of particle size on a two-phase turbulent jet. Exp. Thermal Fluid Sci. 8: 315–327
Shraiber AA, Gavin LB, Naumov VA, Yatsenko VP (1990) Turbulent flows in gas suspension. Hemisphere, New York
Shuen J-S, Solomon ASP, Zhang Q-F, Faeth GM (1985) Structure of particle-laden jets: measurements and predictions. AIAI J. 23: 396–404
Soo SL (1990) Multiphase fluid dynamics. Science Press and Gover Technical, Beijing
Spalding DB (1971) Concentration fluctuation in round turbulent free jet. Chem. Eng. Sci. 26: 95–107
Spalding DB (1976) Development of the eddy-break-up model of turbulent combustion. The Sixteenth Symposium (International) on Combustion, Pittsburgh, Pa., pp. 1657–1663
Squires KD, Eaton JK (1990) Particle response and turbulence modulation in isotropic turbulence. Phys. Fluids A. 2: 1191–1203
Sun TV, Faeth GM (1986) Structure of turbulent bubbly jets–1. Methods and centerline properties. Int. J. Multiphase Flow 12: 99–114
Sundaram S, Collins LA (1999) A numerical study of the modulation of isotropic turbulence by suspended particles. J. Fluid Mech. 379: 105–143
Swain LM (1929) On the turbulent wake behind a body of revolution. Proc. R. Soc. London, Ser. A 125: 647–659
Theofanous TG, Sullivan J (1982) Turbulence in two-phase dispersed flow: J. Fluid Mech. 116: 343–362
Townsend AA (1976) The structure of turbulent shear flow. Cambridge University Press, Cambridge
Tsuge S, Sagara K (1978) Arrhenius' law in turbulent media and an equivalent tunnel effect. Combust. Sci. Technol. 18: 179–189
Tsuji Y, Morikawa Y (1982) LDV measurements of an air-solid two-phase flow in a horizontal pipe. J. Fluid Mech. 120: 385–406
Tsuji Y, Morikawa Y, Shiomi H (1984) LDV measurements of an air-solid two-phase flow in a vertical pipe. J. Fluid Mech. 139: 417–434

Tsuji Y, Morikawa Y, Tanaka T, Karimine K, Nishida S (1988) Measurements of an axisymmetric jet laden with coarse particles. Int. J. Multiphase Flow 14: 565–574

Uberoi MS, Freymuth P (1970) Turbulent energy balance and spectra of the axisymmetric wake. Phys. Fluids 13: 2203–2210

Varaksin AYu (1998) To questions about fluctuated velocity and temperature of the non-Stokesian particles moving in the turbulent flows. Heat Transfer. In: Lee JS (ed) The 11th IHTC, Kyongju, Korea, 2, pp. 147–150

Varaksin AYu, Kurosaki Y, Satoh I, Polezhaev YuY, Polyakov AF (1998) Experimental Study of the Direct Influence of the Small Particles on the Carrier Air Turbulence Intensity for Pipe Flow. The Third International cConference on Multiphase Flow. ICMF 98, Lyon, France, June 8–12.

Vulis LA (1960) Towards the role of turbulent fluctuations in turbulent combustion. The 3rd All-Union Conference on Combustion Theory. (in Russian) AN SSSR, Moscow, pp. 86–90

Vulis LA (1961) Thermal regimes of combustion. McGraw-Hill, New York

Vulis LA (1972) On turbulent burning velocity. Combust. Explosion Shock Waves 8: 1–5

Vulis LA, Ershin ShA, Yarin LP (1968) Foundations of the theory of gas torch. (in Russian) Energia, Leningrad

Wang SK, Lee SJ, Jones OSJr, Lahey RTJr (1987) 3-D turbulence structure and phase distribution measurements in bubbly two-phase flows. Int. J. Multiphase Flow 13: 327–343

Yarin LP, Hetsroni G (1994a) Turbulence intensity in dilute two-phase flows—1. Effect of particle-size distribution on the turbulence of the carrier fluid. Int. J. Multiphase Flow 20: 1–15

Yarin LP, Hetsroni G (1994b) Turbulence intensity in dilute two-phae flows—2. Temperature fluctuations in particle-laden dilute flows. Int. J. Multiphase Flow 20: 17–25

Yarin LP, Hetsroni G (1994c) Turbulence intensity in dilute two-phase flows—3. The particles-turbulence interaction in dilute two-phase flow. Int. J. Multiphase Flow 20: 27–44

Yoshida A, Tsuji H (1978) Measurements of fluctuating temperature and velocity in a turbulent premixed flame. The Seventeenth Symposium (International) on Combustion. The Combustion Institute, Pittsburgh, Pa., pp. 945–956

Yuan Z, Michaelides EE (1992) Turbulence modulation in particulate flow—a theoretical approach. Int. J. Multiphase Flow 18: 779–785

Zel'dovich YaB (1949) Towards the theory of combustion unpremixed gases. J. Tech. Phys. 19: 1199–1210 (in Russian)

Zel'dovich YaB, Barenblatt GJ, Librovich VB, Makhviladze GM (1985) The Mathematical theory of combustion and explosion. Consultants Bureau, New York

Zisselmar R, Molerus O (1979) Investigation of solid-liquid pipe flow with regard to turbulence modulation. Chem. Eng. J. 18: 233–239

2

Combustion wave propagation

2.6 Combustion waves in two-phase media

2.6.1 Preamble

Low-speed combustion waves in homogeneous media. The theory of combustion waves in homogeneous media incorporates a wide class of problems related to the analysis of the physical processes in flames, as well as to the mathematical aspects of solution of the system of nonlinear equations describing combustion wave propagation. A comprehensive discussion of these problems is contained in the monographs and surveys by Frank-Kamenetskii (1969), Williams (1985), Zel'dovich et al. (1985), and Merzhanov and Khaikin (1988, 1992). Referring the reader to these works, we confine our consideration to the calculation of the combustion wave speed.

A consistent thermal theory of combustion wave propagation in a homogeneous mixture was developed by Zel'dovich and Frank-Kamenetskii (1938 a, 1938 b). It accounts for the main features of the process, namely, for the sharp dependence of the chemical reaction rate on temperature for the intensive heat release, for mass and heat transfer due to molecular diffusion and thermal conductivity. The theory agrees with experiments fairly well.

In accordance with the thermal theory the mechanism of the combustion wave propagation in homogeneous media consists of the following. Let a thin layer of thickness δ of an initially cold mixture with temperature T_0 undergo an instantaneous heating up to temperature T_* by means of an external source (Fig. 6.1). (Note that the reaction rate at temperature T_0 $W(T_0)$ is assumed to be zero.) An increase in the mixture temperature triggers chemical reaction within the heated layer. The latter leads to a further heating of the mixture and, under certain conditions, to its ignition. Heat transfer from the high temperature zone to the cold mixture ensures heating and ignition of the neighboring layers, i.e. propagation of a self-sustaining chemical reaction in a reactive medium. At the setting stage of the combustion wave propagation, the process develops under the conditions of a continuous variation of the temperature and concentration fields. Also the speed of

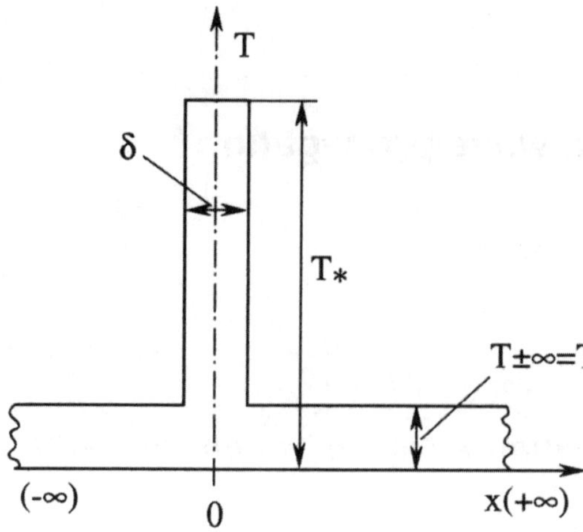

Fig. 6.1 Temperature distribution at $t = 0$

the combustion wave varies up to its value corresponding to the stationary regime of combustion. The survey of the analytical investigations that dealt with the establishment of the stationary regime of combustion wave propagation is given in the monography by Zel'dovich et al. (1985). The numerical study of this problem was carried out by Spalding (1953), Zel'dovich and Barenblatt (1959) and Merzhanov et al. (1969).

Speed of combustion waves. The thermal structure of the combustion wave is shown in Fig. 6.2. At large activation energy of the chemical reaction it includes three distinct characteristic zones: (i) a relatively wide heating zone I filled with the fresh mixture of fuel and oxidizer, (ii) a narrow reaction zone II where the conversion of the initial components into final products occurs, and (iii) a high temperature zone III filled with combustion products.

To estimate the speed of the combustion wave we use the method of the thermal flux (Frank-Kamenetskii 1969). Approximating the real temperature distribution in the combustion wave $T(x)$ by a broken line with linear sections T_o=const and T_m=const connected by a straight line tangent to the curve $T(x)$ at the inflection point, we find the heat flux from the high to low temperature zone as

$$q_* = \lambda \frac{T_m - T_0}{\delta} \tag{6.1}$$

where λ is the thermal conductivity; T_m and T_0 are the temperatures of the combustion products and of the fresh mixture, respectively, and δ is the conditional thickness of the reaction zone.

The heat flux q_* is spent in heating of the fresh mixture with the initial temperature T_0 up to the combustion temperature T_m. Therefore

$$q_* = c_p \rho u_f (T_m - T_0) \tag{6.2}$$

where u_f is the speed of the combustion wave (flame), and c_p is the specific heat.

Comparing Eqs. (6.1) and (6.2), we obtain

$$u_f = \frac{\alpha}{\delta} \tag{6.3}$$

where α is the thermal diffusivity.

Assuming that the thickness of the combustion wave δ is proportional to the product of the speed of the combustion wave u_f to the residence time of the mixture inside the flame, i.e. $\delta = u_f \tau_r$, we obtain

$$u_f = \sqrt{\frac{\alpha}{\tau_r}} = \sqrt{\frac{F \cdot \alpha}{\tau_m}} \tag{6.4}$$

where τ_m is the characteristic time of the chemical reaction at the maximum temperature T_m, $F = \tau_m / \tau_r$ –, $F<1$, since τ_m represents a fraction of the residence time τ_r.

The expression (6.4) can be obtained using dimensional analysis without any assumptions on the structure of the combustion wave. Indeed, taking into account the mechanism of the combustion wave propagation, it is possible to select two groups of parameters: those responsible for the heat transfer, u_f, c_p, ρ, λ, and that for the chemical reaction, τ_m, respectively. Therefore the problem incorporates five parameters having the following dimensionalities: u_f $[LT^{-1}]$, ρ $[ML^{-3}]$, c_p $[L^2T^{-2}\theta^{-1}]$, λ $[LMT^{-3}\theta^{-1}]$, τ $[T]$, (θ is in kelvins). Requiring a product of certain powers of these parameters to be dimensionless

$$u_f^{\beta} \rho^{\gamma} c_p^{\omega} \lambda^{\nu} \tau^{\mu} = \Pi \tag{6.5}$$

we arrive at the four equations for the unknown exponents β, γ, ω, ν and μ

$$\begin{aligned} &\beta - 2\gamma + 2\omega + \nu = 0 \\ &-\beta - 2\omega - 3\gamma + \mu = 0 \\ &\gamma + \nu = 0 \\ &\omega + \gamma = 0 \end{aligned} \tag{6.6}$$

Therefore one of the exponents in the system (6.6) can be chosen arbitrarily. Then we find the dimensionless group characterizing the problem

$$\frac{u_f^2 \tau}{\alpha} = F \tag{6.7}$$

were F has the same dimensionless value as in Eq. (6.4).

The value of the parameter F is a function of the other dimensionless groups corresponding to the problem. They are E/RT_0 and E/RT_m (Frank–Kamenetskii 1969). When the method of the exponent expansion is used, F depends on one dimensionless parameter $\theta_m = [E/(RT_m^2)](T_m - T_0)$, combining the two mentioned above. The expression for the speed of the combustion wave takes the form

$$u_f = \sqrt{\frac{\alpha}{\tau_m} F(\theta_m)} \tag{6.8}$$

The exact form of the dependence of the speed of the combustion wave propagation on the physico-chemical and kinetic parameters is found by solving the energy and diffusion equations. In the frame of reference associated with the combustion front (moving with a constant speed u_f) the equations read:

$$\rho u_f c_p \frac{dT}{dx} = \frac{d}{dx}\left(\lambda \frac{dT}{dx}\right) + qW(C,T) \tag{6.9}$$

$$\rho u_f \frac{dC}{dx} = \frac{d}{dx}(\rho D \frac{dC}{dx}) - W(C,T) \tag{6.10}$$

where D is the diffusivity of the fuel–oxidizer mixture and q is the heat of reaction. At Le$= \alpha/D = 1$ the profiles of the concentration and of the normalized temperature $\tilde{T} = (T - T_0)/(T_m - T_0)$ are similar since they satisfy the identical equations and are subject to identical boundary conditions. That allows one to express the concentration C as a function of temperature and to assume that W(C,T)=W(T). Then the problem reduces to a consideration of the energy equation only

$$\rho u_f c_p \frac{dT}{dx} = \frac{d}{dx}(\lambda \frac{dT}{dx}) + qW(T) \tag{6.11}$$

with boundary conditions

$$x = -\infty,\ T = T_0;\ x = +\infty,\ T = T_m \tag{6.12}$$

The boundary conditions (6.12) can be satisfied only at a certain value of the velocity u_f. The latter is the eigenvalue of Eq. (6.11) which is determined from the solution of the problem itself, since for the infinite interval $-\infty \le x \le \infty$ the solution is always up to a constant, i.e. depends on x+const. Since the value of the constant cannot be found via (6.12), one of the boundary conditions serves, actually, to find u_f. The study of the existence and uniqueness of the solution of the problem (6.9) and (6.10) was carried out by Zel'dovich (1948). The detailed mathematical analysis of this problem is presented in the monograph by Zel'dovich et al. (1985). Below we derive an approximate expression for the speed of the combustion wave. Within the narrow reaction zone II adjoining the zone III (cf. Fig. 6.2) is possible to neglect the contribution of the convective heat flux compared to the contribution of thermal conductivity and the heat release. Then Eq. (6.9) takes the form

$$\frac{d}{dx}\left(\lambda\frac{dT}{dx}\right) + qW(T) = 0 \tag{6.13}$$

Introducing the new variable $z = \lambda(dT/dx)$, we arrive at the equation

$$z\frac{dz}{dT} + \lambda q(W(T) = 0 \tag{6.14}$$

which should be integrated taken into account the boundary conditions

$$z = 0, \quad T = T_m \tag{6.15}$$

Integration of Eq. (6.14) with the boundary condition (6.15) yields the flux from the reaction to the heating zone

$$\lambda_m\left(\frac{dT}{dx}\right)_m = \sqrt{2\lambda_m q\int_T^{T_m} W(T)dT} \tag{6.16}$$

Bearing in mind that the rate of the chemical reaction is negligibly small at low temperatures, it is possible to take T_0 as the lower limit in the integral of Eq. (6.16). The cumulative heat release of the reaction equals the chemical energy contained in the reactive mixture. Therefore

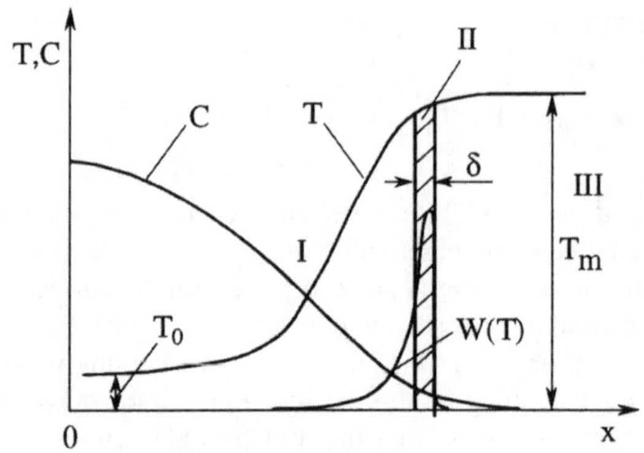

Fig. 6.2 The structure of combustion wave at a certain moment of time. (the wave propagates from right to left). I, heating zone; II, reaction zone; III, high temperature zone

$$\rho_0 u_f C_0 q = \sqrt{2\lambda_m q \int_{T_0}^{T_m} W(T)dT} \tag{6.17}$$

where subscript 0 corresponds to the initial state. Then speed of the combustion wave is given by

$$u_f = \frac{1}{\rho_0 C_0}\sqrt{\frac{2\lambda_m}{q}\int_{T_0}^{T_m} W(T)dT} \tag{6.18}$$

Equation (6.18) shows that the speed of the combustion wave depends on the physico-chemical and kinetic characteristics of the reactive mixture. Since the rate of chemical reaction depends on pressure and temperature as P^n and $\exp(-E/RT_m)$, respectively (n is the order of the reaction), the dependence of the combustion wave speed on P and T has the form

$$u_f \sim P^{\frac{n}{2}-1} \exp\left(-\frac{E}{2RT_m}\right) \tag{6.19}$$

It is seen that the speed of the combustion wave increases sharply with T_m. It does not depend on P for bimolecular reactions with n=2. On the other hand, u_f increases with P for trimolecular reactions (n=3), and decreases for monomolecular reactions (n=1). The comparison of the theoretical predictions with the experimental data on the speed of combustion waves in homogeneous propane–air and kerosene–benzene vapor–air mixtures is presented in Fig. 6.3.

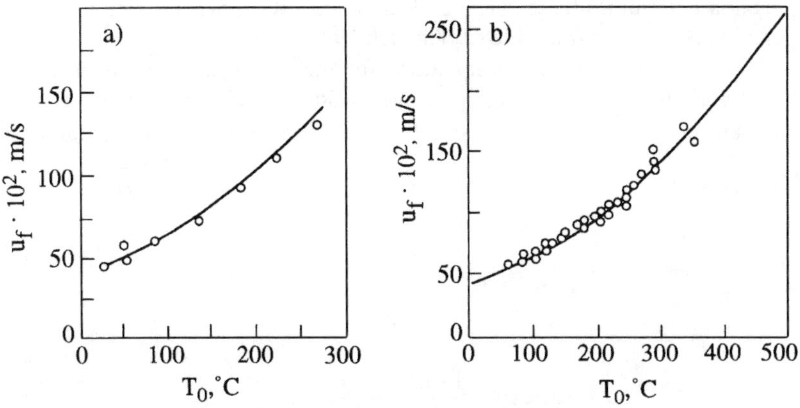

Fig. 6.3 Comparison of the experimental (symbols) and theoretical (lines) dependences for the speed of combustion wave on temperature. Reprinted from Zel'dovich et al. (1985). a) Propane–air mixture. b) Kerosene–benzene vapor–air mixture

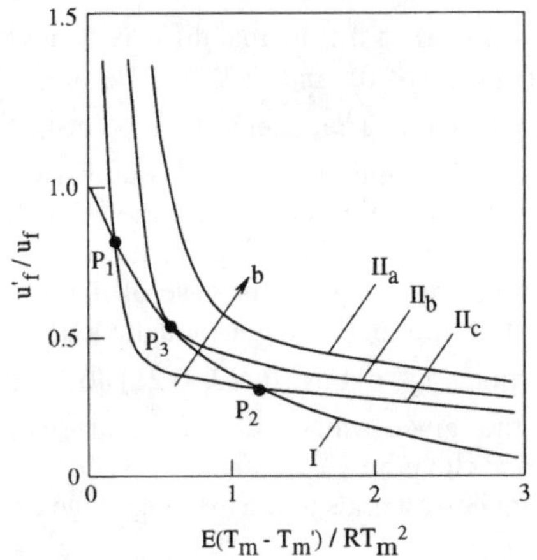

Fig. 6.4 Graphical solution of Eqs. (6.20) and (6.21). Reprinted from Zel'dovich et al. (1985). Curve I corresponds to Eq. (6.21), curves II a, II c correspond to Eq. (6.20), P_1 and P_2 are the upper and lower intersection points corresponding to the stationary (stable P_1 and unstable P_2 regimes), and P_3 is the tangency point

Limits of flame propagation. The estimate of the wave propagation speed u_f given in the previous section corresponds to the adiabatic conditions. The heat losses due to radiation and heat transfer into the ambient medium leads to reduction of the combustion temperature and of the speed of combustion wave. In turn, the latter is accompanied by an increase in the residence time of reactive mixture in the reaction zone and to still larger heat losses. The competition between the heat losses and heat release accompanying combustion wave propagation determines the existence of limits outside of which combustion wave propagation becomes impossible (Zel'dovich 1941).

Using the energy balance equations for the combustion wave propagation with and without heat losses, it is possible to find the relation between the flame temperature under non-adiabatic conditions T'_m and the speed of the combustion wave u'_f

$$T'_m = T_m - \frac{b}{(u'_f)^2} \tag{6.20}$$

$$u'_f = u_f \exp\left[-\frac{E(T_m - T'_m)}{2RT_m^2}\right] \tag{6.21}$$

The expressions (6.20) and (6.21) correspond to flame propagation in a pipe; T_m and u_f corresponds to the adiabatic conditions $b = (4h\alpha(T_m - T_0))/(r_0 c_p \rho_0)$, h is the heat transfer coefficient, α is the thermal diffusivity, and r_0 is the pipe radius.

The solution of Eqs. (6.20) and (6.21) determines possible regimes of combustion wave propagation. The intersection points of the curves $u'_f(T'_f)$ corresponding to Eqs. (6.20) and (6.21) yield stationary (stable and unstable) regimes, whereas the tangency point determines the critical states (Fig. 6.4). The lower intersection point, P_2 is unstable, since at the lower branch of curve I an increase of the heat losses results in an increase of the speed of the combustion wave. As can be seen from Fig. 6.4, at large enough values of the parameter b, the curves $u'_f(T'_f)$ corresponding to Eqs. (6.20, 6.21) lose the intersection points. That means that at the given values of the parameters propagation of the combustion wave is impossible.

The tangency point corresponds to the limiting value of the parameter $b = b_*$

$$b_* = \frac{u'_{f*} T_m^2 R}{E} \tag{6.22}$$

The corresponding largest possible decrease of the combustion temperature is given

$$T_m - T'_{m*} = \frac{RT_m^2}{E} \tag{6.23}$$

The speed of the wave at the combustion limit is

$$u'_{f*} = \frac{u_f}{\sqrt{e}} \tag{6.24}$$

It is smaller that the adiabatic by the factor $2/\sqrt{e} \simeq 0.607$.

2.6.2 Thermal structure of combustion waves in gas–particle mixtures

In the previous section it was shown that propagation of combustion waves in homogeneous media is fully determined by the physico-chemical characteristics of the reactive mixtures. In contrast with that, the flame propagation in a two-phase medium also depends on its heterogeneity which determines the dynamic, thermal and diffusion non-equilibrium of the reactive system.

Consider qualitatively the structure of a combustion wave in a two-phase medium. We confine our consideration to the analysis of the effect of particle size on the thermal structure of a combustion wave propagating in a monodisperse mixture loaded with non vaporizing particles. As was noted earlier (see Chapter 1.1), dynamic interaction of continuous and dispersed phases is characterized by the dynamic relaxation time proportional to the square of the particle diameter. In accordance with that, the effect of sliding of phases is negligibly small in flows loaded with fine particles. In these flows the particles' Reynolds number is small enough so that the Nusselt and Sherwood numbers are constant (Nu=Sh=2). Accordingly, the heat and mass transfer coefficients are inversely proportional to the particle diameter: $h \sim 1/d$, and $h_m \sim 1/d$.

Under the conditions of intensive heat and mass transfer between a particle and the surrounding fluid the process kinetics is the main factor governing burning of fine particles. In this case the rate of combustion essentially depends on the particle temperature. At large values of the heat transfer coefficient the particle temperature is close to the temperature of the carrier fluid. It is possible to use the one-velocity and one-temperature approximation for the description of combustion waves in mixtures loaded with fine particles. A typical thermal structure of such a wave is presented in Fig. 6.5. As can be seen the temperature distribution in the flame of fine particles is similar to the temperature distribution in the flame of a homogeneous gaseous mixture. It is emphasized that such a similarity is observed not only in the shape of curves T(x) but also in a number of the other important characteristics of the flame. For example, in both cases the thickness of the heating zone is larger than the thickness of the reaction zone; the speed of the flame front depends exponentially on the combustion temperature, etc.

The rate of combustion of the coarse particles depends slightly on the particle temperature. It is determined mostly by the oxidizer diffusion.

In gaseous mixtures loaded with coarse particles the change of the continuous and dispersed phase temperatures in the combustion waves depends on the mechanism of heat transfer from the high temperature reaction zone to the

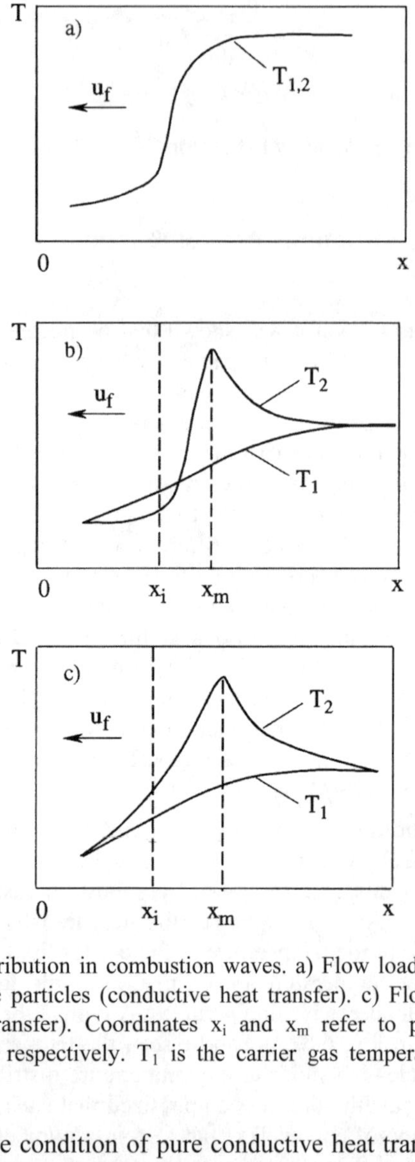

Fig. 6.5 Temperature distribution in combustion waves. a) Flow loaded by fine particles. b) Flow loaded by coarse particles (conductive heat transfer). c) Flow loaded by coarse particle (radiative heat transfer). Coordinates x_i and x_m refer to particle ignition and maximum temperature, respectively. T_1 is the carrier gas temperature, and T_2 is the particle temperature

fresh mixture. Under the condition of pure conductive heat transfer, the heat flux from the high temperature combustion products is spent primarily in heating of the carrier fluid in front of the reaction zone. Within the region $-\infty < x < x_i$ (Fig. 6.5b) the particle heating occurs by means of: (i) the heat transfer from the carrier fluid, and (ii) the heat release due to the heterogeneous reaction at the particle surface. The contributions of the heat release and heat transfer in the thermal balance of particle heating are different. They depend (the other conditions being fixed) only on the distance from the flame front to a given cross-section of the combustion wave. At low enough temperatures corresponding to the kinetic regime, the heat release is exponentially small. Accordingly, within the region $-\infty < x < x_i$ the heat transfer from the carrier fluid to particles is dominant. In this case the structure of the heating zone depends on the ratio of the particle residence time to the characteristic time of heat transfer in the heating zone

$$z = \frac{\tau u_f^2}{\alpha} \tag{6.25}$$

where τ is the particle residence time, u_f is the rate of combustion, α is the thermal diffusivity. When $z \gg 1$ (fast combustion), the temperatures of the particles and carrier gas in the heating zone are different, whereas when $z \ll 1$ (slow combustion), they are close to each other (Rumanov and Khaikin 1971).

Closer to the cross-section $x=x_I$, the influence of the heat release due to the chemical reaction increases significantly. At $x=x_i$ the intensities of the heat release and the heat transfer are equal

$$qW(T_{2.i}) = hs(T_{1.i} - T_{2.i}) \tag{6.26}$$

where $W(T_{2.i})$ is the rate of chemical reaction at $T_2=T_{2.i}$, q is the heat of reaction, subscripts 1 and 2 correspond to the carrier gas and particle, i corresponds to the cross-section $x=x_i$, and s is the specific interface surface.

The growth of particle temperature in the vicinity of the cross-section $x=x_i$ (at $x \approx x_i$ or $x > x_i$) is extremely sharp. That allows for treatment of Eq. (6.26) as the condition determining the particle ignition by hot gas. The sharp rise of the particle temperature is accompanied by a change of the combustion regime from the kinetic to the diffusion regime. Within the reaction zone the particle temperature essentially exceeds the temperature of the surrounding gas. The rate of chemical reaction varies only slightly across the reaction zone. The thickness of this zone is larger than that in the combustion wave propagating in a homogeneous mixture.

The pattern of the thermal structure of combustion waves in gas–particle mixtures is depicted in Fig. 6.6, where the results of the numerical calculations by Gurevich et al. (1972) are presented. It corresponds to propagation of a combustion wave in a motionless, monodisperse air mixture loaded with carbon particles. These data show that there are two characteristic zones within the combustion wave. In the first of these $-\infty < \overline{x} < \overline{x}_1$, the heating of the dispersed phase is due to the heat transfer from the carrier gas, whereas in the second zone, the particles return thermal energy to the gaseous phase. The change in the heat flux direction occurs at the cross-section $\overline{x} = \overline{x}_1$, where $\overline{T}_1 = \overline{T}_2$ and $d\overline{T}_1/d\overline{x} < d\overline{T}_2/dx$.

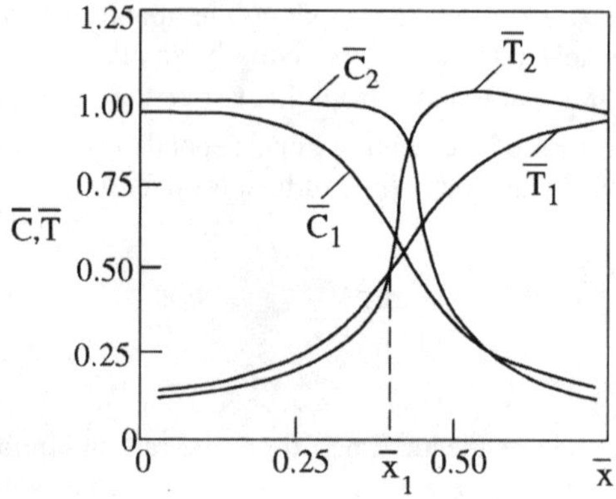

Fig. 6.6 Structure of the combustion wave in motionless air–carbon particle mixture. Reprinted from Gurevich et al. (1972). $\bar{T},\bar{C}$ and $\bar{x}$ are the dimensionless temperature and concentration, and the longitudinal coordinate, respectively; subscripts 1 and 2 refer to the carrier gas and particles. The carrier gas and particle temperatures are rendered dimensionless by the combustion products temperature, while the concentrations C_1 and C_2 by their values in a fresh mixture. The longitudinal coordinate is rendered dimensionless by the combustion wave thickness

The effect of the flow hydrodynamics on the thermal structure of combustion waves in gas–particle mixtures was studied by Vainshtein and Nigmatulin (1971). Using the two-temperature and two-velocity model of the medium, the authors obtained detailed data on the temperature and velocity distribution within the combustion wave in a carbon particle–air mixture. The comparison of the results corresponding to flame propagation in motionless (Fig. 6.6) and moving (Fig. 6.7) media shows that the existence of phase sliding does not change the character of the temperature distribution in the combustion wave. In both cases there are two characteristic zones with opposite directions of the heat fluxes within the combustion wave. In the first zone, the rate of chemical reaction is negligible and the particle heating occurs, mainly, due to the transfer from the surrounding gas. In the second zone, after particle ignition the particle temperature significantly exceeds the carrier gas temperature. Therefore, in the second zone the heat flux is directed from the dispersed to the continuous phase.

The velocity distribution within the combustion wave propagating in a gas–particle mixture is shown in Fig. 6.8. It is seen that the velocity of the gaseous phase significantly increases near the reaction zone. An acceleration of the gaseous phase in the combustion wave is due to the thermal expansion of the carrier gas resulting from its heating. The particle motion is completely determined by the carrier gas motion and depends on the interphase friction force.

As can be seen in Fig. 6.8, even in the mixture loaded with relatively small particles (d~50 μm) a noticeable difference between the phase velocities exists. The existence of interphase sliding significantly affects the combustion process and manifest itself more clearly in that diffusion regime of particle burning. It leads to a change in the chemical reaction rate (Fig. 6.9), as well as in the speed of combustion wave propagation.

The thermal structure of the combustion wave with radiative heat transfer is depicted in Fig. 6.10. In this case the heat flux from the reaction zone is mostly spent in heating of the dispersed phase and only slightly in heating of the carrier gas. Accordingly, the particle temperature exceeds the gas temperature within the heating and reaction zones.

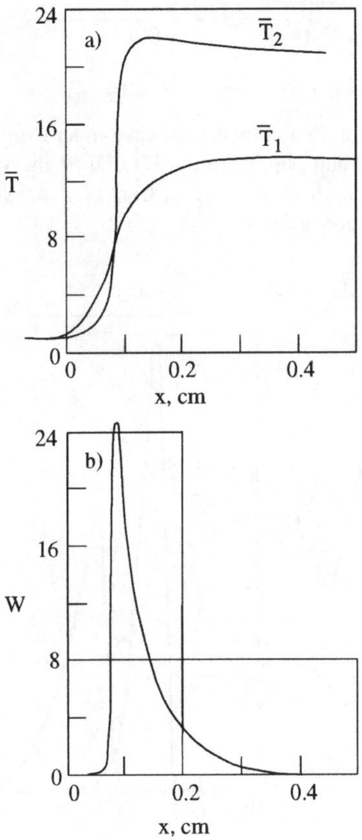

Fig. 6.7 Structure of the combustion wave in a moving oxygen–carbon particle mixture. Reprinted from Vainshtein and Nigmatulin (1971). a) The temperature distribution within the combustion wave. b) The chemical reaction rate distribution within the combustion wave. $\overline{T}$ is dimensionless temperature, $\overline{W}$ is rate of chemical reaction; subscripts 1 and 2 refer to the carrier gas and particles. The carrier gas and particle temperatures are rendered dimensionless by the fresh mixture temperature. The chemical reaction rate is normalized to the mass flow rate of carrier gas in the reaction zone

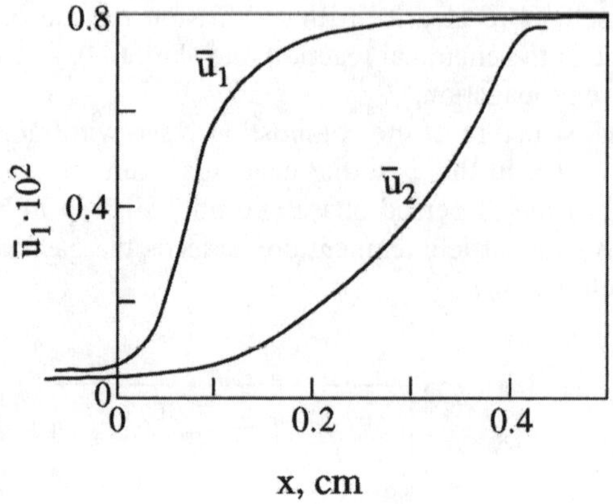

Fig. 6.8 Velocity distribution in a combustion wave in an oxygen–carbon particle mixture Reprinted from Vainshtein and Nigmatulin (1971). $\bar{u}$ is the velocity is rendered dimensionless by the sound velocity in carrier gas in front of combustion wave, and subscripts 1 and 2 refer to carrier gas and particles

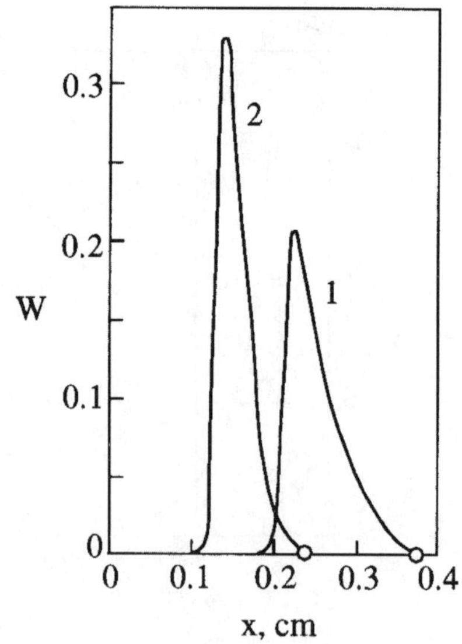

Fig. 6.9 Rate of chemical reaction in the combustion wave. Reprinted from Vainshtein and Nigmatulin (1971). Curve 1 corresponds to the one-velocity approximation; curve 2 corresponds to the two-velocity approximation. The chemical reaction rate is normalized to the mass flow rate of carrier gas in the reaction zone

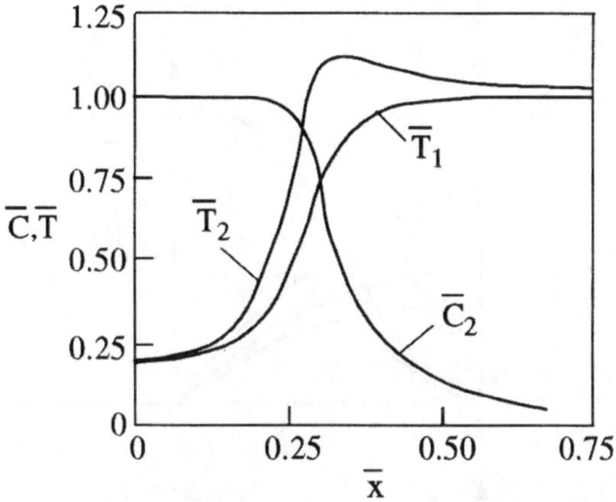

Fig. 6.10 Structure of the combustion wave in a gas–carbon particle mixture (radiative heat transfer). Reprinted from Ozerov (1980). The carrier gas and particle temperatures T_1 and T_2 are rendered dimensionless by the combustion products temperature and the concentration C_2 by its value in the fresh mixture. The longitudinal coordinate is rendered dimensionless by the combustion wave thickness

The temperature distribution in the flame of polydispersed mixtures possesses certain characteristics due to the different scale times of relaxation processes for fine and coarse particles. When the conductive heat transfer is dominant, the ignition of the fine particles is accompanied by heating of the coarse ones, temperature growth in the heating zone and shortening of its length. At the same time, burning out of the fine particles and a decrease of the oxidizer concentration in the reaction zone lead to a decrease in the rate of combustion of the coarse particles, as well as to an increase in their burning time and in the thickness of the reaction zone.

The physico-chemical conversion of the dispersed phase, i.e. pyrolysis, evaporation and devolatilization significantly affect the temperature distribution within the combustion wave. The influence of these processes on the temperature field depends on their rate, succession and the accompanying thermal effects. In particular, in the spray flame the temperature of the condensed phase varies within a rather narrow range $T_0<T_2<T_b$ (T_0 and T_b are the initial and boiling temperature), whereas the gas temperature increases up to the combustion temperature. It is emphasized that the droplet temperature changes, as a rule, only in the heating zone and remains practically constant within the zones were kinetic and diffusion burning are dominant.

A similar character of the temperature distribution is also observed in flames of polydisperse sprays. The existence of the mixture of droplets of different sizes affects not only the value of the gas temperature but also the lengths of the heating and combustion zones .

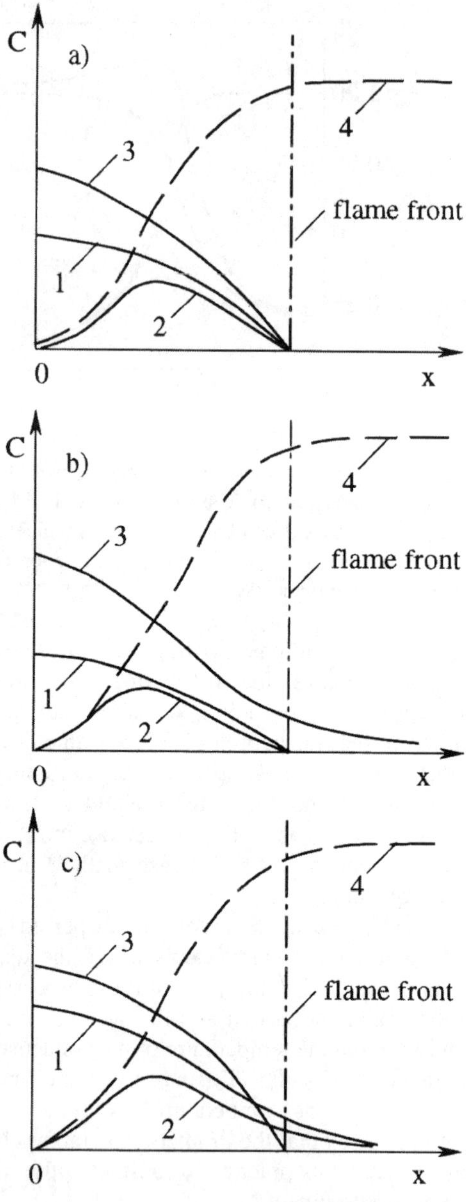

Fig. 6.11 Distributions of the fuel and oxidizer concentrations in the combustion wave propagating in gas–droplet mixture. Curves 1 show the fuel (liquid and gaseous component); curves 2 the vapor concentration; curves 3 oxidizer concentration, curves 4 the combustion products. a) Stoichiometric mixture. b) Lean mixture. c) Rich mixture

The distributions of the fuel and oxidizer concentrations in the combustion wave that is spreading in a two-phase mixture are shown in Fig. 6.11. It is seen that the fuel and oxidizer concentrations decrease monotonically from their initial values corresponding to the fresh mixture and approach the values corresponding to burning out of both ($\gamma_{ox} = 1$) or one ($\gamma_{ox} > 1$, or $\gamma_{ox} < 1$) of the reagents ($\gamma_{ox} = \rho_{1.0}/(\rho_{2.0}\sigma)$ is the excess of the oxidizer, σ is stoichiometric oxidizer-to-fuel mass ratio). The concentration of the combustion products increases from zero at the front boundary of the combustion zone up to the maximal value, corresponding to full burning out of one or both reagents.

The distribution of the concentrations of gaseous components of the reactive mixture (vapor of liquid fuel, or volatile of solid fuel) is more complicated. As can be seen in Fig. 6.11, the vapor (volatile) concentration increases in the forepart of the combustion wave and then monotonically decreases, approaching the values corresponding to full burning out of the stoichiometric mixture or full burning out of the gaseous oxidizer.

The structure of the combustion wave which propagates in a gas–droplet mixture depends essentially on its composition (excess or lack of fuel), droplet size, as well as the initial temperature T_0. At a fixed T_0, the distributions of the temperature and concentration across the flame are determined by the stoichiometry of the two-phase mixture (the initial fuel–oxidizer ratio) and the droplet diameter. Lin et al. (1988) proposed a classification of possible spray burning regimes that is based on two parameters: the critical droplet size and the stoichiometry of the fresh mixture. When the droplet diameter is less than the critical value, the liquid fuel vaporization is fully completed within the prevaporized zone which is located in the forepart of the heating zone, so that in the forepart of the reaction zone only gaseous vapor–oxidizer mixture is present. When the droplet diameter exceeds a critical value, partially prevaporized burning takes place. In this case the vapor–oxidizer mixture is loaded with coarse droplets that penetrate the flame front. Evaporation and combustion of a two-phase vapor–oxidizer–droplet mixture occur within the region located behind the flame front.

2.6.3 Stationary combustion waves in two-phase media

Regimes of combustion wave propagation. Combustion wave propagation in a two-phase reactive mixture is due to the heat transfer from the high to low temperature zone of the flame by means of thermal conductivity and radiation. The intensity of conductive and radiative heat transfer depends on thermophysical and optical properties of continuous and dispersed phases, concentration of solid particles or droplets, as well as their sizes.

The mutual importance of the radiative and conductive heat transfer is determined by the following parameter

$$N = \frac{\tilde{\sigma} T_f^3 d}{\lambda Nu} \tag{6.27}$$

where $\tilde{\sigma}$ is the Stefan–Boltzmann constant, T_f is the flame temperature, d is the particle diameter, λ is the thermal conductivity, and Nu is the Nusselt number.

When N<1 (a mixture loaded with fine particles), conductive heat transfer significantly affects the flame propagation. In contrast with that, the flame propagation in the mixture loaded with coarse particles (N>1) is determined by radiation. The structure of the heating zone depends on the ratio of the characteristic heat transfer time to the particle residence time in the heating zone

$$N_C = \frac{\tau_c}{\tau_r} \tag{6.28}$$

where τ_c is the characteristic heat transfer time ($\tau_c \sim d^2$ and $\tau_c \sim d$ for the conductive and radiant heat transfer, respectively), $\tau_r = \alpha / u_f^2$ is the residence time in the heating zone, α is the thermal diffusivity, and u_f is the combustion wave speed.

The parameter N_C determines possible regimes of wave propagation. When $N_C<1$, particle heating to the ignition temperature occurs by means of convective heat transfer from the carrier gas at $T_2 \approx T_1$. When $N_C>1$, self-acceleration of the chemical reaction at the particle surface occurs at $T_2>T_1$. The cases $N_C<1$ and $N_C>1$ correspond to slow and fast combustion, respectively, (Rumanov and Khaikin 1971).

Under the conditions of radiative heat transfer, the heating of the carrier gas occurs by means of convective heat transfer from the particles. In this case the temperature in the heating zone of the combustion wave depends on the parameter N_R corresponding to the ratio of the characteristic time of radiative heat transfer to the particle residence time in the heating zone

$$N_R = \frac{\tau_R}{\tau_r} \tag{6.29}$$

where $\tau_R = \delta / u_f$, δ is the thickness of the heating zone proportional to the free path length of radiation $\ell = 1/nr^2$, n is the number of particles in unit volume, and r is the particle radius.

Depending on the value of the parameter $N_R>1$, $N_R<1$ two the regimes corresponding to particle ignition in fast $(T_2>T_1)$ and slow $(T_2 \approx T_1)$ combustion occur.

Problem formulation. Combustion wave propagation in a two-phase reactive medium is described by the system of mass, momentum and energy balance equations for the phases and components of the gaseous mixture. The set of the governing equations also includes the equations of state for the phases and components, the balance equation for the number of particles, as well as the relations for calculation of the physical-chemical parameters of the reactants and combustion products. In order to close the system of equations, it should be supplemented by the correlations determining the interfacial interactions: the intensity of the momentum and energy transport, the rate of devolatilization or evaporation, etc.

To calculate the speed of the combustion wave, various models of two-phase media are used to simplify significantly the mathematical problem. One such model is the model of interpenetrating and interacting continua which allows for application of the methods of mechanics of multiphase systems in studies of combustion wave propagation in two-phase media (Nigmatulin 1991). Such an approach was used by Vainshtein and Nigmatulin (1971, 1973) for the analysis of combustion wave propagation in gas–particle and gas–droplet mixtures and by Sukhov and Yarin (1981 a, 1981 b) for studying combustion waves in bubbly media. At the same time a number of additional assumptions are used for the problem simplification. The most important are: (i) the combustion wave propagates in an infinite medium, (ii) two-phase flow is one-dimensional so that the flame front is planar, (iii) the process of flame propagation is stationary, (iv) the two-phase mixture is monodisperse, (v) physical characteristics of the reactive medium are constant, (vi) the Lewis number equals one, (vii) the temperature distribution inside the particles is uniform.

In the coordinate frame associated with the flame front the problem reduces to the solution of the system of the one-dimensional balance equations containing an unknown speed of combustion wave. The boundary conditions corresponding to the stationary flame propagation with conductive heat transfer are

$$x \to -\infty \quad u_1 = u_2 = u_*; \quad \frac{dT_1}{dx} = \frac{dT_2}{dx} = 0 \tag{6.30}$$

$$x \to +\infty \quad u_1 = u_2 = u_{**}; \quad \frac{dT_1}{dx} = \frac{dT_2}{dx} = 0 \tag{6.31}$$

where subscripts 1 and 2 refer to the carrier gas and particles, respectively; single and double asterisks denote the states far ahead of and behind the flame front. Under the conditions of radiaive heat transfer, the radiative heat flux q_R far from the reaction zone is given as (Vainshtein 1973)

$$x \to -\infty \qquad q_R = 0 \tag{6.32}$$

$$x \to +\infty \qquad q_R = 0 \tag{6.33}$$

As in the case of combustion wave propagation in a homogeneous medium, the equations and the boundary conditions corresponding to combustion wave propagation in a two-phase mixture can be satisfied only at a certain speed of the flame front (which is an eigenvalue of the problem) which is found from the solution of the problem on the combustion wave structure.

Speed of combustion wave. The speed of combustion waves propagating in two-phase media was the subject of a number of experimental investigations (Burgoyine and Cohen 1954, Cassel et al. 1956, Cekalin 1962, Mizutani and Nishuimoto 1972, Polymeroulos and Das 1975, Hayashi and Kumagi 1975, Myers and Lefebver 1986 and Richards and Lefebver 1989). The majority of these works deal with the study of the flame speed in sprays. Great attention in these works was paid to the effect of droplet size, fuel–air ratio, volatiles contained in the fuels, interdroplet distance and flow turbulence on the combustion wave speed. The influence of these factors on the combustion process was studied in detail. It was found that the combustion wave speed in a mixture with coarse droplets ($d > 30\ \mu m$) is inversely proportional to the Sauter mean diameter of the droplets, whereas at $d < 10\ \mu m$ it tends to the speed of the combustion wave in the corresponding premixed gaseous mixture. The combustion wave speed increases linearly as the fuel–air ratio increases. The less volatile fuels have a lower flame speed. They are also more sensitive to changes in the Sauter mean diameter. The interdroplet distance, as well as the flow turbulence significantly affect the speed of combustion waves. At high content of vapor in droplet–vapor–air mixture an increase of the interdroplet distance is accompanied by a gradual decrease of the flame speed, whereas at a low vapor content the flame speed decreases rapidly with an increase of the interdroplet spacing. An increase in the turbulence intensity leads to an increase of the combustion wave speed.

For an order of magnitude estimate of the combustion wave speed in two-phase media, extremely simplified approaches are used. They are based on some elementary assumption on the structure of the combustion wave, as well as on an approximate solution of the energy equation with the temperatures (velocities, etc.) of phases being considered identical.

For pure conductive heat transfer the combustion wave speed is estimated using the condition that the heat flux from the reaction zone is fully spent in heating the two-phase mixture from its initial temperature to the ignition temperature

$$(\rho_{1.0}c_1 + \rho_{2.0}c_2)u_f(T_i - T_0) = \lambda_1\left(\frac{dT_1}{dx}\right)_f \tag{6.34}$$

where c is the specific heat, T is the temperature, $(dT_1/dx)_f$ is the temperature gradient, λ is the thermal conductivity, subscripts 1 and 2 refer to continuous and dispersed phases, respectively, and subscripts 0 and i correspond to the parameters ahead of (far enough) the reaction zone, and to the ignition point, respectively.

Assuming that the temperature distribution within the reaction zone is linear, we present the gradient $(dT_1/dx)_f$ as

$$\left(\frac{dT_1}{dx}\right)_f \sim \frac{T_m - T_0}{\delta} \tag{6.35}$$

where T_m is the combustion temperature, $\delta = u_f t_b$ is the thickness of reaction zone, and t_b is the time of particle burning.

Equations (6.34) and (6.35) yield the following estimate of the combustion wave speed

$$u_f \sim \frac{\lambda}{(\rho_{1.0}c_1 + \rho_{2.0}c_2)}\left(\frac{T_m - T_0}{T_i - T_0}\right)\frac{1}{\delta} \tag{6.36}$$

Bearing in mind that burning time is proportional to the particle diameter in the kinetic regime of the combustion, and to the square of the particle diameter in the diffusion regime, we find the dependence of the combustion wave speed on the particle size (Leipunskii 1960)

$$u_{f.k} \sim \frac{1}{\sqrt{d}}\ ;\ u_{f.d} \sim \frac{1}{d} \tag{6.37}$$

where $u_{f.k}$ and $u_{f.d}$ are the combustion wave speeds in the kinetic and diffusion regimes and d is the particle diameter.

Another approximation is based on the presentation of the combustion wave in the form of two characteristic zones (the heating and reaction zones) separated by the boundary passing through the inflection point of the temperature profile. This approximation yields the following estimate of the combustion wave speed (Ozerov 1980).

$$u_f \sim \frac{1}{c_1\rho_{1.0}}\left(\sqrt{\lambda_1 q\left[\frac{dW(T)}{dT}\right]_{T=T_i}}\right) \tag{6.38}$$

where q and $W(T)$ are the heat of combustion and the rate of chemical reaction.

To estimate the combustion wave speed due to radiative heat transfer (mixture with coarse particles) some additional assumptions are used. Gol'tsiker et al. (1977) obtained the following estimate of the combustion wave speed assuming that the radiative heat flux from the reaction zone is fully spent for particle heating up to the ignition temperature

$$u_{f.R} \sim \frac{6q_R}{c_{p2}\rho_2 \pi d^3 n(T_i - T_0)} \tag{6.39}$$

where q_R is the radiative heat flux.

The expression (6.39) corresponds to an infinitely small intensity of the interfacial heat transfer (Nu=0) and determines the maximum speed of the combustion wave. Taking into account the heat transfer between the particles and carrier gas, as well as the absorption of radiation energy, etc., leads to more complicated relations for the combustion wave speed. A number of models of flame propagation in gas–particle (coal dust) mixtures were proposed by Essenhigh and Csaba (1963), Bhaduri and Bandyopadhyay (1971), Vainshtein and Nigmatulin (1971), Gurevich et al. (1972), Ozerova and Stepanov (1973), Vainshtein (1973), Slezak et al. (1985), Seshadri et al. (1992) and Krishenik et al. (1994). The theory of flame propagation in sprays was developed by Williams (1959), Polymeropoulos (1974), Ballal and Lefebvre (1981), Lin and Sheu (1991), Lin and Lin (1991), Silverman et al. (1993), Greenberg et al. (1996) and Sirignano (1999). The results of the theoretical predictions of the speed of combustion waves propagating in gas–particle and gas–droplet mixtures agree fairly well with experiments.

Effect of polydispersity. Combustion wave propagation in premixed polydisperse sprays was studied by Silverman et al. (1993). The numerical calculations of the width of the combustion wave, its speed and mass flux and the Sauter mean diameter (SMD) of droplets at the flame front were carried out for nine different initial size distributions with the same SMD. Some results of these calculations are presented in Fig. 6.12, where all the characteristic parameters are rendered dimensionless by their values corresponding to a monodisperse admixture. As can be seen from Fig. 6.12, an increase of the coarse fraction in the spray leads to a significant expansion of the burning zone and a reduction of the flame front speed. These effects are not unexpected. They result from the changes in the evaporation rate on changing the fractional compositions of the reactive mixture. A decrease of the evaporation rate in a spray of coarse droplets is accompanied by a decrease in the droplet mass fluxes and SMD at the flame front.

Effect of turbulence. The influence of turbulent fluctuations on the combustion wave speed in gas–droplet mixtures was studied by Mizutani and Nishimoto (1972), Mizutani (1972) and Cekalin (1962). Mizutani and Nishimoto proposed the following empirical relation for the turbulent flame velocity

Size distribution	δ / δ_m	u_f / u_{fm}	G / G_{fm}	d_f / d_{fm}
	1	1	1	1
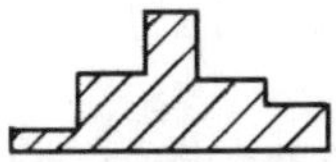	1.54	0.94	1.05	1.18
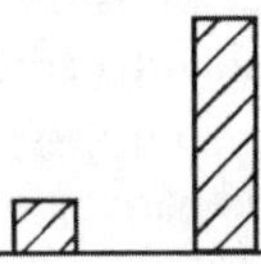	3.44	0.52	1.33	2.4

Fig. 6.12 Effect of polydispersity on the characteristics of the combustion wave. Reprinted from Silverman et al.(1996), with permission. δ is the thickness of the reaction zone, u_f is the combustion wave speed, u_{fm} corresponds to monodisperse admixture, g is the mass flux of the droplets at the flame front, and d is the Sauter mean diameter at the flame front

$$u_{f.T} = \frac{1}{d} A(\phi - 0.012) Tu^{1.15} \tag{6.40}$$

based on measurements performed for a wide variation of the droplet diameter, the fuel-to-air mass ratio, as well as the initial temperature of liquid fuel.

In Eq. (6.40) $u_{f.T}$ is the turbulent flame velocity, $T_u = \sqrt{\overline{u'^2}}/\overline{u}$ is the turbulence intensity, u' and $\overline{u}$ are the fluctuation and average velocities, ϕ is the fuel-to-air mass ratio, A is the correction factor depending on the type of liquid fuel (A=$6.8 \cdot 10^3 \left[m^2/s \right]$ and $4.3 \cdot 10^3 \left[m^2/s \right]$ for kerosene and light diesel sprays, respectively), and d is the Sauter mean droplet diameter (in microns).

An extremely strong dependence of the turbulent flame velocity on the turbulence intensity as in Eq. (6.40) is instructive. It is noteworthy, that such a dependence is stronger than that for homogeneous mixture. Indeed, under the conditions of large scale turbulence, the speed of a turbulent flame in a homogeneous mixture is proportional to T_u (Khitrin 1957 and Williams 1985), whereas in a turbulent spray it is proportional to $T_u^{1.15}$. The difference is probably due to the different mechanisms of influence of large scale fluctuations in homogeneous and heterogeneous flames. In the first case the large scale fluctuations cause the flame front to undulate and increase its surface, whereas in the second case they also affect the rate of droplet evaporation.

To estimate the combustion wave velocity in a turbulent gas–droplet flow Mizutani (1972) used a model which accounts for the molecular and turbulent heat transfer. Assuming that the flow is uniform and one-dimensional in the direction normal to the flame front, he used the following energy balance equation:

$$\frac{d}{dx}(Gh) = \frac{d}{dx}\left[(\lambda + \rho c_p \varepsilon)\frac{dT}{dx}\right] + Q \qquad (6.41)$$

where $G=\rho_1 u_{f.T}$ is the mass flow rate of gas passing through a unit area of the flame front, Q is the volumetric rate of heat release, ε is the eddy thermal diffusivity defined as $\varepsilon = \ell u'_u$, and ℓ is the Lagrangian turbulence scale.

The set of the governing equations used for estimating the combustion wave velocity also includes the equations for (i) heating up and burning of droplets, (ii) heat release rate and (iii) size distribution function.

The predictions of Mizutani's theory agree qualitatively with experimental data. The calculations confirm the existence of a strong dependence of the turbulent flame velocity on the turbulence intensity. They also demonstrate that the initial temperature of a liquid fuel significantly affects the combustion wave speed, which rapidly increases when the temperature approaches the evaporation temperature.

2.6.4 Non stationary combustion wave propagation

The initiation of a chemical reaction in a layer of gas–particle mixture leads to the formation of a local ignition spot and the inception of an unsteady combustion wave. Depending on the properties of the reactive mixture and the conditions of its ignition, different scenarios of development of the combustion process can be realized. When the power of the thermal spot is so high that the heat flux from the layer of burning particles can ignite the neighbouring ones, the unsteady combustion wave the transforms into a steady wave spreading in a gas–particle mixture with a constant velocity. In contrast with that, at low power of the thermal spot the unsteady combustion wave degenerates after a long time and disappears.

Propagation of the unsteady combustion waves in gas–particle mixtures is accompanied by a complex evolution of the velocity, temperature and concentration fields. It may be described by the system of unsteady transport equations for a two-velocity and two-temperature reactive medium.

In order to estimate the behavior of the unsteady combustion waves it is possible to use a simplified formulation of the problem which is based on the assumption that the wave propagates in a motionless monodispersed mixture of an invariable composition. In particular, in the case of radiative heat transfer the problem reduces to the solution of the energy and radiation transfer equations subject to the initial and boundary conditions corresponding to the ignition of a

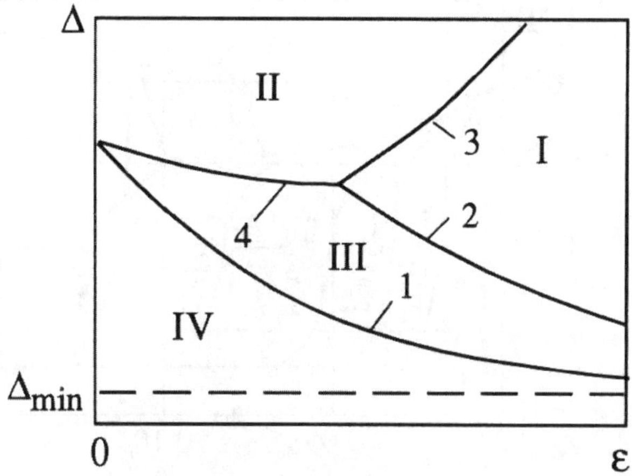

Fig. 6.13 Map of the possible regimes of ignition and combustion of a gas–particle mixture. Reprinted from Gol'tsiker et al. (1977)

two-phase mixture by a layer of high temperature combustion products (Gol'tsiker et al. 1977). The heat flux from this layer depends on its thickness, the particle sizes, the temperature and concentration, the difference between the ignition and initial temperatures of the gas–particle mixture, as well as on the thermo-physical properties of the carrier gas and dispersed fuel.

To classify the possible regimes of wave propagation in a gas–particle mixture it is expedient to use the following dimensionless variables:

$$\Delta = \frac{d\tilde{\sigma}T_2^4}{8\lambda_1(T_i - T_o)} \tag{6.42}$$

$$\varepsilon = \frac{\pi n c_{21}\rho_{21} d^3}{6} \tag{6.43}$$

$$\chi = \frac{8\alpha_1 n S_1 \tau_R}{\varepsilon d} \tag{6.44}$$

where T_i and T_0 are the ignition and initial temperature, C is the mass concentration, $c_{21}=C_2/C_1$, $\rho_{21}=\rho_2/\rho_1$, n is numerical particle concentration, S is the radiative surface of a particle, and τ_R is the characteristic time of radiative heating of a particle.

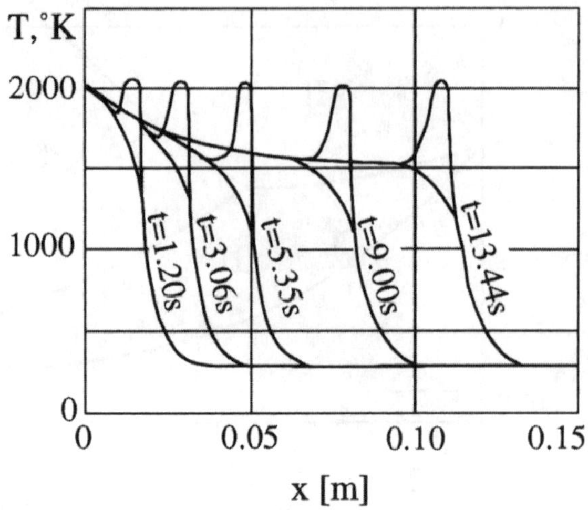

Fig. 6.14 Structure of a non stationary flame front in a coal particle–air mixture. Reprinted from Ozerov (1980)

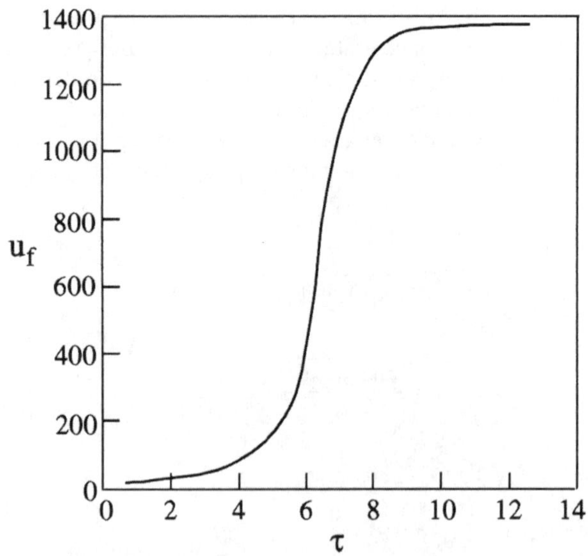

Fig. 6.15 Velocity of propagation of the combustion front in a dust–air mixture (u_f versus time u_f max $(1,400) \approx 5.5\,\text{m}/\text{s}$, period of induction t_{ind}~3.5 s). Reprinted from Krishenik et al. (1994), with permission

The map of the possible regimes of ignition and combustion of a gas–particle mixture is plotted in Fig. 6.13. Curves 1–4 correspond to limiting states. These curves separate the parametric plane $\Delta - \varepsilon$ (at $\chi = \text{const}$) into four domains. Domain I is characterized by the regimes at which ignition of the gas–particle mixture always results in the formation of a stationary combustion wave. This corresponds to large values of the parameters Δ and ε, i.e. to systems with intensive radiative heat transfer and high particle concentration. Domain II corresponds to the states in which of both stationary and oscillatory waves can occur. In domain III ignition of the gas–particle mixture is accompanied by quenching of the combustion process. Under the conditions corresponding to domain IV ignition of the gas–particle mixture is impossible.

The dispersion of solid fuel and its concentration affect significantly the development of combustion in a gas–particle mixture. An increase of the dispersion of the solid phase (a decrease of the parameter Δ) promotes the quenching of combustion. An increase of the particle concentration (an increase of the parameter ε) leads to the formation of a stationary combustion wave.

Unsteady flame propagation also depends on details of the ignition of the gas–particle mixture. This problem was studied by Stepanov (1976) for conductive or radiative mechanisms of heat transfer. The numerical calculations revealed the effect of the oxidizer concentration in a gas–particle mixture and of the details of ignition on the processes of establishment of a stationary flame. At $\gamma_{ox} > 1$ the ignition of the gas–particle mixture by a hot wall leads to relatively fast formation of a stationary combustion wave that propagates with a speed determined by the convective mechanism of heat transfer (Fig. 6.14). In the case of ignition of agas–particle mixture by a half-space filled with hot combustion products containing non burning particles ($\gamma_{ox} < 1$), the flame front initially accelerates and then slows down and stabilizes at a speed corresponding to the radiative mechanism of heat transfer.

The temporal evolution of a combustion wave propagating in a gas–carbon particle mixture was considered by Krishenik et al. (1994). They used a two-dimensional nonstationary model accounting for the difference between the temperatures of the solid and gaseous phases, as well as for the radiative and conductive heat transfer. Transition from the low-speed conductive regime to the high-speed radiative regime was considered in detail. The results characterizing such a transformation are presented in Fig. 6.15 in the form of the dependence of the flame speed on time $u_f(\tau)$. As can be seen in Fig. 6.15 there are two distinct regimes of combustion wave propagation. During the initial period when heat transfer occurs mainly by conduction, the low-speed regime takes place. Then at higher τ a steep (practically explosive) increase of the speed of the combustion wave is observed. The cause of the transition to this combustion regime is in the change in the structure of the combustion wave accompanied by expansion of the reaction zone, as well as by the change of the contributions of the conductive and radiative heat transfer components.

References

Ballal DR, Lefebvre AH (1981) Flame propagation in heterogeneous mixtures of fuel droplets, fuel vapor and air. The Eighteenth Symposium (International) on Combustion. The Combustion Institute, Pittsburgh, Pa., pp. 321–328

Bhaduri D, Bandyopadhyay S (1971) Combustion in coal dust flames. Combust. Flame 17: 15–24

Burgoyne JH, Cohen L (1954) The effect of droplet size on flame propagation in liquid aerosols. Proc. Rl Soc. London Ser. A 225: 375–392

Cassel HM, Liebman I, Mock WK (1957) Radiative transfer in dust flames. The Sixth Symposium (International) on Combustion. The Combustion Institute. Reinhold, New York, pp. 602–605

Cekalin EK (1962) Propagation of flame in turbulent flow of two-phase fuel-air mixture. The Eighth Symposium (International) on Combustion. The Combustin Institute Willliams and Wilkins, Md., Baltimore pp. 1125–1129

Essenhigh RH, Csaba J (1963) The thermal radiatin theory for plane flame propagation in coal dust clouds. The Ninth Symposium (International) on Combustion. The Combustion Institute. Academic Press, New York pp. 111–125

Frank-Kamenetskii DA (1969) Diffusion and heat transfer in chemical kinetics. 2nd edn. Plenum, New York

Gol'tsiker AD, Todes OM, Chiviliatin CA (1977) Theory unstationary flame propagation in aerodispersed systems. In: Combustion and Explosion, Nauka, Moscow pp. 300–306 (in Russian)

Greenberg JB, Silverman I, Tambour YA (1996) New heterogeneous burning velocity formula for the propagation of a laminar flame front through a polydisperse spray of droplets. Combust. Flame 104: 358–368

Gurevich MA, Ozerova GE, Stepanov AM (1972) The calculation of flame speed in dust of solid fuel. In: Combustion and Explosion. Nauka, Moscow, p. 198 (in Russian)

Hayshi S, Kumagai S (1975) Flame propagation in fuel droplet-vapor-air mixture. The Fifteenth Symposium (International) on Combustion. The Combustion Institute., Pittsburgh, Pa., pp. 445–452

Khitrin LN (1957) Physics of combustion and explosion. (in Russian) Moscow University, Moscow

Krishenik PM, Rumanov EN, Shkadinskii KG (1994) Modeling of combustion wave propagation in a carbon dust/gas mixture. Combust. Flame 99: 713–722

Leipunskii OI (1960) About the dependence of combustion rate of black powder on pressure. J. Phys. Chem. 34: 114–121 (in Russian)

Lin TH, Sheu YY (1991) Theory of laminar flame propagation in near stoichiometric dilute sprays. Combust. Flame 84: 333–342

Lin TH, Law CK, Chung SH (1988) Theory of laminar flame propagation in off-stoichiometric dilute sprays. Int. J. Heat Mass Transfer 31: 1023–1034

Liu C-C, Lin T-H (1991) The interaction between external and internal heat losses on the flame extinction of dilute sprays. Combust. Flame 85: 468–478

Merzhanov AG, Khaikin BI (1988) Theory of combustion caves in homogeneous media. Prog. Energy. Combust. Sci. 14: 1–98

Merzhanov AG, Khaikin BI (1992) Theory of combustion waves in homogeneous media. (in Russian) AN SSSR, Chernoglovka

Merzhanov AG, Khaikin BI, Shkadinskii KG (1969) Establish of stationary flame propagation at gas ignitin by overheated surface. Prikl. Mech.Tech.Phys. 5: 42–48 (in Russian)

Mizutani Y (1972) Turbulent flame velocities in premixed sprays. Part II. Theoretical analysis. Combust. Sci. Technol. 6: 11–21

Mizutani Y, Nishimoto T (1972) Turbulent flame velocities in premixed sprays. Part I. Experimental study. Combust. Sci. Technol. 6: 1–10

Myers GD, Lefebvre AH (1986) Flame propagation in heterogeneous mixtres of fuel drops and air. Combust. Flame 66: 193–210

Nigmatuilin RI (1991) Dynamics of Multiphase Media. v. 1 and 2. Hemisphere. London

Ozerov ES (1980) Foundations of the theory of combustion of gas-dispersed systems. (in Russian) L.P.J., Leningrad

Ozerova GE, Stepanov AM (1973) Effect of radiation on flame propagation through a gas suspension of solid fuel particles. Combust. Explos. Shock Waves 9: 543–549

Polymeropoulos CE (1974) Flame propagation in a one-dimensional liquid fuel spray. Combust. Sci. Technol. 9: 197–207

Polymeropoulos CE, Das S (1975) The effect of droplet size on the burning velocity of kerosene-air-sprays. Combust. Flame 25: 247–257

Richards GA, Lefebvre AH (1989) Turbulent flame speeds of hydrocarbon fuel droplets in air. Combust. Flame 78: 299–307

Rumanov EN, Khaikin BI (1971) Towards the flame propagation in gas/particle mixtutre. Sov. Phys. Dokl. 201: 144–147

Rumanov EN, Khaikin BI (1972) Regimes of flame propagation in gas/particle mixture. Combkustion and Explosion. The Third All Union Symposium on Combustion and Explosion. Nauka, Moscow pp. 161–165

Seshadri K, Berlad AL, Tangirala V (1992) The structure of premixed particle-cloud flames. Combust. Flame 89: 333–342

Silverman I, Greenberg JB, Tambour Y (1993) Stoichiometry and polydisperse effects in premixed spray flames. Combust. Flame 93: 97–118

Sirignano WA (1999) Fluid dynamics and transport of droplets and sprays. Cambridge University Press, Cambridge

Slezak SE, Buckius RO, Krier H (1985) A model of flame propagation in rich mixtures of coal dust in air. Combust. Flame 59: 251–265

Spalding DB (1953) Theoretical aspect of flame stabilizatin. An approximate graphical method for the flamespeed of mixed gases. Aircraft Eng. 25: 264–276

Stepanov AM (1976) The theoretical study of regimes of ignition and flame propagation in gas-dispersed stems. Ph.D. Thesis. (in Russian) Leningrad Polytech. Institute

Sukhov GS, Yarin LP (1981a) Combustion waves in bubbly media. Sov. Physics Dokl. 256: 376–380

Sukhov GS, Yarin LP (1981b) Laws of combustion of bubbled media. Combust. Explos. Shock Waves 17: 251–257

Vainshtein PB (1973) Radiative frame front in gas/particle mixture. Prikl. Mech. Tech. Phys. 3: 83–91 (in Russian)

Vainshtein PB, Nigmatulin RI (1971) Combustion of gas/particle mixture. Prikl. Mech.Tech. Phys. 4: 19–33 (in Russian)

Vainshtein PB, Nigmatulin RI (1973) Towards the theory of flame propagation in gas/droplet mixture. Prikl. Mech.Tech.Phys. 4: 101–108 (in Russian)

Williams FA (1959) Spray combustion theory. Combust. Flame 3: 215–228

Williams FA (1985) Combustion theory. 2nd edn. Benjamin/Cummings, Menlo Park, Calif.
Zel'dovich YaB (1941) The theory of limit of slow flame propagation. J. Exp. Theor. Phys. 11: 159–168
Zel'dovich YaB (1948) Towards a theory of flame propagation. J. Phys. Chem. 22:.27–48 (in Russian)
Zel'dovich YaB, Barenblatt GI (1959) Theory of flame propagation. Combust. Flame. 3: 61–74
Zel'dovich YaB, Frank-Kamenetskii DA (1938a) The theory of thermal flame propagation. J. Phys. Chem. 12: 100–105 (in Russian)
Zel'dovich YaB, Frank-Kamenetskii DA (1938b) Towards the theory of uniform flame propagation. Sov. Phys. Dokl. 19: 693–695
Zel'dovich YB, Barenblatt GI, Librovich VB, Makhviladze GM (1985) Mathematical theory of combustion and explosion. Plenum, New York.

2.7. Combustion wave propagation in bubbly media

2.7.1. Process mechanism

Bubbly media belong to a vast class of gas–liquid systems in which numerous gas inclusions are located within the carrier fluid. They occur in various industrial systems such as boilers, chemical reactors, etc. The existence of small gas inclusions in the bulk has an important influence on the physical properties of bubbly media, in particular, on their rheology (Kraynik 1988 and Reinelt and Kraynik 1996). The thermal features of bubbly media such as heat capacity, thermal conductivity, the temperature rise in the reactive bubbly media at full burn-out of gaseous oxidizer, etc. depend largely on the volumetric content of the gaseous phase (Yarin and Sukhov 1987).

Typically, two limiting kinds of bubbly media are distinguished: (i) gas bubble suspensions and (ii) foams. The structure of such media is depicted in Fig. 7.1. In bubble suspensions gas inclusions are separated by thick layers of fluid, whereas in foams they are separated by thin liquid films. In both cases combustion wave propagation is accompanied by heating and vaporizing of the liquid reagent or inert carrier fluid, as well as by the ignition and combustion of the reactive gaseous mixture. At the same time there are a number of significant distinctions between the combustion of bubble suspensions and foams.

The development of combustion in bubble suspensions may be illustrated as follows. Let the bubbly medium consists of liquid and gaseous reactants, the carrier liquid fuel containing fine bubbles filled with gaseous oxidizer. The partial pressure of the liquid reagent vapor is negligible at low temperature corresponding to the initial state of the bubble suspension. Under these conditions bubbles contain gaseous oxidizer only. Since the rate of oxidation of the carrier fluid by direct interaction of liquid/gaseous reagents is very small, the bubble suspension is practically inert. That means that its composition and temperature remain invariable for physically realistic intervals of time (Fig. 7.2a). The situation changes qualitatively with local heating of the suspension by an external source (Fig. 7.2b). An increase of the temperature within a suspension layer leads to vaporization of liquid reagent and the formation of a reactive vapor–oxidizer mixture inside the bubbles. The initiated chemical reaction in such a mixture leads to further heating of the suspension. At high enough temperature, heat release due to the chemical reaction exceeds heat losses to the surrounding fluid, which results in the progressive heating of the mixture inside the bubbles and its ignition (Fridman et al. 1981).

Chemical reaction initiated inside the bubbles significantly affects the intensity and direction of the interfacial heat transfer. During the first stage of the process when the temperature of liquid reagent within the heated layer exceeds the temperature of the gaseous oxidizer, the heat flux is directed towards the cold bubbles. A sharp temperature rise in the gaseous phase after ignition of the reactive mixture results a change in the direction of the heat flux (from hot bubbles to

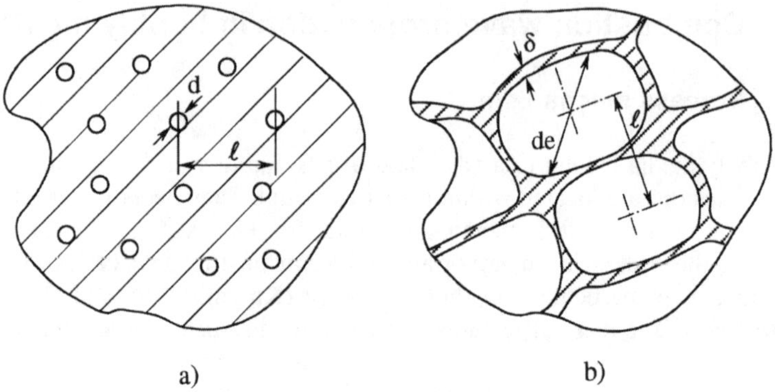

Fig. 7.1 Structure of bubbly media. a) Bubble suspension. b) Foam. d is the bubble diameter, d_e is the volume equivalent diameter of irregular bubbles, ℓ is the interbubble size, and δ is thickness of the liquid film

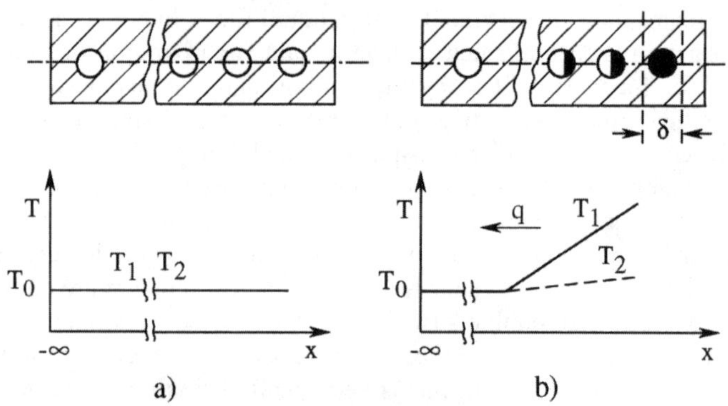

Fig. 7.2 Temperature distribution within bubble suspension. a) The low temperature regime corresponding to the initial state of the medium. b) State of the medium with a heated layer. T_1 and T_2 are the temperatures of the gaseous and liquid phases, T_0 is the initial suspension temperature, q is the heat flux from the heated layer, and δ is the thickness of the heated layer. The degree of bubble shading corresponds to the completeness of combustion

the surrounding fluid), heating of the neighboring layers of liquid reagent, its vaporization and formation of vapor–oxidizer mixture inside the primarily cold bubbles. The ignition and burning of the mixture in these bubbles sustain the combustion wave and its propagation in the bubble suspension. Therefore, the thermal interaction of the system of bubbles with the carrier fluid determines the transfer of the thermal pulse from an initially high temperature region of bubble suspension to a low temperature region (Fig. 7.3), i.e. the development of a macroscopic, multistage process leading to the formation of the combustion wave. Such a process can only occur if the characteristic time of combustion wave propagation in a given region of reactive suspension is much less than the induction time:

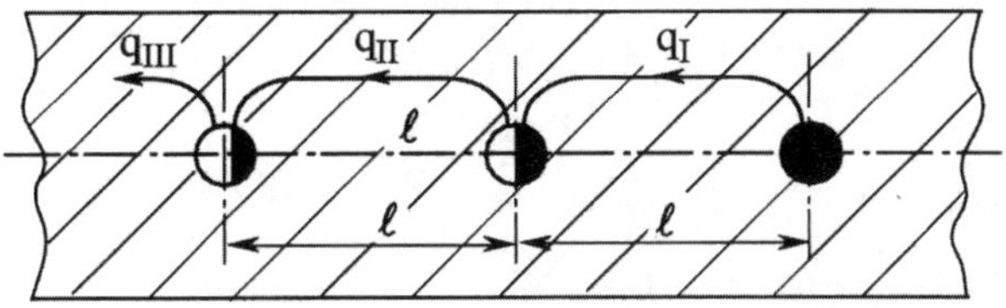

Fig. 7.3 Scheme of the heat transfer in a bubble suspension. The degree of bubble shading corresponds to the completeness of the vapor–oxidizer mixture inside the bubbles

$$t_w \ll t_{ind} \tag{7.1}$$

where $t_w = L/u_f$, L is the extension of a region of the bubble suspension, u_f is the speed of combustion wave, and t_{ind} is the induction time corresponding to the ignition delay inside the bubbles.

From the physical point of view the inequality (7.1) corresponds to the case when the properties of the bubble suspension outside a thin reaction zone remain invariable. In contrast with that in the case when $t_w \approx t_{ind}$, or $t_w > t_{ind}$, burn-out of the fresh mixture outside a reaction zone is essential. Changing of the composition of the fresh mixture in front of the reaction zone makes propagation of a combustion wave with constant speed impossible.

The mechanism of combustion wave propagation in gas–liquid media with high volumetric content of gaseous phase (foams) is noticeably different from that described above. Heating of such media accompanied by changing of their structure leads to the destruction of the liquid matrix and the formation of a gas–droplet reactive mixture. Zamanshchikov and Kakutkina (1993, 1994) and Babkin et al. (1994) showed that these processes significantly affect the speed of combustion wave propagation in water-based foams, as well as its dependence on the volumetric content of liquid phase. It is emphasized that the change of the bubble medium structure manifests itself not only in variation of the absolute value of the combustion speed but also in its dependence on the content of phases.

2.7.2. Combustion waves in bubble suspensions

Modeling of the medium. Combustion waves in bubble suspensions were studied by Sukhov and Yarin (1980, 1981 a, 1981 b, 1987), and the following consideration is based on these studies. In order to capture the main features of the phenomenon the authors explored wave propagation in a medium with non deformable bubbles "frozen" in the carrier fluid. In spite of the simplified character of this model, one can expect that it yields a plausible description of the complex process since the model accounts for the basic features of the phenomenon, namely, the multistage heat transfer from the combustion products to fresh mixture, as well as for the sharp acceleration of the chemical reaction with the temperature rise.

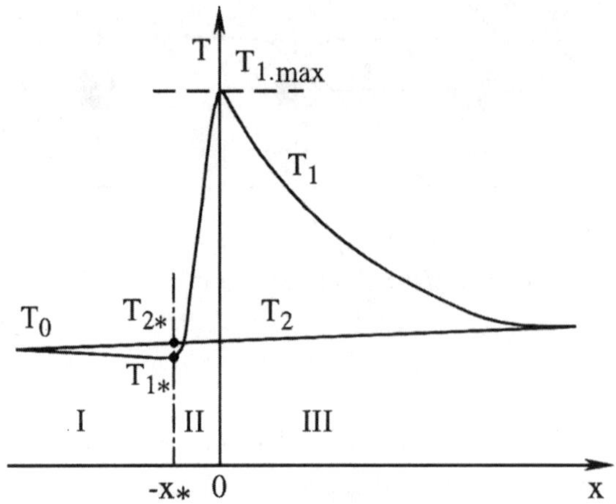

Fig. 7.4 Thermal structure of the combustion wave in bubble suspension. Reprinted from Yarin and Sukhov (1987). I, II and III are heating, reaction and thermal relaxation zones, respectively. T_1 and T_2 are the temperatures of gaseous and liquid phases, respectively. T_{1*} and T_{2*} are the temperatures of the gaseous and liquid phases in the cross-section of the combustion wave in which the rates of chemical reaction and heat transfer are equal to each other.

For brevity and clarity of the explanation, we consider the simplest model of suspension as a liquid reagent with bubbles containing gaseous oxidizer only. We assume that the bubbles are uniformly distributed in the carrier fluid and the spacing between the neighboring bubbles is much larger than their own sizes. This makes it possible to neglect bubble coalescence, as well as the transfer of gaseous phase from one bubble to another. Under these conditions the heat transfer within the combustion wave occurs as follows. First heat from the "hot" bubbles already containing combustion products is transferred through the gas–liquid interface to the surrounding fluid. Then due to thermal conductivity, heat is transferred towards the "cold" bubbles in front of the reaction zone.

The structure of the combustion wave in a bubble suspension is illustrated in Fig. 7.4. There are three characteristic zones in which heating and evaporation of liquid reactant (I), ignition and combustion of reactive vapor–oxidizer mixture (II), and thermal relaxation of the combustion products (III) occur. The temperatures of the gaseous and liquid reactants within the first zone are close to the initial temperature of the bubble suspension. Under these conditions the effect of the heat release due to the chemical reaction is negligible. The heating of bubbles when the reaction zone approaches occurs mainly via heat transfer from the liquid reagent. An increase of the temperature of the vapor–oxidizer mixture promotes chemical reaction, further heating of the gaseous mixture inside the bubble, and a decrease in the intensity of the interfacial heat transfer. At some distance from the reaction zone, the rates of the heat release and heat transfer become equal to each other. The cross-section of a combustion wave in which such an equality is fulfilled can be taken as the boundary that subdivides the heating and the combustion zones. Accordingly, the temperature of the gaseous phase corresponding to this cross-section ($-x_*$ in Fig. 7.4) can be taken as a conditional ignition temperature.

To describe the combustion wave propagation in reactive bubble suspension Sukhov and Yarin (1981 a, 1981 b, 1987) used the model of interpenetrating and interacting continua (Nigmatulin 1991). In accordance with this model the liquid and gaseous phases are presented as continua possessing some effective density ρ_i related to the physical density of a given phase ρ_i^o and its volumetric content ξ_i by means of the following relation:

$$\rho_i = \rho_i^o \xi_i \tag{7.2}$$

In the case of low volumetric content of gaseous oxidizer in a bubble suspension it is possible to omit the conductive heat and diffusive mass transfer in the gaseous phase, as well as the latent heat of evaporation. In this approximation the system of the governing equations for the phases and components of the reactive mixture takes the following form:

$$\frac{\partial \rho_1}{\partial t} = W_e \tag{7.3}$$

$$\frac{\partial \rho_2}{\partial t} = -W_e \tag{7.4}$$

$$\frac{\partial \rho_\ell}{\partial t} = W_\ell \tag{7.5}$$

$$\rho_1 c_1 \frac{\partial T_1}{\partial t} = qW - hs(T_1 - T_2) \tag{7.6}$$

$$\rho_2 c_2 \frac{\partial T_2}{\partial t} = \frac{\partial}{\partial x}\left(\lambda_2 \frac{\partial T_2}{\partial x}\right) + hs(T_1 - T_2) \tag{7.7}$$

where ρ_1 and ρ_2 are the effective density of the gaseous and liquid phases; ρ_ℓ is the effective density of the ℓ-th component of the bubble suspension; T_1 and T_2 are the temperatures of the gaseous and liquid phases; c and λ are the specific heat and thermal conductivity, h is the heat transfer coefficient; s is the specific area of the interfacial surface; q and W are the heat release and the chemical reaction rate; W_e is the rate of evaporation; W_ℓ is the rate of conversion of the ℓ-th component of the reactive mixture; subscripts $\ell = a$, b_1, b_2, c and d correspond to oxidizer, vapor of reactive liquid, liquid reagent, combustion products and an inert admixture, respectively.

The rate of conversion of the ℓ-th component is expressed as

$$W_a = -\sigma W \;; \quad W_{b_1} = -W + W_e; \tag{7.8}$$

$$W_{b.2} = -W_e; \quad W_c = (1+\sigma)W, \quad W_d = 0$$

where σ is the stoichiometric oxidizer-to-fuel mass ratio.

Assuming that chemical reaction of zero order leads to a full burn-out of the gaseous oxidizer and the formation of gaseous products only, we take the kinetic function W in the following form:

$$W = \xi_1 z f(\eta) \exp(-\frac{E}{RT_1}) \tag{7.9}$$

where ξ_1 is the volumetric content of the gaseous phase; $\eta = (\rho_{a.0} - \rho_a)/\rho_{a.0}$ is the completeness of combustion; $f(\eta)$ is the burning-out function [$f(\eta) = 1$ when $\eta < 1$, $f(\eta) = 0$ when $\eta = 1$] E is the activation energy; R is the universal gas constant; z is the pre-exponential factor; subscript 0 corresponds to the initial state.

The specific area of the interfacial surface s depends on the bubbles' diameters and concentration. In a suspension containing N bubbles of diameter d in unit volume, the specific area of the interfacial surface equals

$$s = \pi d^2 N \tag{7.10}$$

In some cases it is more convenient to express s as a function of the volumetric content of the gaseous phase ξ_1 and the numerical concentration of bubbles N. Taking into account that

$$\xi_1 = \frac{\pi d^3}{6} N \tag{7.11}$$

we find that s equals

$$s = (36\pi\xi_1^2 N)^{\frac{1}{3}} \tag{7.12}$$

The system of Eqs. (7.3) and (7.10) should be supplemented by the equation of state of the gas, by the equation determining the dependence of vapor pressure on temperature, as well as by the mass balance equation relating the volumetric content of the gaseous and liquid phases.

Assuming that the vapor behaves as a perfect gas and the evaporation process is inquasi-equilibrium, we obtain these equations in the following form:

$$\xi_1 P_1 = \rho_1 R_g T_1 \tag{7.13}$$

$$P_{b.1} = z_b \exp\left(-\frac{q_e}{R_{b.1} T_2}\right) \tag{7.14}$$

$$\xi_1 + \xi_2 = 1 \tag{7.15}$$

where q_e is the latent heat of evaporation, z_b is a constant, and $R_{b.1}$ and R_g are gas constants for the vapor and gaseous mixture, respectively.

Taking into account Eqs.(7.8) and (7.15), we can present the mass balance equations (7.3)–(7.5) as follows:

$$\frac{\partial \rho_1}{\partial t} - \rho_2^o \frac{\partial \xi_1}{\partial t} = 0 \tag{7.16}$$

$$\frac{\partial \rho_{b.1}}{\partial t} - \frac{\partial \rho_1}{\partial t} = W \tag{7.17}$$

$$\rho_{a.s} \frac{\partial \eta}{\partial t} = \sigma W \tag{7.18}$$

It is convenient to associate the reference frame with the combustion front. Introducing the new variables

$$t = t', \qquad \zeta = x + \int_0^t u_f \tag{7.19}$$

and taking into account that $\partial/\partial t = \partial/\partial t' + u_f\,(\partial/\partial\zeta)$; $\partial/\partial x = \partial/\partial\zeta$ we obtain (in the steady state approximation) the following system of equations:

$$\frac{d\rho_1}{d\zeta} - \rho_2^o \frac{d\xi_1}{d\zeta} = 0 \tag{7.20}$$

$$u_f \frac{d(\rho_1 - \rho_{b.1})}{d\zeta} = W \tag{7.21}$$

$$\rho_{a.0} u_f \frac{d\eta}{d\zeta} = \sigma W \tag{7.22}$$

$$c_1\rho_1 u_f \frac{dT_1}{d\zeta} = qW - hs(T_1 - T_2) \tag{7.23}$$

$$c_2\rho_2^o(1-\xi_1)u_f \frac{dT_2}{d\zeta} = \frac{d}{d\zeta}\left(\lambda_2 \frac{dT_2}{d\zeta}\right) + hs(T_1 - T_2) \tag{7.24}$$

where u_f is the combustion wave speed.

The boundary conditions corresponding to the present problem are

$$\zeta = -\infty,\ \ T_1 = T_2 = T_0;\quad \eta = 0;\ \ \xi = \xi_0;\ \ \rho_1 = \rho_{1.0},\ \ W = 0 \tag{7.25}$$

$$\zeta = +\infty \quad \frac{dT_1}{d\zeta} = \frac{dT_2}{d\zeta} = 0,\quad \eta = 1$$

To find a physically realistic solution that corresponds to combustion wave propagation with a constant speed, conditions (7.25) should be supplemented by the condition $\zeta < \zeta_*$, $W = 0$. The latter corresponds to canceling the kinetic function at some temperature $T_1 = T_{1*} > T_0$. That means that the rate of chemical reaction in the gaseous mixture equals zero within the temperature range $T_0 \le T_1 < T_{1*}$. Such a condition is identical to that used in the classical theory of combustion wave propagation in a homogeneous mixture (Zel'dovich et al. 1985). It prevents burning out of the mixture at low temperatures, while the combustion wave approaches from a formally infinite distance.

The ignition of a bubble suspension is determined by the condition that expresses the equality of the intensities of heat release and heat transfer in some cross-section of the combustion wave. It can be written as follows:

$$qz\xi_1 \exp(-\frac{E}{RT_{1*}}) = hs(T_{2*} - T_{1*}) \tag{7.26}$$

where T_{1*} and T_{2*} are the temperatures of the gaseous and liquid phases in the cross-section of the combustion wave in which the rates of chemical reaction and heat transfer are equal to each other.

2.7.3. The thermal structure of a combustion wave

Consider the temperature distribution within a combustion wave which propagates over a bubble suspension. For this aim we use the thermal balance equations (7.23) and (7.24). We first estimate variation of the volumetric content of the gaseous and liquid phases. Integration of Eq. (7.20) yields (taking into account the boundary condition $\zeta = -\infty\ \rho_1 = \rho_{a.0}, \xi_1 = \xi_{1.0})$ the following expression for the current value of the volumetric content of the gaseous phase ξ_1:

$$\xi_1 = \xi_{1.0} + \frac{\rho_1 - \rho_{a.0}}{\rho_2^o} \tag{7.27}$$

Since $\rho_2^o >> \rho_1$ and $\xi_1 + \xi_2 = 1$, we find that

$$\xi_1 \approx \xi_{1.0}, \quad \xi_2 \approx \xi_{2.0} \tag{7.28}$$

Combining Eqs. (7.21) and (7.22) we obtain $\frac{\rho_{a.0}}{\sigma}\frac{d\eta}{d\zeta} = \frac{d(\rho_1 - \rho_{b.1})}{d\zeta}$. Integrating it subject to the condition $\xi = -\infty,\ \eta = 0,\ \rho_{b.1},\ \rho_1 = \rho_{a.0}$ we obtain

$$\rho_1 = \rho_{b.1} + \left(1 + \frac{\eta}{\sigma}\right)\rho_{a.0} \tag{7.29}$$

As a rule, the stoichiometric oxidizer-to-fuel ratio $\sigma >> 1$, whereas the completeness of combustion is always less than or equal to one $\eta \le 1$. Therefore, we obtain the following estimate of the density of the gaseous phase:

$$\rho \approx \rho_{b.1} + \rho_{a.0} \tag{7.30}$$

At large values of the ratio $q_e/(RT_0)$ corresponding to the initial temperature of the bubble suspension, the vapor density is much less than the density of the gaseous oxidizer. In this case it is possible to assume that

$$\rho_1 \approx \rho_{a.0} = \rho_{1.0} \tag{7.31}$$

The estimates (7.30) and (7.31) make it possible to neglect the variation of ρ_1 and ξ_1 and to reduce the problem to solving Eqs. (7.22)–(7.24) to determine the functions T_1, T_2 and η. Combining Eqs. (7.72)–(7.24) and integrating them from the initial to the final state, we obtain the following expression for the final temperature of bubble suspension:

$$T_{fin} = T_0 + \frac{q\ \rho_{a.0}}{\sigma[c_2\rho_2^o(1-\xi_{1.0}) + c_1\rho_{1.0}]} \approx T_0 + \frac{q\rho_{a.0}}{\sigma c_2\rho_2^o} \tag{7.32}$$

where subscript fin corresponds to the final state. The second term on the right in of Eq. (7.32) is much less than the first one ($T_0 \approx 300\,K$, $(q\rho_{a.0})/(\sigma c_2 \rho_2^o) \approx 10\,K$). Therefore, in a bubble suspension, there is only a slight difference between the temperatures corresponding to the initial and the final states of a reactive medium. The latter is due to the high heat capacity of such system, as well as to the low mass content of the gaseous oxidizer.

To calculate the temperature distribution in the characteristic regions of the combustion wave, we assume that the heat capacity of the gaseous and liquid phases, as well as the heat transfer coefficient and specific area at the interfacial surface are constant. We also take into account that the rate of chemical reaction within the heating and thermal relaxation zones is negligible and the interfacial heat transfer does not significantly affect the temperature field in the reaction zone

A. Heating zone. In the case when $W = 0$ the momentum equation for the gaseous phase (7.23) takes the form

$$c_1 \rho_1 u_f \frac{dT_1}{d\zeta} = -hs(T_1 - T_2) \tag{7.33}$$

Taking into account that within the heating zone $-\infty < \zeta_*$ the derivative $dT_1/d\zeta$ does not exceed the derivative $dT_2/d\zeta$ and $c_1\rho_1 \ll c_2\rho_2^o(1-\xi_{1.0})$, it is possible to present the thermal balance equation as follows:

$$c_2 \rho_2^o (1-\xi_{1.0}) u_f \frac{dT_2}{d\zeta} = \lambda_2 \frac{d^2 T_2}{d\zeta^2} \tag{7.34}$$

Integration of Eq. (7.34) yields

$$T_2 = T_0 + (T_{2*} - T_0)\exp\left(\frac{\zeta}{\ell_2}\right) \tag{7.35}$$

where $\ell_2 = (2\lambda_2)/(c_2\rho_2^o(1-\xi_{1.0})u_f)$ is the characteristic liquid heating length.

Equation (7.33) should be integrated using the expression (7.35). The integral of Eq. (7.33) is

$$T_1 = T_0 + g(\zeta)\exp\left(-\frac{\zeta}{\ell_1}\right) \tag{7.36}$$

where

$$g(\zeta) = A + (T_{2*} - T_0)\left\{\frac{\ell_2}{\ell_1 + \ell_2}\exp\left[\frac{\zeta(\ell_1 + \ell_2)}{\ell_1 \ell_2}\right]\right\} \tag{7.37}$$

The characteristic gaseous phase heating length is denoted by $\ell_1 = (c_1 \rho_1 u_f)/hs$.

Bearing in mind that the solution of the problem must be bounded as $\zeta \to -\infty$, we find that constant A in Eq.(7.37) equals zero. Then the general solution of the energy equation for the gaseous phase takes the following form:

$$T_1 = T_0 + (T_{2*} - T_0)\frac{\ell_2}{\ell_1 + \ell_2}\exp\left(\frac{\zeta}{\ell_2}\right) \tag{7.38}$$

To find the temperature corresponding to the boundary that separates the regions of heating and combustion, we assume $\zeta = -\zeta_*$ in Eq. (7.38). Since $T_1(\zeta_*) = T_{1*}$ and $\zeta_*/\ell_2 \ll 1$, we obtain

$$T_{1*} = T_0 + (T_{2*} - T_0)\frac{\ell_2}{\ell_1 + \ell_2} \tag{7.39}$$

This temperature should be considered as the ignition temperature of the reactive mixture inside the bubbles.

B. Reaction zone. The thermal regime in the combustion zone depends on the ratio of heat released due to chemical reaction to the heat losses from burning bubbles to the surrounding liquid reagent. The ratio is expressed as

$$G = \frac{qW}{hs(T_1 - T_2)} \tag{7.40}$$

When the parameter $G \gg 1$ and the activation energy of the chemical reaction is large enough, burn-out of the gaseous oxidizer occurs extremely quickly. In this case the domain of the intensive chemical reaction is localized within a narrow high temperature zone (the combustion front) where the reactants are consumed under almost adiabatic conditions. The situation changes qualitatively when the heat losses are dominant. In this case degenerative low temperature regimes with low rates of reaction are realized. At $G \le 1$ the thickness of the reaction zone becomes significantly extended.

Hereinafter we confine the consideration to a combustion wave with a thin front and determine the temperature of the gaseous phase within the reaction zone. Assuming that $qW \gg hs(T_1 - T_2)$ and using Eqs. (7.22) and (7.23), we arrive at the equation

$$c_1\rho_{1.0}\frac{dT_1}{d\zeta} = q\frac{\rho_{a.0}}{\sigma}\frac{d\eta}{d\zeta} \tag{7.41}$$

Integration of Eq. (7.41) across the reaction zone yields

$$T_{1.m} = T_{1*} + q\frac{\rho_{a.0}}{\sigma c_1 \rho_{1.0}} \tag{7.42}$$

where $T_{1.m}$ is the maximal temperature of the gaseous phase.

C. Relaxation zone. Comparison of Eqs. (7.32) and (7.42) shows that the temperature of the liquid phase varies much less than the temperature of the gaseous phase. That makes it possible to neglect the variation of the liquid phase temperature within the narrow reaction zone and to assume that

$$T_2 = T_{2.m} \tag{7.43}$$

where $T_{2.m}$ is the liquid temperature at $\zeta = 0$.

Since the temperature of the liquid changes only slightly within the relaxation zone, it is possible to assume that variation of the interfacial heat transfer is determined by the change in the temperature of the gaseous phase. Bearing in mind the estimate (7.32), it is possible to assume that $T_2 \approx T_{fin}$ and to recast Eq. (7.6) into the following form:

$$c_1\rho_{1.0}u_f\frac{dT_1}{d\zeta} = -hs(T_1 - T_{fin}) \tag{7.44}$$

Integration of Eq. (7.44) subject the boundary condition $\zeta = 0,\ T_1 = T_{1.m}$ yields

$$T_1 = T_{fin} + (T_{1.m} - T_{fin})\exp\left(-\frac{\zeta}{\ell_1}\right) \tag{7.45}$$

Using this expression we find the first integral of Eq. (7.24) as

$$\frac{dT_2}{d\zeta} = \left\{\frac{hs(T_{1.m} - T_{fin})}{\lambda_2 D}\exp(-D\zeta) + C\right\}\exp\left(\frac{2\zeta}{\ell_2}\right) \tag{7.46}$$

where $D = (2\ell_1 + \ell_2)/(\ell_1\ell_2)$, and C is a constant of integration.

Taking into account the boundary conditions $\zeta \to \infty$, $dT_2/d\zeta \to 0$, we find that $C = 0$.

The second integral of Eq. (7.24) at point $\zeta = 0$ determines the temperature of the liquid phase in the reaction zone. It is expressed as

$$T_{2*} = T_{fin} - \frac{c_1\rho_{1.0}}{c_2\rho_2^o(1-\xi_{i.0})}\frac{T_{1.m} - T_{fin}}{1+\dfrac{\ell_2}{2\ell_1}} \tag{7.47}$$

2.7.4. Speed of the combustion wave

The system of the algebraic Eqs. (7.26), (7.32), (7.30), (7.42) and (7.47) contains five unknowns which determine the structure of the combustion wave and the speed of its propagation in a bubble suspension. When all the other unknowns are excluded, the system reduces to the thermal balance at the ignition point (Eq. 7.26) with the only unknown being the combustion wave speed. Taking into account that $c_1\rho_{1.0} \ll c_2\rho_2^o(1-\xi_{1.0})$, $T_{1*} < T_{fin}$ and $T_{fin} \sim T_0$, we obtain

$$\frac{Se}{\theta_{fin}}\exp\frac{2\theta_{fin}}{(1+\delta U^2)(2+\delta U^2)} = \frac{\delta U^2}{(1+\delta U^2)(2+\delta U^2)} \tag{7.48}$$

where $U = (c_2\rho_2^o(1-\xi_{1.0})u_f)/\sqrt{hs\lambda_2}$ is the dimensionless combustion wave speed, $\theta_{fin} = (E(T_{fin} - T_0))/(RT_0^2) = (\delta\varepsilon)/(\sigma\gamma)$; $\gamma = (c_1RT_0^2)/(qE)$; $\varepsilon = \rho_{a.0}/\rho_{1.0}$; $Se = (z\xi_{1.0}qE)/(hsRT_0^2)\exp(-E/RT_0)$ is the Semenov number; $\delta = (c_1\rho_{1.0})/(c_2\rho_2^o(1-\xi_{1.0}))$.

The transcendental Eq. (7.48) can be solved by the grapho-analytical method. Consider separately the right-hand and left-hand sides of Eq. (7.48). The first of these $Q_I(\delta U^2) = (\delta U^2)/[(1+\delta U^2)(2+\delta U^2)]$ is a nonlinear function of δU^2 having the maximum $Q_{I.max}$=0.17 at the point $\delta U^2 = \sqrt{2}$ and vanishing at $\delta U^2 \to \infty$ (Fig. 7.5). The left-hand side of Eq. (2.48) $Q_{II}(\delta U^2) = (Se/\theta_{fin})\exp[(2\theta_{fin})/((1+\delta U^2)(2+\delta U^2))]$ is a monotonically decreasing function of δU^2.

In the general case the curves $Q_I(\delta U^2)$ and $Q_{II}(\delta U^2)$ intersect at two points A and B which determine the combustion wave speed. The physical explanation of the appearance of the two solutions is as follows. The condition (7.26) corresponding to the ignition of vapor–oxidizer mixture is satisfied not only at the cross-section $\zeta = \zeta_*$, but also at another cross-section $\zeta = \zeta_*'$ (Fig. 7.6), located in the region of relatively low phase temperatures ($T_{1*}' < T_{1*}$, $T_{2*}' < T_{2*}$).

The existence of the second intersection point of the curves $Q_I(\delta U^2)$ and $Q_{II}(\delta U^2)$ is related to the finiteness of the heat release due to chemical reaction at $\zeta \to -\infty$ $(qW \neq 0)$, whereas the interfacial heat flux $hs(T_1 - T_2)$ tends to zero. However, the ignition at the point $\zeta = \zeta_*'$ is not realizable in reality since the heat release due to chemical reaction is not a dominant factor which determines heating of a reactive mixture. Therefore, the solution corresponding to the ignition at the point $\zeta = \zeta_*'$ has no physical meaning and must be discarded. It should be added that this unphysical solution corresponds to point B in the $Q - \delta U^2$ diagram in Fig. 7.5.

The location of the curves $Q_I(\delta U^2)$ and $Q_{II}(\delta U^2)$ on the parametric plane $Q - \delta U^2$ depends on the value of the Semenov number. An increase of Se leads to shifting of the curve $Q_{II}(\delta U^2)$ from the horizontal axis in Fig. 7.5 and approaching of the intersection points A and B to each other. At a certain (critical) value of the Semenov number $Se = Se_{cr}$ these points merge. In the case when $Se > Se_{cr}$ there are no solutions of Eq. (7.48) corresponding to steady-state combustion wave propagation in a bubble suspension. This phenomenon results from the intensification of the chemical reaction at temperatures close to the initial temperature, i.e. it is associated with the transition from the thermal mechanism of propagation of the combustion to a self-accelerating explosive process in the whole volume of the bubble suspension.

At $Se = Se_{cr}$ the curves $Q_I(\delta U^2)$ and $Q_{II}(\delta U^2)$ are tangential at a single point, where the conditions

$$Q_I = Q_{II} \tag{7.49}$$

$$\frac{dQ_I}{dU^2} = \frac{dQ_{II}}{dU^2} \tag{7.50}$$

are satisfied. It is emphasized that Eq. (7.48) is distinct from the ordinary balance equations of the thermal theory of explosion (Zel'dovich et al. 1985). Equation (7.48) does not express the relation between the heat release and heat removal, but it expresses the balance between two components of heat supply to the vapor–oxidizer mixture in the bubbles, namely the chemical reaction and the interfacial heat transfer.

Using the conditions (7.49) and (7.50), we find the critical value of the Semenov number

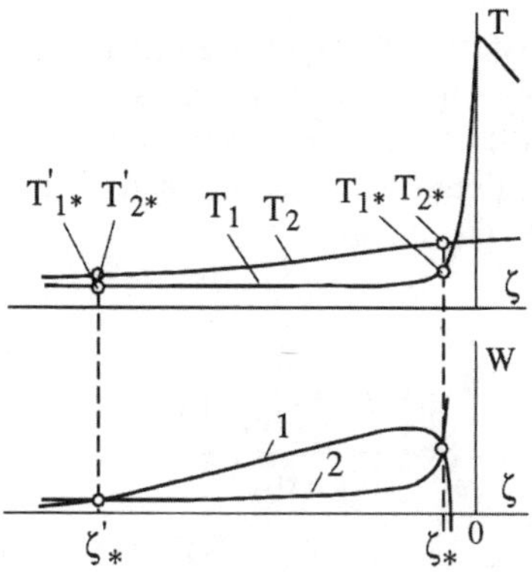

Fig. 7.5 Dependencies $Q_I(\delta U^2)$ and $Q_{II}(\delta U^2)$. Reprinted from Yarin and Sukhov (1987)

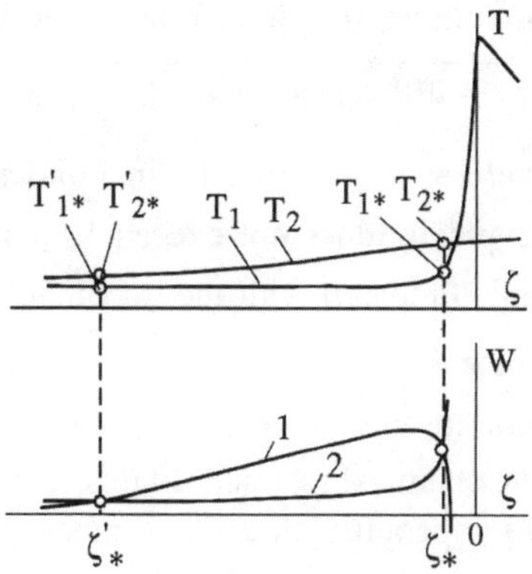

Fig. 7.6 Temperature distribution within the heating zone. Reprinted from Yarin and Sukhov (1987)

$$\mathrm{Se}_{cr} = \frac{\chi\theta_{fin}}{(1+\chi)(2+\chi)}\exp\left[-\frac{2\theta_{fin}}{(1+\chi)(2+\chi)}\right] \tag{7.51}$$

$$(\chi^2-2)(\chi+1)(\chi+2) = 2\chi(3+2\chi)\theta_{fin} \tag{7.52}$$

where $\chi = \delta U_{cr}^2$, and subscript cr corresponds to the critical state. It is seen that the critical Semenov number depends only on θ_{fin}.

Figure 7.7 shows the dependence $\mathrm{Se}_{cr}(\delta)$, which limits from above the region of existence of the combustion regimes in the system of liquid hexane–oxygen bubbles at the pressure $P = 5 \cdot 10^6$ Pa and $T_0 = 400$ K. It should be noted that propagation of combustion in a bubble suspension is not possible at all values of $\mathrm{Se} < \mathrm{Se}_{cr}$. In fact, as Se decreases, the combustion regime degenerates as a result of increasing of heat transfer to the condensed phases. The minimum value of Se at which the combustion is still possible can be estimated as follows. Representing the function G of Eq.(7.40) in the form of a ratio of dimensionless quantities, taking into account that the liquid temperature changes are insignificant within the reaction zone $(\theta_2 \approx \theta_{2*})$, we obtain

$$G = \mathrm{Se}\frac{\exp\left[\dfrac{\theta_1}{1+\beta\theta_1}\right]}{(\theta_1 - \theta_{2*})} \tag{7.53}$$

where $\theta_{2*} = (\varepsilon\delta)/(\gamma\sigma(1+\delta U^2))$ and $\beta = (RT_0)/E$.

In the region $\theta_1 > \theta_{2*}$ the function $G(\theta_1)$ is no monotonic (Fig. 7.8). On the left it is bounded by the asymptote at $\theta_1 = \theta_{2*}$; it has a minimum at the temperature $\theta_1' = [(1-2\beta)/(2\beta^2)]\left[1-\sqrt{(1-4\beta^2)((1+\theta_{2*})/(1-\beta^2))}\right]$ a maximum at the point $\theta_1'' = 1/\beta^2$, and tends to zero at $\theta_1 \to \infty$. The final dimensionless temperature of real bubble suspensions typically does not exceed 10, and $\theta_{2*} < \theta_{fin}$. Therefore, $\theta_1' \approx 1+\theta_{2*}$, i.e. it is small compared with the maximal possible gas temperature in the reaction zone. Since $\theta_{1m} < \beta^{-2}$, the minimum of function $G = f(\theta_1)$ is reached within the temperature interval $\theta_{2*} - \theta_{1m}$, where interfacial heat transfer hinders development of the chemical reaction. In this case heating of the gas phase in the reaction zone to a temperature $\theta_1 > \theta_1'$ is impossible, since at $\theta_1 > \theta_1'$ the heat losses to the liquid are equal to the heat release due to the reaction, as a result of which the temperature ceases to rise. Accordingly, at $\mathrm{Se}=\mathrm{Se}_{min}$ the maximum possible temperatures in the reaction zone are bounded by the small quantity θ_{min}. This means that as $\mathrm{Se} \to \mathrm{Se}_{min}$, combustion degenerates, which is accompanied by a sharp decrease in the chemical reaction rate and a broadening of the reaction zone. The degeneration boundary can be approximately determined using relations (7.48), (7.53) and the expression for θ_1' by selecting the value of Se that yields $G_{min}=1$.

In the parametric plane $\mathrm{Se}-\delta$ (Fig. 7.7) local heating of the liquid within the interval $\mathrm{Se}_{cr}>\mathrm{Se}>\mathrm{Se}_{min}$ gives rise to a self-propagating combustion wave, the process being characterized by high reaction rates, high gas-phase temperature and slight heating of the liquid (Fig. 7.7, domains I and II). The existence of the combustion wave becomes impossible when the mass content of the active mixture is less than a certain limiting value $\delta < \delta_{\ell im}$, and the liquid phase acts as ballast in the system. In the vicinity of the line $\mathrm{Se}=\mathrm{Se}_{min}$ combustion degenerates and at $\mathrm{Se}<\mathrm{Se}_{min}$ and $\delta > \delta_{\ell im}$ (Fig. 7.7) the low temperature regime sets in. Then the process is characterized by low values of the reaction rate.

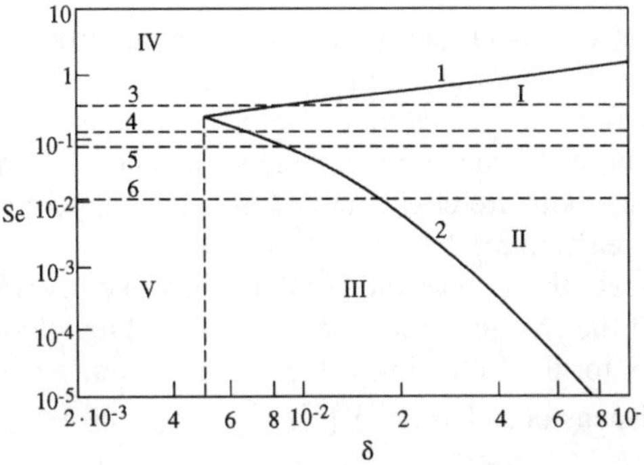

Fig. 7.7 Domains of existence of various reaction regimes in bubble suspensions. Reprinted from Yarin and Sukhov (1987). Curve 1, $Se_{cr}(\delta)$; curve 2, $Se_{min}(\delta)$, curves 3–6, $Se_{\ell im}(\theta_2)$ ($3 - \theta_2 = 0$, $4 - \theta_2 = 1.5$, $5 - \theta_2 = 2$, $6 - \theta_2 = 4$) $\varepsilon = 0.62$, $\sigma = 3.5$, $\beta = 0.027$, $\gamma = 5.33 \cdot 10^{-4}$

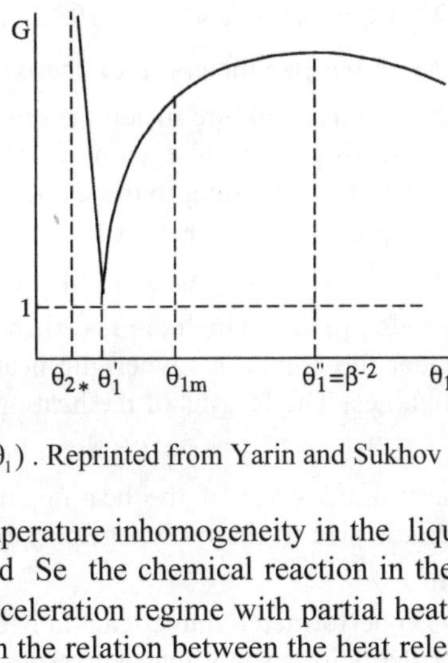

Fig. 7.8 Dependence $G(\theta_1)$. Reprinted from Yarin and Sukhov (1987)

If there is no temperature inhomogeneity in the liquid, then at any values of the parameters δ and Se the chemical reaction in the vapor–oxidizer mixture proceeds in the self-acceleration regime with partial heat transfer to the liquid. In this case, depending on the relation between the heat release and heat transfer, the development of the reaction either ends up in self-ignition or (if the loss of heat to the liquid is considerable) it proceeds slowly at the phase temperatures close to the initial temperature T_0 (slow reaction regime).

Using the approach developed for determining the combustion degeneration boundary and taking into account the fact that in the self-ignition regime $G \gg 1$ and $T_2 \approx T_0$, we can approximately determine the boundary separating the domain of the self-ignition from the domain where only a slow reaction is possible. This boundary corresponds to the limiting value of the Semenov number $Se_{\ell im} = 0.375$. Therefore, in the parametric plane $Se - \delta$ the following domain can be distinguish (Fig. 7.7). Domains I and II, where both stationary combustion waves and non-stationary phenomena of self-ignition coexist, domain II of slow reaction, as well as domains where only non-stationary (self-accelerating processes) are realized in the high-temperature (IV) and low-temperature (V) regimes. Also there is domain III where slow reaction and degenerate combustion regimes exist. In the latter, there is no substantial difference between the slow reaction and degenerate combustion regimes, since both are characterized by weak phase heating and low values of the chemical reaction rate.

Now we evaluate the rate of the combustion wave, as well as the order of magnitude values of the characteristic temperatures and length of heating, reaction and relaxation zones for the following values of the parameters of a monodisperse hexane–oxygen suspension: $P = 5\cdot 10^6\,[\mathrm{N/m^2}]$, T_0=400 K, $N = 4.8\cdot 10^{10}\,\mathrm{m}^{-3}$, $d = 2\cdot 10^{-4}\,\mathrm{m}$, $m_{1.0}$=0.2, $s = 6\cdot 10^{-3}\,\mathrm{m}^{-1}$, $q = 4.2\cdot 10^4\,[\mathrm{kJ/kg}]$, $E = 1.26\cdot 10^5\,[\mathrm{kJ/mol}]$, $z = 4\cdot 10^{12}\,[\mathrm{kg/m^3 s}]$, $c_1 = 3.4[\mathrm{kJ/kg\,K}]$, c_2=1.7 [kJ/kg K], $q_e = 3.2\cdot 10^2\,[\mathrm{kJ/mol}]$, $z_b = 2.6\cdot 10^9\,[\mathrm{N/m^2}]$, $\rho_2^o = 550[\mathrm{kg/m^3}]$, $\rho_{1.0} = 17.7[\mathrm{kg/m^3}]$, $\lambda_2 = 2.1\cdot 10^{-3}\,[\mathrm{kJ/m\,s\,K}]$, $Se = 6\cdot 10^{-3}$, $h = 4.2\cdot 10^{-3}\,[\mathrm{kJ/m^2\,s\,K}]$ and $\delta = 0.08$. The combustion wave speed corresponding to the above parameters is of the order of $10^{-3}\,[\mathrm{m/s}]$. The maximal temperature of the reactive mixture inside the bubbles and the final temperature of the liquid and gaseous phases are equal to 3,980 K and 687 K, respectively. The phase temperatures corresponding to the ignition point are T_{1*} = 472 K and T_{2*} = 520 K. Therefore, preheating of the gaseous mixture $(\Delta T = T_{1*} - T_{in})$ up to its ignition equals $\Delta T = 72\,\mathrm{K}$. The ignition of the bubble suspension occurs at temperatures of the gaseous mixture which are less than the temperature of the surrounding liquid, i.e. under the conditions when the heat flux is directed from the carrier fluid towards bubbles. The lengths of the heating, reaction and thermal relaxation zones are: $\ell_h \approx 2\cdot 10^{-2}\,\mathrm{m}$, $\ell_r \approx 2\cdot 10^{-3}\,\mathrm{m}$, $\ell_{rel.} = 1.4\cdot 10^{-2}\,\mathrm{m}$, respectively. Therefore, the linear dimensions of the heating, reaction and relaxation zones exceed the characteristic size of the inhomogeneity of the system (the bubble diameter) $d = 2\cdot 10^{-4}\,\mathrm{m}$.

Likhachev et al. (1989) performed a numerical study of combustion wave in bubble suspensions. The calculations show that the results of the numerical and analytical solutions are close to each other. For example, the dimensionless rate of combustion wave, the maximal gaseous phase temperature and the final temperature of the bubble suspension equal U=4.15 · 10^{-3}, T_{1m}=3988 K, T_{fin}=520 K and $U = 2.92\cdot 10^{-3}$, T_{1m}=4082 K, T_{fin}=559 K for the analytical and numerical solutions, respectively.

The calculations show that an increase of the chemical reaction rate (an increase of the Semenov number) is accompanied by an increasing of the combustion wave rate (Fig. 7.9). The linear dependence U(Se) corresponding to small values of Se is replaced by a much faster growth of U at large values of the Semenov number. A sharp nonlinear growth of U at Se>1 allows for the assumption that the function U(Se) has an asymptote. The latter manifests the changes of

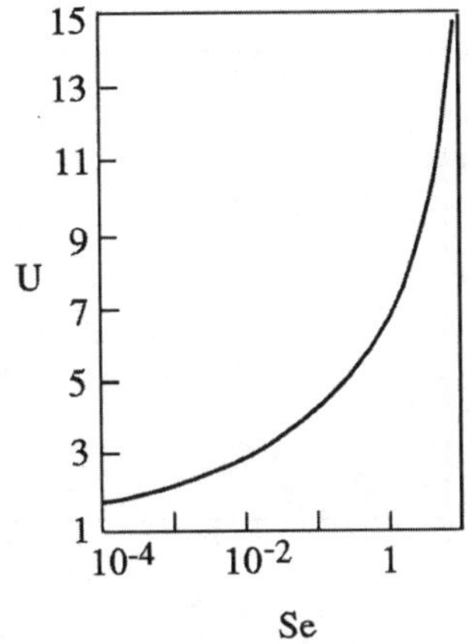

Fig. 7.9 Combustion wave rate versus the Semenov number. Reprinted from Likhachev et al.(1986)

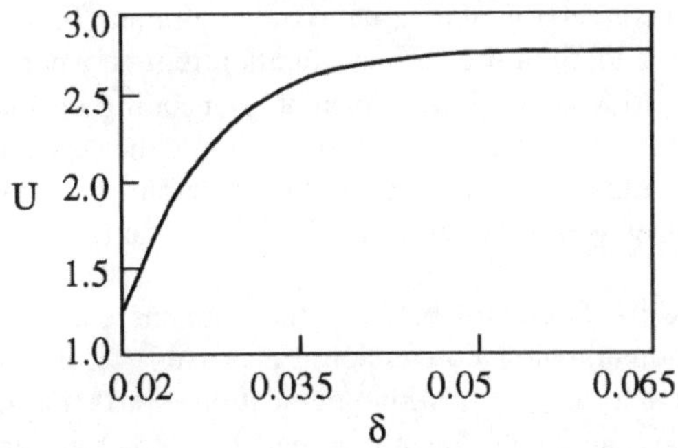

Fig. 7.10 Combustion wave rate versus δ. Reprinted from Likhachev et al.(1989)

the mechanism of combustion at large Se. This result of the numerical calculations is corroborated by the result of the analytical study of the process which shows the existence of the boundary of the oscillatory wave propagation regime in the bubble suspension when the value of Semenov number equals 1.4. The change of the mass content of the gaseous oxidizer practically does not affect the rate of the combustion wave within the range $0.04<\delta<0.065$ (Fig. 7.10). At $\delta<0.02$ the function $U(\delta)$ is sharply decreasing. That corresponds to the break down of the combustion process. The analytical theory determines the value of δ corresponding to this boundary, $\delta=0.02$. The initial content of the gaseous oxidizer in bubbles noticeably affects the rate of the combustion wave. For example, an increase of the parameter ε from 0.2 to 0.8 leads to an increase of the combustion wave rate by a factor of 3.2 (Fig. 7.11).

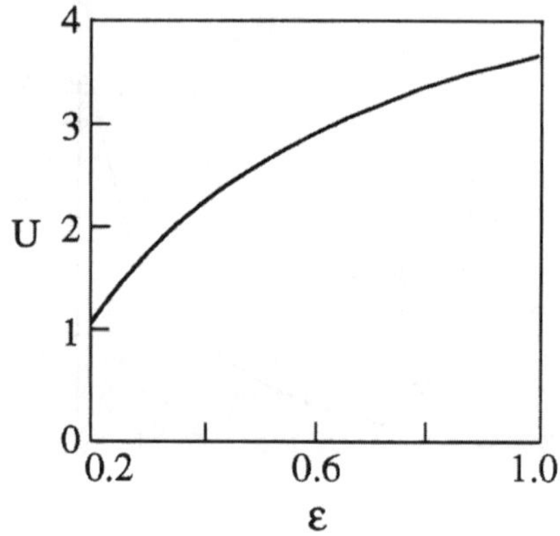

Fig. 7.11 Combustion wave rate versus ε. Reprinted from Likhachev et al.(1989)

2.7.5. Inductional ignition and site ignition

The chemical conversion of a vapor–oxidizer mixture occurs at any values of the parameters Se and δ in the self-acceleration regime when there are no temperature heterogeneities in the liquid reagent. Depending on the intensity of the heat release and heat transfer, the chemical reaction either leads to self-ignition of the vapor–oxidizer mixture inside the bubbles (regime of thermal explosion) or proceeds slowly at temperatures close to the initial one (regime of slow conversion).

To determine the boundary between the domains corresponding to the regimes of self-ignition and slow conversion it is possible to use the same approach as that for the boundary of combustion degradation. For this aimpurpose we consider once again the function G. Assuming that θ_2 is constant (at an early stage of self-heating, when $\theta_1 \ll \theta_{1m}$), we obtain

$$G_{min} = Se_{\ell im} \frac{\exp\left[\dfrac{\theta_1'}{1+\beta\theta_1'}\right]}{\theta_1' - \theta_2} = 1 \tag{7.54}$$

where $\theta_1' = [(1-2\beta)/(2\beta^2)] \left[1-\sqrt{(1-4\beta^2)((1+\theta_2)/(1-2\beta)^2)}\right]$

This relation determines the limiting value of the Semenov number $Se_{\ell im}$ which subdivides the domains corresponding to the high temperature ($Se > Se_{\ell im}$) and low temperature ($Se < Se_{\ell im}$) states. The value of $Se_{\ell im}$ decreases as the liquid temperature increases. In the parametric plane $Se-\delta$ (Fig. 7.7) a family of

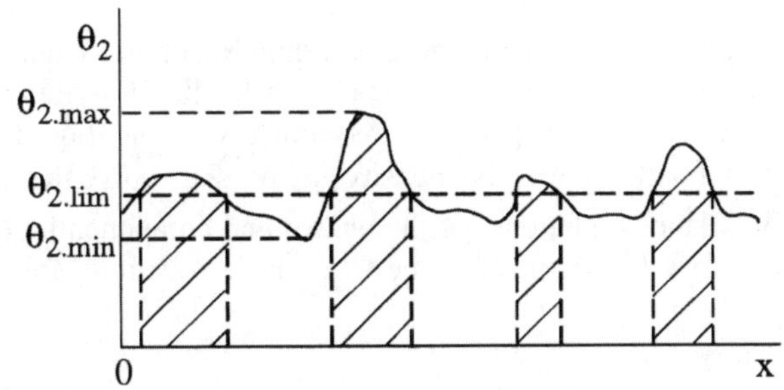

Fig. 7.12 Local ignition of a non uniformly heated bubble suspension. The shaded zones correspond to the self-ignition sites

straight lines parallel to the horizontal axis depicts the dependence of $Se_{\ell im}$ on the liquid temperature θ_2. Figure 7.7 shows that there are two possible ways of development of the combustion process depending on the initial state of bubble suspension: the inductional ignition or the self-propagation process of burning-out of bubbles. Contrary to that, outside the domain between the curves 1 and 2 in Fig. 7.7 either thermal explosion of vapor–oxidizer mixture (the domain where $Se > Se_{\ell im}$) or a slow oxidation at relatively low temperature (the domain where $Se < Se_{\ell im}$) may be realized.

The existence of temperature heterogeneities in bubble suspensions changes the character of the process. In this case domains with different temperatures of the liquid reagent are characterized by their own local values of the limiting Semenov number. The self-ignition of a bubble suspension is possible when $Se_{\ell im}$ corresponding to the local temperature is less than the value of Se corresponding to the bubble suspension as a whole. In other words, self-ignition occurs within the domains (sites) where the liquid temperature exceeds some limiting value of θ_2 corresponding to the condition $Se = Se_{\ell im}$ (Fig. 7.12).

The behavior of a bubble suspension within the "cold" regions corresponding to the spaces between the ignition sites in Fig. 7.12 depends on the values of the parameters δ and Se. Outside the domains where propagation of combustion is impossible there still remain regimes of low temperature oxidation. Within the domains where the combustion process is possible, burning of vapor–oxidizer mixture in combustion waves generated by high temperature zones still occurs. Note that a similar phenomenon takes place during combustion of homogeneous mixtures (Merzhanov et al. 1963).

2.7.6. Effect of bubble expansion

The model of a bubble suspension with non deformable bubbles does not account for the flow of liquid and gaseous phases due to the thermal expansion of bubbles. The aim of the present section is the prediction of the combustion wave speed in a gas–liquid medium containing deformable bubbles capable of expanding. We begin with a phenomenological approach for the description of the combustion process in bubble suspension. Assuming that the liquid and gaseous phases are continua with an effective density ρ_i we shall solve the mass, momentum and energy balance equations for the phases and components of the gas mixture. In the laboratory frame of reference x', t' these equations are (Likhachev et al. 1992)

$$\frac{\partial \rho_1}{\partial t'} + \frac{\partial(\rho_1 u_1')}{\partial x'} = W_e \tag{7.55}$$

$$\frac{\partial \rho_b}{\partial t'} + \frac{\partial(\rho_b u_1')}{\partial x'} = -W + W_e \tag{7.56}$$

$$\frac{\partial \rho_a}{\partial t'} + \frac{\partial(\rho_a u_1')}{\partial x'} = -\sigma W \tag{7.57}$$

$$\rho_2^o \frac{\partial(1-\xi_1)}{\partial t'} + \rho_2^o \frac{\partial}{\partial x'}\left[(1-\xi_1)u_2'\right] = -W_e \tag{7.58}$$

$$\rho_1 \frac{\partial u_1'}{\partial t'} + \rho_1 u_1' \frac{\partial u_1'}{\partial x'} = -K(u_1' - u_2') \tag{7.59}$$

$$\rho_2^o(1-\xi_1)\frac{\partial u_2'}{\partial t'} + \rho_2^o(1-\xi_1)u_2' \frac{\partial u_2'}{\partial x'} = \frac{\partial}{\partial x'}\left[\rho_2^o \nu_2 (1-\xi_1)\frac{\partial u_2'}{\partial x'}\right] + K\left(u_1' - u_2'\right) - \frac{\partial P}{\partial x'} \tag{7.60}$$

$$c_1\rho_1 \frac{\partial T_1}{\partial t'} + c\rho_1 u_1' \frac{\partial T_1}{\partial x'} = qW - hs(T_1 - T_2) + W_e(q_e - c_1 T_1) \tag{7.61}$$

$$c_2\rho_2^o(1-\xi_1)\frac{\partial T_2}{\partial t'} + c_2\rho_2^o(1-\xi_1)u_2' \frac{\partial T_2}{\partial x'} = \frac{\partial}{\partial x'}\left[\lambda_2(1-\xi_1)\frac{\partial T_2}{\partial x'}\right] + hs(T_1 + T_2) - W_e(q_e - c_2 T_2) \tag{7.62}$$

where P is the pressure, K is the phase-interaction coefficient, and u_1' and u_2' are the velocities of the gaseous or liquid phases, respectively.

Note that in the momentum and energy equations the terms accounting for the momentum change (vapor recoil) due to the interfacial mass transfer and dissipation of mechanical energy are omitted because the rate of vaporization and corresponding velocity (Stefan flux) are small. The buoyancy and the forces associated with the added mass are also negligibly small, as well as the work of expansion.

Transition from the laboratory frame of reference to that associated with combustion front corresponds to the transformation

$$t = t', \quad \zeta = x' + \int_0^{t'} u_f dt \tag{7.63}$$

$$u_{1.2} = u'_{1.2} - u_f, \quad \frac{\partial}{\partial x'} = \frac{\partial}{\partial \zeta}; \quad \frac{\partial}{\partial t'} = \frac{\partial}{\partial t} + u_f \frac{\partial}{\partial x'}$$

where u_f is the combustion wave speed (still unknown).

Assuming also a stationary process $(\partial/\partial t = 0)$ we reduce Eqs. (7.55)–(7.62) to the following form:

$$\frac{d(\rho_1 u_1)}{d\zeta} - \frac{d\rho_b u_1}{d\zeta} = W \tag{7.64}$$

$$\rho_{a.0} \frac{d}{d\zeta}\left[(1-\eta)u_1\right] = -\sigma W \tag{7.65}$$

$$\rho_2^o \frac{d}{d\zeta}\left[(1-\xi_1)u_2\right] + \frac{d(\rho_1 u_1)}{d\zeta} = 0 \tag{7.66}$$

$$\rho_1 u_1 \frac{du_1}{d\zeta} = -K(u_1 - u_2) \tag{7.67}$$

$$\rho_2^o(1-\xi_1)u_2 \frac{du_2}{d\zeta} = \frac{d}{d\zeta}\left[\rho_2^o \nu_2 (1-\xi_1)\frac{du_2}{d\zeta}\right] + K(u_1 - u_2) - \frac{dP}{d\zeta} \tag{7.68}$$

$$c_1 \rho_1 u_1 \frac{dT_1}{d\zeta} = qW - hs(T_1 - T_2) + (q_e - c_1 T_1)\frac{d(\rho_1 u_1)}{d\zeta} \tag{7.69}$$

$$c_2 \rho_2^o (1-\xi_1) u_2 \frac{dT_2}{d\zeta} = \frac{d}{d\zeta}\left[\lambda_2 (1-\xi_1)\frac{dT_2}{d\zeta}\right] + hs(T_1 - T_2) - (q_e - c_2 T_2)\frac{d(\rho_1 u_1)}{d\zeta} \tag{7.70}$$

The boundary conditions of the problem are

$$\zeta=-\infty,\ T_1=T_2=T_0,\ u_1=u_2=u_f,\ \xi_1=\xi_{1.0},\ P=P_0, \tag{7.71}$$
$$\rho_a=\rho_{a.0},\ \rho_b=\rho_{b.0},\ \rho_1=\rho_{1.0},\ \eta=0,\ W=0$$
$$\zeta=+\infty,\ T_1=T_2=T_{fin},\ u_1=u_2=u_{fin},\ \xi_1=\xi_{1.fin},P=P_{fin},\ \eta=1.$$

where the parameters of the initial state are determined by the external pressure P_0, the initial temperature T_0 and the gas content $\xi_{1.0}$.

$$\rho_{b.0}=\frac{\xi_{1.0}P_{b.0}}{R_bT_0},\quad \rho_{a.0}=\frac{\xi_{1.0}P_{a.0}}{R_aT_0},\quad \rho_{1.0}=\rho_{a.0}+\rho_{b.0} \tag{7.72}$$
$$P_{a.0}=P_0-P_{b.0},\quad P_{b.0}=z_e\exp\left(-\frac{q_e}{R_bT_0}\right).$$

Here z_e is the pre-exponential factor in the Clapeyron–Clausius equation for $P_{b.0}$.

A. Parameters in the final state. Consider the combustion wave as a succession of three characteristic zones (heating, intensive chemical reaction and thermal relaxation) and integrate the system of governing equations. Multiplying Eq. (7.66) by c_2T_2 and combining the resulting equation with Eqs. (7.65), (7.69) and (7.70), we obtain

$$c_1\frac{d(\rho_1u_1T_1)}{d\zeta}+c_2\rho_2^o\frac{d}{d\zeta}\left[(1-\xi_1)u_2T_2\right]+\frac{\rho_{a.0}q}{\sigma}\frac{d}{d\zeta}[(1-\eta)u_1]= \tag{7.73}$$
$$\lambda_2\frac{d}{d\zeta}\left[(1-\xi_1)\frac{dT_2}{d\zeta}\right]$$

Integration of Eqs. (7.73) and (7.66) from the initial to the final state yields

$$T_{fin}=\frac{u_fq\rho_{a.0}/\sigma+c_1\rho_{1.0}\cdot T_0+c_2\rho_2^o(1-\xi_{1.0})T_0}{c_1\rho_{1fin}+c_2\rho_2^o(1-\xi_{1fin})} \tag{7.74}$$

$$\xi_{1.fin}=1+\frac{\rho_{1.fin}}{\rho_2^o}-\left[(1-\xi_{1.0})+\frac{\rho_{1.0}}{\rho_2^o}\right]\left(\frac{u_f}{u_{fin}}\right) \tag{7.75}$$

Combining Eqs. (7.67) and (7.68) with Eq. (7.66) multiplied by u_2 and taking into account that

$$\rho_1 u_1 \frac{du_1}{d\zeta} + u_2 \frac{d\rho_1 u_1}{d\zeta} = \frac{d\rho_1 u_1^2}{d\zeta} + (u_2 - u_1)\frac{d\rho_1 u_1}{d\zeta} \tag{7.76}$$

we find

$$\frac{d\rho_1 u_1^2}{d\zeta} + (u_2 - u_1)\frac{d\rho_1 u_1}{d\zeta} + \rho_2^o \frac{d}{d\zeta}\left[(1-\xi_1)u_2^2\right] = \frac{d}{d\zeta}\left[\rho_2^o \nu_2 (1-\xi_1)\frac{du_2}{d\zeta}\right] - \frac{dP}{d\zeta} \tag{7.77}$$

where ν_2 is the kinematic viscosity of liquid phase.

Integrating Eq. (7.77) from $-\infty$ to $+\infty$, we obtain

$$\rho_{1fin} u_{fin}^2 - \rho_{1.0} u_f^2 + \rho_2^o (1-\xi_{1fin}) u_{fin}^2 - \rho_2^o (1-\xi_{1.0}) u_f^2 + {} + P_{fin} - P_0 = \int_{-\infty}^{+\infty} (u_1 - u_2)\frac{d\rho_1 u_1}{d\zeta} d\zeta \tag{7.78}$$

To evaluate the integral on the right-hand side in Eq. (7.78), we estimate the values of the characteristic parameters in the heating and thermal relaxation zones. Since $\eta = 0$ and W=0 within the heating zone, it follows from Eq. (7.65) that $\rho_{a.0} u_1 = \rho_{a.0} u_f$, i.e . $u_1 = u_f$ = constant. Accordingly, it follows from Eq. (7.67) that $u_2 = u_1$=const.

Integration of Eq. (7.64) over the thermal relaxation zone (W=0) yields

$$u_1(\rho_1 - \rho_b) = u_{fin}(\rho_{1fin} - \rho_{bfin}) = \text{const} \tag{7.79}$$

Since $\rho_a = 0$ and $\rho_1 = \rho_b + \rho_c$, within this zone Eq. (7.79) takes the form

$$\rho_c u_1 = \rho_{c.fin} u_{fin} \tag{7.80}$$

In the present case $\rho_c = \rho_{c.fin}$. Hence $u_1 = u_{fin}$=const and it follows accordingly from Eq. (7.67), that $u_2 = u_{fin}$=const. Thus, the difference $u_1 - u_2 = 0$ in the interval from $-\infty$ to $-\zeta_*$ and from 0 to $+\infty$. Therefore, the integral on the right-hand side of Eq. (7.78) equals

$$\int_{-\infty}^{+\infty} (u_1 - u_2)\frac{d\rho_1 u_1}{d\zeta} d\zeta = \int_{-\zeta_*}^{0} (u_1 - u_2)\frac{d\rho_1 u_1}{d\zeta} d\zeta \tag{7.81}$$

For activated reactions, when the thickness of the reaction zone is sufficiently small, this integral can be taken as zero. Then Eq. (7.78) takes the following form:

$$\rho_{1.fin} u_{fin}^2 - \rho_{1.0} u_f^2 + \rho_2^o (1 - \xi_{1fin}) u_{fin}^2 - \rho_2^o (1 - \xi_{1.0}) u_f^2 + P_{fin} - P_0 = 0 \tag{7.82}$$

Solution of Eq. (7.82) for u_{fin} yields

$$\left(\frac{u_{fin}}{u_f} \right)^2 = \frac{\rho_{1.0} + \rho_2^o (1 - \xi_{1.0}) - (P_{fin} - P_{1.0}) / u_f^2}{\rho_{1.fin} - \rho_2^o (1 - \xi_{1.fin})} \tag{7.83}$$

Using the equation of state of the gas

$$\xi_1 P = \rho_1 R_g T_1 \tag{7.84}$$

we find (in the approximation $R_{g.0}$=$R_{g.fin}$) the final value of the volumetric content of the gaseous phase:

$$\frac{\xi_{1.fin}}{\xi_{1.0}} = \frac{\rho_{1.fin} T_{fin} P_0}{\rho_{1.0} T_0 P_{fin}} \tag{7.85}$$

Equations (7.74), (7.75), (7.83) and (7.85) define the final values of the parameters T, ξ_1, u, P as a function of the initial value of the parameters, as well as the rate of the combustion wave (unknown a priori). To find u_f, the thermal balance equation at the point of ignition, (Eq. 7.26) should be used.

B. Thermal structure of the characteristic zones. Reaction zone. To determine the maximal gaseous phase temperature we use Eq. (7.69). Bearing in mind that the combustion process proceeds within the reaction zone under almost adiabatic conditions, it is possible to omit the term corresponding to heat losses in Eq. (7.69). The integration of Eq. (7.69) over the thickness of the reaction zone yields (taking into account Eq. (7.65)) the following relation for the maximal gaseous phase temperature $T_{1.m}$:

$$T_{1.m} = \frac{u_f \rho_{1.0}}{u_{fin} \rho_{1.fin}} \left[T_0 + \frac{\rho_{a.0} q}{c_1 \sigma \rho_{1.0}} - \frac{q_e}{c_1} \left(\frac{\rho_{1.fin} u_{fin}}{\rho_{1.0} u_f} - 1 \right) \right] \tag{7.86}$$

The sum of the first two terms in the square brackets in Eq. (7.86) corresponds to the maximal gaseous phase temperature in a bubble suspension with non deformable bubbles, the third term and the factor before the square brackets take into account the bubble deformation resulting from the of thermal expansion of the gas.

Heating zone. Integration of Eq. (7.66) over the heating zone (from $-\infty$ to ζ) yields

$$\rho_2^o(1-\xi_1)u_2 + \rho_1 u_1 = \rho_2^o(1-\xi_{1.0})u_f + \rho_{1.0}u_f \tag{7.87}$$

The velocities of the liquid and gaseous phases are equal to each other within the heating zone, i.e. $u_1=u_2=u_f$. Then Eq. (7.87) is transformed to

$$\xi_1 = \xi_{1.0}\frac{1-\rho_{1.s}^o/\rho_2^o}{1-\rho_1^o/\rho_2^o} \tag{7.88}$$

At moderate pressures when $\rho_{1.0}^o << \rho_2^o$ and $\rho_1^o << \rho_2^o$, it is possible to assume that $\xi_1 = \xi_{1.0}$ and s=const. Then the integration of Eq. (7.73) from $-\infty$ to ζ yields the following expression:

$$u_f[c_1\rho_1 T_1 + c_1\rho_2^o(1-\xi_{1.0})T_2] = \lambda_2(1-\xi_{1.0})\frac{dT_2}{d\zeta} + \tag{7.89}$$
$$+[c_1\rho_{1.0}T_0 + c_2\rho_2^o(1-\xi_{1.0})]u_f$$

Since $T_1 \approx T_2$ and $c_1\rho_1 \ll c_2\rho_2^o$, Eq. (7.89) can be reduced to the from

$$\ell_*\frac{dT_2}{dx} = T_2 - T_0 \tag{7.90}$$

where $\ell_* = \lambda_2/(c_2\rho_2^o u_f)$

From Eq. (7.90) we obtain (taking into account the condition $\zeta = -\zeta_*, T_2 = T_{2*}$) the following relation for the liquid phase temperature:

$$T_2 = T_0 + (T_{2*} - T_0)\exp\left[\frac{\zeta+\zeta_*}{\ell_*}\right] \tag{7.91}$$

In the case of a thin reaction zone, Eq. (7.91) takes the form

$$T_2 = T_s + (T_{2*} - T_0)\exp\left(\frac{\zeta}{\ell_*}\right) \tag{7.92}$$

Now we integrate Eq. (7.69) taking into account the expression (7.92). As a result, we obtain the following expression for the temperature distribution of the gaseous phase within the heating zone:

$$T_1 = T_0 + \frac{hs(T_{2*} - T_0)}{c_1\rho_{1.0}u_f}\left(\frac{1}{\ell_*} + \frac{hs}{c_1\rho_0 u_f}\right)^{-1} \exp\left(\frac{\zeta}{\ell_*}\right) \tag{7.93}$$

In the case of sufficiently small thickness of the reaction zone $(\zeta_*/\ell_* \ll 1)$, the following expression for the gaseous phase temperature at the ignition point is valid:

$$T_{1*} = T_0 \frac{hs(T_{2*} - T_0)}{c_1\rho_{1.0}u_f}\left(\frac{1}{\ell_*} + \frac{hs}{c_1\rho_{1.0}u_f}\right)^{-1} \tag{7.94}$$

Thermal relaxation zone. The integral of Eq. (7.66) is

$$\rho_2^o(1-\xi_{1.fin})u_{fin} - \rho_2^o(1-\xi_1)u_2 + \rho_{1.fin}u_{fin} - \rho_1 u_1 = 0 \tag{7.95}$$

Since within the thermal relaxation zone $u_1{=}u_2{=}u_{fin}$, the following relation for the volumetric content of the gaseous phase is valid:

$$\xi_1 = \xi_{fin}\frac{1-\rho_{1fin}^o/\rho_2^o}{1-\rho_1^o/\rho_2^o} \approx \xi_{1.fin} \tag{7.96}$$

In this case (cf. Eq. 7.66) $d\rho_1 u_1/d\zeta = 0$ and $\rho_1 u_1 = \rho_{1fin}u_{1fin}$. Then Eq. (7.69) takes the form

$$c_1\rho_{1fin}\frac{dT_1}{d\zeta} = -hs_{fin}(T_1 - T_2) \tag{7.97}$$

where $s_{fin} = (36\pi N)^{1/3}\xi_{1.fin}^{2/3}$.

The liquid temperature varies only slightly within the thermal relaxation zone. That allows for the assumption that $T_2{\sim}T_{fin}$ in the expression Eq. (7.92). Then the solution of Eq. (7.97) subject to the condition $\zeta{=}0$, $T_1{=}T_{1.m}$ becomes

$$T_1 = T_{fin} + (T_{1.m} - T_{fin})\exp\left(-\frac{hs_{fin}}{c_1\rho_{1.fin}u_{fin}}\zeta\right) \tag{7.98}$$

Using this relation we recast Eq. (7.78) into the following form

$$\frac{d^2T_2}{d\zeta^2}-\frac{c_2\rho_2^o u_{fin}}{\lambda_2}\frac{dT}{d\zeta}+\frac{hs_{fin}}{\lambda_2(1-\xi_{i.fin})}(T_{1.m}-T_{fin})\exp\left(-\frac{hs_{fin}}{c_1\rho_{1.fin}u_{1.fin}}\zeta\right)=0 \tag{7.99}$$

Integration of equation (7.99) yields the following expressions for the current temperature of the liquid reagent, as well as the temperature at the ignition point:

$$T_2=T_{fin}-\frac{(T_{1.m}-T_{fin})c_1\rho_{1.fin}u_{fin}}{(1-\xi_{1.fin})[c_2\rho_2^o u_{fin}-\lambda_2 r_*]}\exp(-r_*\zeta) \tag{7.100}$$

$$T_{2*}=T_{fin}-\frac{(T_{1.m}-T_f in)c_1\rho_1 u_{fin}}{(1-\xi_{1.fin})[c_2\rho_2^o u_{fin}+\lambda_2 r_*]} \tag{7.101}$$

where $r_*=(hs_{fin})/(c_1\rho_{1fin}u_{fin})$.

Solving Eqs. (7.94) and (7.101) for T_{1*} and T_{2fin}, and taking into account the relation (7.86), we find

$$T_{1*}=\frac{(1-a)T_0+a(1+b)T_{fin}-aQ}{1-\bar{u}ba} \tag{7.102}$$

$$T_{2*}=\frac{(1+b)T_0-b\bar{u}(1-a)T_0-Q}{1+\bar{u}ba} \tag{7.103}$$

where

$$a=\frac{hs_0}{c_1\rho_{1.0}u_f}\left(\frac{1}{\ell_*}+\frac{hs_0}{c_1\rho_{1.0}u_f}\right),\quad b=\frac{c_1\rho_{1.fin}u_{fin}}{(1-\xi_{1.fin})[\bar{c}_2\rho_g^o u_{fin}+\lambda_2 r_*]},\quad \bar{u}=\frac{\rho_{1.0}u_f}{\rho_{1.fin}u_{fin}},$$

$$Q=b\bar{u}\left[\frac{\rho_{a0}q}{c_1\sigma\rho_{1.0}}+\frac{q_e}{c_1}\left(\frac{1}{\bar{u}}-1\right)\right]$$

The substitution of the expressions (7.102) and (7.103) in Eq. (7.26) yields an equation which determines the speed of the combustion wave.

From Eqs. (7.64) and (7.65) we have

$$\frac{d}{d\zeta}[\rho_1 u_1-\rho_b u_1+\rho_{a.0}\frac{(1-\eta)u_1}{\sigma}]=0 \tag{7.104}$$

Integrating this equation from the initial to the final state, we obtain

$$\frac{u_{fin}}{u_f} = \frac{\rho_0 - \rho_{b.0} + \rho_{a0}/\sigma}{\rho_{1.fin} - \rho_{b.fin}} \tag{7.105}$$

Taking into account expressions (7.72) we recast Eq. (7.105) to the form

$$\frac{u_{fin}}{u_f} = \frac{1 + \frac{\rho_{a.0}}{\sigma\rho_{1.0}} - \frac{\xi_{1.0}}{\rho_{1.0}}\frac{z_e}{R_b T_0}\exp\left(-\frac{q_e}{R_b T_0}\right)}{\frac{\rho_{1fin}}{\rho_{1.0}} - \frac{\xi_{1fin} z_e}{\rho_{1fin} R_b T_{fin}}\exp\left(-\frac{q_e}{R_b T_{fin}}\right)} \tag{7.106}$$

The system of Eqs. (7.74), (7.75), (7.83), (7.85), (7.26) and (7.105) contains six parameters determining the final temperature T_{fin}, the velocity u_{fin} and the pressure P_{fin} of the liquid and gaseous phases, density of the gaseous phase $\rho_{1.fin}$, as well as speed of the combustion wave u_f.

Rendering all the parameters involved dimensionless by the following scales

$$\beta = \frac{RT_0}{E}, \quad \varepsilon = \frac{\rho_{a.0}}{\rho_{1.0}}, \quad \delta = \frac{c_1\rho_{1.0}}{c_2\rho_2^o(1-\xi_{1.0})}$$

$$\varsigma_0 = \frac{P_0 c_1 c_2 \rho_2^o(1-\xi_{1.0})}{hs_0\lambda_2}, \quad U = c_2\rho_2^o u_f\sqrt{\frac{1-\xi_{1.0}}{hs_0\lambda_2}},$$

$$a = (1+\delta^2)^{-1}, \chi = \frac{\xi_{1.fin} z_e}{\rho_{1.fin} R_b T_{fin}}\exp\left(-\frac{q_e}{R_b T_0}\right),$$

$$b = \delta\frac{\rho_{1.fin}(1-\xi_{1.fin})}{\rho_{1.0}(1-\xi_{1.0})}\left(1 + \frac{\rho_{1.0} s_{fin} u_f^2)}{\delta U^2 \rho_{1.fin} s_0 u_{fin}^2}\right),$$

we obtain Eqs. (7.74), (7.75), (7.83), (7.85), (7.26) and (7.105) in the dimensionless form:

$$\bar{T}_{fin} = \bar{u}_f \frac{\beta\varepsilon/\sigma + (1+\delta)/\delta}{\bar{\rho}_{fin} + (1-\xi_{1.fin})/\delta(1-\xi_{1.0})} \tag{7.107}$$

$$\xi_{1.fin} = 1 + \bar{\rho}_{1.fin}\frac{\delta(1-\xi_{1.0})}{c_{1.2}} - \bar{u}_f(1 + \frac{\delta}{c_{1.2}})(1-\xi_{1.0}) \tag{7.108}$$

$$(\bar{u}_f)^{-2} = \frac{1 + c_{1.2}\delta - \varsigma_{fin}/(\delta U^2)}{\rho_{1.fin}\rho_{1.0} + c_{1.2}(1-\xi_{1.fin})/[\delta(1-\xi_{1.0})]} \tag{7.109}$$

$$\overline{\rho}_{1.\mathrm{fin}} = \frac{\overline{\xi}_{1.\mathrm{fin}}}{\overline{T}_{\mathrm{fin}}}\left(\frac{\varsigma_{\mathrm{fin}}}{\varsigma_0}+1\right) \tag{7.110}$$

$$(\mathrm{Se})\exp\left(\frac{A\overline{q}}{1+ab\overline{u}}\right) = \delta U^2\left(\frac{A\overline{q}}{1+ab\overline{u}}\right) \tag{7.111}$$

$$(\overline{u}_f)^{-1} = \frac{1+\dfrac{\varepsilon}{\sigma}-\chi}{\overline{\rho}_{1.\mathrm{fin}} - \chi\exp[-\psi(\overline{T}_{\mathrm{fin}}^{-1}-1)]\overline{\xi}_{1.\mathrm{fin}}/\overline{T}_{\mathrm{fin}}} \tag{7.112}$$

where $\overline{u}_f = u_f/u_{\mathrm{fin}}$, $\overline{T}_{\mathrm{fin}} = T_{\mathrm{fin}}/T_0$, $\overline{\xi}_{1.\mathrm{fin}} = \xi_{1.\mathrm{fin}}/\xi_{1.0}$, $\overline{\rho}_{1.\mathrm{fin}} = \rho_{1.\mathrm{fin}}/\rho_{1.0}$.

Characteristics of the process. The dependences of the parameters corresponding to the final state $\xi_{1.\mathrm{fin}}$, $\overline{T}_{\mathrm{fin}}$ and $\overline{\rho}_{\mathrm{fin}}$ on the initial gas content of the gaseous phase $\xi_{1.0}$, as well as on the initial temperature (via the parameter β) are depicted in Fig. 7.13. It is seen that an increase in $\xi_{1.0}$ is accompanied by a monotonic, practically linear increase in the final temperature of the bubble suspension. This is associated, first of all, with the increase in the total heat release because of the increase in the oxidant content in the bubble medium and, correspondingly, the increase in the reagent mass. It is also associated with the decrease in the specific heat of the system. An increase in the initial gas content also leads to an increase in the final value of the volumetric content of the gaseous phase $\xi_{1.\mathrm{fin}}$. The rate of change of $\xi_{1.\mathrm{fin}}$ is considerably larger than the rate of change of $\overline{T}_{\mathrm{fin}}$. This is explained by the dependence of $\xi_{1.\mathrm{fin}}$ not only on the thermal expansion of the gas due to the temperature rise but also on the intensity of evaporation of the liquid reagent, which increases sharply with increase in $\overline{T}_{\mathrm{fin}}$. Considering the influence of the initial temperature T_0 on $\overline{T}_{\mathrm{fin}}$, $\xi_{1.\mathrm{fin}}$ and P_{fin}, it is evident that an increase in β leads to an increase in $\xi_{1.\mathrm{fin}}$ and $\overline{T}_{\mathrm{fin}}$ and a decrease in P_{fin} (cf. Fig. 7.13).

The dependence of the maximal temperature on the initial temperature of the bubble suspension and the gas content in it is depicted in Fig. 7.14 for two different models of bubble media: with non deformable and deformable (expanding) bubbles. The comparison of the curves $\theta_{\max}(\beta)$ and $\theta_{\max}(\xi_{1.0})$ shows that they are significantly different for different models. This is because the heat released during combustion in the medium with non deformable bubbles is consumed solely in changing the internal energy of the gaseous phase. In the model with deformable (expanding) bubbles, the relative reduction in the temperature $\theta_{\max}$ in this case is associated with the additional heat loss from the reaction zone at rate $u_{\mathrm{fin}}>u_f$.

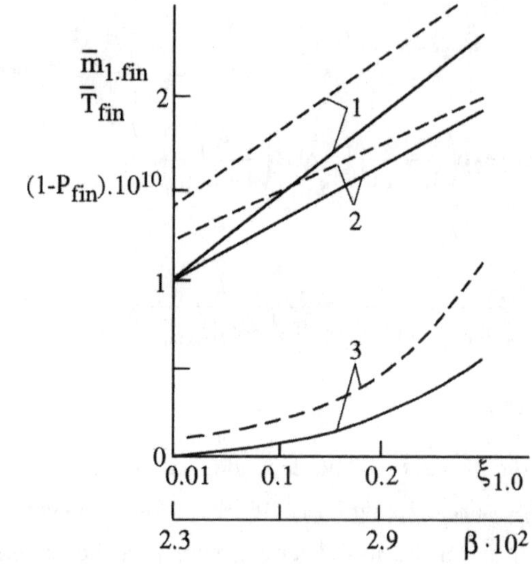

Fig. 7.13 Dependence of the final parameters of a bubble suspension on the initial volumetric content of the gaseous phase (solid lines) and the initial temperature (dashed lines). Reprinted from Likhachev et al.(1992). Curve 1, $\overline{\xi}_{1.fin} = \xi_{1.fin}/\xi_{1.0}$; curve 2, $\overline{T}_{fin} = T_{fin}/T_0$; curve 3, $(1-\overline{P}_{fin})\cdot 10^{10}$, $\overline{P}_{fin} = P_{fin}/P_0$

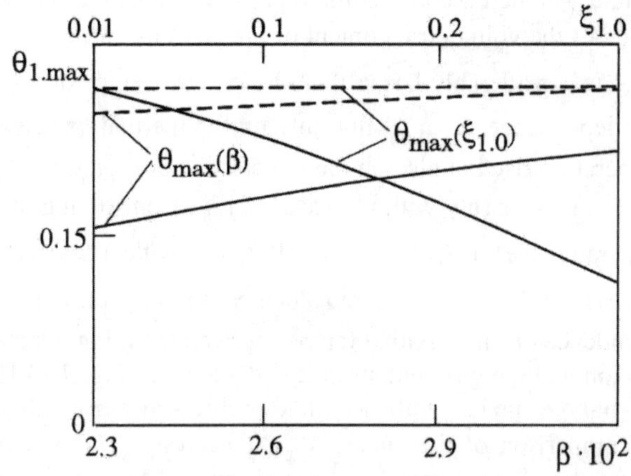

Fig. 7.14 Dependence of the maximal temperature on the initial temperature and volumetric gaseous phase content: dashed lines correspond to the bubble suspension with non deformable bubbles, and solid lines correspond to the medium with deformable (expanding) bubbles. Reprinted from Likhachev et al.(1992)

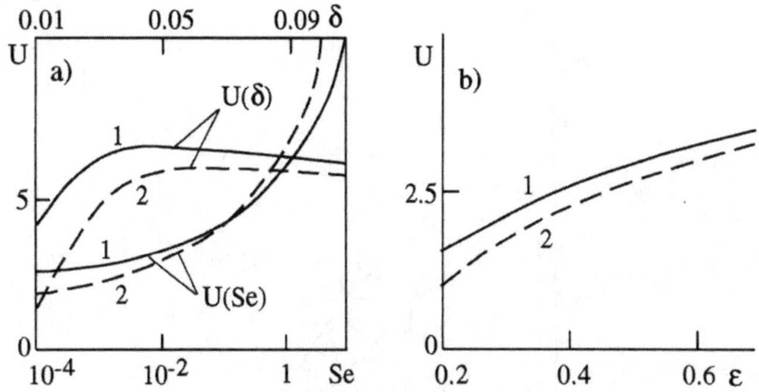

Fig. 7.15 The rate of combustion waves. Reprinted from Likhachev et al.(1992). a) The dependencies U(Se), U(δ). b) The dependence U(ε) The solid lines correspond to the model with deformable (expanding) bubbles. The dashed lines correspond to the model with non deformable bubbles

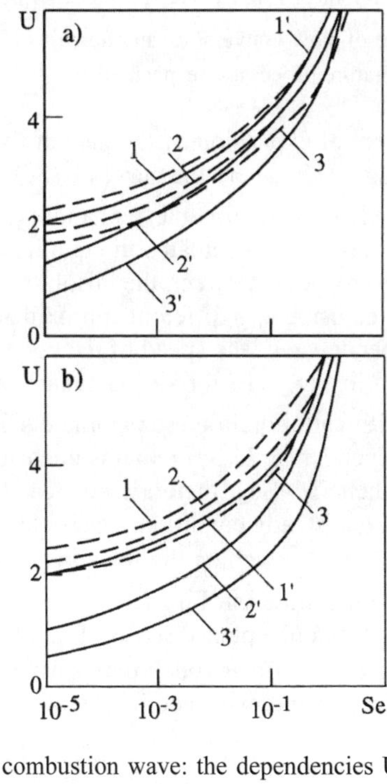

Fig. 7.16 The rate of the combustion wave: the dependencies U(Se). Reprinted from Likhachev et al.(1992), with permission. a) $\varepsilon = 0.7$ curve (1′), 0.5 (2′), 0.3 (3′); $\gamma = 5.3 \cdot 10^{-1}$ curve (1), $2 \cdot 10^{-1}$ (2), $1 \cdot 10^{-1}$ (3). b) $\chi = 1 \cdot 10^{-3}$ curve (1′), $1 \cdot 10^{-2}$ (2′), $2 \cdot 10^{-2}$ (3′); $\delta = 0.03$ curve (1), 0.05 (2), 0.08 (3)

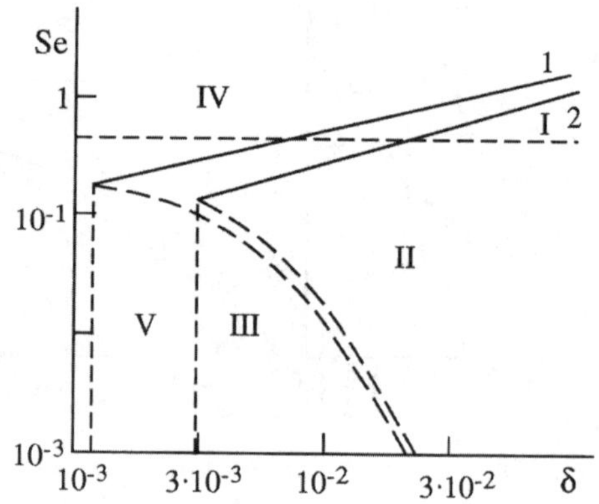

Fig. 7.17 Domains corresponding to various reaction regimes. Reprinted from Likhachev et al. (1992), with permission. Curve 1, $Se_{cr}(\delta)$ for deformable (expanding) bubbles; curve 2, $Se_{cr}(\delta)$ for non deformable bubbles. I and II are the domains of existence of steady combustion wave, III the domain of slow conversion, and domains IV and V correspond to unsteady high- and low- temperature processes, respectively

The influence of thermal expansion of the gas on the speed of the combustion wave is evident in Fig. 7.15, where the curves U(Se), $U(\delta)$ and $U(\varepsilon)$ for the media with non deformable and deformable (expanding) bubbles are shown. In both cases, the shape of these curves remains the qualitatively similar. However, there is a considerable discrepancy between the absolute values of the rate of the combustion wave calculated using two different approximations.

The data on the dependence of the speed of the combustion wave on the Semenov number are shown in Fig. 7.16 for various values of ε, χ, γ and δ. The calculations show that, with variation in these parameters over a sufficiently broad range, the shape of the dependence U(Se) remains unchanged: U increases with an increase in Se, and when $Se > Se_{cr}$ there are no steady solutions for U. This indicates that, when $Se > Se_{cr}$, steady conditions cannot be realized.

The dependence $Se_{cr}(\delta)$, as well as the domains corresponding to different combustion conditions are illustrated in Fig. 7.17, where the characteristic border lines are plotted on the parametric plane $Se - \delta$. Figure 7.17 shows that taking into account the hydrodynamics effects (bubble expansion) leads to an expansion of the region corresponding to the existence of steady conditions of combustion wave propagation.

The comparison presented in Figs. 7.14, 7.15 and 7.17 indicates a pronounced quantitative discrepancy between the results of the models of bubble suspensions with non deformable and deformable (expanding) bubbles. Nevertheless, the basic trends remain qualitatively similar, which means that the simpler model of the bubble media with non deformable bubbles is adequate for the qualitative investigation of the complex phenomenon.

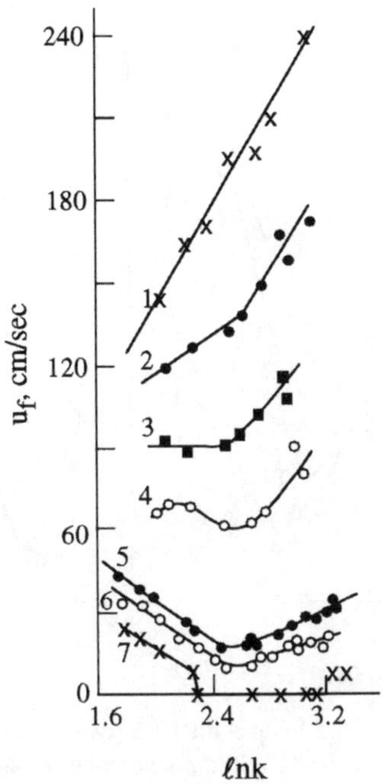

Fig. 7.18 Dependence $u_f(k)$. Reprinted from Zamashchikov and Kakutkina (1993). Curves (1–4): hydrogen–air mixture (hydrogen concentration 15% for curve 1, 13% for curve 2, 11.6% for curve 3, 10% for curve 4) ; curves (5-7): methane-air mixture (methane concentration 9.5% for curve 5, 8% for curve 6, 7% for curve7)

2.7.7. Combustion waves in media with a high volumetric content of gaseous phase

In the previous sections of the present chapter considerable attention was paid to the combustion wave propagating in bubbly media with low volumetric content of gaseous oxidizer in liquid reagent. The large spacing between the neighboring bubbles mostly determines the character of the combustion process, as well as the succession of its stages. At the same time, in principle, combustion wave propagation in bubbly media of some other types is possible, in particular, in media consisting of the inert liquid and bubbles filled with a combustible gaseous mixture. Usage of solutions with surface-active substances allows one to enhance stability of such media at high volumetric content of gaseous phase.

The works dealing with systematic experimental and theoretical investigation of combustion wave propagation in gas-liquid media with high volumetric contents of gaseous phase are scarce. A number of important results concerning the laws of combustion wave propagation in water-based foams was obtained by Zamashchikov et al. (1988, 1993, 1994). Their measurements, carried out for a wide variation of composition of gaseous mixture, average bubble diameters, as

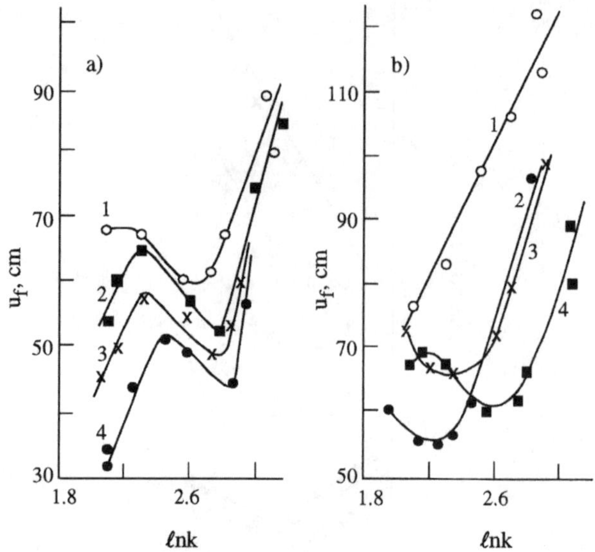

Fig. 7.19 Dependence $u_f(k)$ for the foam with 10% hydrogen–air mixture. Reprinted from Zamashchikov and Kakutkina.(1993). a) Sulfanole content in the foaming solutions (3% for curve1, 1% for curve2, 0.5% for curve 3, 0.2% for curve 4). b) Glycerin content in foaming solution (30% for curve 1, 15% fir curve 2, 7.5% for curve 3, 0 % for curve 4)

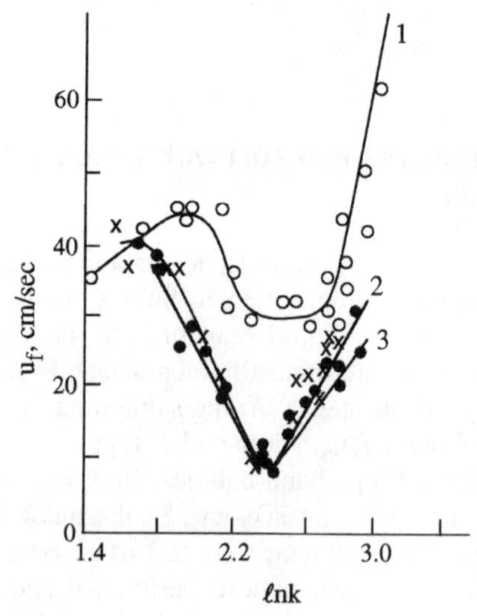

Fig. 7.20 Dependence $u_f(k)$ for the foams with 9.5% methane–air mixture. Reprinted from Zamashchikov and Kakutkina (1993). (The average bubble diameter d_{av} [mm] 4 is for curve 1, 2.5 for curve 2, and 2 for curve 3)

well as the foam strength, revealed some general features of the phenomenon. It was shown that the dependence of the combustion wave rate on the degree of aeration k $(k = \xi_2^{-1})$ is non monotonic (Fig. 7.18). The shape of this curve remains qualitatively similar on changing the composition of the gaseous mixture, the average diameter of bubbles and the concentration of various additives in aqueous solution (Figs. 7.19 and 7.20). The non monotonic character of the dependence $u_f(k)$ is due to the change in the foam structure on its heating and combustion. These processes are accompanied by destruction of the liquid matrix, disintegration of liquid films in the foam, and formation of a gas–droplet mixture in front of the reaction zone. The question of the development of a model suitable for the description of combustion of gas–liquid media with a high volumetric content of gaseous phase remains open.

An attempt to account for the characteristics of combustion of water-based foams was undertaken by Zamashchikov et al. (1994) and Babkin et al. (1994) in the framework of the one-dimensional continuous model. In accordance with these works, the combustion of foams is modeled as a combustion of the gaseous mixture in which a part of the heat released by chemical reaction is spent in the destruction of the liquid membranes and the heating and evaporation of droplets. Such an approach leads to a fairly good agreement between the theoretical predictions and experimental data, if some empirical information on changes of the combustion surface and heat spent in destruction of the liquid matrix is employed.

References

Babkin VS, Kakutkina NA, Zamashchikov VV (1994) Characteristics of water-base foam combustion. The Twenty-Fifth Symposium (international) on Combustion. The Combustion Institute, Pittsburgh, Pa., pp. 1627–1627

Fridman NB, Kitain MM, Shteinberg AS, Merzanov AG (1981) On the mechanism of bubble ignition. Sov. Phys. Dokl. 258: 961–965

Kraynik AM (1988) Foam flows. In: Lumey Jl, Van Dyke M, Reed HL (eds) Annu. Rev. Fluid Mech. 20: 325–357

Likhachev VN, Sukhov GS, Yarin LP (1986) On some regularities of combustion wave propagation in bubble media. Chemical physic of combustion and explosion, combustion processes, In: combustions condensed systems. OIKhF. of AN SSSR, Chernogolovka, pp. 120–122

Likhachkev VN, Sukhov GS, Yarin LP (1992) Combustion of bubble media. Combust. Explos. Shock Waves 28: 120–129

Merzhanov AG, Barzykin VV, Gontkovski VV (1963) The problem of seated thermal explosion. Sov. Phys. Dokl. 148: 380–383

Nigmatulin RI (1991) Dynamics of multiphase media. Hemisphere, London. vols. 1 and 2.

Reinelt DA, Kraynik AM (1996) Simple shearing flow of a dry Kelvin soap foam. J. Fluid Mech 311: 327–343

Sukhov GS, Yarin LP (1980) Combustion wave propagation in bubbly media. Chemical Physics of Problems of Combustion and Explosion. Combustion of Condensed and Heterogeneous Systems. AN SSSR, Chernogolovka, pp. 108–111

Sukhov GS, Yarin LP (1981) Combustion waves in bubbly media. Sov. Phys. Dokl. 256: 376–380

Sukhov GS, Yarin LP (1981) Laws of combustion of bubbled media. Combust. Explos. Shock Waves 17: 251–257

Yarin LP, Sukhov GS (1987) Fundamentals of the theory of two-phase media combustion. (in Russian) Energoatomizdat, Leningrad

Zamashchikov VV, Kakutkina NA (1993) Experimental studies of the combustion mechanism of water-base foams filled with fuel gases. Combust. Explos. Shock Waves 29: 142–147

Zamashchikov VV, Kakutkina NA (1994) Effect of interphase heat transfer on the combustion velocity of foams. Combust. Explos. Shock Waves 30: 772–780

Zamashchikov VV, Babkin VS, Tikhomolov EM, Golulbushkin LM, Sofilkanich OK, Kann KB, Shreiber IR (1988) Experimental studies hydrocombustible foam. Combust. Explos. Shock Waves 24: 413–415

Zel'dovich YaB, Barenblatt GI, Librovich VB, Makhviladze GM (1985) Mathematical theory of combustion and explosion. Plenum, New York

2.8. Filtration combustion

2.8.1. Definition, method and a process analysis

General comments. Filtration combustion is defined as the process in which a self-sustaining exothermal reactive wave propagates over a porous reagent by means of gaseous (liquid) oxidizer filtration through a solid matrix towards the reaction zone. This process is typical of many natural phenomena and industrial technologies such as smoldering combustion of powdered media, combustion of coal seams and oil-bearing strata, high temperature synthesis of refractory materials, etc.

Characteristics of filtration combustion determine a number of factors reflecting the kinetics of chemical reaction of solid–gaseous reagents, the thermal and filtration properties of the system, the existence of heat losses, as well as the scheme of oxidizer transport to the reaction zone. Depending on the flame front and oxidizer flow velocities direction, the countercurrent, concurrent and two-side filtration regimes occur (Fig. 8.1).

There are two causes determining oxidizer filtration through a porous matrix: (i) its consumption by chemical reaction, (ii) the existence of a given pressure drop between the opposite sides of the porous layer. Their contribution to the filtration flux depends on the physical properties of the reagents and combustion products, the kinetics of chemical reaction, the thermal regime of combustion and the pressure distribution over the external surface of the porous layer. When oxidizer consumption due to chemical reaction is dominant, the regime of natural filtration takes place. The forced filtration corresponds to the conditions in which the effect of external pressure is dominant. In the general case, when the contribution of both factors is significant, the mixed filtration regime occurs.

At high activation energy and exothermal chemical conversion, filtration combustion develops as a self-propagation combustion wave. It includes two characteristic zones: (i) a narrow chemical reaction zone, (ii) a relatively broad heating zone (Fig. 8.2). The rate of this wave, as well as its structure, depend on a number of interconnected processes determining the intensity of heat transfer, chemical reaction, filtration and diffusion of the components of the gaseous phase. Although some aspects of filtration combustion were discussed more than 50 years ago (Khitrin 1957, Kantorovich 1958) a consistent theory of this process has been developed comparatively recently by Aldushin et al. (1974) who applied the asymptotic methods of combustion theory to this phenomenon. The authors derived a relation for the rate of the combustion wave which accounts for the fundamental factors determining filtration combustion: kinetics, thermo-physical and filtration properties of reactants, as well as the regime parameters of the process. They showed that depending on the permeability of the solid matrix in countercurrent filtration two different regimes of filtration combustion can be realized. At low hydrodynamic resistance of the porous layer and high intensity of filtration flux, the regime of total conversion of solid the reagent takes place.

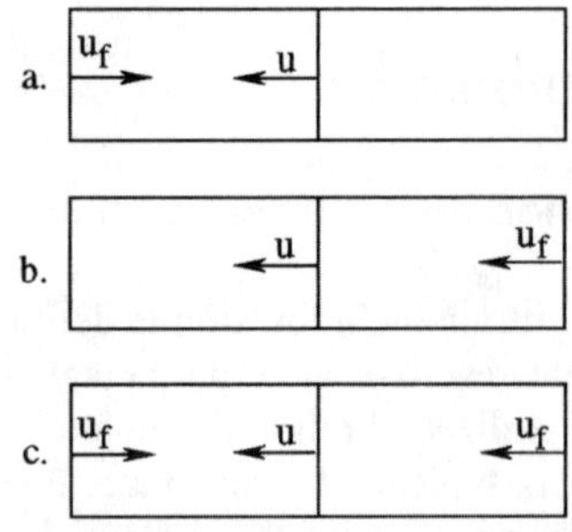

Fig. 8.1 Schemes of filtration. a),b) and c) correspond to countercurrent, concurrent and two-side filtration; u and u_f are the flame and the filtration flux velocities; bold line corresponds to flame front

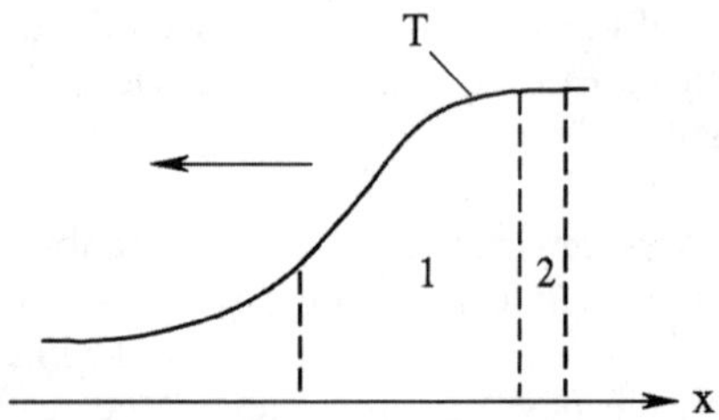

Fig. 8.2 Structure of filtration combustion wave: temperature distribution. Reprinted from Yarin and Sukhov (1987) 1, Heating zone; 2, Reaction (chemical conversion) zone, the arrow shows the direction of combustion wave propagation

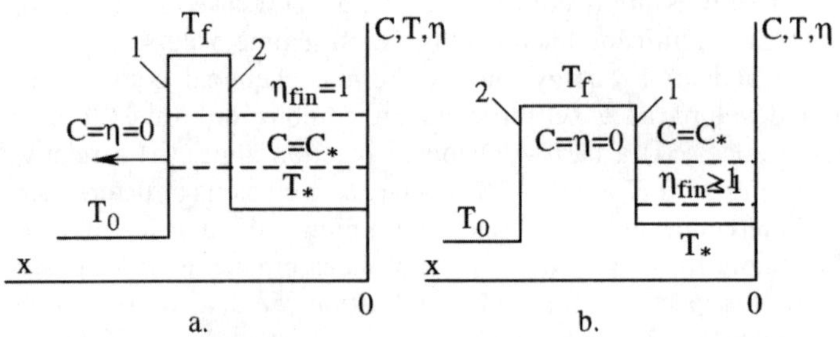

Fig. 8.3 Structure of the combustion wave in porous layer. Reprinted from Yarin and Sukhov (1987). a) Normal structure $C_* > (C_g\sigma_g)/(C_p\sigma_p)$. b) Inversion structure $C_* < (C_g\sigma_g)/(C_p\sigma_p)$. C_g and C_p are the gaseous oxidizer and combustion products concentrations, σ_g and σ_p are the stoichiometric coefficients defined by the gaseous oxidizer and combustion products, respectively, T_* and T_O are the temperature of the gas supplied through the hot butt-end of the porous layer and the initial temperature of porous medium ; 1 and 2 correspond to the combustion and thermal relaxation fronts, respectively

In contrast, in a dense matrix the combustion process develops under the conditions of partial conversion of the porous reagent.

The generalized pattern of filtration combustion corresponding to the countercurrent, concurrent and two-side filtration was explored by Aldushin et al. (1980). The numerical calculations performed for different schemes of filtration revealed some important characteristics of the process. In particular, it was shown that under the conditions of two-side filtration, the combustion wave disintegrates into two fronts which move in the same direction with different velocities.

When a gaseous mixture containing oxidizer and an inert admixture is blown through a porous layer, the thermal regime of filtration combustion depends markedly on the intensity of the gas flux due to oxidizer consumption in the chemical reaction, as well due to the pressure drop ΔP between the external boundaries of the porous layer. With countercurrent blowing it is possible to regulate the characteristics of filtration combustion (the velocity of the flame front, its temperature and the completeness of combustion) up to quenching of the process by changing the flow rate of the gaseous mixture. A more complicated phenomenon takes place with concurrent blowing through the porous layer. Supplying cold gas through the hot butt-end of the porous layer leads, under certain conditions, to local cooling of the substance and the formation of a cooling wave in the combustion product zone. Accordingly, the temperature profile becomes non uniform. It has a characteristic maximum which is located within the section confined by the combustion and cooling fronts (Fig. 8.3a). Heat transfer from the combustion product zone to the flame front is accompanied by an increase in temperature which exceeds the adiabatic one. When the oxidizer concentration in gaseous mixture is decreased, the velocity of the cooling wave exceeds the velocity of the combustion front. In this situation the cooling wave overtakes the reaction zone and decreases its temperature and the completeness of conversion of the solid reagent. The oxidizer flow removes the heat of reaction into the zone located in front of the flame front. This process is accompanied by inversion of the wave structure (Fig. 8.3b) (Aldushin and Seplyarskii 1979).

At present the theory of filtration combustion covers a wide family of the problems related to combustion of porous media. They concern filtration combustion under the conditions of natural, forced and mixed filtration (Aldushin and Seplyarskii 1977, 1979, Aldushin et al. 1975, 1977, Aldushin and Merzhanov 1988); stability of filtration combustion (Lebedev et al. 1976, 1977, Sukhov and Yarin 1980,1987; Aldushin and Kasparyan 1981, Boody and Matkowsky 1991); smolder waves propagation (Ohlemiller and Lucca 1983, Schult et al. 1995, 1996 and Ohlemiller 1985); filtration combustion in dissociating system and porous deformable condensed materials (Aldushin 1993, Shkadinskii et al. 1992); interaction of gasless and filtration combustion (Aldushin et al. 1994 a, 1994 b), etc.

Governing equations: one-dimensional approximation. Consider filtration combustion in a porous layer confined by adiabatic gas-impenetrable side walls (Fig. 8.4). The right butt-end of this layer is closed, whereas the left one is open. At moderate pressure the density of the gaseous phase contained in the porous reagent is less than the density of solid material. Accordingly, at typical values of the stoichiometric oxidizer/solid reagent mass ratio ($\sigma \sim 1$) the total mass of the

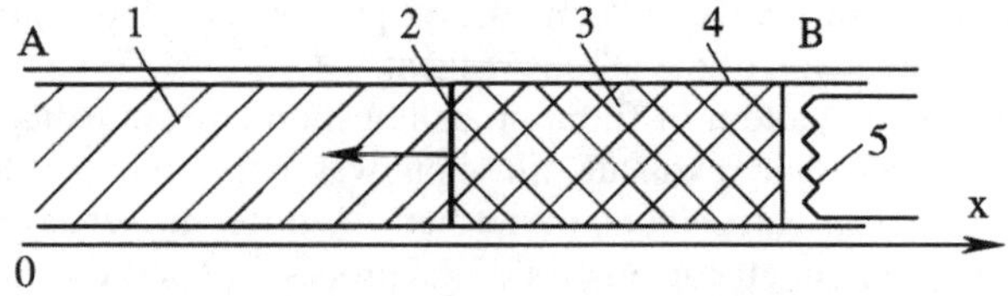

Fig. 8.4 Scheme of combustion of porous layer of finite length. Reprinted from Yarin and Sukhov (1987). 1, solid reagent; 2, flame front; 3, combustion product; 4, envelope of porous layer; 5, igniter; A and B are the open and closed butt-ends of the porous layer

oxidizer in the porous substance is less than that needed for development of combustion with high completeness of chemical transformation. Under these conditions such a process occurs only when gaseous oxidizer is supplied from the environment into the porous material through the open butt-end. The mass flux needed for total conversion of solid reagent is

$$\rho_g \mathbf{u}_f = \sigma \rho_{m.o} \mathbf{u} \tag{8.1}$$

where ρ_g and $\rho_{m.0}$ are the densities of the gaseous oxidizer and porous reagent; u_f and u are the velocities of filtration and the combustion wave, respectively; subscript 0 corresponds to the initial state. The equality (8.1) shows that the filtration velocity is larger than the velocity of the combustion wave since $\rho_{m.0}/\rho_g \gg 1$.

To describe filtration combustion we use the mass, momentum (in the form of the Darcy's law) and energy equations. In the laboratory frame of reference the equations read

$$\frac{\partial \rho_g}{\partial t} + \frac{\partial \rho_g \mathbf{u}_f}{\partial x} = -\sigma W \tag{8.2}$$

$$\rho_{m.0} \frac{\partial \eta}{\partial t} = W \tag{8.3}$$

$$\frac{dP}{dx} = -\frac{u_f}{k_f} \tag{8.4}$$

$$\frac{\partial}{\partial t}\left\{(c_g \rho_g + c_m \rho_m + c_p \rho_p)T\right\} + c_g \frac{\partial}{\partial x}(\rho_g u_f T) = \lambda \frac{\partial^2 T}{\partial x^2} + qW \tag{8.5}$$

where ρ, P and T are the density, pressure and temperature, respectively, W is the chemical reaction rate; $\eta = (\rho_{m.0} - \rho_m)/\rho_{m.0}$ is the completeness of conversion of the solid reagent; c_m and c_p are the specific heat of the solid reagent and product, c_g is the isochoric specific heat of the gaseous oxidizer, q is the reaction heat released, k_f is the filtration coefficient equal to the permeability of the porous matrix divided by the gas viscosity, λ is the thermal conductivity, subscripts g, m and p refer to the gaseous phase, solid reagent and combustion product, respectively.

The system of Eqs. (8.2) – (8.5) is supplemented by the equation of state of the gaseous phase assumed to be an ideal gas

$$mP = \rho_g R_g T \tag{8.6}$$

by the macrokinetic law expressing the reaction rate

$$W = zP^{\nu} f(\eta) \exp\left[-\frac{E}{RT}\right] \tag{8.7}$$

by the relation for specific heats of the solid reagent, combustion product and gaseous oxidizer, which corresponds to the assumptions that the specific heats of the reagents and products are constant and the heat released is independent of temperature

$$c_m + \sigma c_g = (1+\sigma) c_p \tag{8.8}$$

by the definition of the completeness of combustion of the solid reagent

$$\eta = \frac{(\rho_{m.0} - \rho_m)}{\rho_{m.0}} \tag{8.9}$$

and by the stoichiometric relations

$$\rho_m = \rho_{m.0}(1-\eta) \tag{8.10}$$

$$\rho_p = (1+\sigma)\rho_{m.0}\eta \tag{8.11}$$

In Eq. (8.7) $f(\eta) = (\eta_o + \eta)^{-n}$ is the burning-out function $\left(f(\eta) = 1,\ f(1) = 0\right)$, z, ν, n and $\eta_o << 1$ are the kinetic coefficients, E is the activation energy, R is the universal gas constant; R_g is the gas constant in Eq. (8.6); m is the medium's porosity.

Quasi-stationary problem. The system of the non-linear Eqs. (8.2) – (8.11) can be solved numerically. However, it is possible to obtain an approximate solution of the problem under the assumption that combustion of the porous layer is quasi-stationary. This assumption is valid when the combustion front is located far from the cold butt-end of the porous layer and its slow displacement does not affect significantly the filtration conditions. Whith such an approach the given process can may be considered as a succession of stationary states which differ only in the position of the combustion wave relative to the boundaries of the porous layer.

For the frame of reference ζ, τ associated with the combustion front

$$\zeta = x - x_o + \int_{t_0}^{t} u(t)dt; \quad \tau = t \tag{8.12}$$

under the assumption of a stationary process when $\partial/\partial\tau = 0$ the governing equations reduce to

$$\frac{d(\rho_g u_f)}{d\zeta} = -\sigma W \tag{8.13}$$

$$\rho_{m.0} u \frac{d\eta}{d\zeta} = W \tag{8.14}$$

$$\frac{dP}{d\zeta} = -\frac{u_f}{k_f} \tag{8.15}$$

$$\frac{d}{d\zeta}\left\{\left[(c_g\,\rho_g + c_m\rho_m + c_p\rho_p)u + c_g\rho_g u_f\right]T\right\} = \lambda\frac{d^2T}{d\zeta^2} + qW \tag{8.16}$$

It should be noted that when the ratios ρ_g/ρ_m, u/u_f and ε/L (ε and L are the width of the combustion zone and the length of the porous layer) have the order $\delta \ll 1$, the derivatives $\partial/\partial t$ have the order δ and δ^2, whereas the $\partial/\partial\zeta$ derivatives have the order δ^0.

As was noted above, filtration combustion with total conversion of the porous reagent occurs only when gaseous oxidizer is supplied from the environment. The intensity of the oxidizer flux towards the reaction zone is determined by the hydraulic resistance of the cold part of the porous layer, R_h. At large R_h the intentsity of the flux is not enough for total conversion of the solid reagent. In this case the residual completeness of combustion $\eta_{fin} = 1$ and the oxidizer pressure P_{fin}=0.

These conditions correspond to the so-called filtration regime of combustion. In contrast, at small R_h the intensity of the oxidizer flux supplied towards the reaction zone exceeds that needed for total conversion of the solid reagent. In this case the residual completeness of conversion of solid reagent $\eta_{fin} = 1$ and the residual oxidizer pressure $P_{fin} > 0$. These conditions correspond to the kinetic regime of combustion. Aldushin et al. (1974) formulated a general condition corresponding to both the kinetic and filtration regimes of combustion of porous media

$$P_{fin}(1 - \eta_{fin}) = 0 \tag{8.17}$$

Filtration combustion in a porous layer of finite length. This problem was studied in the seminal work by Aldushin et al. (1974). Below, we consider the solution of the problem, as well as the general results related to it. To describe the process of filtration combustion we use the system of Eqs. (8.6)–(8.16).The boundary conditions of the problem are

$$\zeta = -\zeta_{fl},\ P = P_0,\ T = T_0,\ \eta = 0 \tag{8.18}$$

$$\zeta = L - \zeta_{fl},\ \frac{dP}{d\zeta} = \frac{dT}{d\zeta} = 0 \tag{8.19}$$

where ζ_{fl} is the distance from the flame front to the open butt-end of the porous layer.

Let us transform the system of the governing equations. First of all, we reduce the number of the unknown parameters by eliminating the filtration velocity u_f from Eqs. (8.13), (8.15) and (8.16). Combining Eqs. (8.13) and (8.14) and integrating the resulting equation, we arrive at the relation

$$\rho_g u_f = \sigma \rho_{m.0} u(\eta_{fin} - \eta) \tag{8.20}$$

Using the Darcy law (8.15) and the relation (8.20), we obtain the filtration equation

$$\frac{dp^2}{d\zeta} = -2\sigma \frac{R_g \rho_{m.0}}{m k_f} uT(\eta_{fin} - \eta) \tag{8.21}$$

Bearing in mind Eqs. (8.6), (8.8), (8.10), (8.11) and (8.18), we transform the energy equation (8.16) to the following form

$$c_m \rho_{m.0} u(1+\tau\eta_{fin})\frac{dT}{d\zeta} = \lambda \frac{d^2T}{d\zeta^2} + q\rho_{m.0} u \frac{d\eta}{d\zeta} \tag{8.22}$$

where $\tau = (\sigma c_g)/c_m$.

Integrating Eq. (8.22) taking into account conditions (8.18), we obtain

$$\lambda \frac{dT}{d\zeta} = c_m \rho_{m.0} u(1+\tau\eta_{fin})(T-T_0) - q\rho_{m.0} u\eta \tag{8.23}$$

Taking into account the conditions (8.18), we determine from Eq. (8.23) the final temperature T_{fin}

$$T_{fin} = T_0 + \frac{q\eta_{fin}}{c_m(1-\tau\eta_{fin})} \tag{8.24}$$

Thus, the problem reduces to the system of three first-order differential equations (8.13), (8.21) and (8.23) containing three unknowns P, T and η. The number of the equations can be reduced to two by transition from the spatial variable ζ to T.

$$\frac{dP^2}{dT} = -\frac{2\sigma R_g \lambda T(\eta_{fin} - \eta)}{mk_f[c_m(1+\tau\eta_{fin})(T-T_{fin}) - q\eta]} \tag{8.25}$$

$$\frac{d\eta}{dT} = \frac{z\lambda P^{\nu} \exp[-E/(RT)]}{\rho_{m.0}^2 u^2 (\eta_o + \eta)^n [c_m(1+\tau\eta_{fin})(T-T_0) - q\eta]} \tag{8.26}$$

Solutions of Eqs. (8.25) and (8.26) are subject to the boundary conditions

$$T = T_0, \quad P = P_0, \quad \eta = 0 \tag{8.27}$$

$$T = T_{fin}, \quad P = P_{fin}, \quad \eta = \eta_{fin}, \quad P_{fin}(1-\eta_{fin}) = 0 \tag{8.28}$$

where T_{fin} and η_{fin} are related by Eq. (8.24).

The velocity of the combustion wave u, which is contained as a parameter in Eq. (8.26), is determined as a function of the physico-chemical characteristics of the porous medium and the regime parameters via integration of Eq. (8.26). In the case of an activated reaction when the process of chemical conversion of solid reagent is localized at the final temperature T_{fin}, the integration of Eq. (8.26) for η yields

$$u^2 = \frac{z\lambda}{q\rho_{m\cdot 0}^2 \psi(\eta_{fin})} \int_{T_0}^{T_{fin}} P^{\nu}(T)\exp\left[-\frac{E}{RT}\right] dT \tag{8.29}$$

where

$$\psi(\eta_{fin}) = \frac{c_m(1+\tau\eta_{fin})(T_{fin}-T_0)\eta_{fin}^{n+1}}{(n+1)q} - \frac{\eta_{fin}^{n+2}}{n+2}$$

To evaluate the integral in Eq. (8.29) we should determine the dependence of the oxidizer pressure on temperature. We present the function P(T) as the linear part of its Taylor series

$$P^2(T) = P_{fin}^2 + (\frac{dP^2}{dT})_{fin}(T - T_{fin}) \tag{8.30}$$

The expression (8.30) determines the pressure as a function of temperature, the parameters T_{fin} and P_{fin}, and the expansion coefficient $\left(dP^2/dT\right)_{fin}$. The last quantity can be found from Eq. (8.25) as a limit at $\eta \to \eta_{fin}$, $T \to T_{fin}$

$$\left(\frac{dP^2}{dT}\right)_{fin} = -2\frac{\sigma R_g}{m\kappa_f q}\lambda T_{fin}\eta_{fin} \tag{8.31}$$

The parameter P_{fin} is determined via the approximate integration of Eq. (8.21). Bearing in mind that the pressure drop in the porous layer is determined, mostly, by the filtration resistance of its cold part, we integrate Eq. (8.21). Assuming that the length of the cold part of the porous layer equals ℓ, and the oxidizer temperature is close to the initial temperature, we obtain

$$P_{fin}^2 = P_0^2 = 2\sigma R_g \rho_{m.0} u T_0 \eta_{fin} \frac{\ell}{m\kappa_f} \tag{8.32}$$

Evaluating the integral in Eq. (8.29) taking into account Eqs. (8.30) and (8.31), we arrive at

$$u^2 = \frac{z\lambda P_0^{\nu} T_{fin}}{\rho_{m.0}^2 \psi(\eta_{fin}) \Pi^{\nu/2} qE} \Gamma\left(1+\frac{\nu}{2}, \Pi\frac{P_{fin}^2}{P_0^2}\right) \exp\left(\Pi\frac{P_{fin}^2}{P_0^2} - \frac{E}{RT_{fin}}\right) \tag{8.33}$$

where

$$\Pi = \frac{m\kappa_f P_0^2 qE}{2\sigma R_g \lambda R T_{fin}^3}; \; \Gamma\left(1+\frac{\nu}{2}, \Pi\frac{P_{fin}^2}{P_0^2}\right) = \int_{\Pi\frac{P_{fin}^2}{P_0^2}}^{\infty} t^{\frac{\nu}{2}} \exp(-t)dt$$

and Γ is the incomplete Γ function.

Analysis of the solution. The system of Eqs. (8.17), (8.24), (8.32) and (8.33) determines the velocity of the combustion wave, and the final values of pressure, temperature and completeness of combustion of the solid reagent depending on the position of the combustion front in the porous layer. Using the grapho-analytical method developed by Stolyarova et al. (1980), we study the possible solutions of the problem. Introducing the dimensionless variables $\pi_{fin} = P_{fin}/P_0$ and $\theta_{fin} = (E(T_{fin} - T_0))/(RT_{fin}^2)$ and eliminating the combustion wave velocity, we arrive at

$$\gamma\theta_{fin} = \frac{\eta_{fin}}{1+\tau\eta_{fin}} \tag{8.34}$$

$$\eta_f^n = (n+1)(n+2)Af(\pi_{fin}, \theta_{fin}) \tag{8.35}$$

where

$$f(\pi_{fin}, \theta_{fin}) = (1+\beta\theta_{fin})^{2+\frac{3\nu}{2}} (1-\pi_{fin}^2)^{-1} \Gamma(1+\frac{\nu}{2}, \Pi\pi_{fin}^2) \exp\left\{\Pi\pi_{fin}^2 + \frac{\theta_{fin}}{1+\beta\theta_{fin}}\right\},$$

$$\gamma = \frac{c_m RT_0^2}{qE}; \; \beta = \frac{RT_0}{E}; \; A = \left\{\frac{z\lambda P_0^{\nu} T_0^2}{qE}\right\} \left(\frac{2\sigma T_0 \ell}{m\kappa_f P_0^2}\right)^2 \cdot \left(\frac{2\sigma R_g \lambda R T_0^3}{m\kappa_f P_0 qE}\right)^{\frac{\nu}{2}} \cdot \exp\left[-\frac{E}{RT_0}\right]$$

are the parameters accounting for the combined effect of filtration, heat transfer and chemical kinetics on combustion of the porous layer.

Let us explore the solutions of the system of Eqs. (8.34) and (8.35) subject to the condition $\pi_{fin}(1-\eta_{fin}) = 0$. For this aim we reveal the dependence of the functions of heat release q_1 and heat transfer q_2 on temperature

$$q_1 = \frac{\eta_{fin}}{1+\tau\eta_{fin}} \tag{8.36}$$

$$q_q = \gamma\theta_{fin} \tag{8.37}$$

Bearing in mind the possibility of both the total and partial conversion of solid reagent, we consider separately two cases of the chemical kinetics: (i) when reaction is hindered by the layer of combustion product that covers the particles (n>0) and (ii) when such hindering is absent (n=0).

At n>0 and $\eta_{fin} < 1$, the dimensionless final pressure $\pi_{fin} = 0$ (the filtration regime), and the function of heat release takes the characteristic S-shape as a consequence of the exponential dependence of $f(0, \theta_{fin})$ on temperature. In the general case the curves $q_1(0, \theta_{fin})$ and $q_2(\theta_{fin})$ have three intersection points (points 1, 2, and 3 in Fig. 8.5) which correspond to the possible steady states. The limiting values of the functions $q_1(\theta_{fin})$ and $q_2(\theta_{fin})$ and θ_{fin} are also depicted in Fig. 8.5. On the parametric $q-\theta_{fin}$ plane they form a rectangle with the sides $1/(1+\tau)$ and $1/(\gamma(1+\tau))$ which confine the domain of the physically realistic solutions.

The position of the curve $q_1(\theta_{fin})$ on the parametric plane $q-\theta_{fin}$ is determined by the parameter A. An increase in A is accompanied a displacement of $q_1(\theta_{fin})$ to the left in Fig. 8.5. That leads to the approach of the steady state 2 to state 1 and turns the former state into the low temperature state. Simultaneously point 3 is displaced towards point 4 in Fig. 8.5, i.e. the limiting state corresponding to the total conversion of the solid reagent occurs ($\pi_{fin} = 0,\ \eta_{fin} = 1,\ \theta_{fin} = \theta_{max}$). The value of the parameter A corresponding to this state is

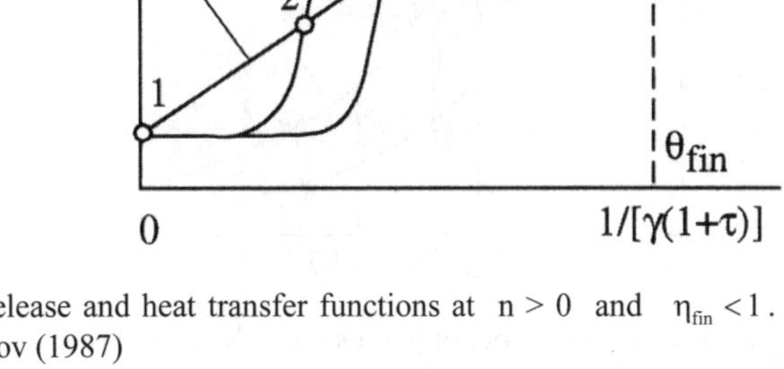

Fig. 8.5 Heat release and heat transfer functions at $n > 0$ and $\eta_{fin} < 1$. Reprinted from Yarin and Sukhov (1987)

$$A_{\ell im} = [(n+1)(n+2)f(0,\theta_{max})]^{-1} \tag{8.38}$$

When $A > A_{\ell im}$ the state corresponding to point 3 loses its physical meaning and we have the unique solution. The decrease of the parameter A leads to the approach of points 2 and 3. At some critical value of $A = A_{cr}$ points 2 and 3 merge. At $A<A_{cr}$ the high temperature regimes with partial conversion of solid reagent are absent.

Under the conditions corresponding to the kinetic regime of filtration combustion the system of Eqs. (8.34) and (8.35) has a single solution. In this case the process proceeds at the temperature $\theta_{fin} = \theta_{max}$ and the residual pressure of the gaseous oxidizer in the combustion product is determined by Eq. (8.35). The following range of the parameter A, $0 \le A \le A_{\ell im}$, corresponds to the kinetic regime. Therefore, the ranges of the existence of the kinetic and filtration regimes supplement each other, so that within the range $A_{\ell im} \le A < A_{cr}$ and $A_{cr} \le A \le A_{\ell im}$ a single or multiple solutions exist. If the physico-chemical properties of the porous layer are constant, a decrease of the parameter A results in shortening of the distance between the cold butt-end and the combustion front during its propagation over the porous layer. In the general case combustion of the porous layer begins after its ignition under the condition $A(L) > A_{\ell im}$ and develops into the filtration regime until the critical cross-section where it inevitably transforms into the kinetic regime. The character of the process remains the same up to the moment when the combustion front reaches the cold butt-end (A=0). This process is represented schematically in Fig. 8.6 (curve 1).

The inevitably transition between the combustion regimes is due to the existence of the critical state which is determined by the following conditions

$$q_1 = q_2; \quad \frac{dq_1}{d\theta_{fin}} = \frac{dq_2}{d\theta_{fin}} \tag{8.39}$$

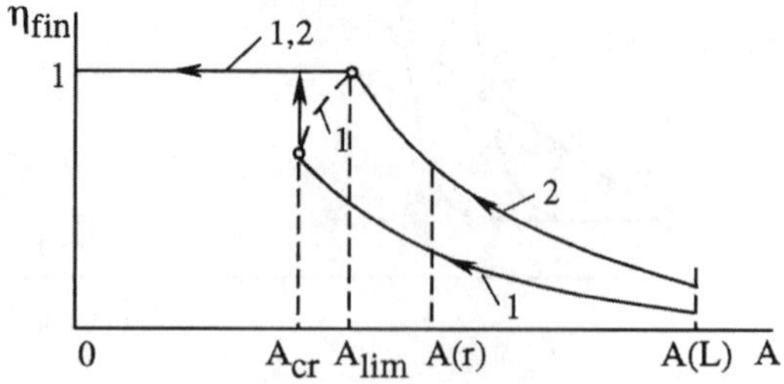

Fig. 8.6 Possible regimes of burning out of porous layer. Reprinted from Yarin and Sukhov (1987)

In accordance with Eq. (8.39), the critical temperature θ_{cr} is expressed as

$$1-\frac{\theta_{cr}(1-\tau\gamma\theta_{cr})}{n(1+\beta\theta_{cr})}\{1+\frac{\theta_{cr}^{\frac{\nu}{2}}(1-\beta\theta_{cr})}{(1+\beta\theta_{cr})^{1+\nu}\Gamma(1+\frac{\nu}{2})\exp\frac{\theta_{cr}}{(1+\beta\theta_{cr})^2}}+ \tag{8.40}$$

$$\beta(2+\frac{3\nu}{2})(1+\beta\theta_{cr})\}=0$$

The analysis of Eq. (8.40) shows that the critical states $(\theta_{cr} < \theta_{max})$ are possible at $n>0$ when the parameter τ exceeds some limiting value of τ_*. For example, at n=1, $\nu = 0.5$, $\gamma = 0.01$, and $\beta = 0.04$ the limiting value of the parameter τ equals $\tau_* = 5.9$. When $\tau < \tau_*$ the critical states are absent $(\theta_{cr} > \theta_{max})$. In this case transition from the filtration to the kinetic regime of combustion occurs monotonically (curve 2 in Fig. 8.6) in the limiting cross-section where $\eta_{fin} = 1$, and $\pi_{fin} = 0$. The transition between the combustion regimes occurs when $A(L)/A_{\ell im} = R_* > 1$. In the case when $R_* < 1$, combustion in the porous layer always proceeds with total conversion of the solid reagent.

Consider the case when the kinetics of chemical reaction is described by a linear law: n = 0. In accordance with Eq. (8.40), the critical states are absent in the filtration regime of combustion, and the process in the porous layer develops monotonically. At n=0 Eqs. (8.24), (8.32) and (8.33) determine the heat release and heat transfer functions as follows (the functions are depicted in Fig. 8.7)

$$q_1 = \frac{\gamma\theta_{fin}}{0.5 + Af(0.\theta_{fin})} \tag{8.41}$$

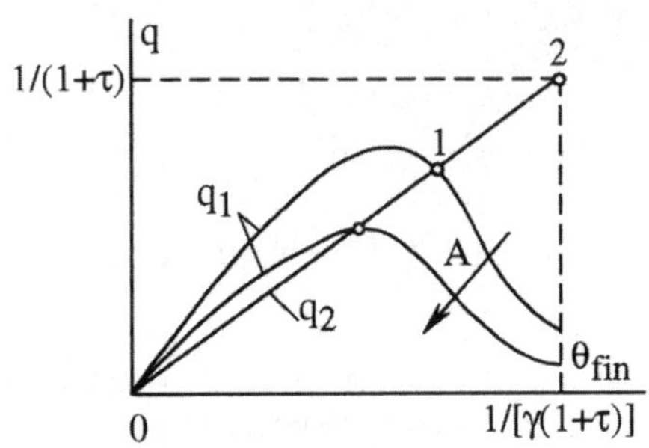

Fig. 8.7 Heat release and heat transfer functions at n = 0 and $\eta_{fin} = 1$. Reprinted from Yarin and Sukhov (1987)

$$q_2 = \gamma\theta_{fin} \tag{8.42}$$

At $A > A_{\ell im}$ the intersection point of the curves q_1 and q_2 corresponds to a single high temperature state (point 1). At $A \to A_{\ell im}$ it tends to point 2 where $\pi_{fin} = 0$ and $\eta_{fin} = 1$. Within the range $0 \le A \le A_{\ell im}$ combustion occurs in the kinetic regime at a maximal temperature and a non-zero residual pressure behind the combustion front.

At n=0, as in the case n>0, the parameter R_* subdivides different regimes of the process. At $R_*<1$ the burning out of a layer proceeds until total conversion of the solid reagent, whereas at $R_*>1$ it proceeds until a transition from the filtration to the kinetic regime (curve 2 in Fig. 8.6).

The current values of the pressure and temperature are determined (at known η_{fin}, π_{fin} and θ_{fin}) by the following relations

$$\pi^2 = \pi_{fin}^2 - 2\sigma R_g \rho_{m.0} u T_0 \eta_{fin} \frac{\zeta}{mk_f P_0^2} \tag{8.43}$$

$$\theta = \theta_{fin} \exp\{c_m \rho_{m\cdot 0} u(1+\tau\eta_{fin})\frac{\zeta}{\lambda}\} \tag{8.44}$$

Bearing in mind that the reaction zone is narrow, it is possible to assume that the profile of the completeness of combustion is uneven

$$\zeta < 0,\ \eta = 0;\ \zeta > 0,\ \eta = \eta_{fin} \tag{8.45}$$

The approximate analytical analysis of the problem performed by Aldushin et al. (1975) reflects the qualitative pattern of the process and yields the results close to those of the numerical solution of Eqs. (8.1)–(8.11) under the following conditions

$$t > 0 \begin{cases} x = 0, & T = T_i,\ (t < t_i) \\ x = 0, & \dfrac{dT}{dx} = 0,\ \dfrac{dP}{dx} = 0\ \ (t > t_i) \\ x = L, & P = P_o,\ T = T_o\ \ (t > t_i) \end{cases} \tag{8.46}$$

The conditions assume the ignition of the porous layer by a thermal impulse with amplitude T_i (hot wall which is assumed to be adiabatic after ignition) and duration t_I, whereas the initial conditions are

$$t = 0,\ 0 \le x \le L,\ P = P_0,\ T = T_0;\ \eta = 0 \tag{8.47}$$

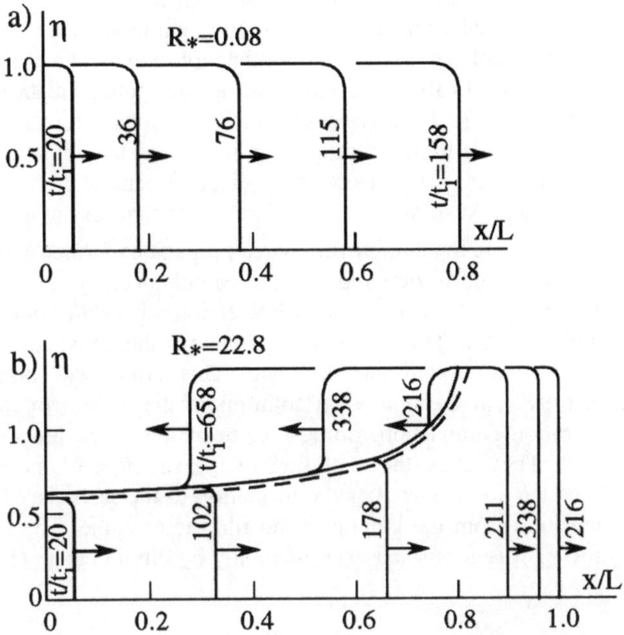

Fig. 8.8 Dynamics of burning out of the porous layer. Reprinted from Aldushin et al. (1975). numerical calculation, -- analytical solution of Aldushin et al. (1975) (n=0, $\tau = 1$, $\nu = 1/2$, $(c_g\rho_g)/(c_m\rho_{m0}) = 10^{-3}$, $\Pi(t_{max}) = 10^2$, $\beta = 0.018$, $\gamma = 0.0015$)

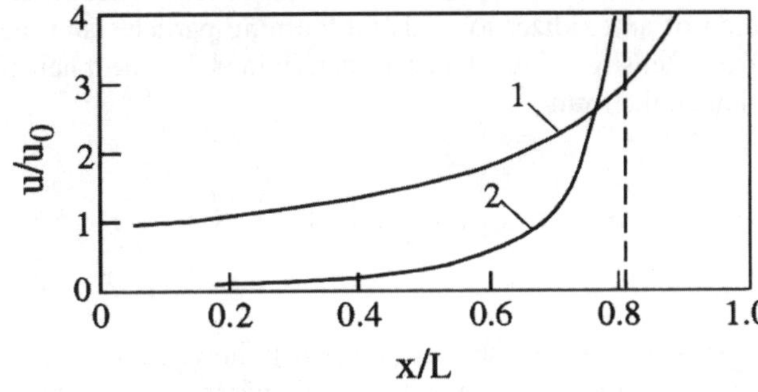

Fig. 8.9 Velocity of the initial (1) and secondary combustion wave at $R_* = 22{,}5$. Reprinted from Aldushin et al. (1975). u_0 is the characteristic velocity determined by known parameters of the problem

As was noted above the parameter R_* determines the regimes of filtration combustion. A small R_* corresponds to the kinetic regime in which conversion of the solid regent is completed within the combustion wave. The velocity of the flame front remains, practically, invariable during the motion of the flame front over porous media. At large R_* the process is more complicated. During the initial stage combustion proceeds in the filtration regime in which completeness of conversion of the solid reagent increases with the displacement of the flame front towards the cold butt-end. At some cross-section the completeness of combustion η reaches the value one and the combustion wave propagates further under the conditions of total conversion of the solid reagent. Simultaneously, the second wave (the burning-out wave) is formed in the cross-section where the transition of the combustion regimes occurs. This wave propagates over the initial substance in the opposite direction relatively to the first one. This effect was discovered by Aldushin et al. (1977) in the numerical solution of the system of the transient equations of filtration combustion. Some results of this work are illustrated in Figs. 8.8 and 8.9. They show the evolution of the structure of the combustion wave at small and large R_*, as well as the formation of the secondary burning-out wave in the transition from the kinetic to the filtration regime. The existence of such a wave at $R_*>1$ was confirmed experimentally by Pityulin et al. (1979).

2.8.2. Heterogeneous model of combustion of porous media

Preliminary comments. In the previous sections, the processes of heat and mass transfer in a heterogeneous medium were considered within the framework of a quasi-homogenous model. According to this model, on the one hand, the chemically active phases are represented as interpenetrating continua, and on the other hand, account is taken of the factors inherent in real heterogeneous systems, such as filtration of an oxidizer towards the burning particles and the interfacial interaction. The intensity of the latter is determined by the phenomenological macrokinetic law of the form

$$W = zP^{\nu} f(\eta) \exp\left[-\frac{E}{RT}\right] \tag{8.48}$$

Such a model proposed by Aldushin et al. (1974) allowed for understanding of a number of specific features characteristic of the given phenomenon, namely, of the regime of stationary flame propagation and of the regime with self-sustained oscillations, combustion with the complete and incomplete conversion of the solid reagent, waves of complete burning etc. The existence of these features was subsequently corroborated experimentally by Pityulin et al. (1979).

The multi-stage character of the process of the oxidizer supply to the reaction zone is characteristic of combustion of porous media. Along with filtration, it includes the sorption stage, i.e. the adsorption of the oxidizer by the surface of the solid phase and its diffusion through the growing film of the product. As was noted by Frank-Kamenetskii (1969), in many practically important cases (combustion of coarsely ground particles with the formation of compact films of the condensed product) the diffusion stage plays the dominant role in the mechanism of thermal flame spread. This role, however, cannot be elucidated within the framework of the quasi-homogeneous model employing a set of the empirical constants in the macrokinetic law (8.48), which are not explicitly related to the sorption characteristics of the medium and reflect only the combined effect of adsorption, diffusion and the chemical reaction on the rate of product formation. At the high temperatures inherent in combustion, the chemical reactions occur in the solid phase, as a rule, very rapidly and the rate of conversion at the combustion front is practically determined by the diffusion kinetics. A heterogeneous model, which includes, along with filtration, the subsequent stages of adsorption and diffusion of an oxidizer in the solid phase, must be employed for the detailed analysis of the process. This approach, developed by Yarin and Sukhov (1989), in contrast to the quasi-homogeneous model, makes it possible to determine explicitly the combustion rate as a function of the standard characteristics of the dispersion medium.

Scheme of combustion and governing equations. Consider the quasi-stationary problem of flame propagation in a layer of porous material with adiabatic side walls impermeable to gas and a hot butt-end. For the analysis, use is made of the representation of a heterogeneous medium as an ordered set of layers of different phases (Fig. 8.10). In this case the structure of the solid and gaseous phases is characterized by the linear sizes ℓ_1 and ℓ_2, which characterize the effective values of the half-widths of the pores and solid slices.

In accordance with the reaction–diffusion model, outlined by Hauffe (1955), the chemical reaction is assumed to be localized at the product–metal interface where a product is formed and the reaction heat is released under conditions similar to the equilibrium ones. Neglecting subsequently the solubility of the gas in the solid reagent, we identify the reaction front with the moving boundary of the oxide film within which the region of diffusion transfer is confined. The thermal propagation of the flame in the present case results from the combined effect of the heat transfer, diffusion, adsorption and filtration. Therefore, the combustion rate can only be determined from the simultaneous solution of the corresponding boundary value problems. Consider the mechanism of diffusion transfer in the solid phase. Its intensity is directly dependent on the value of the diffusion coefficient which is a function of the concentration of diffusing material and temperature

$$D = zf(C)\exp\left[-\frac{E}{RT}\right] \tag{8.49}$$

where D is the diffusion coefficient, C is the concentration, E is the diffusion activation energy.

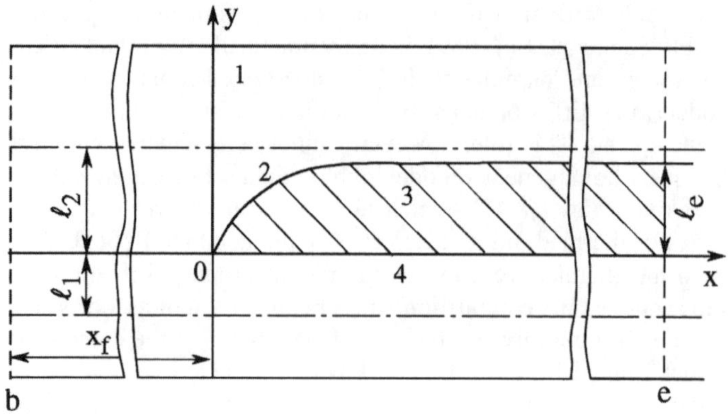

Fig. 8.10 Model medium for the study of combustion wave propagation. Reprinted from Yarin and Sukhov (1989), with permission. 1, solid reagent; 2, combustion front

In what follows we shall confine ourselves to the consideration of the case where the temperature dependence of the diffusion coefficient plays a dominant role. This circumstance enables us to neglect the functional dependence f(C) in Eq. (8.49), compared to the exponential one, and to use the method of "truncation" of the diffusion kinetics at some temperature T_0 close to the final one T_e.

In the frame of reference associated with the "truncation" point (Fig. 8.10) the diffusion problem can be formulated as

$$u\frac{\partial C}{\partial x}=\frac{\partial}{\partial x}\left(D\frac{\partial C}{\partial x}\right)+\frac{\partial}{\partial y}\left(D\frac{\partial C}{\partial y}\right) \tag{8.50}$$

$$y=0,\ C=C_*;\ \ y=y(x),\quad C=C_s$$

$$x=x_e,\qquad \frac{\partial C}{\partial x}=\frac{\partial C}{\partial y}=0$$

where u is the rate of combustion wave propagation, x and y are the longitudinal and transverse coordinates, and subscripts *, s, b and e correspond to the interface, the reaction front and the initial and final states at the cold and hot butt-ends of the layer of porous material.

In comparison with the diffusion stage, the adsorption and desorption processes, which precede it, occur very rapidly, and the state at the surface of the solid phase $y = 0$ is considered to correspond to thermodynamic equilibrium. In the general case, the concentration of gas dissolved at the surface of the solid phase is determined, according to Landau and Lifshitz (1968), by the function

$$C_* = P^n b(T_*) \tag{8.51}$$

where $b(T_*)$ is the solubility constant; the exponent $n = 1$ corresponds to the case when the gas passes into the solid solution without disintegration of the gas molecules, whereas n=0.5 corresponds to diatomic gases in which the molecules are dissociated during the dissolution process.

The oxidizer concentration C_s at the combustion front can be calculated by the thermodynamic method in the equilibrium approximation as a function of pressure and temperature

$$C_s = C(T_s, P_s) \tag{8.52}$$

Describing the process of heat transfer in a layer of the condensed phase, we use the thermal balance equation with the corresponding boundary conditions

$$\begin{gathered} u\frac{\partial T}{\partial x} = \frac{\partial}{\partial x}\left(\alpha_j \frac{\partial T}{\partial x}\right) + \frac{\partial}{\partial y}\left(\alpha_j \frac{\partial T}{\partial y}\right) \\ x = x_b,\ T = T_b;\ x = x_e \quad \frac{\partial T}{\partial x} = \frac{\partial T}{\partial y} = 0 \\ y = 0,\ \ T = T_*;\ \ y = y_s(x) \quad T = T_s; \\ y = \ell_2,\ \frac{dT}{dy} = 0 \end{gathered} \tag{8.53}$$

Here α_j is the thermal diffusivity; j = m, p, the subscripts m and p correspond, respectively, to the solid reagent and the condensed product; ℓ_2 is the half-width of the solid phase layer and u is the flame speed.

According to the adopted model, the filtration process occurs in laminated pores of width $2\ell_1$, assumed to be one-dimensional and characterized by a sorption loss through the interface. Using the Darcy approximation, at moderate oxidizer pressures $(\rho \ll \rho_p)$ we obtain

$$\frac{dP}{dx} = -\frac{u_f}{k_f} \tag{8.54}$$

$$\ell_1 \frac{d(\rho u_f)}{dx} = -\phi(x) \tag{8.55}$$

where ρ is the oxidizer density; u_f is the filtration rate, k_f is the filtration coefficient (which is assumed to be constant); $\phi(x)$ is the specific rate of adsorption transfer.

Using the relationship between the state parameters $P = \rho R_g T$ (R_g is the gas constant), we obtain from the Eqs. (8.54) and (8.55) the filtration equation

$$\frac{d}{dx}\left(\frac{1}{T}\frac{dP^2}{dx}\right) = \frac{2R_g}{\ell_2 k_f}\phi(x) \tag{8.56}$$

Its solution is a subject to the boundary conditions

$$x = x_b, \ P = P_b; \ x = x_e, \ \frac{dP}{dx} = 0 \tag{8.57}$$

which correspond to filtration towards the combustion wave.

The parameters u_f, T_s, $\phi(x)$ and $y_s(x)$ involved in the problem formulation are unknown a priori and can be determined from the mass and thermal balances at the combustion front and at the interface. In the approximation $\rho \ll \rho_b$ at the interface surface $y_s(x)$ the fluxes of the components are related by

$$\left(\frac{\partial C}{\partial x}\frac{dy_s}{dx} - \frac{\partial C}{\partial y}\right)_s = \frac{C_s \rho_p + \sigma \rho_m}{\rho_p D_s}\frac{dy_s}{dx} \tag{8.58}$$

where σ is the stoichiometric coefficient of the reaction.

Neglecting the conductive heat transfer in the gas dissolved in the solid phase and the difference between the values of the specific heat of the components, we obtain the thermal balance equation at the front

$$\alpha_m\left(\frac{\partial T}{\partial x}\frac{dy_s}{dx} - \frac{\partial T}{\partial y}\right)_{m.s} - (1+\sigma)\alpha_p\left(\frac{\partial T}{\partial x}\frac{dy_s}{dx} - \frac{\partial T}{\partial y}\right)_{p.s} = \frac{uq}{c_h}\frac{dy_s}{dx} \tag{8.59}$$

where c_h is the specific heat of solid phases, and q is the reaction heat.

The assumption of the quasi-stationary character of the process and the ideal character of thermal contact between the gas in the pores and the surface of the solid phase imply the following conditions:

$$\phi(x) = -\rho_p D_* \left(\frac{\partial C}{\partial y}\right)_* \tag{8.60}$$

$$\lambda_p \left(\frac{\partial T}{\partial y} \right)_* + q_a \phi(x) = \ell_1 c_h \frac{d}{dx}(\rho u_f T_*) + \ell_1 \lambda \frac{d^2 T_*}{dx^2} \tag{8.61}$$

where q_a is the specific heat of adsorption and λ is the coefficient of thermal conductivity of gas.

For further analysis it is convenient to turn to dimensionless variables, taking as the characteristic scales of time, longitudinal and transverse lengths, velocity, pressure, diffusion coefficient and thermal diffusivity the following quantities: $t^o = \ell_e^2 / D^o$, $\ell_x^o = u^o t^o$, ℓ_2, $u^o = \sqrt{\alpha^o / t^o}$, $D^0 = z^0 \exp[-E/(RT_b)]$, and $\alpha = \alpha_m$.

In the dimensionless variables $z = x/\ell_x^o$, $\eta = y/\ell_2$, $U = u/u^o$, $\theta = (E(T - T_b))/(RT_b^2)$ and $\Pi = P/P_b$ the equations and boundary conditions of the problem take the form

$$\begin{cases} U\dfrac{\partial C}{\partial z} = \delta \dfrac{\partial}{\partial z}\left(\overline{D} \dfrac{\partial C}{\partial z} \right) + \eta_e^2 \dfrac{\partial}{\partial \eta}\left(\overline{D} \dfrac{\partial C}{\partial \eta} \right) \\ \eta = 0,\ C = C_*;\ \eta = \eta_s,\ C = C_s;\ z = z_e,\ \dfrac{\partial C}{\partial \eta} = \dfrac{\partial C}{\partial z} = 0 \end{cases} \tag{8.62}$$

$$\begin{cases} U\dfrac{\partial \theta}{\partial z} = \dfrac{\partial}{\partial z}\left(\overline{\alpha}_j \dfrac{\partial \theta}{\partial z} \right) + \dfrac{\eta_e^2}{\delta} \dfrac{\partial}{\partial \eta}\left(\overline{\alpha}_j \dfrac{\partial \theta}{\partial \eta} \right) \\ z = z_b,\ \theta = 0;\ z = z_e,\ \dfrac{\partial \theta}{\partial z} = \dfrac{\partial \theta}{\partial \eta} = 0;\ \eta = 0,\ \theta = \theta_* \\ \eta = \eta_s,\ \theta = \theta_s;\ \eta = 1,\ \dfrac{\partial \theta}{\partial \eta} = 0 \end{cases} \tag{8.63}$$

$$\begin{cases} \dfrac{d}{dz}\left(\dfrac{1}{1 + \beta\theta_*} \dfrac{d\Pi^2}{dz} \right) = \omega \dfrac{\overline{\phi}}{\eta_e^2} \\ z = z_b,\quad \pi = 1,\quad z = z_e,\quad \dfrac{d\Pi}{dz} = 0 \end{cases} \tag{8.64}$$

$$\left(\delta \frac{\partial C}{\partial z} \frac{d\eta_s}{dz} - \eta_e^2 \frac{\partial C}{\partial \eta} \right)_s = \frac{[\sigma + (1+\sigma) C_s]}{(1+\sigma) D_s} \frac{U d\eta_s}{dz} \tag{8.65}$$

$$\begin{cases} \left(\frac{\partial\theta}{\partial z}\frac{d\eta_s}{dz} - \frac{\eta_e^2}{\delta}\cdot\frac{\partial\theta}{\partial\eta}\right)_{m.s} - (1+\sigma)\bar{\alpha}_p\left(\frac{\partial\theta}{\partial z}\frac{d\eta_s}{dz} - \frac{\eta_e^2}{\delta}\frac{\partial\theta}{\partial\eta}\right)_{p.s} = \frac{U}{\gamma}\frac{d\eta_s}{dz} \\ z=0,\ \eta_s = 0 \end{cases} \tag{8.66}$$

$$\bar{\phi} = -\bar{D}_*\left(\frac{\partial C}{\partial\eta}\right)_* \tag{8.67}$$

$$\bar{\alpha}_p\left(\frac{\partial\theta}{\partial\eta}\right)_* = \left(\frac{1}{\gamma_a} + \frac{1+\beta\theta_*}{\beta}\right)\delta\bar{D}_*\left(\frac{\partial C}{\partial\eta}\right)_* + \sqrt{\delta}\frac{\bar{G}_f}{\eta_e}\frac{d\theta_*}{dz} + \frac{\delta}{\eta_e^2}\varepsilon\bar{\alpha}\frac{d^2\theta_*}{dz^2} \tag{8.68}$$

where $\bar{D} = D/D^\circ = \exp([\theta/(1+\beta\theta)] - [\theta_e/(1+\beta\theta_e)])$, $\delta = D^\circ/\alpha_m$, $\beta = (RT_b)/E$, $\bar{\phi} = (\ell_2\phi)/(\rho_p D^\circ)$, $\omega = (2R_g T_b \alpha_m \ell_2)/(\kappa_f P_b^2 \ell_1)$, $\gamma = (c_h RT_b^2)/(qE)$, $\bar{G}_f = (\ell_1\rho u_f)/(\alpha_m\rho_p)$, $\varepsilon = (\ell_1\rho)/(\rho_p\ell_2)$, $\gamma_a = (c_n RT_b^2)/(q_a E)$, $\bar{\alpha} = \alpha/\alpha_m$ is the dimensionless thermal diffusivity of the gas, $\eta_s = y_s/\ell_2$ is the completeness of combustion of the solid reagent, and $0 \le \eta_s \le \eta_b \le 1$.

Integral relations. The parameter δ, equal to the ratio of the scale values of the diffusion coefficient and the thermal diffusivity appears in the system of equations (8.62) – (8.68). Its value is significantly less than unity since heat transfer in a condensed medium occurs much more intensively than diffusion. This circumstance, together with the analysis of combustion of condensed systems in the paper by Aldushin and Khaikin (1974), makes it possible to simplify significantly the initial formulation of the problem by reducing the thermal conduction equation to the form containing only derivatives of temperature with respect to the coordinate. In fact, in the limit $\delta = 0$ Eq. (8.63) yields $\theta = \theta(z)$. The latter means that at sufficiently small but nonzero values of δ, the temperature $\theta(z,\eta)$ together with its derivatives is only slightly dependent on η. Taking this into account, we integrate otherwise Eq. (8.63) in the intervals $0 \le \eta \le \eta_s$ and $\eta_s \le \eta \le 1$. Combining the results and using the thermal balance conditions (8.66) and (8.68), we obtain an approximate relation similar to the thermal conduction equation

$$\left[(1+\sigma)\eta_s U + (1-\eta_s)U + \frac{1+\sigma}{\sqrt{\delta}}\eta_e\bar{G}_f\right]\frac{d\theta}{dz} = \frac{d}{dz}\left(\alpha_\Sigma\frac{d\theta}{dz}\right) + \frac{U}{\gamma}\frac{d\eta_s}{dz} \tag{8.69}$$

which expresses in the quasi-one-dimensional form the characteristics of heat transfer in a heterogeneous system. In Eq. (8.69) $\alpha_\Sigma = (1+\sigma)\bar{\alpha}_p\eta_s + 1 - \eta_s$ is the effective thermal diffusivity.

Equation (8.69) can be simplified using the assumption that the factor in square brackets in front of the derivative $d\theta/dz$ is proportional to the total mass flux $\rho_m u \ell_2 (1+\sigma\eta_e)$ in a layer of width $\ell_1 + \ell_2$. This assumption approximately holds when small values of the order of ρ/ρ_m are neglected. Then the problem of the flame speed reduces to integrating the thermal conduction equation

$$(1+\sigma\eta_e)U\frac{d\theta}{dz} = \frac{d}{dz}(\alpha_\Sigma \frac{d\theta}{dz}) + W \tag{8.70}$$

with the boundary conditions

$$z = z_b,\ \theta = 0; \quad z = z_e,\ \frac{d\theta}{dz} = 0$$

The source function $W = (U/\gamma)(d\eta_s/dz)$ can be found by solving simultaneously the diffusion (in the approximation $\delta = 0$) and filtration problems

$$\left[\begin{array}{l} U\dfrac{\partial C}{\partial z} = \eta_e^2 \dfrac{\partial}{\partial \eta}(\overline{D}\dfrac{\partial C}{\partial \eta}) \\ \eta = 0,\ C = C_*;\ \eta = \eta_s,\ C = C_s;\ z = z_e;\ \dfrac{\partial C}{\partial z} = 0 \end{array}\right. \tag{8.71}$$

$$\left[\begin{array}{l} \dfrac{\partial \eta_s}{dz} = \dfrac{(1+\sigma)\overline{D}\eta_e^2}{[\sigma+(1+\sigma)C_s]U}\left(\dfrac{\partial C}{\partial \eta}\right)_s \\ z = 0,\ \eta_s = 0 \end{array}\right. \tag{8.72}$$

$$\left[\begin{array}{l} \dfrac{d}{dz}\left(\dfrac{1}{1+\beta\theta}\dfrac{d\Pi^2}{dz}\right) = -\omega\eta_e^2\overline{D}\left(\dfrac{\partial C}{\partial \eta}\right)_* \\ z = z_b,\ \Pi = 1;\ z = z_e,\ \dfrac{d\Pi}{dz} = 0 \end{array}\right. \tag{8.73}$$

To find the function W we turn to Eq. (8.71). Assuming the process of diffusion transfer to be activated ($D=D(\theta)$), using the new variables $\varphi = \eta/\sqrt{\zeta}$, $\zeta = (\eta_e^2/U)\int_0^z \overline{D}(\theta)dz$ and taking into account that the transverse concentration gradient in the product film is much larger than the longitudinal one, we transform the diffusion equation (8.71) to the form

$$2\frac{d^2\overline{C}}{d\varphi^2}+\varphi\frac{d\overline{C}}{d\varphi}=0 \tag{8.74}$$

where $\overline{C}=(C-C_s)/(C_*-C_s)$.
The solution of the Eq. (8.74) is subject to the boundary conditions

$$\varphi=0,\ \ \overline{C}=1,\ \ \varphi=\eta_s,\ \ \overline{C}=0$$

The solution for the oxidizer concentration in the solid phase is

$$C=C_*-(C_*-C_s)\frac{\operatorname{erf}(\frac{\varphi}{2})}{\operatorname{erf}(\frac{\varphi_s}{2})} \tag{8.75}$$

where

$$\operatorname{erf}(\frac{\varphi}{2})=\frac{2}{\sqrt{\pi}}\int_0^{\frac{\varphi}{2}}\exp(-t^2)dt$$

is the error function. Using the solution (8.75), we calculate the derivative

$$\frac{d\eta_s}{d\zeta}=\frac{(1+\sigma)(C_*-C_s)\eta_e^2\overline{D}\exp(-\frac{\varphi_s^2}{4})}{\sqrt{\pi\zeta}U[\sigma+(1+\sigma)C_s]\operatorname{erf}(\frac{\varphi}{2})} \tag{8.76}$$

Let us simplify the derived expression. We assume the temperature dependence $\overline{D}(\theta)$ in it to be dominant and neglect the terms containing C_s, (since according to the experimental data of Hauffe (1955), as a rule, $C_s \ll C_*$). Then Eq. (8.76) reduces to

$$\frac{d\eta}{d\zeta}=\frac{(1+\sigma)C_{*0}\exp(-\frac{\varphi_s^2}{4})}{\sigma\sqrt{\pi\zeta}\operatorname{erf}(\frac{\varphi_s}{2})} \tag{8.77}$$

where $C_{*0}=b(\theta_e)\Pi_o^n$ is the characteristic surface concentration within the reaction zone.

The solution of Eq. (8.77) is

$$\eta_s = \varphi_s \sqrt{\zeta} \tag{8.78}$$

where φ_s is a constant which satisfies the relation

$$\varphi_s \operatorname{erf}(\frac{\varphi_s}{2}) = a \exp(-\frac{\varphi_s^2}{4}) \tag{8.79}$$

Here $a = \dfrac{2(1+\sigma)C_{*o}}{\sigma\sqrt{\pi}}$

Numerical solution of Eq. (8.79) shows that the function $\varphi_s(a)$ increases monotonically at $a > 0$ from the value $\varphi_s(0) = 0$. For $\varphi_s(0) \ll 1$ the following approximations $\exp\left(-\varphi_s^2/4\right) \simeq 1$ and $\operatorname{erf}\left(\varphi_s/2\right) \approx \varphi_s/2$ are valid, and Eq. (8.79) reduces to

$$\varphi_s^2 = -\frac{4(1+\sigma)C_{*o}}{\sigma\sqrt{\pi}} \tag{8.80}$$

Equqtion (8.80) means that φ_s is constant in the combustion zone. Therefore, differentiating Eq. (8.78), we obtain

$$\frac{d\eta_s}{d\zeta} = \frac{\varphi_s^2}{2\eta_s} \tag{8.81}$$

We now consider the energy equation (8.70). Introducing the variable ζ, we recast Eqs. (8.70) and (8.71) into the following form

$$\begin{cases} (1+\sigma\eta_e)U^2 \dfrac{d\theta}{d\zeta} = \dfrac{d}{d\zeta}\left(\alpha_\Sigma \eta_e^2 \overline{D} \dfrac{d\theta}{d\zeta}\right) + \dfrac{U^2}{\gamma}\dfrac{d\eta_s}{d\zeta} \\ \zeta = \zeta_b,\ \theta = \eta_s = 0;\ \zeta = \zeta_e,\ \dfrac{d\theta}{d\zeta} = 0 \end{cases} \tag{8.82}$$

The first integral of Eq. (8.82) reads

$$\frac{d\theta}{d\zeta}=\frac{U^2}{\alpha_\Sigma\gamma\bar{D}\eta_e^2}\left[\eta_e-\eta_s-\gamma(1+\sigma\eta_e)(\theta_e-\theta)\right] \quad (8.83)$$

Converting to the coordinate η_s with the aid of Eq. (8.81) , we obtain

$$\frac{d\theta}{d\eta_s}=\frac{2\eta_s U^2[\eta_e-\eta_s-\gamma(1+\sigma\eta_e)(\theta_e-\theta)]}{\alpha_\Sigma\varphi_s^2\gamma\bar{D}\eta_e^2} \quad (8.84)$$

The repeated integration of Eq. (8.84) within the combustion zone determines [after linearizing the exponent in expression (8.49)] the speed of the flame as an explicit function of the thermal and filtration characteristics (which determine the pressure P_o), as well as the sorption and structural characteristics of the medium

$$U^2=\frac{\gamma\eta_e^2\varphi_s^2(1+\beta\theta_e)^2}{2f(\eta_e)} \quad (8.85)$$

where

$$f(\eta_e)=\eta_e\Delta^{-2}\left[\Delta\eta_e-\ln(\Delta\eta_e+1)\right]-\Delta^{-3}\left[\frac{(\Delta\eta_e+1)^2}{2}-2\Delta\eta_e-\frac{1}{2}+\ln(\Delta\eta_e+1)\right]$$

is a weak function η_e, and $\Delta=(1+\sigma)\bar{\alpha}_p-1$.

The dimensional form of Eq. (8.85) is

$$u^2=\frac{2z^0\alpha_m(1+\sigma)b(T_e)P_o^n c_h RT_e^2\exp\left[-\frac{E}{RT_e}\right]}{\sqrt{\pi}\sigma\ell_2^2 qEf(\eta_e)} \quad (8.86)$$

According to Eq. (8.86) the combustion rate will be determined if the parameters P_0, T_e and η_e are known. The value of the final temperature of the process is determined by the first integral of Eq. (8.82) at the point $\theta=0$ and $\eta=0$ as

$$\theta_e=\frac{\eta_e}{\gamma(1+\sigma\eta_e)} \quad (8.87)$$

The values of P_0 and η_e can only be found taking into account the details of the filtration process. Consider the filtration equation (8.73). We introduce the variable ζ instead of z in the external derivative on its left-hand side. We then calculate the derivative $(\partial C/\partial \eta)_*$ on its right-hand side using Eq. (8.75). Taking also into account the previously used estimates $\varphi_s \ll 1$, $C_s \ll 1$, we obtain the first integral of Eq. (8.73) in the form

$$\frac{d\Pi^2}{d\zeta} = -\frac{\sigma\omega}{1+\sigma} U(1+\beta\theta)(\eta_e - \eta_s) \tag{8.88}$$

Equation (8.88) is integrated once again within the reaction zone. We assume that the filtration process in unaffected by heat released due to chemical reaction. As a result, we find the relation between the characteristic pressure

$$\Pi_o^2 = \Pi_e^2 + \left(\frac{\sigma U}{1+\sigma}\right)^2 \frac{\sqrt{\pi\omega}\eta_e(1+\beta\theta_e)}{12C_{*_o}} \tag{8.89}$$

On the other hand, the integration of Eq. (8.88) over the whole width of the layer, neglecting the filtration resistance of the combustion wave, determines the final pressure depending on the location of the flame front

$$\Pi_e^2 = 1 - \frac{\sigma\omega\eta_e}{1+\sigma} U\zeta_f \tag{8.90}$$

where $\zeta_f = -\zeta_b$ is the distance between the combustion front and the cold butt-end of the layer.

When the filtration combustion develops, as was first noted by Aldushin et al. (1974), two possible cases may be realized. In the first case the factor limiting the rate of the process is the reaction kinetics; in the present case it is the diffusion (the kinetic regime). In the second case filtration of the oxidizer towards the combustion zone limits the rate of the process (the filtration regime). Accordingly, the process of growth of the product film is completed either due to the complete combustion of the solid reagent a with a residual oxidizer $(\eta_e = 1;\ \Pi_e > 0)$ or as a result of the insignificant intensity of the oxidizer filtration ($\eta_e < 1,\ \ \Pi_e = 0$). This alternative is expressed by the relation $\Pi_e(1-\eta_e) = 0$ which forms a closed system of equations together with Eqs. (8.85), (8.87), (8.89) and (8.90). Its solution makes it possible to determine all the characteristic parameters of the process, u, η_e, θ_e, Π_0 and Π_e and to understand its main features.

Analysis of the process. For the following analysis it is convenient to present Eqs. (8.85), (8.89) and (8.90) in the form

$$U^2 = A\Pi_0^n,\ \Pi_e^2 = 1 - BU\zeta_f,\ \Pi_0^2 = \Pi_e^2 + \frac{CU^2}{\Pi_0^n} \tag{8.91}$$

where A, B and C are the known functions of θ_e and η_e:

$$A = \frac{2\gamma(1+\sigma)b(\theta_e)\eta_e^2(1+\beta\theta_e)^2}{\sqrt{\pi}\sigma f(\eta_e)},\ B = \frac{\omega\sigma\eta_e}{1+\sigma},\ C = \frac{\sqrt{\pi}\omega\sigma^2(1+\beta\theta_e)\eta_e}{12(1+\sigma)^2 b(\theta_e)}$$

Combining Eqs. (8.91), we obtain the relation for the final pressure Π_e

$$\left(\frac{1-\Pi_e^2}{B\zeta_f}\right)^2 = A(\Pi_e^2 + AC)^{\frac{n}{2}} \tag{8.92}$$

Denoting the left- and right-hand sides of Eq. (8.92) by $S_1(\Pi_e^2, \zeta_f)$ and $S_2(\Pi_e^2)$, respectively, and analyze the behavior of these functions in the interval $0 \le \Pi_e^2 \le 1$ which corresponds to the kinetic regime of combustion. In this case $\eta_e = 1$ and A, B and C are given by the previously found constants. The function S_1 is concave and monotonically decreases with increasing Π_e^2 and ζ_f.

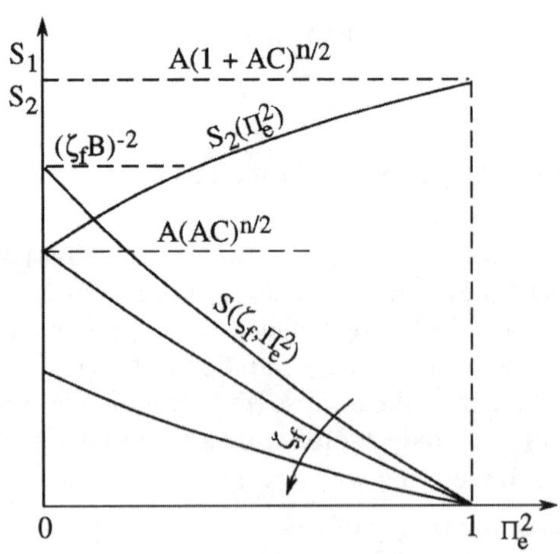

Fig. 8.11 Grapho-analytical analysis of the solution of Eq. (8.92). Reprinted from Yarin and Sukhov (1989), with permission

Also $S_1(0)=1/\sqrt{B\zeta_f}$ and $S_1(1)=0$ (Fig. 8.11). The function S_2 is, in contrast, convex and monotonically increasing. Also $S_2(0)=A(AC)^{n/2}$. The intersection of the curves S_1 and S_2 determines the value of the final pressure Π_e. Since $S_1 \sim 1/\sqrt{\zeta_f}$, a sufficiently small distance $\zeta_f \neq 0$ can always be found at which $\Pi_e^2 > 0$ at the intersection point. With increasing ζ_f the value of Π_e decreases. At some critical value of $\zeta_f = \zeta_f^0$ the curves S_1 and S_2 become tangential at the point $\Pi_e = 0$ which corresponds to the limiting stage $\eta_e = 1$, $\Pi_e = 0$ in which

$$\zeta_f^0 = [AB^2(AC)^{\frac{n}{2}}]^{-\frac{1}{2}} \tag{8.93}$$

A further increase in $\zeta_f > \zeta_f^0$ leads to separation of the curves S_1 and S_2, which means the disappearance of the solution at $\eta_e = 1$ and the transition to the filtration regime of combustion. Thus, during the propagation over the layer of the porous reagent, the combustion wave successively passes through the filtration and kinetic stages. In this case its velocity increases continuously. In fact, taking into account the relationship (8.87) between θ_e and η_e and selecting in Eq. (8,92) the strongest dependence on η_e, we obtain for the filtration regime of combustion the estimate $\zeta_f \sim [b(\theta_e)\eta_e^{4+\frac{3}{2}n}]^{-\frac{1}{2}}$. Since the solubility constant usually increases with temperature, according to the estimate, the approach of the combustion wave to the cold end of the layer should be accompanied by an increase in η_e and θ_e. The latter, according to Eq. (8.86), increases the rate of combustion.

In the kinetic regime, from Eq. (8.91) for the rate of combustion, we obtain the equation $U^{\frac{4}{n}} + A^{\frac{2}{n}}B\zeta_f U - A^{\frac{2}{n}}(1+AC) = 0$. Differentiating it, we obtain

$$\frac{dU}{d\zeta_f} = -\frac{A^{\frac{2}{n}}BU}{\frac{4}{n}U^{\frac{4}{n}-1} + BA^{\frac{2}{n}}\zeta_f} < 0 \tag{8.94}$$

which also corresponds to the acceleration of the combustion wave as it approaches the cold end of the layer.

The results obtained above are consistent with the conclusions of the works by Aldushin et al. (1974, 1977, 1980) devoted to the analytical and numerical investigation of filtration combustion using the quasi-homogeneous approach.

At the same time it is of interest to elucidate the development of combustion affected by change in the sorption and structural properties of the medium, i.e. by those parameters whose role cannot be accounted for in the framework of the quasi-homogeneous model. In particular, according to Eq. (8.86) the enhancement of the gas absorbing capacity of the solid phase (increase in the solubility constant) should be accompanied by an increase in the rate of combustion. The same effect is also achieved in the case of enhancement of the diffusion permeability of the product film due to the increase in the diffusion coefficient. The influence of the characteristic size ℓ_2 on the rate of combustion is more complicated. According to Eq. (8.91)

$$u \sim \frac{\sqrt{A\Pi_o^n}}{\eta_e \ell_2} \exp\left[-\frac{E}{2RT_e}\right] \tag{8.95}$$

If a porous medium consist of a conglomerate of solid particles, different packing densities are possible, which can result in two different limiting stages of combustion.

In a layer consisting predominantly of coarsely divided particles the permeability of the medium is sufficiently high and combustion occurs in the kinetic regime when $\Pi_o \approx 1$. Since in this case $\eta_e = 1$ and the final temperature irrespective of the particle size is $T_e = T_b + [q/(c_h(1+\sigma))]$, Eq. (8.95) yields $u \sim 1/\ell_2$.

On the other hand, in densely packed media, consisting of fine particles weak filtration transfer limits the intensity of combustion and completeness of combustion of the solid reagent. In this case the oxidizer pressure in the reaction zone is reduced to zero. Therefore, the parameter $\Pi_0 \ll 1$ and its value is insignificant for the calculation of the combustion rate. The parameters T_e and η_e involved in Eq. (8.95) are determined by the filtration regime and, consequently, are functions of ℓ_2. All these factors complicate the character of the dependence $u(\ell_2)$. It can be simplified, however, by using the relations $A\Pi_0^n = (B\zeta_f)^{-2}$ following from Eq. (8.91) at $\Pi_e = 0$. Substituting this into Eq. (8.95), and neglecting the dependence on η_e^2, we obtain

$$u \sim \left(\frac{mk_f}{\ell_2}\right) \exp\left[-\frac{E}{2RT_e}\right] \tag{8.96}$$

Here $m \approx \ell_1/\ell_2$ is the porosity of the medium.

Based on the data of Zabrodskii (1963) and Aravin and Numerov (1953), we assume approximately that $m \sim \ell_2^s$ and the filtration coefficient $\kappa_f \sim \ell_2^2$ where

$s<1$. Consequently, $m\kappa_f \sim \ell_2^{2+s}$. Bearing this in mind and taking into account the estimate $m\kappa_f \sim uT_e^2$ obtained via the integration of the filtration equation (8.56), we transform Eq. (8.96) into the following form:

$$u \sim \ell_2^{1+s} \exp\left(-\frac{E\sqrt{u}}{2R\ell_2^{1+\frac{s}{2}}}\right) \tag{8.97}$$

which is indicative of an increase in the value of $u(\ell_2)$ in the finely divided media.

From the physical point of view the result obtained expresses the dominant role of the different stages of mass transfer during combustion and depends on the degree of dispersion of particles in the layer. In dense, finely divided bulk systems, where the rate of combustion is determined by the intensity of the oxidizer filtration, an increase in the particles' sizes stimulates the filtration transfer and increases the speed of flame propagation. However, the transport of the oxidizer towards the reaction zone is simultaneously hindered at the diffusion stage. Under these conditions a decrease in the rate of flame propagation related to a further increase of the particle size occurs due to the dominant influence of the build-up of the diffusion resistance of the product film The latter agrees with the observation of flame propagation in the systems titanium–nitrogen $T_i + N_2$, with variable effective size d of the solid reagent particle in the range $4 \cdot 10^{-6}$ to $3.25 \cdot 10^{-4}$ m, Fig. 8.12 (Yarin and Sukhov 1987).

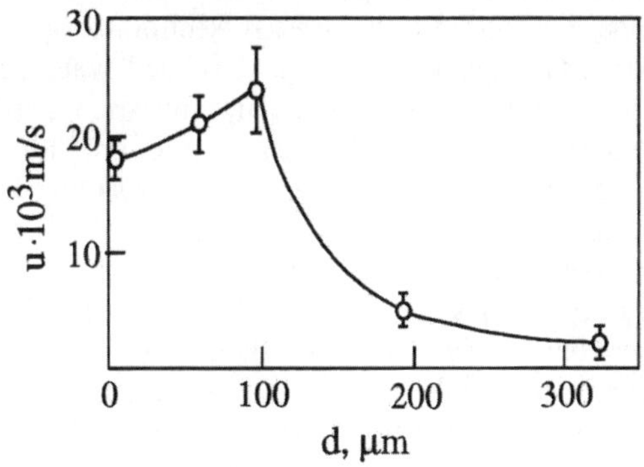

Fig. 8.12 Dependence of the velocity of flame propagation on the particle size. Reprinted from Yarin and Sukhov (1987)

Similarly to the rate of combustion, the width of the zone of complete combustion is an important parameter which expresses the dynamics of chemical transformation in a layer of finite length. Its value is significantly affected not only by filtration but also by the sorption characteristics of the solid phase.

In fact, an increase in the diffusion permeability of particles, which corresponds to a decrease in E according to Eq. (8.49), leads to an increase in the coefficient A at constant B and C. According to Eq. (8.93), this results in a decrease in ζ_f^0.

The enhancement of the gas absorbing capacity of the solid phase, which is associated with an increase in the solubility constant, is accompanied by an increase in A at constant B and C. It also leads to a decrease in the width of the zone of complete combustion. In contrast, the expansion of the latter is promoted by an increase in the particle size. In fact, with an increase in $\ell_2 \sim \omega^{-1/(2+s)}$ the coefficients B and C decrease, A is constant and ζ_f^0 increases.

An additional feature elucidated by the estimate of the ratio of the width of the reaction zone ℓ_x^0 to the linear scale ℓ_2 should be noted. The ratio is sufficiently large $(\ell_x^o/\ell_2 \sim 1/\sqrt{\delta} \gg 1)$ that the use of the quasi-homogeneous model of the medium, taking into account the restrictions inherent in it discussed above, is justifiable in the analysis of combustion.

The capabilities of the heterogeneous model of combustion are not limited to the description of the processes of flame propagation. A similar approach is applicable to the process occurring in a layer of an immobile catalyst according to the mechanism of the interporous catalytic interaction proposed by Zel'dovich (1984). In this case, in the space between the catalyst particles the gas motion should be assumed to be driven by filtration, whereas in the micropores of the particles it should be assumed to be driven by diffusion which develops under the effect of the concentration gradients of the reagents and reaction products.

Thus, the model introduced in the present section is capable of describing a number of very important features of several related heterogeneous processes, which are driven by macroscopic (filtration) and microscopic (diffusion transfer) phenomena. At the same time, the model's capabilities are limited because it does not account for the detailed features of the process, resalting from the statistical character of the porous medium structure. According to Zel'dovich and Sokolov (1985), fractals display similar structures and transfer processes. Transfer processes on fractals can be adequately described by a system of functional dependences which nowhere have derivatives. With the development of the mathematical formalism of the theory of fractals, the solution of such problems will undoubtedly yield more comprehensive information on the character of the processes of heat and mass transfer during combustion of heterogeneous media.

2.8.3. Stability of filtration combustion

Brief survey of the problem. As was noted above, filtration combustion is a complex multistage process developing under the conditions of strong interaction of hydrodynamic, thermal, diffusion and kinetic factors. The analytical study of this process is performed, as a rule, in the frame of the stationary or quasi-stationary approximations. Such solutions should be considered as limiting for the solutions of the unsteady problems for infinite time intervals. However, the approaches ignoring the dynamics of the transient processes leading to steady states should be supplemented by stability considerations. Indeed only stable steady states can become attractors of the transient processes. This make stability studies of the limiting steady states extremely important.

Approaches to the study of the stability of combustion wave propagation in homogenous gaseous mixtures were developed by Landau and Lifshitz (1959), Darrieus (1938), Markstein (1951, 1953), Barenblatt et al. (1962), Istratov and Librovich (1966 a, b). A comprehensive analysis of this problem can be found in monographs by Zel'dovich et al. (1985), Merzhanov and Khaikin (1992), as well as in the surveys by Istratov and Librovich (1966 a, b), Sivashinsky (1983) and Buckmaster (1983).

In accordance with the Landau–Darrieus theory a plane unperturbed flame front separating two uniform flows is a borderline between a uniform flow of reactive mixture entering the flame and a uniform flow of combustion products leaving the flame. The normal component of the flow velocity, its density and temperature undergo a jump at the flame front, whereas the pressure remains continuous. In the perturbed state the flame front takes the form of a surface perturbed by a periodic perturbation. Also the pressure and velocity fields on both sides of the flame front are perturbed with the same wavelength of perturbation. To determine the temporal evolution of the perturbations the continuity and momentum equations are used. At the flame front the tangential velocity component and pressure perturbation are continuous, and the flame front (taking the form of mathematical surface) follows the kinematically perturbed gas flow.

The Landau–Darrieus theory assumes that perturbations do not affect the flame velocity. The latter is assumed to be constant and known from the solution for the flame propagation in an unperturbed medium (Chapter 2.7). These assumptions are valid for perturbations with wavelengths λ_w significantly exceeding the width of the reaction zone δ. At $\lambda_w/\delta \gg 1$ perturbations practically do not disturb the structure of the combustion wave and, accordingly, its velocity. The Landau–Darrieus theory predicted an absolute instability of laminar flames since it found that small perturbations grow in time. This does not agree with numerous experimental observations. The contradiction between the theory and experiment is related to the assumption of a constant velocity of the combustion wave and its independence of flow perturbations. In accordance with the Landau–Darrieus theory the perturbation growth rate is inversely proportional to its wavelength. Therefore, the instability of the flame front is determined by the shortest wavelengths of the perturbation spectrum and thus the problem is ill posed. However, the shortest perturbations should affect the flame structure and its velocity. This circumstance motivates the study of the response of the flame structure to the external perturbations. The analysis of the stability of the flame front taking into account such "small scale" factors as diffusion, heat transfer and viscous forces eliminates the above-mentioned contradiction between the theory and experiment (Istratov and Librovich 1966 a, b). The methods developed to study the stability of the flame front in homogeneous gaseous mixtures and condensed mixed system are used in the stability studies of filtration combustion. A number of important results related to these problems were obtained by Lebedev et al. (1976, 1977), Sukhov and Yarin (1980), Yarin and Sukhov (1987), Aldushin and Kasparyan (1981), and Boody and Matkowsky (1991). Below we consider the stability of filtration combustion under the conditions of isothermal and nonisothermal filtration. The present analysis is mostly based on the results of Lebedev et al. (1976, 1977), Yarin and Sukhov (1980), and Yarin and Sukhov (1987).

Isothermal filtration. Consider the stability of a plane flame front neglecting the influence of perturbations on the structure of the combustion wave in the same approximation as in the Landau–Darrieus theory. We assume the activation energy to be high enough and the rate of chemical reaction to be zero within the domains located outside of the combustion front. The latter corresponds to an infinitely thin flame front separating the domains of fresh reagents and combustion products (Fig. 8.13). The filtration process develops at temperatures T_{in} and T_{fin} corresponding to the initial and final states of reactive mixture.

We restrict our consideration to the kinetic regime of combustion $(\eta_{fin} = 1)$. The system of the governing equations in the frame of reference x, y associated with the unperturbed flame front (cf. Fig. 8.13) is

$$\frac{\partial \rho_g}{\partial t} + \frac{\partial \rho_g u_f}{\partial x} + \frac{\partial \rho_g v_f}{\partial y} + \sigma \rho_{m.1}\left(\frac{\partial \eta}{\partial t} + u\frac{\partial \eta}{\partial x}\right) = 0 \tag{8.98}$$

$$\frac{\partial P}{\partial x} + \frac{u_f - u}{k_f} = 0 \tag{8.99}$$

$$\frac{\partial P}{\partial y} + \frac{v_f}{k_f} = 0 \tag{8.100}$$

$$mP = \rho_g R_g T \tag{8.101}$$

where u is the velocity of combustion wave, $\rho_{m.1}$ is the density of porous reagent in the initial state, and subscript g corresponds to gaseous phase. Integration of Eq. (8.98) for the undisturbed state (where $\partial/\partial t = \partial/\partial y = 0$) yields

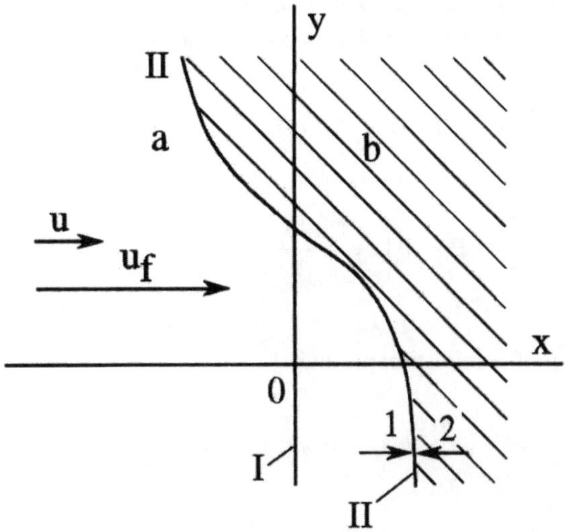

Fig. 8.13 Combustion front in undisturbed (I) and disturbed (II) states. Reprinted from Sukhov and Yarin (1980), with permission. Filtration combustion with a non-iniform temperature distribution. Domain a corresponds to the fresh mixture, and domain b to the combustion products. The reference frame is associated with the unperturbed flame, and u and u_f are the flame and the oxidizer filtration speed, respectively

$$(\rho_g u_f)_a - (\rho_g u_f)_b = \sigma\rho_{m.1} u \tag{8.102}$$

where subscripts a and b refer to the domains a and b corresponding to the fresh mixture and combustion products, respectively (cf. Fig. 8.13). Since $(\rho_g u_f)_a > \sigma\rho_{m1} u$, we obtain the following estimates:

$$\frac{u}{u_{fa}} < \frac{\rho_{ga}}{\rho_{m.1}\sigma} << 1; \qquad \frac{u}{u_{fb}} = \frac{\rho_{gb}}{\sigma\rho_{m.1}(\alpha_{ox} - 1)} \tag{8.103}$$

where $\alpha_{ox} = (\rho_g u_f)_a / (\sigma\rho_{m.1} u)$. When the value of α_{ox} is not close to 1 the ratios u/u_a and u/u_b are less than unity. That allows us to neglect the velocity of the combustion wave in comparison with the filtration velocity in Eq. (8.99). Then Eq. (8.99) takes the form

$$\frac{\partial P}{\partial x} + \frac{u_f}{k_f} = 0 \tag{8.104}$$

The first integrals of Eqs. (8.98) and (8.99) determine the gas density and pressure variation in the basic (undisturbed) state for the region of solid reagent

$$(\rho_g u_f)_a = \rho_{g1} u_{f1} \tag{8.106}$$

$$P_a = P_1 \left(1 - \frac{2u_{f1}x}{k_{f1}P_1}\right)^{\frac{1}{2}} \tag{8.107}$$

and for the region of combustion product

$$(\rho_g u_f)_b = \rho_{g2} u_{f2} \tag{8.107}$$

$$P_b = P_2 \left(1 - \frac{2u_{f2}x}{k_{f2}P_2}\right)^{\frac{1}{2}} \tag{8.108}$$

Subscripts 1 and 2 refer to the parameters at the combustion front in the domains a and b, respectively (cf. Fig. 8.13).

In the disturbed case the governing parameters consist of a sum of their basic values plus small perturbations. The linearized equations for the small perturbation of pressure P' and components of filtration velocity u' and v' are obtained from Eq. (8.98) and (8.100) in the following form:

$$\frac{\partial P'}{\partial t} + u_f \frac{\partial P'}{\partial x} + P\frac{\partial v'}{\partial y} + P'\frac{du_f}{dx} + P\frac{\partial u'}{\partial x} + u'\frac{dP}{dx} = 0 \tag{8.109}$$

$$\frac{\partial P'}{\partial x} + \frac{u'}{k_f} = 0 \tag{8.110}$$

$$\frac{\partial P'}{\partial y} + \frac{v'}{k_f} = 0 \tag{8.111}$$

In the linear approximation all the parameters can be presented as a Fourier mode of wave number **k** and amplitude $L_0 \exp(\Omega t)$ dependent on time

$$L = L_0 \exp(i\mathbf{k} \cdot \mathbf{r} + \Omega t) \tag{8.112}$$

we find from the system of Eqs. (8.109)–(8.111) the characteristic equation of the problem

$$\varepsilon r^2 - 2u_f r - (\Omega + \varepsilon k^2 + \frac{u_f^2}{\varepsilon}) = 0 \tag{8.113}$$

where L=P'; u' and v', $\mathbf{k}=k_x\mathbf{e}_x+k_y\mathbf{e}_y$ is the wave vector, and $\mathbf{r}=x\mathbf{e}_x+y\mathbf{e}_y$ is the radius vector; Ω is the complex frequency (the real part of Ω is the rate of perturbation growth in the case of a possible instability), $\varepsilon = \kappa_f P$ is the piezoconductivity, $r=ik_x$, $k=k_y$, and $\mathbf{e}_x$ and $\mathbf{e}_y$ are unit vectors in the x and y directions, respectively.

From Eq. (8.113), we obtain

$$r_{1,2} = \frac{u_f}{\varepsilon} \pm \sqrt{2\left(\frac{u_f}{\varepsilon}\right)^2 + k^2 + \frac{\Omega}{\varepsilon}} \tag{8.114}$$

Since perturbations vanish far from the flame front, we should take plus for domain a and minus for domain b in Eq. (8.114). Then the fluctuations of pressure and velocity are expressed as

$$P_a' = A\exp(iky + \Omega t) \tag{8.115}$$

$$u_1' = -k_{f1} r_1, A\exp(iky + \Omega t) \tag{8.116}$$

$$v_a' = -ikk_{f1} A\exp(iky + \Omega t) \tag{8.117}$$

$$P_b' = B\exp(iky + \Omega t) \tag{8.118}$$

$$u_b' = -k_{f2} r_2 B\exp(iky + \Omega t) \tag{8.119}$$

$$v_b' = -ik \cdot k_{f2} B\exp(iky + \Omega t) \tag{8.120}$$

where A and B are the amplitudes of the pressure fluctuations in domains a and b, respectively.

The position of the perturbed flame front is given by

$$\ell = C\exp(iky + \Omega t) \tag{8.121}$$

where C is the perturbation amplitude, and ℓ is the flame displacement in the x direction (cf. Fig. 8.13).

Displacement of the combustion front in a domain with a non uniform of distribution of parameters is accompanied by additional disturbances. The total fluctuations of pressure $\widetilde{P}'$ and velocity $\widetilde{u}'$, $\widetilde{v}'$ may be represented as a superposition of the periodic disturbances (8.115)–(8.120) and additional disturbances that are proportional to the displacement of the flame front

$$\widetilde{P}_1' = (A - \frac{u_{f1}}{k_{f1}} C)\exp(iky + \Omega t) \tag{8.122}$$

$$\widetilde{u}_1' = (-k_{f1} r_1 A + \frac{u_{f1}}{\varepsilon_1} C - \Omega C)\exp(iky + \Omega t) \tag{8.123}$$

$$\widetilde{v}_1' = ik(-k_{f1} A + u_{f1} C)\exp(iky + \Omega t) \tag{8.124}$$

$$\widetilde{P}_2' = (B - \frac{u_{f2} C}{k_{f2}})\exp(iky + \Omega t) \tag{8.125}$$

$$\widetilde{u}_2' = (-kf_2 r_2 B + \frac{u_{f2} C}{\varepsilon_2} - \Omega C)\exp(iky + \Omega t) \tag{8.126}$$

$$\widetilde{v}_2' = ik(-k_{f2} B + u_{f2} C)\exp(iky + \Omega t) \tag{8.127}$$

At the flame front the disturbances (8.122)–(8.127) are connected by a certain balance relation which can be found by integrating the system of Eqs. (8.98)–(8.101) in the vicinity of disturbed and undisturbed fronts and assuming that the width of reaction zone is limited to zero. Combining the resultant integrals we arrive at the following relations:

$$\left(u_{f1} \frac{m_1}{R_g T_1} - u_{f2} \frac{m_2}{R_g T_2} \right) \widetilde{P}_1' + \frac{m_1}{R_g T_1} P_1 \widetilde{u}_1' - \frac{m_2}{R_g T_2} P_2 \widetilde{u}_2' - \sigma \rho_{m.1} u_*' = 0 \tag{8.128}$$

$$\widetilde{P}_1' = \widetilde{P}_2' \tag{8.129}$$

$$\widetilde{v}_1' k_{f2} = \widetilde{v}_2' k_{f1}' \tag{8.130}$$

where $u_*' = -d\ell/dt = -\Omega C \exp(iky + \Omega t)$.

Substitution of the expressions for the pressure and velocity fluctuations in Eqs. (8.128)–(8.130) yields a system of algebraic equations containing the amplitudes of fluctuations with frequency Ω and wave number k. From the first equation of the system (8.128)–(8.130) it follows that

$$AW_1 + BW_2 + CW_3 = 0 \tag{8.131}$$

where

$$W_1 = \frac{\sigma\rho_{m1}u - \varepsilon_1\rho_{g1}r_1}{P_1},\quad W_2 = \frac{\varepsilon_2\rho_{g2}r_2}{P_1},\quad W_3 = \frac{\rho_{g2}u_{f2}W_4}{P_1} + \Omega(\rho_{g2} - \rho_{g1} + \sigma\rho_{m1}),$$

$$W_4 = \frac{u_{f1}}{k_{f1}} - \frac{u_{f2}}{k_{f2}}$$

The second and third equations of the system (8.128)–(8.130) are not independent and yield a single relation

$$A–B–CW_4 = 0 \tag{8.132}$$

Thus, to determine the amplitudes A, B and C it is necessary to employ an additional equation. This can be obtained from the assumption that long wavelength disturbances do not affect the combustion wave structure. The latter means that the heat released is the same at the disturbed and non disturbed flame fronts:

$$\rho_{g1}\tilde{u}_1' + \frac{\rho_{g1}u_{f1}}{P_1}\tilde{P}_1' + \frac{c_m}{c_g}\rho_{m.1}u_*' = 0 \tag{8.133}$$

where c_m and c_g are the specific heat capacity of the solid and gaseous reagents, respectively.

Substitution of the expressions for the fluctuations $\tilde{u}_1'$, $\tilde{P}_1'$ and u' in Eq. (8.133) yields the relation

$$AW_s - CW_6 = 0 \tag{8.134}$$

where $W_5 = u_{f1} - r_1\varepsilon_1$, $W_6 = \Omega\left(1 + (c_m\rho_{m1})/(c_g\rho_{g1})\right)\cdot P_1$

Dispersion equation and stability of combustion. The requirement of the existence of non trivial solutions of Eqs. (8.130), (8.131) and (8.133) determines the form of the dispersion equation:

$$\Omega\left\{(\sigma + \frac{c_m}{c_g})\sqrt{2\frac{u_{f1}^2}{\varepsilon_1^2} + k^2 + \frac{\Omega}{\varepsilon_1}} + \frac{c_{m1}k_{f2}\rho_{g2}}{c_g k_{f1}\rho_{g1}}\sqrt{2\frac{u_{f2}^2}{\varepsilon_2^2} + k^2 + \frac{\Omega}{\varepsilon_2}}\right\} + \tag{8.135}$$

$$+\frac{\rho_{g2}}{\rho_{m.1}}\sqrt{\left(2\frac{u_{f1}^2}{\varepsilon_1^2} + k^2 + \frac{\Omega}{\varepsilon_1}\right)\left(2\frac{u_{f2}^2}{\varepsilon_2^2} + k^2 + \frac{\Omega}{\varepsilon_2}\right)}\cdot k_{f2}\left(\frac{u_{f1}}{k_{f_1}} - \frac{u_{f2}}{k_{f2}}\right) = 0$$

Equation (8.135) can be simplified significantly for a system with dense combustion products where the difference $1-(u_{f2}k_{f1})/(u_{f1}k_{f2})$ is small enough. In this case Eq. (8.135) takes the following form:

$$\Omega = -\frac{\rho_{g2}}{\rho_{m.1}}\frac{ab(1-M)}{F} \tag{8.136}$$

where

$$F = (\sigma + \frac{c_m}{c_g})a + \frac{c_m}{c_g}\frac{k_{f2}}{k_{f1}}\frac{\rho_{g2}}{\rho_{g1}}b,\ a = 2\frac{u_{f1}^2}{\varepsilon_1^2} + k^2,\ b = 2\frac{u_{f2}^2}{\varepsilon_2^2} + k^2,\ M = \frac{u_{f2}}{u_{f_1}}\frac{k_{f1}}{k_{f2}}$$

The growth rate Ω is real from Eq. (8.136). Its sign depends on the value of parameter M which manifests itself the as ratio of the pressure gradient in the region of the combustion products to that in the region of the initial reagent. Equation (8.136) shows that the conditions $M < 1$ and $M > 1$ correspond to the stable and unstable states of the combustion front since they yield $\Omega < 0$ and $\Omega > 0$, respectively. The stability boundary corresponds to $M = 1$.

To study the effect of the properties of a porous medium on the stability of filtration combustion it is convenient to present parameter M in the following form:

$$M = \left(\frac{\alpha_{ox} - 1}{\alpha_{ox}}\right)\frac{k_{f1}m_1T_2}{k_{f2}m_2T_1} \tag{8.137}$$

The coagulation of a solid material at high temperature is accompanied by a change in its characteristics. As a result the porosity and filtration coefficient of the combustion product, as a rule, are smaller than that of the initial reagent. Since $T_2/T_1 \sim 10$ and $(k_{f1}m_1)/(k_{f2}m_2) > 1$, the ratio $(k_{f1}m_1T_2)/(k_{f2}m_2T_1) > 10$. Taking into account that the stable and unstable states correspond to $M < 1$ and $M > 1$, respectively, we obtain the following estimates for the excess of oxidizer α_{ox} for the stable

$$\alpha_{ox}\left(1 + \frac{m_2k_{f2}T_1}{m_1k_{f1}T_2}\right) < 1 \tag{8.138}$$

and for the unstable regimes of filtration combustion.

$$\alpha_{ox}\left(1 + \frac{m_2k_{f2}T_1}{m_1k_{f1}T_2}\right) > 1 \tag{8.139}$$

An increase in the excess of oxidizer α_{ox} promotes (with the other parameters being fixed) the combustion instability. It should be noted that at fixed values of pressure in domains a and b, the excess of oxidizer α_{ox} depends on the porosity and filtration coefficients of the initial reagent and combustion product. Compression of porous material due to chemical reaction leads to decrease α_{ox}. Thus, if compression of the finite product is significant the filtration combustion is stable.

Consider now the case when the value of the term $(1-k_{f2}m_2T_1)/(k_{f1}m_1T_2)$ is arbitrary. First we find the relations determining the value of Ω. In the present case Ω is expected to be a complex number. We split it into the real and imaginary parts φ and ψ, as per

$$\Omega = \varphi + i\psi \tag{8.140}$$

Then Eq. (8.135) yields the following relations

$$\begin{aligned}&\varphi A_1\sqrt{\frac{1}{2}\alpha_1(N_1+1)}+\varphi A_2\sqrt{\frac{1}{2}\alpha_2(N_2+1)}-\varphi A_1\sqrt{\frac{1}{2}\alpha_1(N_1-1)}-\\&\psi A_2\sqrt{\frac{1}{2}\alpha_2(N_2-1)}+\frac{1}{2}A_3\sqrt{\alpha_1\alpha_2}\left\{\sqrt{(N_1+1)(N_2+1)}-\sqrt{(N_1-1)(N_2-1)}\right\}=0\end{aligned} \tag{8.141}$$

$$\begin{aligned}&\left\{A_1\sqrt{\frac{1}{2}\alpha_1(N_1-1)}+A_2\sqrt{\frac{1}{2}\alpha_2(N_2-1)}\right\}(\varphi+\psi)+\frac{1}{2}A_3\sqrt{\alpha_1\alpha_2}\times\\&\left\{\sqrt{(N_1+1)(N_2-1)}+\sqrt{(N_1-1)(N_2+1)}\right\}=0\end{aligned} \tag{8.142}$$

where

$$\alpha_{1.2}=2\left(\frac{u_f}{\varepsilon}\right)^2_{1.2}+\kappa^2+\frac{\varphi}{\varepsilon_{1.2}},\quad \beta_{1.2}=\frac{\psi}{\varepsilon_{1.2}},\qquad N_{1.2}=\sqrt{1+\left(\frac{\beta}{\alpha}\right)^2_{1.2}},\qquad A_1=\sigma+\frac{c_m}{c_g},$$

$$A_2=\frac{c_m k_{f2}\rho_{g_2}}{c_g k_{f1}\rho_{g1}},\quad A_3=k_{f2}\left(\frac{u_{f1}}{k_{f1}}-\frac{u_{f2}}{k_{f2}}\right)\frac{\rho_{g2}}{\rho_{m.1}}$$

Note that Eq. (8.142) at $\psi=0$ is satisfied identically at any value of increment φ. Therefore the solution Eq. (8.141) can be found at $\psi=0$. In this case Eq. (8.141) is simplified significantly and takes the form

$$A_1\varphi\sqrt{a_1+\frac{\varphi}{\varepsilon_1}}+A_2\varphi\sqrt{a_2+\frac{\varphi}{\varepsilon_2}}+A_3\sqrt{(a_1+\frac{\varphi}{\varepsilon_1})(a_2+\frac{\varphi}{\varepsilon_2})}=0 \tag{8.143}$$

where $a_{1.2}=2\left(u_f/\varepsilon\right)^2_{1.2}+k^2$.

Since the coefficients A_1 and A_2 in Eq. (8.143) are positive, the sign of increment φ depends on the sign of the parameter A_3. In typical cases where $u_{f1}>0$ and $u_{f2}=0$ (countercurrent filtration in a porous layer with closed high temperature butt-end) and $u_{f1}=0$, $u_f<0$ (concurrent filtration in a porous layer with closed cold butt-end) the coefficient A_3 is positive. In these cases the growth rate of the fluctuations $\Omega=\varphi$ is negative and filtration combustion is stable.

The numerical study of the problem performed by Aldushin et al. (1980) and experiments by Pityulin et al. (1979) supported the conclusion that there exist unstable regimes of filtration combustion. They manifest themselves as oscillatory or spin combustion. Their results also show an inadequacy of the isothermal models which do not account for the effects of heat transfer in the combustion wave like the model considered in the previous two subsections. The latter can be taken into account with an improved model of the combustion wave including both an infinitely thin flame front and a relatively broad relaxation zone (Sukhov and Yarin 1980, Yarin and Sukhov 1987). Following the above-mentioned works, we consider the stability of the filtration combustion under the conditions of nonuniform temperature and pressure distribution within the relaxation zone.

Nonisothermal filtration. The scheme of the combustion wave with a non uniform temperature distribution within the relaxation domain is presented in Fig. 8.14. An infinitely thin flame front separates the initial mixture (domain a) and the final product (domain b). The medium outside the flame front is assumed to be inert and its state is described by the following system of equations:

$$\frac{\partial\rho_g}{\partial t}+\nabla\cdot(\rho_g\mathbf{v}_f)=0 \tag{8.144}$$

$$\nabla P+\frac{\mathbf{v}_f}{\kappa_f}=0 \tag{8.145}$$

$$(c_g\rho_g+c_m\rho_m+c_p\rho_p)\frac{\partial T}{\partial t}+c_g\rho_g\mathbf{v}_f\cdot\nabla T-\lambda\nabla^2 T=0 \tag{8.146}$$

$$\rho_m=\rho_{m.0}(1-\dot{\eta}) \tag{8.147}$$

$$\rho_p=\rho_{m.0}(1+\sigma)\eta \tag{8.148}$$

$$mP=\rho_g R_g T \tag{8.149}$$

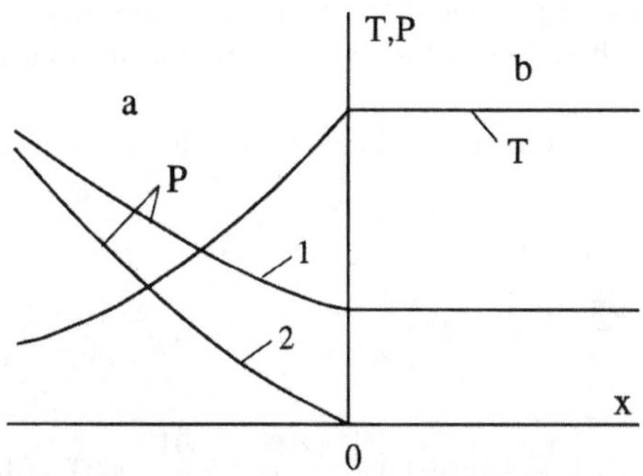

Fig. 8.14 Limited model of combustion wave structure in an undisturbed state. Reprinted from Yarin and Sukhov (1987). Curve 1, kinetic regime; curve 2, filtration regime. The reference frame is associated with the unperturbed front (at x=0)

where c is the specific heat of the components, λ is the thermal conductivity, subscript p corresponds to the combustion product, and 0 is related to the initial state.

We assume that the thermophysical and filtration characteristics of the medium, as well as the heat released by the chemical reaction are constant. In particular, we neglect any coagulation effects. Accordingly the specific heats of the components are related to each other by the following relation:

$$c_m + \sigma c_g = (1+\sigma)c_p \tag{8.150}$$

Kinetic regime of combustion. In this case $\eta_2 = 1$ and $P_2 > 0$. In the unperturbed state Eqs. (8.144)–(8.146) can be integrated to determine the temperature and pressure profiles in the unperturbed state:

$$T_a = T_0 + (T_1 + T_0)\exp\left[\frac{u(1+\tau)x}{\alpha}\right] \tag{8.151}$$

$$P_a^2 = P_1^2 + \frac{2\rho_1 u_{f1}}{k_{f1} T_1}\left\{(T_1 - T_0)\frac{\alpha}{u(1+\tau)}\left[1-\exp\left(\frac{u(1+\tau)x}{\alpha}\right)\right] - T_0 x\right\}$$

for domain a (x<0), and

$$T_b = T_2 \tag{8.152}$$

$$P_b = P_2 \tag{8.153}$$

for domain b (x>0), where $\alpha = \lambda/(c_g\rho_{m.o})$, $\tau = (\sigma c_g)/c_m$, and u is the combustion velocity.

The temperature and pressure are constants behind the flame front due to the adiabatic character of the process and absence of filtration in the layer of combustion product.

For small perturbations Eqs. (8.144)–(8.146) after linearization yield the equations

$$f_{a,b}\frac{\partial T'}{\partial t}+u(1+\tau)\frac{\partial T'}{\partial x}-\alpha\nabla^2 T'=0 \tag{8.154}$$

$$\frac{\partial P'}{\partial t}-k_f P\nabla^2 P'+\alpha_*\frac{\partial P'}{\partial x}+\beta P'-\left(\frac{P}{T}\right)\frac{\partial T'}{\partial t}+\gamma\frac{\partial T'}{\partial x}+\xi T'=0 \tag{8.155}$$

where

$$f_a=1;\ \ f_b=1+\tau;\ \ \alpha_*=u+\frac{k_f P}{T}\frac{dT}{dx}-2k_f\frac{dP}{dx},$$

$$\beta=\left(\frac{k_f}{T}\right)\left(\frac{dP}{dx}\right)\left(\frac{dT}{dx}\right)-k_f\nabla^2 P-\left(\frac{u}{T}\right)\frac{dT}{dx};$$

$$\gamma=\left(k_f\frac{P}{T}\right)\frac{dP}{dx}-u\frac{P}{T};\ \ \xi=\left(\frac{u}{T}\right)\frac{dP}{dx}-\left(\frac{k_f}{T}\right)\left(\frac{dP}{dx}\right)^2-\left(\frac{k_f P}{T}\right)\nabla^2 P$$

As in the isothermal model, the perturbations of temperature and pressure consist of Fourier modes with the amplitudes exponentially dependent on time: $\exp(ik\cdot r+\Omega t)$. Then from Eqs. (8.154) and (8.155), we obtain the expressions for the parameter fluctuations on both sides of the disturbed flame front

$$P_a'=[A\exp(r_a x)+B\exp(n_a x)\exp[iky+\Omega t]] \tag{8.156}$$

$$T_a'=\left[\frac{(r^2-\kappa^2)\varepsilon-\Omega-\alpha_* r-\beta}{\gamma r+\xi-P\dfrac{\Omega}{T}}\right]_a A\exp(r_a x+iky+\Omega t) \tag{8.157}$$

$$P_b'=[C\exp(r_b x)+D\exp(n_b x)]\exp(iky+\Omega t) \tag{8.158}$$

$$T_b'=\left[\frac{(r^2-k^2)\varepsilon-\Omega-\alpha_* r}{\gamma r-P\dfrac{\Omega}{T}}\right]_b C\exp(r_b x+iky+\Omega t) \tag{8.159}$$

$$r_{ab}=(1+\tau)\frac{u}{2\alpha}\pm\sqrt{\left[\frac{(1+r)u}{2\alpha}\right]^2+k^2+f_{ab}\frac{\Omega}{\alpha}} \tag{8.160}$$

$$n_{ab} = \left(\frac{\alpha_*}{2\varepsilon}\right)_{ab} \pm \sqrt{\left(\frac{\alpha_*}{2\varepsilon}\right)_{ab}^2 + k^2 + \frac{(\Omega+\beta_{ab})}{\varepsilon_{ab}}} \tag{8.161}$$

where A,B, C and D are the constants of integration.

Similarly to Eq. (8.114) the present Eqs. (8.156)–(8.161) yield the perturbations in front of and behind the flame front.

The perturbation of the flame front in the present case is given by $\ell = F \cdot \exp(iky + \Omega t)$. Therefore, the parameter perturbations at the flame front are

$$\widetilde{P}_1' = \left[A + B - \left(\frac{u_{f1}}{k_{f1}}\right) F\right] \exp(iky + \Omega t) \tag{8.162}$$

$$\widetilde{T}_1' = \left\{\left[\frac{(r^2-k^2)\varepsilon - \Omega - \alpha_* r - \beta}{\gamma r + \xi - P\dfrac{\Omega}{T}}\right]_1 A + \frac{T_1 - T_o}{\alpha}(1+\tau)uF\right\} \exp(iky+\Omega t); \tag{8.163}$$

$$\widetilde{P}_2' = (C+D)\exp(iky+\Omega t) \tag{8.164}$$

$$\widetilde{T}_2' = \left[\frac{(r^2-k^2)\varepsilon - \Omega - \alpha_* r}{\gamma r - P\dfrac{\Omega}{T}}\right]_2 C\exp(iky+\Omega t) \tag{8.165}$$

where

$$\alpha_{*1} = u + 2u_{f1} + \frac{\varepsilon_1(1+\tau)(T_1-T_0)u}{\alpha T_1}, \ \alpha_{*2} = u\,, \ \beta_1 = \frac{u_{f1}^2}{\varepsilon_1} - \frac{(1+\tau)(T_1-T_0)u^2}{\alpha T_1},$$

$$\beta_2 = 0\,, \quad \gamma_1 = -\frac{P_1(u+u_{f1})}{T_1}, \quad \gamma_2 = -\frac{P_2 u}{T_2}, \quad \xi_1 = uu_{f1}P_1(1+\tau)\frac{T_1-T_0}{\alpha T_1^2} - \frac{uu_{f1}P_1}{\varepsilon_1 T_1},$$

$\xi_2 = 0$

The perturbations (8.162) – (8.165) at the flame front are subject to the boundary conditions expressing balances of the thermal and filtration fluxes, and continuity of the temperature and pressure at the flame front

$$\widetilde{P}_1' = \widetilde{P}_2' \tag{8.166}$$

$$\widetilde{T}_1' = \widetilde{T}_2' \tag{8.167}$$

$$\frac{m\varepsilon_1}{R_g T_2}\left(\frac{\partial \tilde{P}'}{\partial x}\right)\Bigg|_1^2 = \sigma\rho_{m.o} u\left(\frac{u'}{u}+\frac{\tilde{T}_2'}{T_2}-\frac{\tilde{P}_2'}{P_2}\right) \tag{8.168}$$

$$c_m \alpha\left(\frac{\partial \tilde{T}'}{\partial x}\right)\Bigg|_1^2 + qu' = 0 \tag{8.169}$$

We supplement the conditions (8.166)–(8.169) with the expression for the velocity of combustion $u \sim \exp[-E/(2RT_1)]$ (Aldushin and Seplyarskii 1977) which is assumed to be the same in both the unperturbed and perturbed cases. Using it, we can relate the flame velocity and temperature fluctuations as

$$u' = -\frac{\partial \ell}{\partial t} = \frac{uE}{2RT_1^2}\tilde{T}_1' \tag{8.170}$$

The assumption that the velocity of the disturbed flame front depends exponentially on temperature is widely used in the studies of stability of gaseous and condensed media combustion (Zel'dovich et al. 1985). However, in the case of filtration combustion of porous media it is not derived rigorously and is based on the intuitive assumption that the Arrhenius function describing the chemical source remains dominant for the heat and mass transfer under transient conditions. Aldushin et al. (1974) derived a correct integral relation for the thermal fluxes at the flame front by integrating the energy equation with a δ-function chemical source.

$$\lambda\left[\left(\frac{\partial T}{\partial x}\right)\Bigg|_1^2 - \left|\frac{\partial T}{\partial x}\right|_2 \left(\frac{\partial T}{\partial x}\right)_2\right] = 2q\rho_{m.0}\exp\left[-\frac{E}{RT_1}\right] \tag{8.171}$$

The solution of the problem based on the conditions (8.170) and (8.171) yield almost identical results (Sukhov and Yarin 1980, Yarin and Sukhov 1987, Aldushin and Kasparyan 1981). This allows for the conclusion that the analysis based on condition (8.170) yields a correct pattern of the instability of filtration combustion.

Substituting in Eqs. (8.166)–(8.170) expressions (8.162)–(8.165) for the total fluctuations of pressure and temperature and their derivatives

$$\left(\frac{\partial \tilde{P}'}{\partial x}\right)_1 = \left\{r_1 A + n_1 B - \left[\frac{u_{f1}P_1}{\varepsilon_1 T_1}(T_1 - T_0)\frac{u(1+\tau)}{\alpha} + P_1\left(\frac{u_{f1}}{\varepsilon_1}\right)F\right\}\exp(iky+\Omega t) \tag{8.172}$$

$$\left(\frac{\partial \widetilde{T}'}{\partial x}\right)_1 = \left\{ r_1 \left[\frac{(r^2 - k^2)\varepsilon - \Omega - \alpha_* r - \beta}{\gamma r + \xi - \frac{P}{T}\Omega} \right]_1 A + \frac{(T_1 - T_0)u^2(1+r)^2}{\alpha^2} F \right\} \exp(iky + \Omega t) \tag{8.173}$$

$$\left(\frac{\partial \widetilde{P}'}{\partial x}\right)_2 = (r_2 C + n_2 D)\exp(iky + \Omega t) \tag{8.174}$$

$$\left(\frac{\partial \widetilde{T}'}{\partial x}\right)_2 = r_2 \left[\frac{(r^2 - k^2)\varepsilon - \Omega - \alpha_* r}{\gamma r - \frac{P}{T}\Omega} \right] C \exp(iky + \Omega t) \tag{8.175}$$

we obtain the system of five algebraic equations for the fluctuation amplitudes. The nontrivial solutions of this system correspond to the dispersion equation

$$W_1 W_2 = 0 \tag{8.176}$$

where

$$W_1 = (1 + \sqrt{1 + \sigma_*^2 + \omega})[2\chi + (1+\tau)\omega] - \chi[4 + (1+\tau)\omega] - \left(1 - \sqrt{1 + \sigma_*^2 + (1+\tau)\omega}\right)\omega(1+\tau),$$

$$W_2 = \omega(1-\delta) + 2\left(1 + \sqrt{1 + \sigma_*^2 + \omega}\right)\left[\frac{1}{1+\tau} + \frac{2\chi_*}{1+\tau} - \delta(1 + \chi_{**})\right] + 4\frac{\frac{\chi_*^2}{\delta(1+\tau)} - \chi_{**}}{1+\tau}, \quad \delta = \frac{\varepsilon}{a},$$

$$\omega = \frac{4\alpha\Omega}{u^2(1+\tau)^2}, \; \chi = \frac{E(T_1 - T_0)}{RT_1^2}, \; \chi_* = \frac{\mu\rho_{m.0}}{\rho_1}, \text{ and } \chi_{**} = \frac{1 - T_0}{T_1}$$

The dispersion equation (8.176) splits in two equations

$$W_1 = 0 \tag{8.177}$$

$$W_2 = 0 \tag{8.178}$$

accounting for the effect of the combustion rate, as well as of filtration and conductive heat transfer, on the process stability. Equation (8.177) is quite similar to the dispersion equation obtained by Makhviladze and Novozhilov (1971) in the analysis of the stability of combustion of condensed homogeneous systems. Actually, in the case $\tau = 0$ (filtration absent) Eq. (8.177) coincides with this equation.

Representing the growth rate Ω in the form $\Omega = \varphi + i\psi$ we obtain from Eq. (8.176) the following expressions for determining $\varphi(a, \tau, \sigma_*)$ and $\psi(a, \tau, \sigma_*)$:

$$\sqrt{\frac{1}{2}(1+\sigma_*^2+\varphi)}\left\{(2\chi+(1+\tau)\varphi)\sqrt{A+1}-(1+\tau)\psi\sqrt{A-1}\right\}$$
$$+(1+\tau)\sqrt{\frac{1}{2}\left(1+\sigma_*^2+(1+\tau)\varphi\right)}\times\left(\varphi\sqrt{B+1}-\psi\sqrt{B-1}\right)-\chi[2+(1+\tau)\varphi]=0 \qquad (8.179)$$

$$\sqrt{\frac{1}{2}(1+\sigma_*^2+\varphi)}\left\{(1+\tau)\psi\sqrt{A+1}+[2\chi+(1+\tau)\varphi]\sqrt{A-1}\right\}+ \qquad (8.180)$$
$$(1+\tau)\sqrt{\frac{1}{2}\left[1+\sigma_*^2+(1+\tau)\varphi\right]}\times\left(\psi\sqrt{B+1}+\varphi\sqrt{B-1}\right)-\chi(1+\tau)\psi=0$$

where

$$A=\sqrt{1+\frac{\psi^2}{(1+\sigma_*^2+\varphi)^2}};\quad B=\sqrt{1+\left[\frac{(1+\tau)\psi}{1+\sigma_*^2+(1+\tau)\varphi}\right]^2}$$

The particular form of Eqs. (8.179) and (8.180) corresponding to $\varphi=0$ determines the critical value of the parameter $\chi^{o}(\tau,\sigma_*)$ and the fluctuation frequency $\psi^{o}(\tau,\sigma_*)$ at the instability onset.

$$\chi^{o}=\frac{1}{2}(1+\tau)\psi^{o}\sqrt{\frac{1}{2}(1+\sigma_*^2)}\frac{\sqrt{A-1}+\sqrt{B-1}}{\sqrt{\frac{1}{2}(1+\sigma_*^2)(A+1)}-1} \qquad (8.181)$$

$$\frac{\sqrt{A-1}+\sqrt{B-1}}{\sqrt{\frac{1}{2}(1+\sigma_*^2)(A+1)}-1}\left[\sqrt{2(1+\sigma_*^2)(A-1)}-(1+\tau)\psi^{o}\right]+2(\sqrt{A+1}+\sqrt{B+1})=0 \qquad (8.182)$$

Similarly, Eq. (8.178) leads to the system of dispersion relations

$$(1-\delta)\varphi+2\left[\frac{1+2\chi_*}{1+\tau}-\delta(1+\chi_{**})\right]\left[1+\sqrt{\frac{1}{2}(1+\sigma_*^2+\varphi)(A+1)}+4\frac{\frac{\chi_*^2}{\delta}-\chi_{**}(1+\tau)}{(1+\tau)^2}\right]=0 \qquad (8.183)$$

$$(1-\delta)\psi+2[1+2\chi_*-\delta(1-\chi_{**})(1+\tau)]\frac{\sqrt{\frac{1}{2}(1+\sigma_*^2+\varphi)(A-1)}}{1+\tau}=0 \qquad (8.184)$$

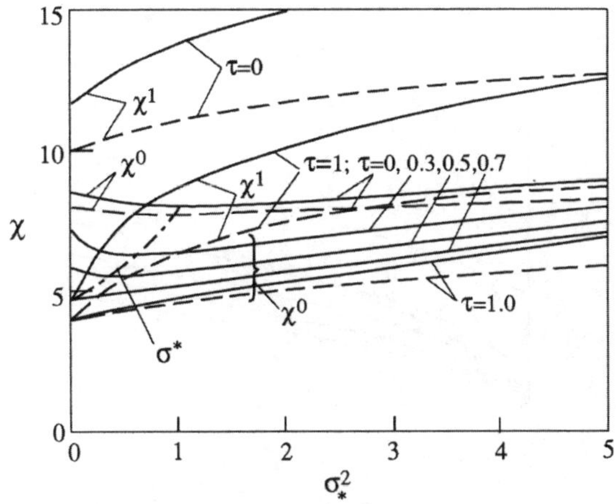

Fig. 8.15 Instability boundaries. Reprinted from Yarin and Sukhov (1987). Solid line in accordance with Yarin and Sukhov (1980). Dotted line in accordance with Aldushin and Kasparyan (1981)

The stability boundary corresponds to the critical values of the parameter $\delta^{\circ}(\tau,\chi_*,\chi_{**},\sigma_*)$ satisfying the equation

$$1+2\chi_*-\delta^{\circ}(1-\chi_{**})(1+\tau)(1+\sqrt{1+\sigma_*^2})+2\frac{\dfrac{\chi_*^2}{\delta^{\circ}}-\chi_{**}(1+\tau)}{1+\tau}=0 \tag{8.185}$$

The solution of the system of Eqs. (8.179) and (8.180) shows that the domains $\chi<\chi^{\circ}$ and $\chi>\chi^{\circ}$ correspond to stable and unstable states, respectively. The instability boundaries $\chi^{\circ}(\tau;\sigma_*)$ (the solid lines in Fig. 8.15) have minima at some values of the parameter $\sigma_*=\sigma_{\rm m}$. The value of $\sigma_{\rm m}$ decreases with increasing τ and tends to zero at $\tau=0.7$. The dependence $\chi^{\circ}(\sigma_*)$ becomes a monotonically increasing function at $\tau>0.7$. The transition to instability is characterized by the frequency $\psi^{\circ}(\tau,\sigma_*)\neq 0$, i.e. it is oscillatory and corresponds to the Hopf bifurcation (Fig. 8.16). Thus, when the contribution of the heat transfer to the filtration flux is insignificant $(\tau<0.7)$, the instability onset is related to the excitement of the traveling waves of growing amplitude with frequencies $\psi^{\circ}(\tau,\sigma_*)$ at the surface of the flame front. Deeper into the domain of instability the spectra of growing perturbations increases. At $\tau>0.7$ transition to the instability results in the emergence of the traveling waves with growing amplitude and frequency $\psi^{\circ}(\tau.0)$. The transition through the critical value of $\chi^{\circ}(\tau,\sigma_*)$ into the instability

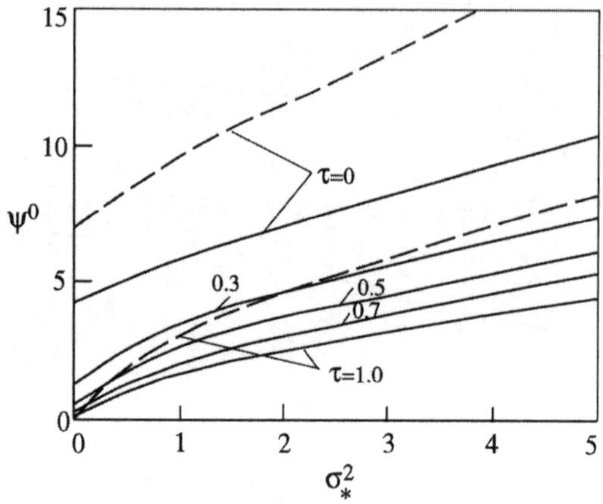

Fig. 8.16 Frequency of fluctuations on the boundary of oscillated instability. Reprinted from Yarin and Sukhov (1987). Solid line in accordance with Yarin and Sukhov (1980). Dotted line in accordance with Aldushin and Kasparyan (1981)

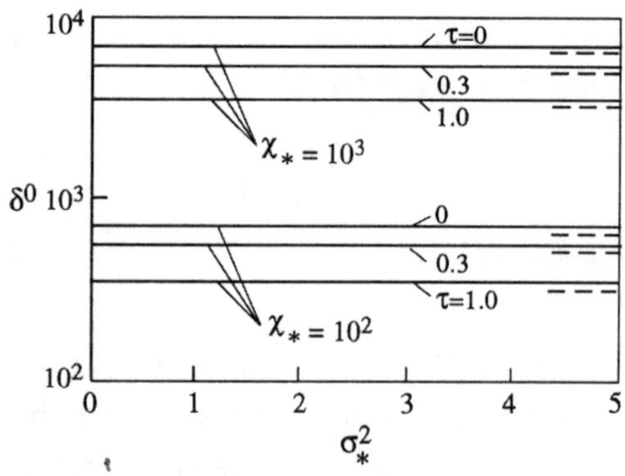

Fig. 8.17 Boundary of instability to parameter δ at $\chi_{**} = 0.718$. Reprinted from Yarin and Sukhov (1987). Dotted line are asymptotes δ^o at $\sigma_*^2 \to \infty$

domain is accompanied by a decrease of the traveling wave frequency up to zero at some boundary $\chi'(\tau,\sigma_*)$, where the oscillatory instability ($\psi^o \neq 0$) is replaced by the non oscillatory one ($\psi^o = 0$). The boundary $\chi'(\tau,\sigma_*)$, is shown in Fig. 8.17.

Equation (8.185) determines the critical value δ^o of the parameter δ affecting the stability of combustion. At $\delta > \delta^o$ the flame front is stable, whereas at $\delta < \delta^o$ it is unstable (Fig.8.17). The instability is non oscillatory ($\psi = 0$). With increasing σ_*, the function $\delta^o(\tau,\chi_*,\chi_{**},\sigma_*)$ decreases slightly and approaches asymptotically the limit $\delta^{o}_{\infty} = (1+2\chi_{*})/((1+\tau)(1-\chi_{**}))$ as $\sigma_{*} \to \infty$. As a result, with decreasing δ^{o} it is possible to expect the development of the instability first in the form of exponentially increasing large scale perturbations and then small scale perturbations of the flame front. In this case the intensification of filtration heat and mass transfer (an increase of τ and a decrease of) χ_{*} expands to a domain of stable combustion. An increase of the heat of combustion (heat release) leads to a contraction of the stability domain.

Filtration regime of combustion. In this case $\eta_2 < 1$, $P_2 = 0$ and the process of chemical conversion proceeds under the condition of a defeat of oxidizer and incomplete conversion of the solid reagent. At an infinitely small filtration resistance of the flame front, pressure $P_1 = 0$. That means that the pressure fluctuation at the flame front equals zero. In this case the study can be restricted to the analysis of the temperature fluctuations via the thermal balance equation (8.154) only. The filtration equation has, in the given case, a trivial solution corresponding to infinitely small (or zero) perturbation amplitudes. The condition $\eta_2 < 1$ allows for the possible existence of fluctuations of the completeness of combustion at the flame front

$$\eta_2' = D\exp(iky + \Omega t) \tag{8.186}$$

Following the approach developed above we find the total temperature fluctuations and their derivatives at the disturbed flame front

$$\tilde{T}_1' = [A + u(1+\tau\eta_2)(T_1 - T_o)\frac{F}{\alpha}]\exp(iky + \Omega t) \tag{8.187}$$

$$\left(\frac{\partial \tilde{T}'}{\partial x}\right)_1 = [r_1.A + u^2(1+\tau\eta_2)^2(T_1 - T_o)\frac{F}{\alpha^2}]\exp(iky + \Omega t) \tag{8.188}$$

$$\tilde{T}_2' = B\exp(iky + \Omega t) \tag{8.189}$$

$$\left(\frac{d\tilde{T}'}{dx}\right)_2 = r_2 B\exp(iky + \Omega t) \tag{8.190}$$

where

$$r_{1.2} = \frac{u(1+\tau\eta_2)}{2\alpha} \pm \sqrt{\left[\frac{u(1+\tau\eta_2)}{2\alpha}\right]^2 + k^2 + f_{1.2}\frac{\Omega}{\alpha}}$$

and $f_1=1$, $f_2=1+\tau\eta_2$.

The total temperature fluctuations are subject to the following boundary conditions at the flame front:

$$\tilde{T}_1' = \tilde{T}_2' \tag{8.191}$$

$$c_m \alpha \left(\frac{\partial \tilde{T}'}{\partial x} \right) \Big|_1^2 + q(u\eta_2' + u'\eta_2) = 0 \tag{8.192}$$

$$u_{f1}\tilde{\rho}_{g1}' - \sigma\rho_{m.o}(u\eta_2' + u'\eta_2) = 0 \tag{8.193}$$

Equations (8.191)–(8.193) express the continuity of the temperature, thermal balance and mass balance at the flame front. In Eq. (8.193) the term $u_{f_1}\tilde{\rho}_{g_1}$ characterizing the fluctuation of the filtration flux is undetermined since as $\tilde{\rho}_{g1} \to 0$, the filtration velocity $u_{f1} \to \infty$. To find the value of the product $u_{f1}\tilde{\rho}_{g1}$, we assume that the reaction proceeds within a narrow zone of width Δx. Integrating Eqs. (8.2) and (8.3) over the reaction zone and using the equation of state we obtain

$$\rho_{g1} u_{f1} = \sigma\rho_{m.o} u\eta_2 \tag{8.194}$$

$$m\tilde{P}_1' = R_g(\tilde{\rho}_{g1}' T_1 + \rho_{g1}\tilde{T}_1') \tag{8.195}$$

Combining these relations, we obtain (as $\Delta x \to 0, \tilde{P}_1' \to 0)$ the following expression:

$$\tilde{\rho}_{g1}' u_{f1} = -\sigma\rho_{m.o} u\eta_2 \frac{\tilde{T}_1'}{T_1} \tag{8.196}$$

In accordance with Eq. (8.196), the Eq. (8.193) takes the form

$$u\eta_2 \tilde{T}_1' + T_1\eta_2 u' + T_1 u\eta_2' = 0 \tag{8.197}$$

Equations (8.170), (8.191), (8.192) and (8.197) yield the dispersion equation

$$S\left(1 - \sqrt{1+\sigma_*^2+\omega}\right) - \omega\left(\sqrt{1+\sigma_*^2+\omega} + \sqrt{1+\sigma_*^2+(1+\tau\eta_2)\omega} + 2\chi_{**}\right) = 0 \tag{8.198}$$

where $S = (2E(T_1 - T_0))/(RT_1^2(1+\tau\eta_2))$. The solution of Eq. (8.198) within the range of increment $\varphi > -(1-\sigma_*^2)/(1+\tau\eta_2)$ corresponds to aperiodic perturbations $(\psi = 0)$ with the growth rate φ satisfying the relation

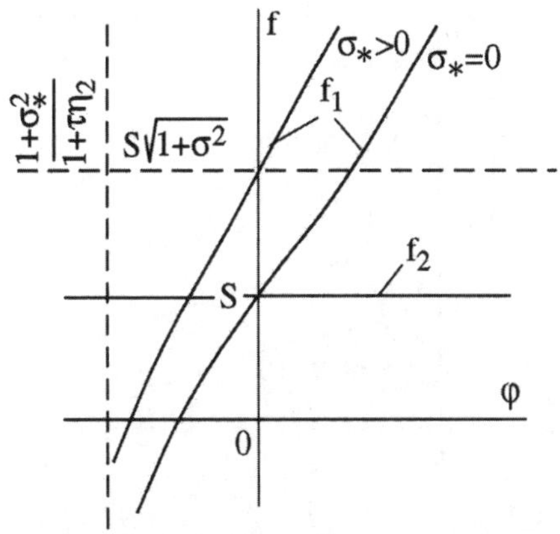

Fig. 8.18 Stability of the flame front in the filtration regime. Reprinted from Yarin and Sukhov (1987)

$$(S+\varphi)\sqrt{1+\sigma_*^2+\varphi}+\varphi(\sqrt{1+\sigma_*^2+(1+\tau\eta_2)\varphi}+2\chi_{**})=S \qquad (8.199)$$

The possible solutions of Eq. (8.199) can be found by using the grapho-analytical method. Representing the left- and right-hand sides of Eq. (8.199) as $f_1(S,\ \sigma_*,\varphi)$ and $f_2=S$, we consider the dependences of f_1 and f_2 on φ (Fig. 8.18). The function f_1 increases with φ at any positive values of σ_* and S. At $\varphi=0$ the function f_1 equals $S\sqrt{1+\sigma_*^2}\geq f_2$. This means that at $\sigma_*>0$ the curves f_1 and f_2 intersect only at $\varphi<0$.

Concluding remarks. The investigations of the stability of filtration combustion considered above elucidate the main features of the complex process that develops under the conditions of strong interaction of the kinetic, filtration and thermal factors. At the same time, the results obtained correspond to a particular model based on a number of simplifying assumptions. The results of the theoretical analysis can differ from those discussed above when other models of filtration combustion are employed (Boody and Matkowsky 1991). For example, the results of the works by Sukhov and Yarin (1980), Yarin and Sukhov (1987) and Aldushin and Kasparyan (1981) corresponding to the countercurrent filtration ($u_{fa}\neq 0,\ u_{fb}=0$) are not fully identical. For example, the oscillatory instability boundary corresponding to χ^o is practically the same in the kinetic regime (Fig. 8.19), whereas the transition to the aperiodic instability at the boundary $\delta=\delta^o$ was not revealed in the analysis of Aldushin and Kasparyan. The latter is due to over-simplification of the mechanism of the heat and mass transfer by Aldushin and Kasparyan (1981): their filtration equation does not take account of the effect of the temperature on the pressure distribution, etc. In accordance with Sukhov and Yarin (1980), Yarin and Sukhov (1987), the flame front is

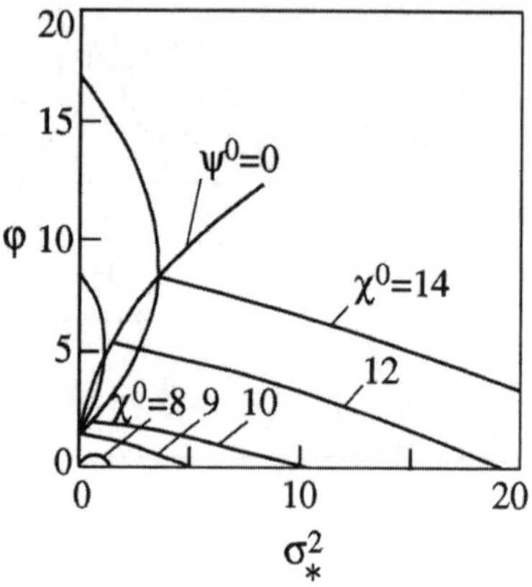

Fig. 8.19 Spectra of increments in the kinetic regime of filtration combustion (natural counter filtration), $\tau = 0$. Reprinted from Aldushin and Kasparyan (1981), with permission. The curve $\psi^0 = 0$ corresponds to the boundary transition from oscillation to exponential instability

exponentially unstable at $\sigma_* > 0$, whereas the study by Aldushin and Kasparyan (1981) predicts the development of aperiodic instability in the supercritical region of the parameter χ: $\chi > \chi^o(\sigma_*)$. These results correspond to the case where gaseous products of combustion are absent and the oxidizer does not contain any inert admixture. The oxidizer concentration in the solid reagent and the combustion product is assumed to be equal to unity and zero, respectively. In reality, even at infinitely small content of oxidizer in the product, its concentration remains identically equal to unity. Taking account of this fact in the dispersion equation of Aldushin and Kasparyan (1981) leads to the same results as those described the previous subsections. The numerical study of filtration combustion in a porous layer with open side surface performed by Ivleva et al. (1980) is devoted to the stability of a two-dimensional flame front in the filtration regime. The results lead to the conclusion that oscillatory instability of the flame front exists in the kinetic regime of combustion. The instability predicted for the filtration regime agrees satisfactorily with the numerical results for one-dimensional unsteady filtration combustion obtained by Aldushin et al. (1980).

References

Aldushin AP (1993) New results in the theory of filtration combustion. Combust. Flame 94: 308–320

Aldushin AP, Kasparyan SG (1981) Stability of stationary filtrational combustion waves. Explos. Shock Waves 17: 615–625

Aldushin AP, Khaikin BI (1974) Combustion of mixtures forming condensed reaction products. Combust. Explos. Shock Waves 10: 273–280

Aldushin AP, Merzhanov AG (1988) Theory of filtration combustion: General representations and the state of research. In: Matros YuSh (ed) Propagation of thermal waves in heterogeneous media. (in Russian) Nauka, Novosibirsk

Aldushin AP, Seplyarskii BS (1977) Theory of filtration combustion of porous metal samples. (in Russian) Preprint. ANSSSR, Chernogolovka

Aldushin AP, Seplyarskii BS (1978) Propagation of waves of exothermal reaction in porous medium during gas blow-through. Sov. Phys. Dokl. 241: 72–75

Aldushin AP, Seplyarskii BS (1979) Inversion structure of combustion wave in porous media at gas blowing. Sov. Phys. Dokl. 249: 585–589

Aldushin AP, Ivleva TP, Merzhanov AG, Khaikin BI, Shkadinskii KG (1975) Spread of combustion front in porous metal samples. (in Russian) In: Merzhanov AG (ed) Processes of combustion in chemical engineering and metallurgy. A.N.SSSR, Chernogolovka, pp. 245–252

Aldushin AP, Matkovsky BJ, Volpert VA (1994a) Interaction of gasless and filtration combustion. Combust. Sci. Technol. 99: 75–103

Aldushin AP, Matkowsky BJ, Volpert VA (1994b) Enhancement of gasless combustion synthesis by counterflow gas filtration. Combust. Sci. Technol. 103: 1–20

Aldushin AP, Merzhanov AG, Khaikin BI (1974) The patterns of the layer-by-layer filtration combustion in porous metals. Sov. Phys. Dokl. 215: 612–616

Aldushin AP, Merzhanov AG, Seplyarskii BS (1977) Theory of filtration combustion of metals. Combust. Explos. Shock Waves 12: 285–294

Aldushin AP, Seplyarskii BS, Shkadinskii KG (1980) Theory of filtration combustion. Combust. Explos. Shock Waves 16: 33–39

Aravin VI, Numerov SN (1953) The theory of motion of liquids and gases in nondeformed porous medium. (in Russian) Gostekhizdat, Moscow

Barenblatt GJ, Zel'dovich YaB, Istratov AG (1962) On diffusional-thermal stability of a laminar flame. Zh. Prikl. Mekh. i Tekh. Fiz. (J. Appl. Mech. Tech. Phys.) 4: 21–26

Boody MR, Matkowsky BJ (1991) On the stability of counter flow filtration combustion. Combust. Sci. Technol. 80: 231–264

Buckmaster J (1993) The structure and stability of laminar flames. In: Lumley L, Van Dyke M, Reed HL (eds) Annu. Rev. Fluid Mech. 25: 21–53

Darrieus G (1938) Propagation d'un front de flamme:assai de theorie des vitesses anomales de deflagration par developpement spontane de la turbulence. Presented at the 6th Int. Congr. Appl. Mech. Paris, 1946

Frank-Kamenetskii DA (1969) Diffusion and heat transfer in chemical kinetics. 2nd edn., Plenum, New York

Hauffe K (1955) Reaktionen in und an festen stoffen. Springer. Berlin, Gőtingen, Heidelberg

Istratov AG, Librovich VB (1966a) Stability of flames. (in Russian) VINITI AN SSSR

Istratov AG, Librovich VB (1966b) The influence of transport processes on the stability of a plane flame front. Prikl. Mat. Mekh. (Appl. Math. Mech.) 30: 451–456 (in Russian)

Ivleva TP, Merzhanov AG, Shkadinskii KG (1980) The surface combustion condensed substances with condensed products. In: Combustion of condensed systems. (in Russian) AN SSSR, Cherkonoglovka, pp. 99–103

Kantorovich BV (1958) Foundations of the theory of combustion and gasification of solid fuel. (in Russian) AN SSSR, Moscow

Khitrin LN (1957) The physics of combustion. (in Russian) Moscow University

Landau LD, Lifshitz AM (1968). Statistical physics. Pergamon, New York

Landau LD, Lifshitz EM (1959) Fluid mechanics. 2nd edn., Pergamon, London

Lebedev AD, Sukhov GS, Yarin LP (1976) Stability of filtration combustion. Combust. Explos. Shock Waves 12: 775–779

Lebedev AD, Sukhov GS, Yarin LP (1977) Theory of filtration combustion. Combust. Explos. Shock Waves 13: 7–11

Makhviladze GM, Novozilov BV (1971) Two-dimensional instability combustion condensed system. Zh. Prikl. Mekh. Tekh. Fiz. 5: 51–59

Markstein GH (1951) Experimental and theoretical studies of flame front stability. J. Aeronaut. Sci. 18: 199–209

Markstein GH (1953) Instability phenomena in combustion waves. The Fourth Symposium (International) on Combustion (Combustion and Detonation waves). Williams and Wilkins, Md., Baltimore pp. 44–59

Merzhanov AG, Khaikin BI (1992) The theory of combustion Wave in homogenous media. (in Russian) Russian Academy, Chernogolovka

Ohlemiller TJ (1985) Modeling of smoldering combustion propagation. Prog. Eng. Combust. Sci. 11: 277–310

Ohlemiller TJ, Lucca DA (1983) An experimental comparison of forward and reverse smolder propagation in permeable fuel beds. Combust. Flame. 54: 131–147

Pityulin AN, Shcherbakov VA, Borovinskaya IP, Merzhanov AG (1979) Laws and mechanism of diffusional surface burning of metals. Combust. Explos. Shock Waves 15: 432–437

Schult DA, Matkowsky BJ, Volpert VA, Fernandez-Pello AC (1995) Propagation and extinction of forced opposed flow smolder waves. Combust. Flame 101: 471–490

Schult DA, Matkowsky BJ, Volpert VA, Fernandez-Pello AC (1996) Forced forward smolder combustion. Combust. Flame 104: 1–26

Sivashinsky GI (1983) Instabilities, pattern Formulation, and turbulence in flames. Annu. Rev. Fluid Mech. 15: 179–199

Shkadinskii KG, Shkadinskaya GV, Matkovsky BJ, Volpert VA (1992) Self-compaction or expansion in combustion synthesis of porous materials. Combust. Sci. Technol. 88: 271–292

Stolyarova NN, Sukhov GS, Yarin LP (1980). Theory of filtration reactor with a stabilized combustion front. Combust. Explos. Shock Waves 16: 174–180

Sukhov GS, Yarin LP (1980) Two-dimensional instability of the combustion of porous substances in a gaseous oxidizer. Combust. Explos. Shock Waves 16: 275–280

Yarin LP, Sukhov GS (1987) Fundaments of combustion theory of two-phase media. (in Russian) Energoatomizdat, Leningrad

Yarin LP, Sukhov GS (1989) A heterogeneous model of combustion of porous media. Combust. Sci. Technol. 64: 67–80

Zabrodskii SS (1963) Hydrodynamics and heat exchange in a fluidized bed. Gosenergoizdat, (in Russian) Moscow Leningrad

Zel'dovich YaB (1984) Towards the theory of reaction on porous or powder-like material. J. Phys. Chem. (1939), 13: 163–169 Selected Papers. Chemical Physics and Gas Dynamics. Moscow, pp. 65–70 (in Russian)

Zel'dovich YaB, Sokolov DD (1985) Fractals, self-similarity and intermediate asymptotics. Uspekhi Fiz. Nauk. 146: 434–506 (in Russian)

Zel'dovich YaB, Barenblatt GI, Librovich VB, Makhviladze GM (1985) Mathematical theory of combustion and explosion. Plenum, New York

2.9 Turbulent heterogeneous flames

2.9.1. General characteristics

Turbulent heterogeneous flames are formed after atomization and ignition of liquid fuel sprays injected into a stagnant or moving oxidizer, as well as in the combustion of particle-laden jets propagating in space filled with oxidizer or combustion products (Fig. 9.1). Such flames are widely used in engineering and technology: in particular, they are encountered in industrial furnaces, gas turbines, jet engines, internal-combustion engines, etc. Data on turbulent heterogeneous flames and their application can be found in the monographs by Smoot and Smith (1985) and Stambuleanu (1976). Below we discuss briefly the main features of turbulent heterogeneous flames.

Turbulent heterogeneous flames can be subdivided in three groups depending on the ratio of the initial mass fluxes of fuel and oxidizer χ_0: lean flames $(\chi_0 < 1)$, rich flames $(\chi_0 > 1)$ and stoichiometric flames $(\chi_0 = 1)$, where $\chi_0 = G_{2.0}/(G_{1.0}\sigma)$, $G_{2.0}$ and $G_{1.0}$ are the fuel and oxidizer mass fluxes at the nozzle exit, respectively, and σ is the oxidizer-to-fuel stoichiometric mass ratio.

Fresh premixed lean and stoichiometric flames contain a surplus or a sufficient amount of oxidizer needed for the full burning-out of the dispersed fuel. Thus, transport of an additional oxidizer from the environment is not required for combustion in such flames. In this case the length of the turbulent flame, the intensity of heat release and the other characteristics are determined by the combustion time of an individual particle, which is equal to the sum of the heating and burning times.

In lean and stoichiometric flames loaded with fine particles fuel is fully consumed due to combustion already within the near field of the turbulent torch, i.e. within its core with the plug velocity profile. In this case the reaction zone is located close to the inner boundary of the mixing layer similarly to the flame front in homogeneous gaseous flames (Fig. 9.2a)

In contrast to the lean and stoichiometric flames, combustion in rich heterogeneous flames is possible only if an additional oxidizer is supplied from the environment. Accordingly, the length of the reaction zone increases significantly. It is located close to the outer boundary of the turbulent jet within the near and far fields of the torch similarly to the flame fronts in diffusion gas flames (Fig. 9.2b). In heterogeneous flames loaded with a polydisperse admixture, fine particles are consumed within the near flow field, whereas the coarse ones are consumed in the far field of the torch. In this case the formation of two reaction zones located within the near and far fields of the flow is possible (Fig. 9.2c). The particle size distribution also affects the reaction zone thickness, which increases for larger particle diameters.

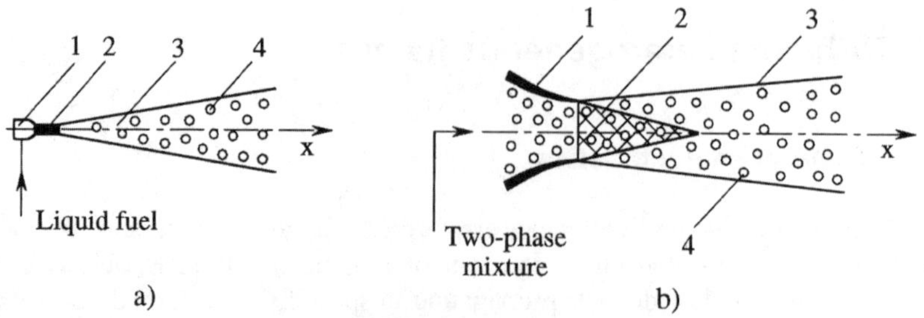

Fig. 9.1 Turbulent two-phase jets. a) Liquid fuel spray: 1, the injector; 2, the intact part of the jet (before fuel atomization); 3, the spray after the atomization has been completed; 4, the droplets. b) Jet of the premixed particle/oxidizer mixture: 1, the nozzle; 2, the inner boundary of the turbulent mixing layer; 3, the outer boundary of the turbulent boundary layer; 4, the particles. The shaded domain corresponds to the jet core with a plug flow

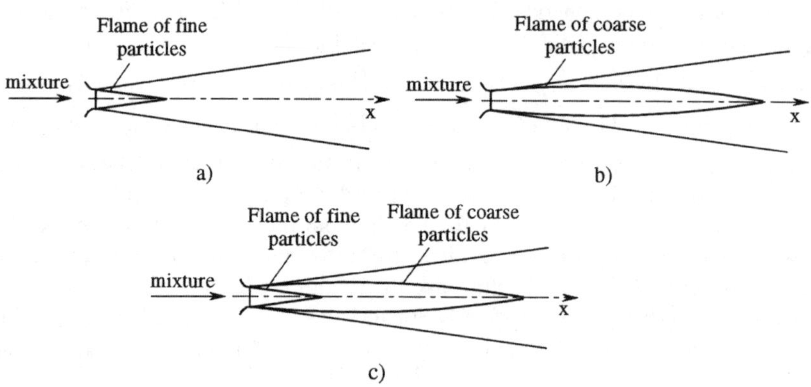

Fig. 9.2 Heterogeneous flames. a) Flame of fine particles. b) Flame of coarse particles. c) Flame of polydisperse particles

2.9.2. Aerodynamics of two-phase jets

The presence of a heavy admixture affects the velocity decay along the axis of two-phase turbulent jets, their ejection features, the intensity of heat and mass transfer and other characteristics. In relation to that we consider briefly the effect of a heavy admixture on the aerodynamics of particle-laden turbulent jets.

The earliest investigations (Laats 1966, Goldschmidt and Eskinazi 1966, Laats and Frishman 1970, Popper et al. 1974) showed that a heavy admixture significantly affects the average characteristics of particle-laden jets. That manifests itself in an increase of the long range of jets, their narrowing, etc. A number of unexpected effects associated with the particle mass concentration in turbulent jets loaded with fine and coarse particles were also observed. In particular, the effect is associated with an increase in particle mass concentration along the axis of a turbulent jet loaded with fine particle, whereas the scattering effect results from a steep attenuation of mass concentration along the axis of a turbulent jet loaded with coarse particles (Laats and Frishman 1970 and Navoznov et al. 1979).

The first attempt to explain the influence of a heavy admixture on the aerodynamics of turbulent jets was apparently undertaken by Abramovich (1963). He considered particle-laden jets as gaseous jets with an equivalent effective density. Accordingly, he attributed an increase in the long range of particle-laden jets, as well as their narrowing to the difference of the densities in the jet and its environment. Although a such model leads to plausible results, it does not account for a number of essential features of the phenomenon and only qualitatively agrees with experiment. In the subsequent investigations detailed data on the turbulent structure of particle-laden jets were obtained (Hetsroni and Sokolov 1971, Levy and Lockwood 1981, Modaress et al. 1984a, 1984b, Shuen et al. 1985, Fleckhaus et al. 1987, Parthasarathy and Faeth 1987, Tsuji et al. 1988, Mostafa et al. 1989, Barlow and Morrison 1990, Sheen et al. 1994, Prevost et al. 1996).

The properties of an air jet laden with fine $(d = 13\,\mu m)$ droplets were studied by Hetsroni and Sokolov (1971). The jet issued from a 25- mm round nozzle with velocity of $50–60$ [m/s]. The intensity of turbulence $\sqrt{u'^2}/u_m$ (u_m is the maximal longitudinal velocity in a cross-section) was uniformly reduced by the small particles (Fig. 9.3). Moreover it decreased almost proportionally to the droplet loading. The turbulence spectra shown in Fig. 9.4 indicate that the particles cause a decrease of the spectral components in the high frequency part of the spectrum. Note that in these experiments the droplets were rather small and the loading was relatively light.

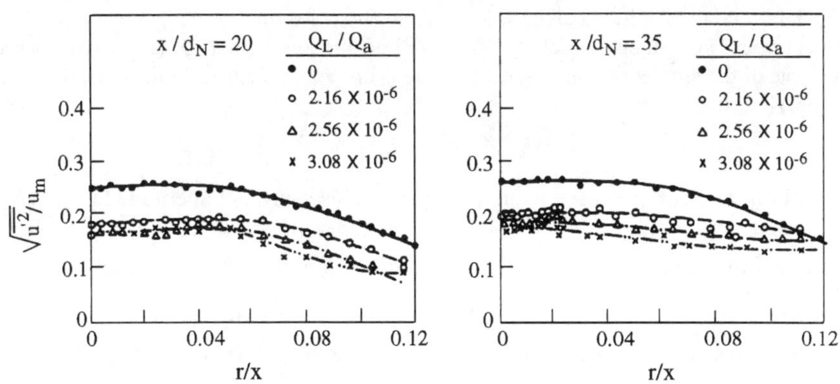

Fig. 9.3 Distribution of the intensity of turbulence across the jet from the distance $x/d_N = 20$ and 35 from the jet origin for several droplets concentrations. Reprinted from Hetsroni and Sokolov (1971), with permission. Q_L and Q_a are the volumetric flow rate of the droplet and air, respectively, and d_N is the nozzle diameter

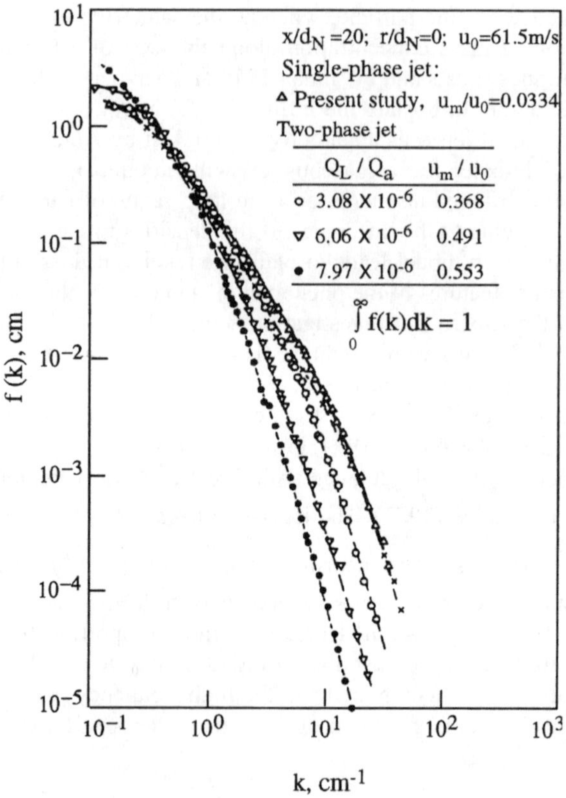

Fig. 9.4 Spectra of the longitudinal velocity fluctuations for single- and two-phase jets. Reprinted from Hetsroni and Sokolov (1971), with permission. Q_L and Q_a are the volumetric flow rate of the droplet and air, respectively, d_N is the nozzle diameter, and u_0 is velocity at the nozzle exit

Tsuji et al. (1988) used a one-dimensional laser-droplet anemometer, (LDA), to measure the motion of 170–1400 μm particles in an air jet. Their data show the effect of coarse particles on the turbulence in the jet. The turbulence intensity (defined as $T_u = \sqrt{\overline{u'^2}}/u_{1.0}$, u_0 is velocity at the nozzle exit), at the axis of the jet loaded by particles of three sizes, is shown in Fig. 9.5. The effect of the 170- μm particles is to decrease the turbulence intensity. There is a very small effect of the larger particles–primarily because the loading was very light—but it seems that they increase the turbulence intensity in at least some parts of the jet, i.e. from $x/d_N \approx 10$, were d_N is the nozzle diameter.

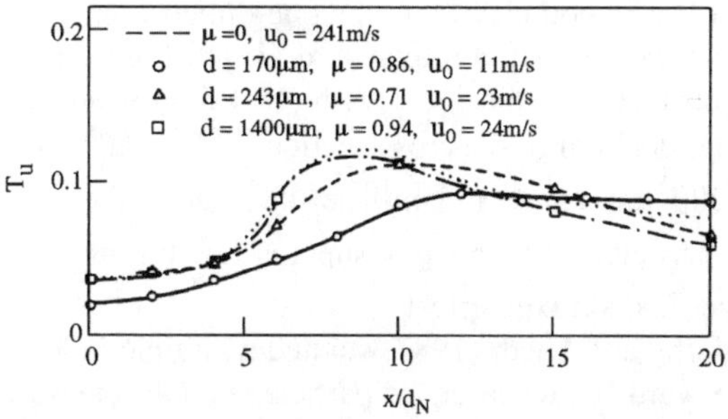

Fig. 9.5 The centerline turbulence in an air jet loaded with particles of three different sizes. Reprinted from Tsuji et al. (1988), with permission. μ is the mass loading ratio, u_0 is velocity at the nozzle exit, and d_N is the nozzle diameter

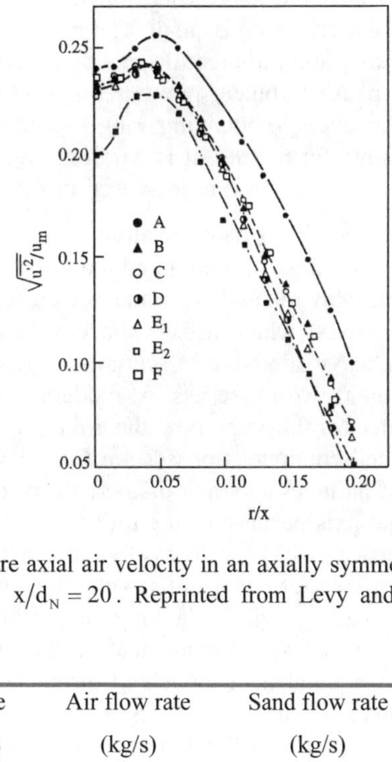

Fig. 9.6 Root mean square axial air velocity in an axially symmetric jet loaded with particles of various sizes at $x/d_N = 20$. Reprinted from Levy and Lockwood (1981), with permission

Code	Particle size range (μm)	Air flow rate (kg/s)	Sand flow rate (kg/s)	Sand-to-air ratio
A	850–1200	$4.29 \cdot 10^{-3}$	$15.0 \cdot 10^{-3}$	3.5
B	600–850	$3.62 \cdot 10^{-3}$	$8.78 \cdot 10^{-3}$	2.42
C	380–700	$3.62 \cdot 10^{-3}$	$8.12 \cdot 10^{-3}$	2.24
D	300–500	$3.62 \cdot 10^{-3}$	$4.42 \cdot 10^{-3}$	1.22
E_1	180–250	$3.62 \cdot 10^{-3}$	$4.13 \cdot 10^{-3}$	1.14
E_2	10–250	$3.62 \cdot 10^{-3}$	$8.45 \cdot 10^{-3}$	2.33
F	clean gas	$4.29 \cdot 10^{-3}$		

Levy and Lockwood (1981) used a one-dimensional LDA to measure the effect of sand particles on a free downward air jet. Figure 9.6 illustrate their data. Here too, the turbulence intensity T_u is plotted for various sizes of sand particles versus the radial position in the cross-section $x/d_N = 20$. Clearly, the larger particles of 850–1200 μm cause a significant increase in the turbulent level, whereas the smaller of particles 180–250 μm suppress the turbulence. The particles of the in-between sizes had a mixed effect.

Parthasarathy and Faeth (1987) studied a particle-laden water jet. They used a two-beam forward-scatter laser for particle velocity measurements. The particle signals were discriminated from the signals of the natural seeded water, based on their amplitude. The results of the measurements are presented in Fig. 9.7. The effect of the turbulence modulation by glass particles loads (particle density and diameter $\rho_p = 2{,}450\,kg/m^3$, $d = 505\,\mu m$, respectively) corresponds to the increased turbulence level near the axis, where turbulence production due to the conventional single-phase mechanism is small. The phenomenon did not appear to influence the overall mixing and turbulent dispersion of the flow, since effects of particles on continuous phase turbulent properties are probably limited to wave numbers higher than the energy-containing range of the turbulence spectrum, which is largely responsible for mixing. It is worth noting that, while the kinetic energy k of the turbulence is somewhat increased in the two-phase jet, as compared to the single-phase one, the cross-correlation $\overline{u'v'}$ is decreased. The latter may point to the fact that, because of inertia and the effect of crossing trajectories the particles fall from one eddy to another, so that the correlation between particle and fluid velocities decreases, which affects the cross-correlation of the fluid. Thus, the existence of a heavy admixture essentially effects the average and fluctuation structure of turbulent two-phase jets. At moderate dispersion of and sufficiently high enough content of the admixture, the effects associated with the interaction of the dispersed and continuous phases are fully evident. For example, an increase in the density of particles and their diameter leads to a decrease of the velocity decay along the jet axis because of the influence of particles moving with large velocities in the carrier gas. This effect is found at different values of the initial phase sliding, in particular at equal initial velocities of the gaseous and dispersed phases. In the last case growth of the long range of the turbulent jet occurs under the conditions of a breakdown of dynamical equilibrium of phases.

At present there are a number of models of turbulence modulation in particle-laden turbulent jets loaded with fine or coarse particles (Abramovich 1970, Al-Tawell and Landau 1977, Hetsroni 1989, Gore and Crowe 1989, Yarin and Hetsroni 1994, Crowe 2000). At the same time analytical and numerical methods of calculation of the aerodynamics of a turbulent two-phase jet have been developed. They allow for a description of the detailed structure of flows (Abramovich et al. 1984, Mostafa et al. 1989, Danon et al. 1977, Frishman et al. 1997).

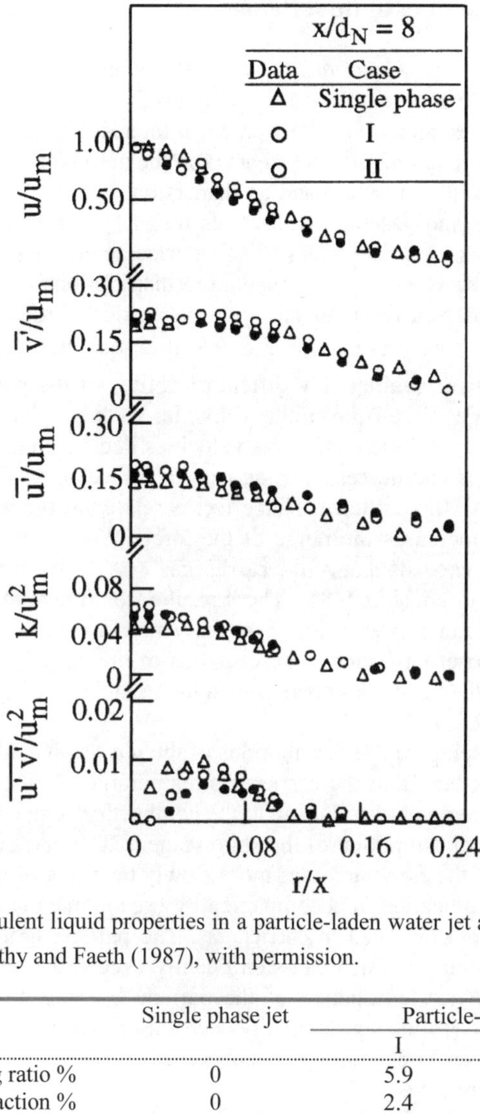

Fig. 9.7 Mean and turbulent liquid properties in a particle-laden water jet at $x/d_N = 8$. Reprinted from Parthasarathy and Faeth (1987), with permission.

Flow	Single phase jet	Particle-laden jet	
		I	II
Mass loading ratio %	0	5.9	11.8
Particle volume fraction %	0	2.4	4.8
Initial average velocity m/s	1.61	1.66	1.72
Jet Reynolds number	8,530	8,795	9,115

u_m is average velocity at jet axis, k is kinetic energy of the turbulence, and $\overline{u}'$, $\overline{v}'$, $\overline{u}'_p$ and $\overline{v}'_p$ are root-mean-square fluctuating velocities

2.9.3. Turbulent coal dust flames

Qualitative characteristics. Coal dust flames are two-phase reactive jets loaded with coal particles. They are widely used in industry, first of all, in various industrial furnaces, and are of great practical interest. The hydrodynamic and thermal structures of coal dust flames depend on the aerodynamic scheme of the flow, the content of solid admixture and its dispersion, the physico-chemical properties of coal particles and gaseous oxidizer, as well as on the velocities of the carrier gas and particles, and the intensity of interfacial heat transfer, etc. Some data characterizing the velocity and temperature distributions along the axis of a coal dust torch at different ratios of the initial velocities of the phases ($n=1$, $n>1$ and $n<1$; $n=u_{2.0}/u_{1.0}$) are presented in Fig. 9.8. It is seen that, depending on the value of the parameter n, qualitatively different profiles of the particle and carrier gas velocities emerge. When the initial velocities of both phases are equal to each other, the particle and the carrier gas velocities decrease monotonically along the torch (Fig. 9.8a). The decrease of the carrier gas velocity along the torch axis is faster than that of the particles. This effect is related to the inertia of the solid admixture which increases the range of the torch. When the initial velocity of the solid admixture exceeds that of the carrier gas ($n>1$), the profile of the gas velocity u_1 has a maximum (Fig. 9.8b). These results from the acceleration of the carrier gas by particles moving with larger speeds. Another pattern is observed at $n<1$. In this case the carrier gas velocity exceeds that of the particles within the near field of the flow but then becomes smaller than the particle velocity because of the inertia effect (Fig. 9.8c).

The axial temperature distributions of the phases in coal dust flame is shown in Fig. 9.9. It is seen that the carrier gas temperature T_1 increases monotonically from the initial temperature T_{10} up to the combustion temperature T_m, which corresponds to full consumption of the solid reagent. The particle temperature within the near field of the flow increases more slowly than that of the carrier gas. In this zone the particle heating is accompanied by escape of volatiles which are mixed with the oxidizer contained in carrier gas. The particle ignition leads to a sharp rise of its temperature which can significantly exceed the temperature of the surrounding gas. After termination of the particle burning, the temperature of the condensed phase (solid slag) gradually decreases to the carrier gas temperature. As far as the oxidizer concentration is concerned, it decreases monotonically along the coal dust flame axis.

In flames loaded with coarse particles the escape of volatiles occurs during all the period of particle motion in the torch volume. Under certain conditions (large volatile content, high initial temperature of the mixture) the volatile concentration can be sufficient for the formation of a combustion front in a homogeneous volatile/oxidizer mixture.

When the processes of particle heating and devolatilization occur in parallel and are completed at the ignition temperature, the structure of the coal dust flame can be schematically presented as follows (Fig. 9.10). There are three

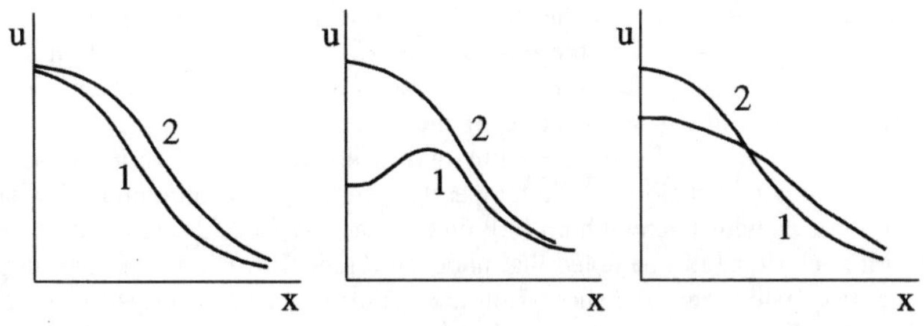

Fig. 9.8 Particle and carrier gas distribution along the axis of a coal dust turbulent flame. Curve 1, carrier gas; curve 2, particle. a) n=1, b) n>1, c) n<1

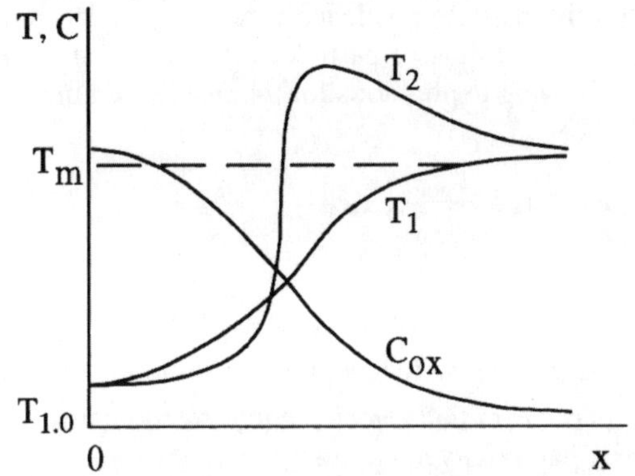

Fig. 9.9 Temperature and oxidizer concentration distributions along the axis of a coal dust turbulent flame

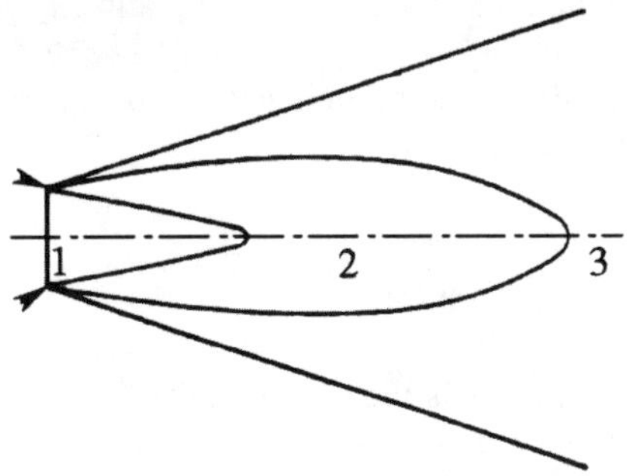

Fig. 9.10 Scheme of a turbulent coal dust torch. 1, heating and evaporation zone; 2, reaction zone; 3, remote zone

characteristic zones in which the particle heating and devolatilization (1), and burning (2) occur, as well as the remote zone (3), where the combustion of the coarse particles takes place. Within the first of them the particle temperature changes from the initial temperature of the reactive mixture up to the ignition temperature T_i. Most of solid fuel (fine and mean fractions) burn in the second zone at a temperature close to the temperature of the diffusion combustion. The largest particles which cannot burn fully in the reaction zone (2) are burnt down in the third zone. It should be noted that under different flow conditions, and different fuel dispersion, etc., the succession and duration of the processes of particle heating, devolatilization and char combustion can vary in a wide range. In particular, during combustion of polydisperse dust, burn-out of char and volatile occurs within the whole flow field of the torch.

Thicknesses of the heating, reaction and dynamic relaxation zones. In order to estimate the thicknesses of the heating and combustion zones of a coal dust flame we determine the particle residence time in the near and far fields of the torch. Assuming that the sliding of particles relative to the carrier gas is negligible, we obtain the following expressions for the residence time

$$\tau_{R.n} = \frac{\ell_{near}}{u_{2.0}}; \quad \tau_{R.f} = \frac{\ell_{far}}{< u_{1.m} >} \tag{9.1}$$

where ℓ_{near} and ℓ_{far} are the characteristic lengths of the near and far fields; $\tau_{R.n}$ and $\tau_{R.f}$ are the particle residence times within the near and far fields of the flame, $u_{2.0}$ is the particle initial velocity equal to the carrier gas issue velocity $u_{1.0}$, $u_{1.m}$ is the carrier gas axial velocity in the flame,

$$< u_{1.m} > = \frac{1}{(x_2 - x_1)} \int_{x_1}^{x_2} u_{1.m}(x)dx$$

is the average velocity within the reaction zone, x_1 and x_2 are the coordinates of the reaction zone boundaries.

The characteristic time of the particle heating is determined by the lumped-capacitance thermal balance equation

$$mc_{p_2} \frac{dT_2}{dt} = hS(T_1 - T_2) \tag{9.2}$$

where m is the particle mass, c_p is the specific heat, h is the heat transfer coefficient, S is the particle surface area, subscripts 1 and 2 correspond to carrier gas and particle, respectively.

The heating time resulting from Eq. (9.2) is

$$\tau_h = \frac{\lambda_{21} d^2}{6\alpha_2 Nu} \tag{9.3}$$

where $\lambda_{21} = \lambda_2/\lambda_1$, λ is the thermal conductivity, α is the thermal diffusivity, $\mathrm{Nu}=(\mathrm{hd})/\lambda_1$ is the Nusselt number, and d is the particle diameter.

The rate of particle burning $u_c\left[\mathrm{kg/m^2s}\right]$ is given by the following formula (Khitrin 1957)

$$u_c = \beta \frac{C_o \rho_1 D_1 \mathrm{Sh}}{d[1 + D_1 \mathrm{Sh}/kd]} \tag{9.4}$$

where C_o is the oxygen concentration, Sh is the Sherwood number, D is the diffusivity, k is the chemical reaction constant, and β is the stoichiometric factor which depends in the general case on the aerodynamic and thermal conditions of the process (within the range 750 °C<T_2<1,500 °C, β=0.375 for coarse particles with d>5 mm and 0.33<β<0.7 for fine particles d<5 mm).

Using the mass balance equation

$$m = u_c S \tau_c \tag{9.5}$$

we find the particle combustion time τ_c. Assuming that we are dealing with the diffusion regime of combustion, $(D_1\mathrm{Sh})/(kd) \ll 1$ and using Eqs. (9.4) and (9.5), we arrive at the following expression for the particle combustion time

$$\tau_c = \frac{\rho_{21} d^2}{6\beta C_o D_1 \mathrm{Sh}} \tag{9.6}$$

where $\rho_{21} = \rho_2/\rho_1$.

Using the expression (9.3) and (9.6) for the characteristic heating τ_h and combustion τ_c times, we obtain the lengths of the heating and reaction zones $\ell_h = \tau_h u_{1.0}$ and $\ell_{react} = \tau_c < u_{1.m} >$

$$\ell_h = \frac{\lambda_{21} d^2 u_{1.0}}{6\alpha_2 \mathrm{Nu}} \tag{9.10}$$

$$\ell_{react} = \frac{\rho_{21} d^2 < u_{1m} >}{6\beta C_o D_1 \mathrm{Sh}} \tag{9.11}$$

Bearing in mind the following expression for $u_{1.m}$ (Abramovich 1963),

$$u_{1.m} = A\frac{u_{1.0}d_N}{x} \tag{9.12}$$

we determine the average velocity within the far flow field

$$< u_{1.m} >= A\frac{u_{1.0}d_N}{x_2 - x_1}\ln\frac{x_2}{x_1} \tag{9.13}$$

where A is a dimensionless constant.

Using Eqs. (9.11) and (9.13) and the equalities $x_2 - x_1 = \ell_{react}$ and $x_2/x_1 = 1 + (\ell_{react}/\ell_h)$, we find the relation between the length of the combustion and heating zones

$$\ell_{react} = \left(\frac{A}{6}\frac{\rho_{21}\,Re_0\;Sc}{\beta C_o hS}\ln(1+\frac{\ell_{react}}{\ell_h})\right)^{\frac{1}{2}} d \tag{9.14}$$

where $Re_0 = (u_{1.0}d_N)/\nu_1$, Sc is the Schmidt number, ν_1 is the kinematic viscosity of the carrier gas.

Introducing dimensional variables $\overline{\ell}_{react} = \ell_{react}/\ell_h$, $\overline{\ell}_h = \ell_h/d_N$, $\overline{d} = d/d_N$, we recast Eqs. (9.10) and (9.14) into the form

$$\overline{\ell}_h = \frac{1}{6}\frac{\lambda_{21}}{\alpha_{21}}\frac{Re_0\,Pr_1}{Nu}\overline{d}^2 \tag{9.15}$$

$$\frac{\overline{\ell}_{react}}{[\ln(1+\overline{\ell}_{react})]^{\frac{1}{2}}} = \left(\frac{A}{6}\frac{\rho_{21}Sc}{\beta C_o\,Re_0\,Pr_1\,hS}\right)^{\frac{1}{2}}\frac{6Nu}{\overline{d}} \tag{9.16}$$

where $\alpha_{21} = \alpha_2/\alpha_1$ and Pr_1 is the Prandtl number of the carrier gas.

Now we estimate the effect of the particle sliding. For this aim we determine the dynamic relaxation length $\overline{\ell}_R$ corresponding to the particle motion in stagnant gas. For a small spherical particle subject to the Stokes force the dependence $\overline{\ell}_R(Re)$ is reads (Boothroyd 1974)

$$\bar{\ell}_R = \frac{\rho_{21}\bar{d}^2}{18}\mathrm{Re}_0 \tag{9.17}$$

where $\ell_R = \tau_R u_{1.0}$ and the dynamic relaxation time $\tau_R = (\rho_{21}d^2)/(18\nu_1)$.

The calculation of the thicknesses of the combustion, heating and dynamic relaxation zones yields, for the characteristic coal dust flame parameters ($d = 10^{-4}$ m, $d_N = 10^{-1}$ m, $u_{1.0} = 20$[m/s], $\lambda_1 = 2.6\cdot10^{-2}$ [W/mK], $\lambda_2 = 0.186$ [W/mK], $c_{p_1} = 10^3$ [J/kgK], $c_{p2} = 1.31\cdot10^3$ [J/kgK], $\nu_1 = 1.5\cdot10^5$ [m^2/s], $\mathrm{Pr}_1 = \mathrm{Sc}_1 = 0.7$), the following estimates

$$\frac{\ell_h}{\ell_R} \sim 1.6\cdot10^3, \quad \frac{\ell_{react}}{\ell_h} \sim 4 \tag{9.18}$$

Thus, the thicknesses of the combustion, heating and dynamic relaxation zones satisfy the inequalities

$$\ell_R \ll \ell_h < \ell_{react} \tag{9.19}$$

The inequalities (9.19) show that under realistic conditions it is possible to neglect the effect of the particle sliding relative to the carrier gas and assume that Nu = 2. The calculations also show that the thickness of the combustion zone is comparable with the boundary layer thickness. The latter makes it impossible to use the infinitely thin flame model for studying the aerodynamics and thermal regimes of coal dust flames.

The one-dimensional models. The engineering calculation of coal dust torches is based on a number of assumptions concerning the structure of the two-phase turbulent flow, the heat and mass transfer in the particle-laden jets, the mechanism of particle ignition and combustion, etc, (Rezn'akov et al. 1968, Kantarovich 1958, Smoot and Smith 1985, Stambuleanu 1976). As a rule, it is assumed that: (i) the volumetric content of solid admixture is small enough and its influence on turbulent flow is insignificant, (ii) the time of particle burning essentially exceeds the time of devolatilization, (iii) the diffusion regime of particle burning holds, etc.

In the frame of the one-dimensional approximation the system of equations describing stationary combustion in a polydisperse coal dust is

$$G_1 c_{p_1}\frac{dT_1}{dx} = -qG_2\frac{dC_2}{dx} - [h_f(T_1 - T_w) + \tilde{\sigma}(T_1^4 - T_w^4)]s_f - Q_\ell \tag{9.20}$$

$$G_2 c_{p_2}\frac{dT_2}{dt} = \left[qu_c - h_p(T_2 - T_1) + \tilde{\sigma}(T_2^4 - T_w^4)\right]s_f \tag{9.21}$$

$$\rho_1 u_1 \frac{dC_1}{dx} = -\sigma W \tag{9.22}$$

$$\rho_2 u_2 \frac{dC_2}{dx} = -W \tag{9.23}$$

$$u_c = \frac{1}{2}\rho_2 \frac{d(d)}{dt} \tag{9.24}$$

$$\rho_1 T_1 = \text{const} \tag{9.25}$$

$$P(d) = \exp(-bd^n) \tag{9.26}$$

where W and q are the specific volumetric rate of chemical reaction and heat release, h_2 and h_f are the particle and flame heat transfer coefficients, G_1 and G_2 are the mass flow rate of the gas and solid reagents, c_{p1} is the specific heat, Q_ℓ are the heat losses related with the incompleteness of particle burning, s_f is the specific flame surface, C is the concentration, $P(d)$ is the particle-size distribution (b and n are constants), σ is the stoichiometric oxidizer-to-fuel mass ratio, $\tilde{\sigma}$ is the Stefan–Boltzman constant, subscripts 1, 2 and w correspond to the gaseous and solid phases and the furnace wall, the gas is perfect and the pressure is assumed to be constant.

The system of Eqs. (9.20) – (9.26) allows for calculation of the temperature and concentration distributions along the torch axis. It also allows for calculation of the completeness of the combustion and some other characteristics. A more consistent approach to the engineering calculation of a coal dust torch (in the near field) is described by Babii and Kuvaev (1986). It is based on using the Patankar–Spalding (1970) semi-empirical method of calculation of a turbulent boundary layer. This model accounts for a number of factors that characterize combustion of a coal dust torch (for the particle-size distribution and radiative heat transfer, as well as for NO_x formation mechanism dating back to Zel'dovich (1947)). It represents a fairly good approximation for the calculation of real phenomena.

2.9.4. Turbulent flames in liquid fuel sprays

Qualitative characteristics. Turbulent torches in liquid fuel sprays consist of turbulent reactive jets loaded with droplets of liquid reagent. They are formed by the injection of liquid jets into an environment containing gaseous oxidizer, moving or of rest (Fig. 9.1a). The mechanism of combustion in such flames consists of primary liquid jet atomization due to strong interaction with the surrounding gas, secondary atomization of large drops, heating and evaporation of droplets, mixing of liquid fuel vapor with the oxidizer contained in the initial mixture or supplied from the environment, chemical reaction in the vapor–oxidizer mixture, as well as mixing of the combustion products with fresh gas–droplet mixture.

Combustion of liquid fuel sprays depends on a number of factors affected by the aerodynamic scheme of the flow, initial content of liquid fuel in combustible mixture, droplet sizes, ambient and fresh mixture temperatures, etc. Depending on the conditions of formation of the combustible mixture rich, lean or stoichiometric gas–droplet mixtures are formed and different regimes of spray combustion occur.

The structure of turbulent flames loaded with fine droplets is similar to the structure of turbulent gas flames. The experiments by Onuma and Ogasawara (1974) showed that heterogeneous mixtures containing many droplets burn within the lowest part of the vertical flame adjoining the nozzle. In the middle and the upper regions of the flame only combustion of vapor–oxidizer mixture occurs. The latter results from the fast evaporation of fine droplets and transformation of the two-phase gas–droplet flow into a single-phase vapor–oxidizer flow within the near-field region of the flame. Accordingly, in both fine sprays and gas flames chemical reaction proceeds in the gaseous mixture formed by turbulent mixing of the gaseous reagents.

Combustion regimes. The limiting conditions corresponding to the gas-phase regime of combustion of a gas–droplet flame are determined by the equality of the characteristic time of droplet vaporization and the characteristic residence time of droplets moving within the constant velocity core of the turbulent two-phase jet.

To estimate the boundary of the domain of the gas-phase regime of combustion in gas–droplet flames we use the known data on the aerodynamics of turbulent two-phase jets and rate of droplet evaporation. The length of the constant velocity core in two-phase jets depends on the initial content of liquid phase in the mixture (Abramovich 1984). It may be presented as follows:

$$\bar{L} = \bar{L}_* \varphi(\chi_0) \tag{9.27}$$

where $\bar{L} = L/d_N$, $\bar{L}$ and $\bar{L}_*$ are the length of the constant velocity core in two-and-single-phase jets, $L_*=ad_N$, with a being dimensionless constant, $\varphi(\chi_0) = 1$, at $\chi_0 = 0$, χ_0 is the initial content of liquid phase.

Assuming that droplets sliding relative to the carrier gas is negligible within the near field of the flow, we obtain the following expression for the droplet resident time

$$\tau_{res} = \frac{L_* \varphi(\chi_0)}{u_{1.0}} \tag{9.28}$$

The droplet evaporation time τ_{ev} is

$$\tau_{ev} = \frac{d_0^2}{k} \tag{9.29}$$

where d_0 is initial droplet diameter, $k = [(8\rho_1 D)/\rho_1]\ln(1+B)$ is the evaporation constant, B is the Spalding transfer number (see Chapter 1.2). Equating the right-hand sides of Eqs. (9.28) and (9.29), we find the limiting initial content of liquid phase corresponding to full droplet evaporation within the constant velocity core of the torch, i.e. to the regime of gas-phase combustion

$$\chi_0 = f_1\left(\frac{\overline{d}_0^2}{a}, \frac{u_{1.0} d_N}{k}\right) \tag{9.30}$$

Similarly it is possible to estimate another limiting value of the initial liquid content χ_o characterizing the state at which the droplet temperature remains less than the boiling temperature within the near field of the flow. Under such conditions, droplet evaporation occurs in a far field of the gas–droplet flame. Bearing in mind that the time needed for droplet heating from the initial temperature T_0 to the boiling temperature T_b is (see Chapter 1.2)

$$\tau_b = \tau_* \ln\frac{T_\infty - T_b}{T_\infty - T_o} \tag{9.31}$$

we find that

$$\chi_0 = f_2(\rho_{21} c_{21} \frac{\overline{d}_o^2}{a} \cdot \frac{u d_N}{\alpha_1} \ln\frac{T_\infty - T_b}{T_\infty - T_o}) \tag{9.32}$$

where τ_* is the thermal relaxation time, $\rho_{21} = \rho_2/\rho_1$, $c_{21} = c_2/c_1$, c is the specific heat, and T_0, T_∞ and T_b are initial, ambient and boiling temperature.

The curves $\chi_0 = f_1(\overline{d}^2)$ and $\chi_0 = f_2(\overline{d}^2)$ subdivide the parametric plane $\chi_0 - \overline{d}_0^2$ into a number of domains that correspond to different conditions of flame combustion (Fig. 9.11). The domains I_a and I_b correspond to the values of the parameters $d_0, u_{1.0}, d_N$, k and a at which the droplet evaporation is fully completed within the near field of the gas–droplet torch. Depending on the initial content of liquid fuel in the fresh mixture, formation of lean or rich vapor–oxidizer mixtures is possible. That means that within the flow field of a turbulent two-phase reactive jet a homogeneous ($\chi_0 < 1/\sigma$), or a diffusion ($\chi_0 > 1/\sigma$) turbulent flame can be formed. The domains IIa and IIb correspond to the regimes in which partial droplet evaporation within the near field of the two-phase flame takes place. Within the domains IIIa and IIIb the processes of liquid evaporation and mixing of liquid vapor with gaseous oxidizer and combustion products occur within the far field of the two-phase torch.

The simplest models. For the description of gas–droplet turbulent flames either a homogeneous model of a turbulent vapor–oxidizer reactive jet (fine droplets) or a heterogeneous model of a two-phase reactive jet loaded with coarse

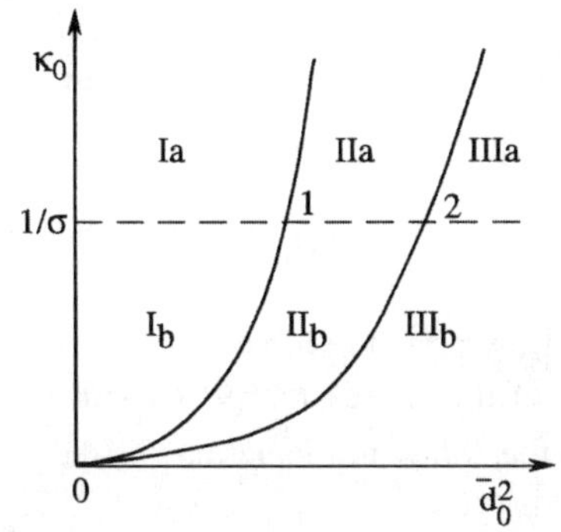

Fig. 9.11 Possible regimes of turbulent combustion of liquid fuel spray. 1, droplets fully evaporate within the near field of the flow; 2, droplets heat up to the boiling temperature within the near field of the flow, the dashed line corresponds to stoichiometric composition of the droplet–oxidizer mixture

liquid fuel droplets should be used. In the first case the general characteristics of the gas–droplet flame can be found by methods similar to those used for calculations of the homogeneous torches (Vulis and Yarin 1978). When the processes of droplet heating and evaporation occur in the torch volume, it is possible to use the integral methods of the theory of turbulent jets. In the case of combustion of rich mixtures in the framework of the integral method, we assume the distributions of dimensionless velocity and fuel concentration to be known functions of the dimensionless variable $\eta = y/y_\delta$ (y_δ is the torch thickness). Then the integral momentum flux and mass flow rate of a two-phase mixture for flow without phase sliding are given by

$$I_x = \rho_1 u_m^2 y_\delta^2 \int_0^1 (1+\chi)\bar{u}_1^2 \eta d\eta \tag{9.33}$$

$$G_x = \rho_1 u_m y_\delta^2 \int_0^1 (1+\chi)\bar{u}_1 \eta d\eta \tag{9.34}$$

where $\bar{u}_1 = u_1/u_m$, u_m is the axial velocity in the torch and $\chi = \rho_2/\rho_1$, ρ_1 and ρ_2 are the effective densities of the gaseous and solid phases, respectively.

From Eqs. (9.33) and (9.34), we obtain

$$u_m = \frac{I_*}{G_x} \frac{\int_0^1 (1+\chi)\bar{u}_1 \eta d\eta}{\int_0^1 (1+\chi)\bar{u}_1^2 \eta d\eta} \tag{9.35}$$

Assuming that combustion of liquid fuel is fully completed in the tip cross-section of the gas–droplet flame $x = \ell_f$, we obtain the mass balance equation in the form

$$G_{f1} = G_{20}(1+\sigma) \tag{9.36}$$

where G_{f1} is the mass flow rate of combustion products in the tip of the torch.

The substitution of the expression (9.36) and $I_x = G_{10}u_{1.0}(1+\sigma\chi_0)$ in relation (9.35) leads to the following equation for the velocity at the tip of a gas–liquid flame.

$$u_{mf} = \frac{u_{1.0}(1+\sigma\chi_o)}{\sigma\chi_o(1+\sigma)} \frac{\int\limits_0^\infty (1+\chi)\bar{u}_1\eta d\eta}{\int\limits_0^\infty (1+\chi)\bar{u}_1^2\eta d\eta} \tag{9.37}$$

Equation (9.37) shows that an increase in the initial content of liquid fuel leads to a decrease in u_{mf} and accordingly to an increase in the length of the gas–droplet flame. The maximal flame length corresponds to the combustion of a homogeneous liquid jet and the minimal velocity at its axis is given by

$$u_{mf} = \frac{u_{1.0}}{\sigma(1+\sigma)} \frac{\int\limits_0^\infty (1+\chi)\bar{u}_1\eta d\eta}{\int\limits_0^\infty (1+\chi)\bar{u}_1^2\eta d\eta} \tag{9.38}$$

The flame length of lean or stoichiometric mixtures loaded with fine droplets can be estimated using the data on speed of the combustion wave propagation in a homogeneous gaseous mixture. In this case the coordinates of the flame front are determined by the differential equation

$$\frac{dy_f}{dx} = \pm\left\{\left(\frac{u_{1.0}}{u_c}\right)^2 - 1\right\}^{\frac{1}{2}} \tag{9.39}$$

where u_c is combustion wave speed, and $y_f(x)$ is coordinate of the flame front.

The integration of Eq. (9.39) under the assumption that u_c is constant yields the following estimate of the flame length:

$$\ell_f \sim \frac{d_N u_{1.0}}{2u_c} \tag{9.40}$$

Bearing in mind that the combustion wave speed in the mixture of fine droplets depends slightly on their diameter ($u_c \approx u_n$, u_n is combustion wave speed corresponding to the vapor–oxidizer mixture) and the combustion wave speed in a mixture loaded with coarse droplets $u_c \sim u_n/d$, we obtain the expressions for the flame length in torches of the fine and coarse droplets

$$\frac{\ell_f}{d_N} \approx \frac{u_{1.0}}{u_n}, \quad \frac{\ell_f}{d_N} \approx \frac{u_{1.0}d}{u_n r} \tag{9.41}$$

where r is the linear size characteristic of the gas–droplet mixture. It can be estimated as $r = N^{-1/3}$, where N is the numerical concentration of droplets.

References

Abramovich GN (1963) The theory of turbulent jet. MTI Press, Cambridge, Mass

Abramovich GN (1970) The effect of an admixture of solid particles or droplets on the structure of turbulent gas jet. Sov. Phys. Dokl. 190: 1052–1055 (in Russian)

Abramovich GN, Girshovich TA, Krasheninnikov SYa, Sekundov A.N, Smirnova IP (1984) Theory of turbulent jets. (in Russian) Nauka, Moscow

Al-Tawell AM, Landau J (1977) Turbulence modulation in two-phase jets. Int. J. Multiphase Flow 3: 341–351

Babii BI, Kuvaev JaF (1986) Combustion of coal dust and coal dust flame calculation. (in Russian) Energoatomizdat, Moscow

Barlow RS, Morrison CQ (1990) Two-phase velocity measurements in dense particle-laden jets. Exp. Fluid 9: 93–104

Boothroyd RG (1974) Following gas–solid suspensions. Charman and Hall, London

Crowe CT (2000) On models for turbulence modulation in fluid-particle flows. Int. J. Multiphase Flow 26: 719–727

Danon H, Wolfschtein M, Hetsroni G (1977) Numerical calculations of two-phase turbulent round jet. Int. J. Multiphase Flow 3: 223–234

Fleckhaus D, Hishida K, Maeda M (1987) Effect of laden solid particles on the turbulent flow structure of a round free jet. Exp. Fluids 5: 323–333

Frishman F, Hussainov M, Kartushinsky A, Mulgi A (1997) Numerical simulation of a two-phase turbulent pipe-jet flow loaded with polydispersed solid admixture. Int. J. Multiphase Flow 23: 765–796

Goldshmidt V, Eskinazi S (1966) Two-phase turbulent flow in a plan jet. J. Appl. Mech. 33: 735–747

Gore RA, Crowe CT (1989) Effect of particle size on modulating turbulent intensity. Int. J. Multiphase Flow 15: 279–285

Hetsroni G (1989). Particles-turbulence interaction Int. J. Multiphase Flow 15: 735–746

Hetsroni G, Sokolov M (1971) Distribution of mass, velocity and intensity of turbulence in two-phase turbulent jet. Trans. ASME, J. Appl. Mech. 38: 315–327

Kantorovich BV (1958) Foundation of the theory of solid fuel combustion and gasification. (in Russian) AN SSSR, Moscow

Khitrin LN (1957) Physics of combustion and explosion. (in Russian) Moscow University, Moscow

Laats MK (1966) Experimental study of the dynamics of an air-dust jet. Inshenerno-Fizicheskii Zh. 10: 11–15

Laats MK, Frishman FM (1970) Assumptions used for the calculation of the two-phase turbulent jet. Fluid Dyn. 5: 333–338

Levy Y, Lockwood FC (1981) Velocity measurements in a particle-laden turbulent free jet. Combust. Flame 40: 333–339

Modarress D, Tan H, Elghobashi S (1984a) Two-component LDA measurement in a two-phase turbulent jet. AIAA J. 22: 624–630

Modarress D, Wuerer J, Elghobashi S (1984b) An experimental study of a turbulent round two-phase jet. Chem. Eng. Commun. 28: 341–354

Mostafa AA, Mongia HC, McDonell VG, Samuelsen GS (1989) Evolution of particle-laden jet flows: a theoretical and experimental study. AIAA J. 27: 167–183

Navoznov SI, Pavel'ev AA, Mulgi AS, Laats MK (1979) Effect of initial slip on admixture dispersion in two-phase jet. In: Laats MK (ed) Turbulent two-phase flows. The III All-Union Conference in Theoretical and Applied Aspects of Turbulent Flows. Part II, pp. 149–157. Tallinn, Moscow (in Russian)

Onuma Y, Ogasawara M (1974) Studies on the structure of a spray combustion flame. In: Chaiken RF (ed) The Fifteenth Symposium (International) on Combustion. The Combustion Institute, Pittsburgh, Pa., pp. 453–463

Parthasarathy RN, Faeth GM (1987) Structure of particle-laden turbulent water jets in still water. Int. J. Multiphase Flow 13: 699–716

Patankar SV, Spalding DB (1970) Heat and mass transfer in boundary layers. 2nd edn., Intertext, London

Popper J, Abuaf N, Hetsroni G (1974) Velocity measurements in two-phase turbulent jet. Int. J. Multiphase Flow 1: 715–726

Prevost F, Boree J, Nuglisch HJ, Charnay G (1996) Measurements of fluid/particle correlated motion in the far field of an axisymmetric jet. Int. J. Multiphase Flow 22: 686–701

Rezn'akov AB, Basina JP, Vdovenko MI, Ustimenko BP (1968) Combustion of fossil fuel. (in Russian) Science, Alma-Ata.

Sheen HJ, Jou BH, Lee YT (1994) Effect of particle size on two-phase turbulent jet. Exp. Thermal Fluid Sci. 8: 315–327

Shuen J-S, Solomon ASP, Zhang Q-F, Faeth GM (1985) Structure of particle-laden jets: measurements and predictions. AIAA J. 23: 396–404

Smoot LD, Smith PJ (1985) Coal Combustion and Gasification. Plenum, New York.

Stambuleanu A (1976). Flame combustion processes in industry. ABACUS Press, Tunbridge Wells, Kent

Sukhov GS, Yarin LP (1986) Calculation of turbulent heterogeneous flames. Combust. Explos. Shock Waves 22: 44–48

Tsuji Y, Morikawa Y, Tanaka K, Karimine K, Nishida S (1988) Measurement of an axisymmetric jet laden with coarse particles. Int. J. Multiphase Flow 14: 565–574

Vulis LA, Yarin LP (1978) Aerodynamic of a torch. (in Russian) Energia, Leningrad

Yarin LP, Hetsroni G (1994) Turbulence intensity in dilute two-phase flows - 3. The particles-turbulence interaction in dilute two-phase flow. Int. J. Multiphase Flow 20: 27–44

Zel'dovich Ya B, Sadovnikov PYa, Frank-Kamenetskii DA (1947) Nitrogen oxidation at combustion. (in Russian) AN SSSR, Moscow

3

High temperature combustion reactor

3.10 Ideally stirred combustion reactor

3.10.1 Preliminary comments

One of the possible ways to accelerate chemical processes is to transfer them into a high temperature regime by using combustion reactors. Such reactors can be divided into two limiting types: ideally stirred reactors and displacement reactors. The first type is characterized by its infinitely high rate of mixing of the fresh reactive mixture with combustion products. At such a high mixing rate uniform temperature and concentration fields in the reactor volume are formed. Accordingly, the completeness of combustion, as well as the process temperature, do not depend on the thermal conductivity and diffusivity of the carrier fluid and are completely determined by the conditions at the reactor inlet and by the intensity of any external heat transfer. In contrast, in the second case (displacement reactors) the rate of mixing of the initial reactants with the final products is negligible. The latter results in the existence of non-uniform temperature and concentration distributions in the reactor volume and in the dependence of process characteristics on the heat and mass transfer between the regions filled with cold reactive mixture and high temperature combustion products.

Ideally stirred reactors are considered below. Such reactors have been an active field of study in recent decades. By the end of the 1950s, the basic approach to the description of possible states of ideally stirred reactors was formulated by Van Heerden (1953), Frank-Kamenetskii (1969), and Zel'dovich et al. (1985). Various aspects of the theory of the stirred combustion reactors have also been examined by Clarke et al. (1965), Evangelista et al. (1968), Bradley et al. (1977), Lignola and Reverchon (1988), Barat (1992), etc.

In most of the theoretical studies, considerable attention has been focused on predicting the stirred combustion reactor regimes by means of the so-called zero-dimensional or ideally stirred model, based on the assumption of ideal mixing of the initial mixture and the combustion products. Vulis (1961) emphasized that such a model is actually a fairly good approximation of the process in combustion chambers with a developed circulation of carrier fluid.

The zero-dimensional model significantly simplifies the theoretical analysis of steady and critical states of combustion reactors. In the framework of such a model, the problem reduces to solving a set of algebraic mass and thermal balance equations containing the terms accounting for heat release and conversion of initial reactants into combustion products as the chemical reaction proceeds. That makes possible a comprehensive analysis of the effects of a variety of factors accounting for physico-chemical properties of reagents and combustion products, e.g. of flow velocity, etc., on the reactor's operating regimes. Aris (1965), Merzhanov and Abramov (1976, 1977) and Perlmutter (1972) obtained a number of important results on steady and unsteady states, parametrical sensitivity and stability of stirred reactors. The concept of ideally stirred reactors developed by Longwell and Weiss (1955) yields detailed data on the chemistry of combustion. In particular, this concept was used by Hottel et al. (1965) to study the kinetics of carbon monoxide and propane combustion; by Williams et al. (1969) to study methane combustion; by Thornton et al. (1987) to investigate pyrolysis and oxidation kinetics of carbon monoxide and n-pentane; as well as by Vaughn et al. (1991) to study combustion of high hydrocarbons in a rich C_2H_4 mixture.

Recently significant attention has also been focused on the study of heterogeneous combustion reactors with separate feed of reacting components in different aggregate states. The developed interfacial area of reacting heterogeneous media determines the unique properties of these reactors (intensive heat and mass transfer, high conversion rate of the initial mixture to finite products, etc.) and their possible applications in modern engineering and technology.

At present, there are a number of models that provide the means for describing some characteristics of a heterogeneous combustion reactor. Genkin et al. (1981) proposed a zero-dimensional model of such a reactor in which the temperature of the carrier fluid (a mixture of gaseous oxidizer with combustion products) containing a multi-temperature admixture of the condensed reagent is uniform in space. Such an approach is based on a very high rate of mixing of gaseous components, as well as on a high thermal conductivity. Subsequently, a similar approach was used by Pushkin et al. (1993) to study steady state ideally stirred gas–droplet reactors. Likhachev et al. (1989, 1991) and Yarin (1990) elaborated the model of ideally stirred bubbly reactors with low volumetric content of gaseous oxidizer. A general approach for exploring states of gas–liquid reactors was considered by Yarin and Hetsroni (1995). The model developed takes into account the principal features of gas–liquid reactive systems, including the multistage character of the process, as well as its dependence on the medium's physico-chemical and structural properties. The instability and self-oscillations during burning of polydisperse fuel in an ideally stirred reactor were examined by Buyevich et al. (1993 a, b).

The dependence of the characteristics of heterogeneous combustion reactors on numerous parameters related to physico-chemical and structural features of the reactive medium, heat losses to reactor walls, etc. makes their theoretical description somewhat difficult. It is worthwhile to use the ideally stirred reactor approximation to study the integral characteristics of the reactor. This approach, however, is accompanied by two difficulties: first, the assumption of ideally stirred components in different aggregate states is unrealistic, and second, due to the random nature of particle motion in the reactor volume, particles with different residence times exist simultaneously. Accordingly, they possess different temperatures and different degrees of combustion completeness. That leads to formation of non uniform temperature (and concentration) field in the carrier fluid at any rate of stirring. Therefore, for ideally stirred heterogeneous combustion reactors one needs first of all to address this point, and justify this assumption.

In accordance with Yarin and Hetsroni (1995) we define an ideally stirred reactor as follows: in the reactor, the properties of the ensemble of particles are homogenous over a scale much larger than that of a single fluctuation (a perturbation introduced by a single particle). The reactor media with such features are globally homogeneous.

3.10.2 Gas–liquid reactor model

Here we deal with the main features of ideally stirred gas–liquid combustion reactors (Fig. 10.1) and their dependence on physico-chemical and structural properties of a reactive two-phase mixture, as well as flow regime parameters. To achieve these goals, the approximate description of an ideally stirred gas–liquid combustion reactor is developed. This approach is based on the thermal combustion theory (Zel'dovich et al. 1985, Williams 1985), as well as on the methods of mechanics of multiphase media (Nigmatulin 1991).

Characteristics of gas–liquid reactive media. Gas–liquid reactive media can be represented as two-phase multi-component, multi-velocity and multi-temperature systems in which one of the phases (the carrier fluid) contains the dispersed particles of the second phase. Examples of such media include gas–droplet and gas–bubble suspensions, reactive foams and some other types of gas–liquid mixtures.

The development of combustion in a gas–liquid mixture depends not only on the total mass content of gaseous and liquid phases and their physico-chemical properties, but also on the dispersed particle sizes which determine the dynamic, thermal and concentration non equilibrium. Here we consider two limiting types of gas–liquid reactive media: gas–droplet medium and bubble medium. These cases are of importance in numerous applications in metallurgy and chemical technology, in particular, production of titanium (Garmata and Gulyanitski 1968), hafnium and zirconium (Barichnikov et al. 1979), acetylene and ethylene (Merzhanov 1973), etc. In spite of a significant difference in the ignition and combustion mechanisms of a single bubble and droplet (Fridman et al. 1981, Yarin and Sukhov 1987, see Chap. 1.3), combustion of gas–droplet and gas–bubble media are similar. In both cases, a chemical reaction occurs between reactants in a gas–vapor mixture which is formed inside bubbles or in the space between droplets. Heating and evaporation of the liquid and the subsequent mixing of the vapor with the gaseous oxidizer are a chain of processes leading to chemical reactions. The ignition and combustion of a reactive mixture is accompanied by heat release.

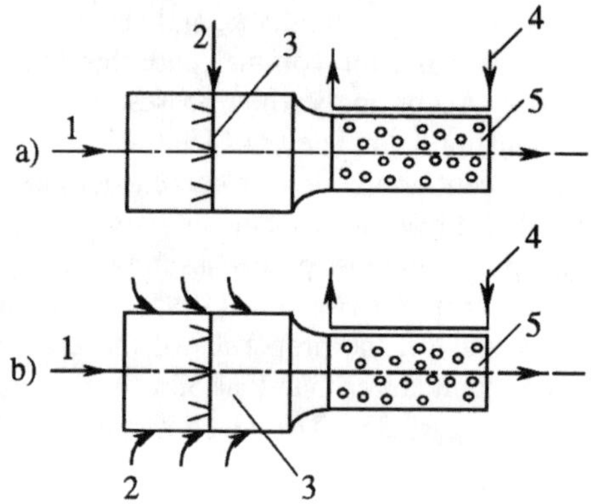

Fig. 10.1 Scheme of gas–liquid combustion reactor. a) Gas–droplet reactor: 1, gaseous oxidizer; 2, liquid reagent; 3, atomizer; 4, cooling system; 5, reactor's channel. b) Bubbly reactor: 1, liquid reagent; 2, gaseous oxidizer; 3, mixer; 4, cooling system; 5, reactor's channel

Heat removal from the carrier fluid to a "cold" mixture ensures heating and ignition of the latter.

Considerable simplification in the description of gas–liquid reactive media may be achieved when the particle's viscous relaxation time is much less than the characteristic hydrodynamic time, i.e.

$$\frac{t_r}{t_h} \ll 1 \tag{10.1}$$

where $t_r = (d^2\bar{\rho})/(18\nu)$ is the viscous relaxation time, d is the particle diameter, $\bar{\rho} = \rho_2/\rho_1$, ρ_1 and ρ_2 are the carrier fluid and particle densities, respectively, ν is the carrier fluid kinematic viscosity, $t_h = V/G_v$ is the characteristic hydrodynamic time, V is the volume under consideration, and G_v is the volumetric flow rate of reactive mixture.

Inequality (10.1) corresponds to the two-phase flow in which the effect of dynamical non equilibrium is negligible. Therefore, we can ignore the relative velocity between the phases and represent the gas–liquid medium as a continuum with "frozen" particles (bubbles or droplets) of the dispersed phase.

At high stirring intensity of a fresh and partially (or fully) reacted mixture, particles with distinct residence time t_r are present in each element of volume of the ideally stirred reactor. Naturally, they have distinct temperatures and completeness of combustion. Due to the finiteness of the thermal conductivity of the carrier fluid, the temperature in the vicinity of different particles is also different. Taking into account the above facts we can represent the gas–liquid medium in the form of gaseous (liquid) cells with temperature $T_{1(i)}$ containing droplets (bubbles) with temperature $T_{2(i)}$ (for the bubbly medium, with bubble temperature $T_{1(i)}$ and cell temperature $T_{2(i)}$), cf. Fig. 10.2. The temperature (concentration) distribution inside a particle and cell is uniform and process of heat (mass) transfer is quasi-stationary when the following conditions are satisfied

$$Bi_i \ll 1 \tag{10.2}$$

$$Fo_i \ll 1 \tag{10.3}$$

where $Bi_i = (h_i d_i)/\lambda_i$ and $Fo_i = d_i^2/(\alpha_i t_h)$ are the Biot and Fourier numbers, λ is the thermal conductivity, h is the heat transfer coefficient, $\alpha = \lambda/(\rho c_p)$ is the thermal diffusivity, c_p is the specific heat, and subscript $i=1, 2$ corresponds to carrier fluid and particles, respectively.

When the thermal conductivity of the carrier fluid is sufficiently large, the reactive gas–liquid medium may be represented as a one-temperature continuum containing particles possessing various temperatures.

Most gas–liquid reactive media are characterized by relatively low volumetric content of dispersed phase (of order of 10^{-4}–10^{-3} for gas–droplet media and 10^{-2}–10^{-1} for bubbly media). The latter determines a number of specific mechanical and thermal properties of those systems: fluidity, thermal conductivity, etc. Below we consider only dilute reactive mixture in which the droplet–droplet (or bubble–bubble) spacing is sufficiently large. In such a case, the effect of the particle interaction on the rate of burning is negligible (Sangiovanni and Kesten 1975, Labowsky 1980, Tsai and Sterling 1990). Taking into account that particle (cell) temperatures and completeness of combustion depend on residence time $t_{r(i)}$, it is possible to consider the substance contained in the i th particles (cells) in the form of an i th continuum. Thus the reactive gas–liquid media may be represented as a system of interpenetrating and interacting continua with different residence times.

Mass and thermal balance equations. The key assumption of the present analysis is that the temperature and concentration distributions in each of the interpenetrating continua in the gas–liquid reactive mixture are homogeneous. The latter permits one to use the zero-dimensional model, i.e. to neglect variation of these parameters in space, averaging them over the whole volume of the reactor. The latter is thought to be a straight cylindrical tube inside which a high temperature exothermal reaction takes place (Fig. 10.1). It is cooled, such that the temperature of the reactor walls is constant. Liquid and gaseous reagents are supplied by pumps to the stirrer at the inlet to the reactor. A homogenous (macroscopically) two-phase mixture is formed in the stirrer. The thermal regime of the reactor is regulated by changing the flow rate of the mixture or varying the heat transfer to the wall.

The equations employed to describe the thermal states of a gas–liquid ideally stirred reactor are obtained from the conservation laws for chemically reacting gas–liquid media. The following simplifying assumptions are made: (i) the thermal conductivity and specific heat of the gaseous and liquid phases are independent of the temperature and composition of the reactive mixture, (ii) reaction heat release is constant, (iii) the volumetric contents of the droplets or bubbles are negligibly small, (iv) the products of the chemical reaction are gaseous, (v) the reaction $a + b_1 \rightarrow c$ (a and b_1 are gaseous reagents, c corresponds to the combustion product) develops according to the following macro-kinetic law

$$W = z\rho_a\rho_{b1}\exp\left(-\frac{E}{RT_1}\right) \tag{10.4}$$

where z is the pre-exponential factor, E is the activation energy, R is the universal gas constant, subscripts a, b_1 correspond to oxidizer and vapor of reactive liquid, and 1 designates the gaseous phase.

We also assume the pressure in the reactor to be constant and the gaseous component of the mixture to be an ideal gas.

Assume that in the gas–liquid mixture there are $n_1, n_2, \ldots n_k$ particles (cells) with residence times equal to $t_{r1}, t_{r2}, \ldots t_{rk}$ respectively. Using the model of interpenetrating continua, we write the expression for an effective density of the ℓ-th component in the i-th cell

$$\rho_{\ell(i)} = \rho^{o}_{\ell}\xi_{\ell(i)} \tag{10.5}$$

where ρ^{o}_{ℓ} is the physical density of the ℓ-th component, $\xi_{\ell(i)}$ is the volumetric content of the ℓ-th component in the i-th cell. The effective densities of the gaseous and liquid phases inside the i-th cell are

$$\rho_{1(i)} = \sum_{\ell_1} \rho_{\ell_{1(i)}}; \quad \rho_{2(i)} = \sum_{\ell_2} \rho_{\ell_2(i)} \tag{10.6}$$

Here we summarize over the gaseous and liquid components inside the i-th cell, subscripts 1 and 2 correspond to gaseous and liquid phases, respectively.

Assume also that the ℓ components are contained in cells from the i-th continuum. Then the reactive gas–liquid media in an ideally-stirred reactor may be represented as a system of interpenetrating and interacting gaseous or liquid continua (for gas–droplet and bubbly media) and continua formed by dispersed material inside bubbles and drops. Since a substance temperature in cells with residence time $t_{r(i)}$ equals $T_{(i)}$, then $T_{(i)}=T^{(i)}$ ($T_{(i)}$ and $T^{(i)}$ are the temperatures of the i-th cell and the i-th continuum, respectively). The effective density of the ℓ-th component in the i-th continua equals

$$\rho^{(i)}_{\ell} = n_i\rho_{\ell(i)} \tag{10.7}$$

where n_i is a number of the i-th cells in the i-th continua.

Consider the reactive gas–liquid medium containing ℓ gaseous and liquid components. The medium represents a mixture of liquid fuel, its vapor, gaseous oxidizer, an inert admixture and combustion products. Using the above-mentioned model we can write the mass conservation equation for the ℓ-th component of the $i_{1.2}$-th continuum. In the zero-dimensional approximation this equation reads

$$\frac{d\rho_\ell^{(i)}}{dt} = W_\ell^{(i)} + \Delta G_\ell^{(i)} + \varepsilon_I \tilde{G}_\ell^{(i)} \tag{10.8}$$

where subscript $\ell = a,\ b_{1,2},\ c$ and d corresponds to the gaseous oxidizer (a), vapor of liquid (b_1), liquid (b_2), combustion product (c), and gaseous inert admixture (d); ε_I is the coefficient accounting for the existence of mass transfer between the i-th and the j-th continua, Δ is the difference between values of the given parameter in initial and final states.

The first term on the right hand side in Eq. (10.8) corresponds to the conversion rate of material (the rate of chemical reaction and rate of evaporation). It may be written as

$$\begin{aligned}
W_a^{(i)} &= -\sigma W^{(i)} \\
W_{b.1}^{(i)} &= -W^{(i)} + W_{e.1}^{(i)} \\
W_{b.2}^{(i)} &= W_{e.2}^{(i)} \\
W_c^{(i)} &= (1+\sigma) W^{(i)} \\
W_d^{(i)} &= 0
\end{aligned} \tag{10.9}$$

where W and $W_{e.2}$ are the volumetric chemical reaction and evaporation rates, respectively, σ is the stoichoimetric oxidizer-to-fuel mass ratio. The terms $W_{e.1}$ and $W_{e.2}$ satisfy the expression

$$W_{e.1} + W_{e.2} = 0 \tag{10.10}$$

The second term on the right-hand side in Eq. (10.8) accounts for convective mass transfer. It is expressed as

$$\Delta G_\ell^{(i)} = G_{\ell.0}^{(i)} - G_\ell^{(i)} \tag{10.11}$$

where $G_{\ell.0}^{(i)} = \rho_{\ell.0}^{(i)} j_0$, $G_\ell^{(i)} = \rho_\ell^{(i)} j$; $j_0 = G_{v.0}/V$ and $j = G_v/V$ are specific volumetric flow rates of a substance that is fed and extracted from the reactor; V is its volume; G_v is the volumetric flow rate of the reactive mixture.

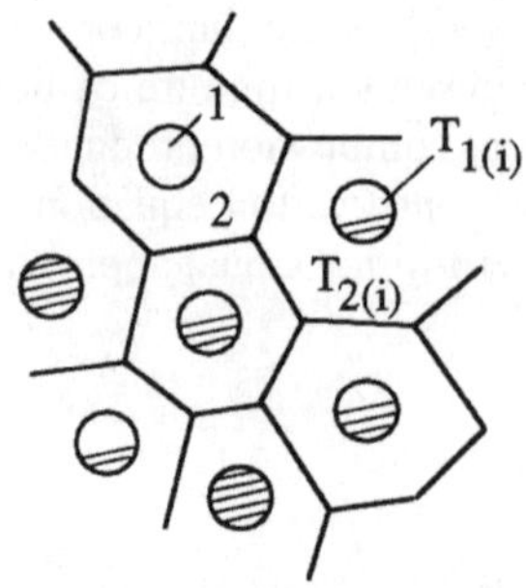

Fig. 10.2 Model of gas–liquid medium. (The degree of cell shading corresponds to the degree of completeness of combustion.) Reprinted from Yarin and Hetsroni (1995), with permission. 1, Particle (bubble or droplet); 2,- carrier liquid cell

The last term on the right-hand side of Eq. (10.8) corresponds to the mass transfer that is due to a distinction of the gaseous components concentrations in the i-th and j-th continua

$$\tilde{G}_{\ell}^{(i)} = \sum_{j=1}^{K} g_{ij}^{(\ell)} \tag{10.12}$$

where $g_{ij}^{(\ell)} = h_{ij}s_{ij}(\rho_{\ell}^{(i)} - \rho_{\ell}^{(j)})$ is the specific mass flux from the i-th to the j-th gaseous continuum, h_{ij} is the mass transfer coefficient, $s_{ij} = S_{ij}/V$, S_{ij} is the surface which separates the i-th and the j-th continua, K is a number of j continua of the ℓ-th component of the gas–iquid mixture.

The coefficient ε_I accounts for the characteristics of the mass transfer between components of the reactive mixture containing in the i-th and the j-th continua. As can be seen from Fig. 10.2 the particles of dispersed phase (bubbles in gas–bubble suspensions, droplets in dilute gas–droplet mixtures) are isolated and do not interact with each other. In accordance with that mass transfer between the dispersed matter in the i-th and the j-th cells, as well as in the i-th and the j-th continua is absent. Bearing this in mind, it can be assumed that $\varepsilon_I = 0$. Since $\sum_{\ell=a,b_1,c,d} \rho_{\ell}^{(i)} = \rho_1^{(i)}$ and $\sum_{\ell=b_2} \rho_{\ell}^{(i)} = \rho_2^{(i)}$, we obtain from Eq. (10.8) the following mass balance equation for gaseous (1) and liquid (2) phases in gas–bubble and gas–droplet media, respectively

$$\frac{d\rho_{1.2}^{(i)}}{dt} = \Delta G_{1.2}^{(i)} + We_{1.2}^{(i)} \tag{10.13}$$

In contrast with that, the continua containing the matter of carrier fluid (liquid in gas–bubble suspensions, gas in dilute gas–droplet mixtures) are contiguous and interact with each other. Under conditions corresponding to intensive stirring of carrier fluid such interaction is very strong. It leads to smoothing of any heterogeneities in the carrier fluid. Thus, the reactive medium in an ideally stirred combustion reactor may be represented as one-temperature (concentration) continuum of carrier fluid containing a number of continua of dispersed phase with different residence times, completeness of combustion, temperatures and concentrations. Note, that in some cases (for example, gas–droplet reactor with initial droplet temperature close to the boiling point) the reactive medium in an ideally stirred reactor may be modeled as a two-phase, two-temperature system.

To determine the temperature of the gaseous and liquid phases we use the thermal balance equation. Neglecting the radiative heat transfer (assume that the optical thickness of the reactive medium is rather small), as well as heat transfer due to mass transfer between the liquid and gaseous continua we obtain

$$qW^{(i)} = \rho_1^{(i)} \frac{di_1^{(i)}}{dt} + \Delta jJ_1^{(i)} + Q_{1-2}^{(i)} + \varepsilon_{II}(Q_{w.1}^{(i)} + Q_{ij.1}^{(i)}) \tag{10.14}$$

$$Q_{1-2}^{(i)} = \rho_2^{(i)} \frac{di_2^{(i)}}{dt} + \Delta jJ_2^{(i)} + q_e W_{e.2}^{(i)} + \varepsilon_{III}(Q_{w.2}^{(i)} + Q_{ij.2}^{(i)}) \tag{10.15}$$

where q is the heat release, q_e is the latent heat of evaporation, $J = \rho i$, i is the enthalpy ($i = c_p T$, c_p is the specific heat), $\Delta jJ^{(i)} = jJ_1^{(i)} - j_o J_{1.0}^{(i)}$; $\Delta jJ_2^{(i)} = jJ_2^{(i)} - j_o J_{2.0}^{(i)}$.

The terms $Q_{1-2}^{(i)}$, $Q_{w.1}^{(i)}$, $Q_{w.2}^{(i)}$ and $Q_{ij.1}^{(i)}$, $Q_{ij.2}^{(i)}$ account for specific interfacial heat flux from the i-th gaseous to the i-th liquid continuum, heat flux from gaseous and liquid phases to the reactor's wall and heat flux from the i-th to the j-th continua, respectively. They are expressed as

$$Q_{1-2}^{(i)} = h_{1-2} s_{1-2}^{(i)} (T_1^{(i)} - T_2^{(i)}) \tag{10.16}$$

$$Q_{w.1}^{(i)} = h_{w.1} s_w^{(i)} (T_1^{(i)} - T_w) \tag{10.17}$$

$$Q_{w.2}^{(i)} = h_{w.2} s_w^{(i)} (T_2^{(i)} - T_w) \tag{10.18}$$

$$Q_{ij.1}^{(i)} = \sum_{j=1}^{K} q_{ij.1} \tag{10.19}$$

$$Q_{ij.2}^{(i)} = \sum_{j=1}^{K} q_{ij.2} \tag{10.20}$$

where $q_{ij.1} = h_{ij.1}s^{(i)}_{ij.1}(T^{(i)}_1 - T^{(j)}_1)$, $q_{ij.2} = h_{ij.2}s^{(i)}_{ij.2}(T^{(i)}_2 - T^{(j)}_2)$, $h_{1\text{-}2}$, $h_{w.1}$, $h_{w.2}$, $h_{ij.1}$, $h_{ij.2}$ are heat transfer coefficients, $s^{(i)}_{1-2} = S^{(i)}_{1-2}/V$, $s^{(i)}_w = S^{(i)}/V$, $S^{(i)}_{1-2}$ and S_w are the surface area which separates the gaseous and liquid continua and the surface area of the reactor walls, respectively, T_w is the wall temperature.

For the bubbly and gas–droplet media coefficients $\varepsilon_{II} = 0;\ 1$ and $\varepsilon_{III} = 1;\ 0$, respectively $(\varepsilon_{II} + \varepsilon_{III} = 1)$.

Equations for the average parameters. To describe the states of the ideally stirred gas–liquid combustion reactor we use the thermal and mass balance equations for the ensemble averaged characteristics. The average value is defined as

$$<\Pi>= N^{-1}\sum_{i=1}^{K} \Pi^{(i)} \tag{10.21}$$

where $N = \sum_{i=1}^{K} n_i$ is the total number of particles in the reactor, $< \ >$ is an operator indicating the ensemble averaging.

Equations for the ensemble averaged density and temperature are derived by multiplying Eqs. (10.8), (10.13)–(10.15) by N^{-1} and subsequent summation of the resulting equations from i=1 to i=K. Since the expressions

$$g_{ii} = 0,\quad q_{ii} = 0,\quad g_{ij} = -g_{ji},\quad q_{ij} = -q_{ji}, \tag{10.22}$$
$$\sum_{i=1}^{K}\left(\sum_{j=1}^{K} g_{ij}\right) = 0\ ;\quad \sum_{i=1}^{K}\left(\sum_{j=1}^{K} q_{ij}\right) = 0$$

we obtain the following mass and thermal balance equations for the components and phases of gas–liquid reactive mixture

$$\frac{d<\rho_\ell>}{dt} =< W_\ell > +\Delta < G_\ell > \tag{10.23}$$

$$\frac{d<\rho_{1.2}>}{dt} = \Delta < G_{1.2} > + < W_{e1.2} > \tag{10.24}$$

$$q < W >=< \rho^{(i)}_1 \frac{di^{(i)}_1}{dt} > +\Delta < jJ_1 > + < Q_{1-2} > +\varepsilon_{II} < Q_{w.1} > \tag{10.25}$$

$$< Q_{1-2} >=< \rho^{(i)}_2 \frac{di^{(i)}_2}{dt} > +\Delta < jJ_2 > +q_e < W_{e.2} > +\varepsilon_{III} < Q_{w.2} > \tag{10.26}$$

Summation of Eqs. (10.25) and (10.26) yields the thermal balance equation for the gas–liquid reactive mixture

$$q < W > = < \rho_1^{(i)} \frac{di_1^{(i)}}{dt} + \rho_2^{(i)} \frac{di_2^{(i)}}{dt} > + \Delta < jI > + < Q_w > + q_e < W_{e.2} > \tag{10.27}$$

where $I = J_1 + J_2$, Q_w is equal to $Q_{w.1}$ and $Q_{w.2}$ for gas–droplet and bubbly media, respectively.

Assuming the time derivatives in Eqs. (10.23)–(10.27) to be zero, we obtain the system of algebraic equations describing stationary states of a gas–liquid combustion reactor. This system is not closed, since the dependence of the average chemical reaction rate and other average characteristics on single particle parameters is yet unknown. Therefore, the system of Eqs. (10.23)–(10.27) should be supplemented by the additional equations describing a single particle behavior.

We supplement the system of Eqs. (10.23)–(10.27) by the particle and cell mass balance equation

$$\frac{dm_{1.2}}{dt} = w_{e_{1.2}} \tag{10.28}$$

By the particle and cell thermal balance equation

$$m_{1.2} \frac{di_{1.2}}{dt} = \varepsilon_{IV} q w + \varepsilon_V q_e w_{e_{1.2}} + \varepsilon_{VI} q_{1-2} \tag{10.29}$$

by the equation of state of the gas

$$P_1 = \rho_1 R_g T_1 \tag{10.30}$$

and by the equation governing the evolution of the distribution of the residence times of the ensemble of particles

$$\frac{d\varphi}{dt} = f(t, \quad t_r) \tag{10.31}$$

Here $m_{1.2} = \rho_{1.2}^0 \left[d_{1.2}^3 \cdot (\pi/6) \right]$ is the mass of particles or material inside a cell; $q_{1-2} = h^* S(T_1 - T_2)$ is the interfacial heat flux; h^* and S are heat transfer coefficient and particle surface area, respectively (for very fine particles $h^* = 2\lambda/d$);

R_g is the gas constant; φ is the distribution function; the coefficients $\varepsilon_{IV} = 0$, $\varepsilon_V = -1$, $\varepsilon_{VI} = 1$ for the liquid phase, and $\varepsilon_{IV} = 1$, $\varepsilon_V = 0$, $\varepsilon_{VI} = -1$ for the gaseous one, w and w_e are the chemical reaction and evaporation rates, respectively.

Note, that Eq. (10.30) is an appropriate one for a perfect vapor–gas mixture and for gas–droplet media when the total volumetric content of the liquid phase is small.

Dimensionless variables. To rearrange the above equations to be dimensionless, we take as the characteristic scales of time $t_m = (z\rho_0)^{-1}\exp\left(E/(RT_{1.0})\right)$, density ρ_0, length d_0, specific volumetric flow rate j_0 and temperature $T^* = RT_{1.0}^2/E$. Rendering all the variables dimensionless as per

$$C_\ell = \frac{\rho_\ell}{\rho_0},\quad j^* = \frac{j}{j_0},\quad \tau = \frac{t}{t_m},\quad \bar{t}_r = t_r j_0,\quad \bar{t} = t j_0,\quad \bar{W} = \frac{W t_m}{\rho_0},\quad \bar{w} = \frac{w t_m}{\rho_0 v_0},$$

$$\bar{W}_e = \frac{W_e t_m}{\rho_0},\quad \bar{w}_e = \frac{w_e t_m}{\rho_0 v_0},\quad \theta_1 = \frac{E(T_1 - T_{1.0})}{RT_{1.0}^2},\quad \theta_2 = \frac{E(T_2 - T_{2.0})}{RT_{1.0}^2}$$

we derive the dimensionless equations as follows

$$\frac{dC_\ell^{(i)}}{d\tau} = \bar{W}_\ell^{(i)} + D_a^{-1}\left(C_{\ell.0}^{(i)} - C_\ell^{(i)} j^*\right) + \frac{\Gamma}{\gamma}\sum_{j=1}^{K} S_{e_{ij}}^{(i)^{-1}}(C_\ell^{(i)} - C_\ell^{(j)}) \tag{10.32}$$

$$\frac{dC_{1.2}^{(i)}}{d\tau} = D_a^{-1}\left(C_{1.2(0)}^{(i)} - C_{1.2}^{(i)} j^*\right) + \bar{W}_{\ell_{1.2}} \tag{10.33}$$

$$\begin{gathered}\frac{\beta}{\gamma}\bar{W}^{(i)} = \beta C_1^{(i)}\frac{d\theta^{(i)}}{d\tau} + D_a^{-1}\left[(1+\beta\theta_1^{(i)})j^* C_1^{(i)} - C_{1.0}^{(i)}\right] + \\ \frac{\beta}{\gamma}\left\{Se_{1-2}^{(i)}\left(\theta_1^{(i)} - \theta_2^{(i)}\right) + \varepsilon_{II}\left[Se_w^{(i)^{-1}}(\theta_1^{(i)} - \theta_w) + \sum_{j=1}^{K} S_{e_{ij}}^{(i)^{-1}}(\theta_1^{(i)} - \theta_1^{(j)})\right]\right\}\end{gathered} \tag{10.34}$$

$$\begin{gathered}Se_{1-2}^{(i)^{-1}}\frac{\beta}{\gamma}\left(\theta_1^{(i)} - \theta_2^{(i)}\right) = \beta c_{p2.1}\frac{d\theta_2^{(i)}}{d\tau} + \\ +Da^{-1}c_{p2.1}\left[\left(1+\beta\theta_2^{(i)}\right) j^* C_{2.0}^{(i)} - C_2^{(i)}\right] + \frac{\beta}{\gamma}\{\bar{q}_e \cdot \bar{W}_{e.2}^{(i)} + \\ \varepsilon_{III}[S_{e_w}^{(i)^{-1}}\left(\theta_2^{(i)} - \theta_w\right) + \sum_{j=1}^{K} Se_{ij}^{(i)^{-1}}\left(\theta_2^{(i)} - \theta_2^{(j)}\right)]\}\end{gathered} \tag{10.35}$$

Here

$$\mathrm{Da}=\frac{z\rho_0}{j_0}\exp\left(-\frac{E}{RT_{1.0}}\right)$$

is the Damkohler number;

$$\mathrm{Se}=\frac{qz\rho_0^2}{hS}\frac{E}{RT_{1.0}^2}\exp\left(-\frac{E}{RT_{1.0}}\right)$$

is the Semenov number, $\Gamma=(h_{ij}\rho_0 c_{p1.0})/h_{ij}^*$, ($\Gamma=1$ for Le=1, Le is the Lewis number), $\bar{q}_e=q_e/q$, $c_{p_{1.2}}=c_{p1}/c_{p2}$, $\beta=RT_{1.0}/E$, $\gamma=(c_{p_{1.0}}R\,T_{1.0}^2)/qE$; $S_{ew}^{(i)}$, $S_{e1-2}^{(i)}$, and $Se_{ij}^{(i)}$ are the Semenov number defined for the values $h_w^* s_w^{(i)}$, $h_{1-2}^* s_{1-2}^{(i)}$ and $h_{ij}^* s_{ij}^{(i)}$, respectively, v_0 is the initial particle volume.

The equations for the average characteristics in the dimensionless form are as follows

$$\frac{d<C_\ell>}{d\tau}=<\bar{W}_\ell>+\mathrm{Da}^{-1}\left(<C_{\ell,0}>-<C_\ell>j^*\right) \tag{10.36}$$

$$\frac{d<C_{1.2}>}{d\tau}=\mathrm{Da}^{-1}\left(<C_{1.2(0)}>-<C_{1,2}>j^*\right)+<\bar{W}_{e_{1.2}}> \tag{10.37}$$

$$\begin{aligned}\frac{\beta}{\gamma}<\bar{W}>=\beta<C_1^{(i)}\frac{d\theta_1^{(i)}}{d\tau}>+D_a^{-1}\left[\left(<C_{1.0}^{(i)}>+\beta<C_1^{(i)}\theta_1^{(i)}>\right)j^*-<C_{1.0}^{(i)}>\right]\\+\frac{\beta}{\gamma}\left\{<Se_{1-2}^{(i)^{-1}}(\theta_1^{(i)}-\theta_2^{(i)})>+\varepsilon_{II}<Se_w^{(i)^{-1}}(\theta_1^{(i)}-\theta_w)>\right\}\end{aligned} \tag{10.38}$$

$$\begin{aligned}\frac{\beta}{\gamma}\left(<Se_{1-2}^{(i)-1}(\theta_1^{(i)}-\theta_2^{(i)})>\right)=\beta c_{p_{2.1}}<C_2^{(i)}\frac{d\theta_2^{(i)}}{d\tau}>+\\ \mathrm{Da}^{-1}c_{p_{2.1}}\left[(<C_2>+\beta<C_2^{(i)}\theta_2^{(i)}>)j^*-<C_{2.0}^{(i)}>\right]+\frac{\beta}{\gamma}\{\bar{q}_e<\bar{W}_{e,2}>+\\ \varepsilon_{III}<Se_w^{-1}\left(\theta_2^{(i)}-\theta_w\right)>\}\end{aligned} \tag{10.39}$$

The combustion reactor equation (10.27) has the following form

$$\begin{aligned}\frac{\beta}{\gamma}<\bar{W}>=\beta\left[<C_1^{(i)}\frac{d\theta_1^{(i)}}{d\tau}>+c_{p_{2.1}}<C_2^{(i)}\frac{d\theta_2^{(i)}}{d\tau}>\right]+\\ \mathrm{Da}^{-1}\left\{<C_1^{(i)}j^*(1+\beta_1\theta_1^{(i)})>+c_{p_{2.1}}<C_2^{(i)}j^*(1+\beta\theta_2^{(i)})>-(C_{1.0}+C_{2.0})\right\}+\\ +\frac{\beta}{\gamma}\left\{\bar{q}_e<\bar{W}_{e.2}>+\varepsilon_{II}<Se_w^{(i)^{-1}}(\theta_1^{(i)}-\theta_w)>+\varepsilon_{III}<Se_w^{(i)^{-1}}(\theta_2^{(i)}-\theta_w)>\right\}\end{aligned} \tag{10.40}$$

The additional Eqs. (10.28)–(10.31) in the dimensionless form are as follows

$$\frac{dC_{1.2}}{d\tau} = \overline{w}_{e1.2} \tag{10.41}$$

$$C_{1.2}\frac{d\theta_{1.2}}{d\tau} = \frac{1}{\gamma}\left(\varepsilon_{IV}\overline{w} + \varepsilon_V \overline{q}_e \overline{w}_{e.2} + \varepsilon_{VI}\overline{q}_{1-2}\right) \tag{10.42}$$

$$\overline{P} = C_1\Gamma_2(1+\beta\theta_1) \tag{10.43}$$

$$\frac{d\overline{\varphi}}{d\tau} = \overline{f}(\overline{t}, \overline{t}_r) \tag{10.44}$$

where $\overline{P} = P/(\rho_0 \cdot j_0^2 S_w)$, $\overline{\varphi} = \varphi/j_0$, $\overline{f} = (f/j_0^2)Da^{-1}$, $\Gamma_2 = (R_g T_{1.0})/(j_0^2 S_w)$, $\overline{q}_{1-2} = q_{1-2}/q$.

3.10.3 Gas–droplet reactor regimes

A. *Stationary states.* Consider the thermal regimes of an ideally stirred gas–droplet combustion reactor. We use the following simplifying assumptions: (i) the initial temperatures of the gaseous and liquid phases are equal, (ii) the initial temperature of droplets is close to the boiling temperature of the reactive liquid, (iii) stirring of the two-phase mixture takes place at a very high rate. Under these assumptions the gas–droplet mixture is a two-temperature medium in which $<C_1>= C_1^{(i)} = C_1$, $<W>= W^{(i)} = W$, etc. According to the latter relations we arrive at the equations describing the stationary states of the gas–droplet reactor. Equating the time derivatives in Eqs. (10.36), (10.37) and (10.40) to zero and assuming $j^*=1$ we obtain in the steady state:

$$\overline{W}_\ell + Da^{-1}(C_{\ell.0} - C_\ell) = 0 \tag{10.45}$$

$$\overline{W}e_{1.2} + Da^{-1}(C_{1.2(0)} - C_{1.2}) = 0 \tag{10.46}$$

$$\begin{aligned}\frac{\beta}{\gamma}\overline{W} &= Da^{-1}\left\{C_1(1+\beta\theta_1) + C_{P_{2.1}} < C_2(1+\beta\theta_2) > -(C_{1.0} + C_{2.0})\right\} + \\ &\frac{\beta}{\gamma}\left\{\overline{q}_e \overline{W}_{e.2} + Se_w^{-1}(\theta_1 - \theta_w)\right\}\end{aligned} \tag{10.47}$$

The dimensionless rate of chemical reaction is given by the expression

$$\overline{W} = C_a C_b \exp\left(\frac{\theta_1}{1+\beta\theta_1}\right) \tag{10.48}$$

Since the evaporation in the reactor occurs in boiling liquid, the relation between the liquid temperature θ_2 and the pressure $\overline{P}$ in the reactor corresponds to the liquid–vapor saturation line $\theta_2 = \theta_{2.0} = \theta_2(\overline{P})$, that is determined by the Clapeyron–Clausious equation.

Under the assumption of quasi-stationary character of the evaporation process, the term $\overline{W}_{e.2}$ in Eqs. (10.46) and (10.47) has the form

$$\overline{W}_{e.2} = \frac{3}{2} C_{2.0} \mathrm{Da}^{-1} \int_0^{\overline{t}_*} (\theta_1 - \theta_2) \frac{\overline{r}}{M} \varphi(\overline{t}_r) d\overline{t}_r \tag{10.49}$$

where t_* is the maximum residence time, $M = (3/2)[(q_e E j_0 m_0 N)/(RT_{1.0}^2 h_0^* s_0)]$ is the parameter characterizing the heterogeneity of gas–droplet mixture, m_0, h_0^* and s_o are the particle initial mass, heat transfer coefficient and specific surface area, respectively. $\overline{r} = r/r_0$, r_0 is the initial particle radius, N is the number of particles.

B. *Adiabatic reactor.* First we consider thermal regimes of an ideally-stirred reactor without heat losses to the reactor walls. Of most interest is to predict the effect of structural characteristics of the gas–droplet mixture on the combustion temperature, concentrations of the reactive components etc. Some results of the calculation of the dependence of the combustion temperature θ_1 (which is the temperature of the gaseous phase) on the Damkohler number for various values of the parameter characterizing heterogeneity, M, are plotted in Fig. 10.3. The calculations were done for the following values of parameters: σ=3.5, $C_{2.0}$=0.05, $C_{1.0}$=0.95, c_{p1}=1, $\beta = 0.02$, $\gamma = 0.0005$, $\overline{q} = 0.001$, $\overline{P} = 0.5 \cdot 10^5$, corresponding to combustion of an oxygen–hexane mixture; the distribution function is taken as $\overline{\varphi} = \overline{t}_1^{-1} \exp\left(-\overline{t}_r/\overline{t}_1\right)$ where $\overline{t}_1 = t_1 j_o$; t_1 is a characteristic residence time. The latter expression corresponds to the assumption by Genkin et al. (1981) that in the stationary states the number of particles leaving the reactor is proportional to the number of particles present in it. In the graph the critical states of the reactor corresponding to ignition (I) and extinction (E) of the gas–droplet mixture are also marked. In the inset (in the top right-hand corner of the graph) the schematic curves $\overline{\theta}_1(\mathrm{Da})$ ($\overline{\theta}_1 = \theta_1/\theta_*$, θ_* is the maximal combustion temperature at M=0) and the curves which pass through the points corresponding to the critical states are shown. These curves divide the parametric plane $\theta_1 - \mathrm{Da}$ into the domains where the high (III), intermediate (II) and low temperature (I) states occur. It is seen that at some value of the parameter $M{=}M_{IE}$ (and accordingly $\mathrm{Da}{=}\mathrm{Da}_{IE}$) the

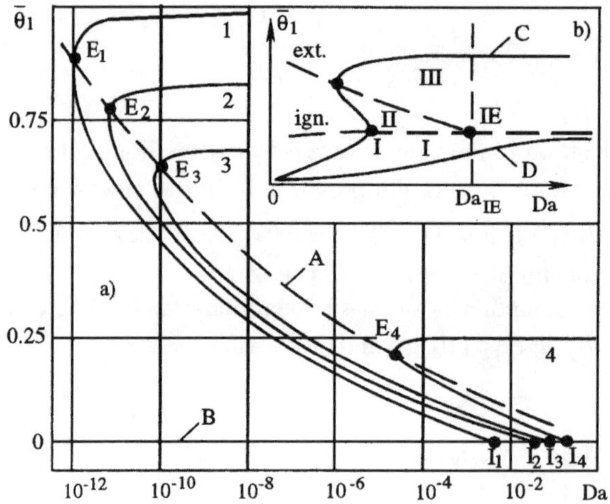

Fig. 10.3 Adiabatic ideally-stirred gas–droplet combustion reactor. Reprinted from Yarin and Hetsroni (1995), with permission. a) Dependence of gaseous phase temperature on the Damkohler number. 1, Combustion of homogeneous gas–vapour mixture (M=0); (2–4), combustion of gas–droplet mixture (2, M=40; 3, M=75; 4, M=150); E_1, E_2, E_3, E_4 are the points of extinction; I_1, I_2, I_3, I_4 are the point of ignition, respectively. b) Dependence of gaseous phase temperature on the Damkohler number of various values of the parameter M. C, curve $\overline{\theta}_1$ (Da) at $M<M_{IE}$; D, curve $\overline{\theta}_1$ (Da) at $M>I_{IE}$; E and I are the points of extinction and ignition, respectively; dashed lines correspond to the extinction and ignition; IE is the point corresponding to the merging of the critical ignition and extinction conditions

curves of ignition and extinction of a gas–droplet mixture intersect. At $M > M_{IE}$ the ignition of a gas–droplet mixture is not possible at any $Da>Da_{IE}$. This domain of the parameter M (large droplet diameter) corresponds to the subcriticial regime with monotonic increase in temperature with the increase of Da. The existence of the limiting values of the heterogeneity parameter M, corresponding to self-ignition of a gas–droplet mixture, results from the decrease in heat released as a consequence of decrease in evaporation surface.

As $M<M_{IE,}$ the curves $\overline{\theta}_1(Da)$ have the characteristic S shape. The upper and lower rising branches of these curves correspond to stable high and low temperature states of gas–droplet reactors, i.e. to the regimes of combustion and low-rate oxidation, respectively. The falling section of the curves $\overline{\theta}_1(Da)$ characterizes the unstable states of the reactor.

A few values of the gas–phase temperature correspond to a single value of Da (within the range $M<M_{IE}$). This corresponds to the existence of an adiabatic hysteresis with characteristic ignition and extinction phenomena. From the physical point of view, the multivalued $\overline{\theta}_1(Da)$ dependences reflect the dominant effect of the chemical reaction and its effects on the combustion rate. The latter is characteristic of the range of $0<M<M_{IE}$.

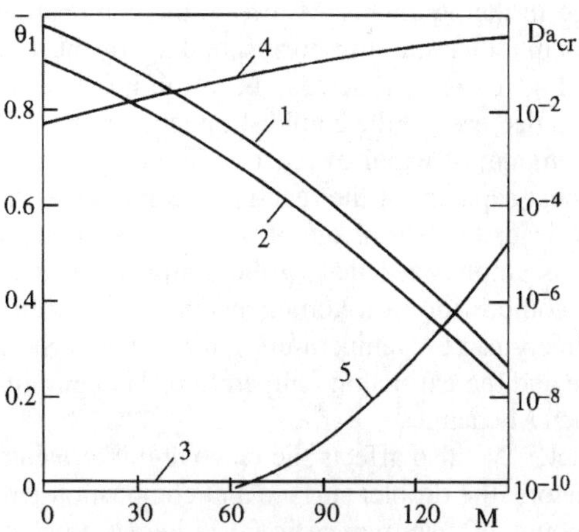

Fig. 10.4 Dependencies of the combustion, extinction and ignition temperature on the parameter M. Curves 1, 2 and 3 corresponds $\bar{\theta}_{1m}$, $\bar{\theta}_{1E}$, $\bar{\theta}_{1I}$, respectively. The dependencies (4) and (5) correspond to the Damkohler number for extinction and ignition. Reprinted from Yarin and Hetsroni (1995), with permission

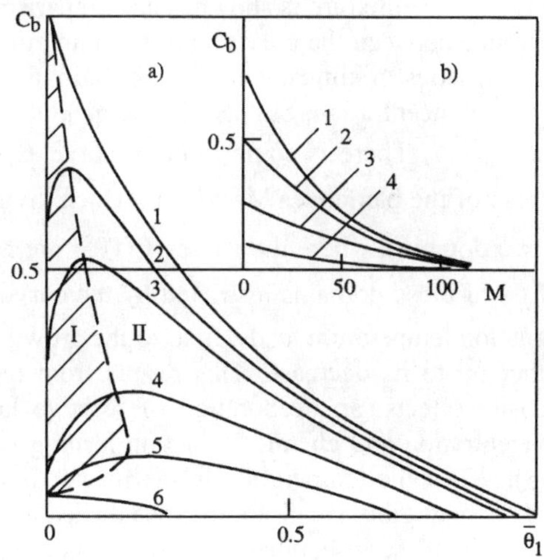

Fig. 10.5 Dependence of reactive liquid vapor concentration on gaseous phase temperature. Reprinted from Yarin and Hetsroni (1995), with permission. a) Curve 1, combustion of homogeneous gas–vapor mixture (M=0); curves 2–6, combustion of gas–droplet mixture (2, M=3; 3, M=10; 4, M=40; 5, M=75; 6, M=150). The dashed line shows the boundary, subdividing the parametric plane C_b - $\bar{\theta}_1$ on the domains where the dominant role is played by evaporation (I) or chemical reaction (II). b) Dependence of reactive liquid vapor concentration on the parameter M. 1, $\bar{\theta}_1$=0.125; 2, $\bar{\theta}_1$ = 0 .25; 3, $\bar{\theta}_1$=0.5; 4, $\bar{\theta}_1$=0.75

The change in the parameter M affects the combustion temperature and the values of the Damkohler number corresponding to the critical states of a gas–droplet reactor (Fig. 10.4). An increase in M (an increase in the initial droplet diameter) leads to a decrease in the combustion temperature. This effect is due to the decrease of the amount of vapor of reactive liquid and a decrease in heat released at a fixed total heat capacity of the reactive medium and heat transfer rate. An increase in M also leads to a shrinkage of the domain corresponding to the combustion regimes. It is emphasized that in the combustion of a gas–droplet mixture, similarly to the combustion of a homogeneous mixture, this regime may be realized in an extremely narrow temperature interval between the maximum combustion temperature and the extinction temperature. The ignition temperature changes very weakly when M changes.

The parameter M also affects the reactive mixture composition. This results from the two factors: the droplet surface and combustion temperature. The former governs the amount of vaporizing liquid, whereas the latter governs both the evaporation and chemical reaction rate. This fact determines the dependence of the vapor concentration on the combustion temperature. The results of the calculation of this dependence are plotted in Fig. 10.5 as the curves $C_b(\overline{\theta}_1)$ corresponding to different values of the parameter M. In comparison, the corresponding curve for the combustion of a homogeneous mixture with physico-chemical properties identical to those of the two-phase mixture is shown. The comparison shows that there is an important difference between the curves corresponding to combustion of the gas–droplet and homogeneous mixtures. This difference manifests itself not only in changes in the vapor concentration, but also in the quantitative changes in the shape of the curves $C_b(\overline{\theta}_1)$. There is some limiting curve $C_b(\overline{\theta}_1)$ (and accordingly the limiting value of the parameter $M = M_{lim}$), which divides the parametric plane $C_b = \overline{\theta}_1$ into two domains, where the curves $C_b(\overline{\theta}_1)$ are monotonic and non monotonic. In the first of these domains restricted by the curves M=0 and M_{lim}, an increase in combustion temperature leads, first, to the growth of the vapor concentration, and further on to its decrease. This results from the competition between the two opposite effects: an evaporation of reactive liquid leading to a growth in vapor concentration and chemical reaction leading to a decrease of C_b (Pushkin et al. 1993). When the temperature in the reactor is relatively low, the chemical reaction rate is negligibly small. In this case the leading role is that of the rate of evaporation. An increase in combustion temperature is accompanied by a significant increase in vapor concentration (see the shadowed domain in Fig. 10.5a). At high temperatures the chemical reaction rate becomes the dominant factor. An increase in $\overline{\theta}_1$ leads in this case to a monotonic decrease in vapor concentration. Such a tendency is maintained in the whole range of temperatures $\overline{\theta}_{1*} < \overline{\theta}_1 < 1$, when the chemical reaction rate exceeds the evaporation rate.

The dependence of vapor concentration on the parameter M is plotted in the inset as Fig. 10.5 b. It is seen that an increase in the degree of heterogeneity of gas–droplet mixture leads to a decrease in the vapor concentration. This effect manifests itself very clearly for relatively low combustion temperatures and the dominant role is played by the evaporation process.

C. *Effect of external heat transfer.* The effect of heat losses to the reactor walls on the combustion process in the gas–droplet reactor is illustrated in Fig. 10.6. In this graph the results of calculations of the combustion temperature for some finite value of the Semenov number are presented. Comparing these results with the results plotted in Fig. 10.3 one notices the characteristics of the combustion process in the adiabatic and non adiabatic reactors. It is seen that the heat losses lead to the appearance of a new pair of critical states (I' and E'). The heat transfer leads to a widening of the temperature range within which stationary states may be realized.

The effect of the flow rate of reactive mixture on the thermal regime of a non-adiabatic reactor is also shown in Fig. 10.6. In a homogeneous mixture, an increase in Da leads to a monotonic decrease in the combustion temperature and to a gradual change from the combustion regime to the regime of low-rate oxidation. This effect is related to a decrease in the total amount of heat release in the reactors at fixed external heat losses. In the combustion of a gas–droplet mixture, the change from the high to the low temperature regime is not gradual. There is a sharp change of temperature, which is characteristic of the critical states corresponding to the ignition and extinction of the gas–droplet mixture. An increase in the parameter M leads to the approach of the points I and I', and at $M = 4$, to their merging. At $M>4$ the states corresponding to the adiabatic and thermal ignition disappear, which shows that there should exist some limiting droplet size from which the self-ignition of a gas–droplet mixture is impossible. In this case, the high temperature states may be attained by using a special ignition device. A further growth in the parameter M leads to a shrinkage of the domain where high temperature regimes exist. At $M>7.5$ such states cannot be realized, due to the low heat release intensity in the coarse droplet mixture with undeveloped interfacial and small mass of reactive liquid vapor. In this case the external heat transfer exceeds the heat released, which causes a decrease in combustion temperature.

The thermal regimes of the non adiabatic gas–droplet combustion reactors are illustrated in Fig. 10.7. In this figure the domains of existence of the hysteresis and critical states are shown in the Da–M plane. Domain I is restricted by the lines 1 and 5, corresponding to the regimes in which hysteresis is absent. At these lines, the high temperature regimes are changed and degrade into low temperature states (by an increase in Da). In domain II, only high temperature states are possible. High and low temperature states corresponding to the adiabatic and non adiabatic hysteresis are attained in domains III and IV, high temperature regimes are localized in domain V. In this domain a high temperature combustion process may be initiated only by an ignition device. In domain VI, only low temperature states exist.

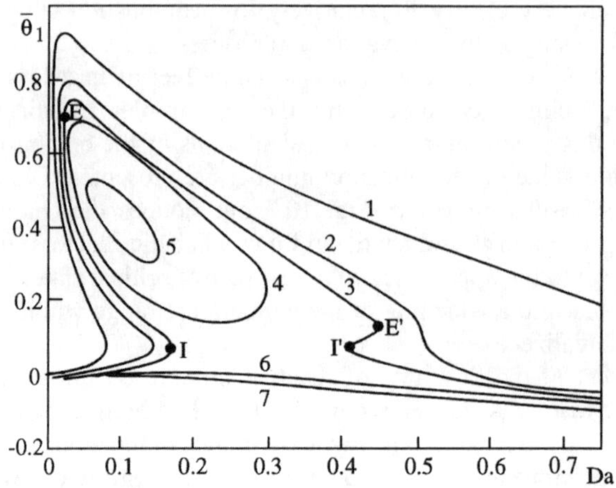

Fig. 10.6 Dependence of the gaseous phase temperature on the Damkohler number in a non adiabatic gas–droplet combustion reactor. Reprinted from Yarin and Hetsroni (1995), with permission. Curve 1, combustion of homogeneous mixture (M=0); curves 2–6, combustion of gas–droplet mixture of Se = 100 (2, M=3; 3, M=3.5; 4, M=5; 5, M=7; 6, M=7.5); the points I and E correspond to the adiabatic ignition and extinction; the points $I^{'}$ and $E^{'}$ correspond to the thermal ignition and extinction

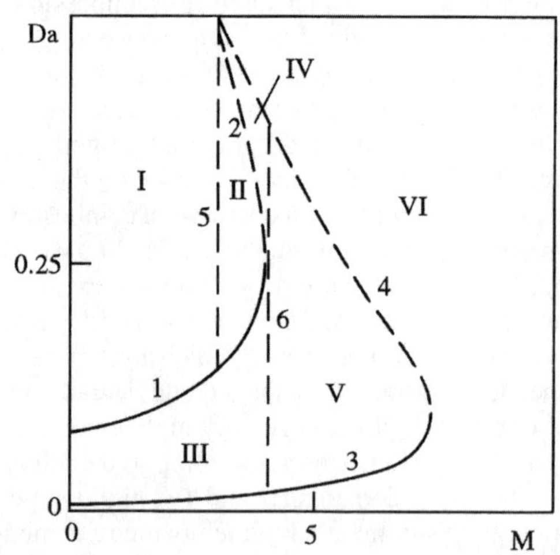

Fig. 10.7 Diagram of the stationary states of non adiabatic gas–droplet combustion reactor. Reprinted from Yarin and Hetsroni (1995). Curves 1 and 2 are the lines of adiabatic and thermal ignition; curves 3 and 4 are the lines of adiabatic and thermal extinction, 5 and 6 are the boundaries where the critical states disappear

3.10.4 Bubbly combustion reactor

A. *The reactor equation.* We now consider the thermal regimes of the gas–liquid combustion reactor where thermal conductivity of the carrier liquid is very small. Such a case is characteristic of various gas–liquid reactive media, for example, oil–gas bubbly media.

At low thermal conductivity, the effect of non uniformity of the temperature field on the combustion process may be critical. In this case it is necessary to take into account the phenomena resulting from the multi-temperature character of the reacting medium. We use here the approach of the thermal combustion theory (Zel'dovich et al. 1985, Willliams 1985) to study this case.

We write the thermal balance equation (10.27) in the standard form of the theory of thermal regimes of combustion (Vulis 1961). Assuming that the derivatives $di_1^{(i)}/dt$ and $di_2^{(i)}/dt$ are equal to zero and also taking into account the approximations $\rho_1^{(i)} \approx \rho_{1.0}^{(i)}$ and $j^* \approx 1$ (Yarin and Sukhov 1987), we reduce Eq. (10.27) to the following form:

$$<Q_1> = q < W > \tag{10.50}$$

$$\begin{aligned} < Q_2 > &= \rho_1 c_{p1} j_0 (< T_1 > - T_{1.0}) + \rho_2 c_{p_2} j_0 (< T_2 > - T_{2.0}) + \\ &+ h_w^* s_w (< T_2 > - T_W) + q_e < W_{e.2} > \end{aligned} \tag{10.51}$$

where $<Q_1>$ and $<Q_2>$ are the functions of heat release and heat removal, respectively.

To rearrange Eqs. (10.50) and (10.51) we find, first, the relation between the average temperatures of liquid and gaseous phases. Assuming the derivative $d\theta_2^{(i)}/d\tau = 0$ in Eq. (10.39) to be zero and $j^*=1$, we find

$$< \theta_2 > = F^{-1}\left[< \theta_1 > - \bar{q}_e S_{e1-2} < \bar{W}_{e.2} > + \alpha_* \theta_w \right] \tag{10.52}$$

where $F = 1 + j_* + \alpha_*$; $j_* = c_{p_{2.1}} (Se_{1-2}/Da)\gamma$, $\alpha_* = Se_{1-2}/Se_w$, $c_{p2.1} = c_{p2}/c_{p1}$; Se_{1-2} is the Semenov number determined by the initial parameters, by the area of the interfacial $S_0 = \left((36\pi\xi_{1.0}^2)/N^2\right)^{1/3}$; N is the numerical bubble concentration.

Note, that here and hereinafter we do not account for hydrodynamic effects, i.e. we assume non deformable bubbles and use the fact that $\rho_2/\rho_0 \sim 1$.

The parameter $F=F(Da, Se^o, Se_w, c_{p2.1})$ is very important in the analysis of the thermal regime of bubbly combustion reactors, because it accounts for the convective transfer, interphase interaction, heat removal to reactor walls, heat release in the chemical reaction and physico-chemical and structural properties of the reactive medium. The value of this parameters is restricted (from below) by the value F=1, which corresponds to the adiabatic reactor without flow.

To calculate the average gaseous phase temperature and the average values of chemical reaction rate, we follow Genkin et al. (1981) and change the averaging from ensemble of bubbles to the time averaging which corresponds to the ergodic hypothesis. Therefore, we write the following expressions, for $<T_1>$ and $<W>$

$$<T_1> = \int_0^{\infty} \varphi(t_r) T_1(t_r) dt_r \tag{10.53}$$

$$<W> = \int_0^{\infty} \varphi(t_r) W(t_r) dt_r \tag{10.54}$$

where $T_1(t_r)$ and $W(t_r)$ are the actual values of gaseous phase temperature and chemical reaction rate, $\varphi(t_r) = (1/t_1)\exp(-t_r/t_1)$ is the distribution function, t_1 is parameter of the distribution function.

To evaluate the integral in the expressions (10.53) and (10.54) we use the approximate form of the temperature dependence $T_1(t_r)$, shown in Fig. 10.8 b replacing the actual temperature profile of Fig. 10.8 a. This form reflects the specific features of the combustion process, namely a slow growth of the gaseous phase temperature during heating and evaporation, and a sharp increase of T_1 after ignition of the gas–vapor mixture. On the curve $T_1(t_r)$ we can distinguish the sections corresponding to different states of the combustion process: (i) heating and evaporation of the liquid and formation of a reactive mixture within the bubbles ($T_1 \sim T_2$, $\eta \ll 1$), (ii) ignition and rapid burning of the mixture inside the bubbles ($T_1 \gg T_2$, $qW \gg h^* s_{1\text{-}2}(T_1 - T_2)$, $\eta \sim 1$). Accordingly, we can approximate the dependence $T_1(t_r)$ by a polygonal line (Figs. 10.8 a and 10.8 b). Then we write the expression for the average temperature of gaseous phase as

$$<T_1> = \int_0^{t_{ind}} \varphi(t_r) T_1^{'}(t_r) dt_r + \int_{t_{ind}}^{\infty} \varphi(t_r) T_1^{''}(t_r) dt_r \tag{10.55}$$

where $T_1^{'}(t_r)$ and $T_1^{''}(t_r)$ are the functions describing the change of gaseous phase temperature during inductive heating and thermal relaxation, respectively, t_{ind} being the induction time.

Since the temperature of the gas–vapor mixture changes slightly during the heating period we assume in calculating $T_1^{'}(t_r)$ that $T_1(t_r) = T_2(t_r)$. Integration of the energy balance equation of the liquid continua (10.39) having regard to the last condition yields (at $\overline{q}_e \ll 1$ and $\rho_2 \approx \rho_{2.0}$) the following expression

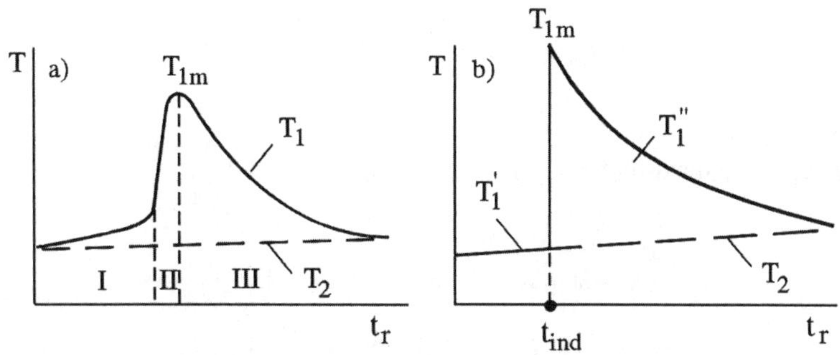

Fig. 10.8 Variations of temperature in liquid and gaseous phase in bubbly reactive medium. Reprinted from Yarin and Hetsroni (1995), with permission. a) Actual profile of $T_1(t_r)$, I is the heating zone, II is the reactive zone, III is the thermal relaxation zone. b) Approximate temperature profile. T_1 and T_2 are the gaseous and liquid phases temperatures, T_1' and T_1'' are the gaseous phase temperatures correspond to heating and thermal relaxation regimes

$$T_2(t_r) = <T_2> - (<T_2> - T_{2.0})\exp\left(\frac{-\bar{t}_r F}{j_*}\right) = T_1'(t_r) \tag{10.56}$$

where $\bar{t}_r = t_r j_o$.

The variation of enthalpy of the gaseous phase during the thermal relaxation period is described by Eq. (10.14) which, after integration (taking into account Eq. (10.56)), yields the following expression for $T_1''(t_r)$

$$T_1''(t_r) = K\exp(-c_{p_{2.1}}\rho_{2.1}\bar{t}_r \cdot j_*^{-1}) + <T_2> - (<T_2> - T_{2.0})\Lambda(\Lambda-1)^{-1}\exp(-\bar{t}_r F j_*^{-1}) \tag{10.57}$$

where

$$K = \exp(c_{p_{2.1}}\rho_{2.1}\bar{t}_{ind} j_*^{-1})\{(T_{i.m} - <T_2>) + (<T_2> - T_{2.0})\Lambda(\Lambda-1)^{-1}\exp(-\bar{t}_{ind} F j_*^{-1})\};$$

$$\Lambda = \frac{c_{p2.1}\rho_{2.1}}{F}; \quad \bar{t}_{ind} = t_{ind} j_o.$$

Using expression (10.55) and the relation for $T_1'(t_r)$ and $T_1''(t_r)$ of Eqs. (10.56) and (10.57), we find the average temperature of the gaseous phase. Assuming the initial phase temperatures to be equal and $\rho_2/\rho_o \sim 1$, we obtain

$$< \theta_1 > = A_2\theta_{1m} + A_3\theta_w + \beta\left(\sum_{i=1}^{3} A_i - 1\right) \tag{10.58}$$

where θ_{1m} is a maximum gaseous phase temperature and

$$A_1 = \psi^{-1}[j_*(\bar{\Omega} - uf) + F(1 - r\bar{e}^{\frac{\omega}{\pi}})$$

$$A_2 = \psi u F e^{-\omega}; \qquad A_3 = \psi^{-1}\alpha_*(\bar{\Omega}^{-1} - uf)$$

$$\psi = \pi^{-1}(F-1) + uf + 1; \quad u = \frac{1+\bar{\Omega}}{\Lambda + \bar{\Omega}}$$

$$r = \frac{\bar{\Omega}}{\Lambda + \bar{\Omega}}; \quad \pi = \frac{\bar{\Omega}}{1+\bar{\Omega}}$$

$$\bar{\Omega} = \frac{\Omega}{F}; \quad \Omega = \frac{j_*}{\bar{t}_1}; \quad \bar{t}_1 = t_1 j_o$$

$$\omega = \frac{t_{ind}}{t_1}; \quad f = e^{-\omega} - \frac{\bar{\Omega}}{1+\bar{\Omega}} e^{-\omega\frac{1+\bar{\Omega}}{\bar{\Omega}}}$$

To determine the maximal gaseous phase temperature θ_{1m}, we use the mass and thermal balance equations of a single bubble (10.28) and (10.29). As a result, we find (at $\rho_1^{(I)} \approx \rho_{a.o}^{(i)}$) that

$$\theta_{1m} = \gamma^{-1}\left(\frac{\eta}{\mu} - \frac{Da}{Se_{1-2}^{o}} \int_0^{\bar{t}_{ind}} (\theta_1 - \theta_2)d\bar{t}_r\right) \tag{10.59}$$

The terms on the right-hand side of Eq. (10.59) correspond to the adiabatic heating of the gas–vapour mixture and to the heat flux into the surrounding liquid during the induction period.

Assuming that at $t = t_{ind}$ the gaseous oxidizer burns out almost completely, we take $\eta = 1$ in Eq. (10.59).

Taking into account that the gaseous phase temperature grows very slowly during the induction period, we can evaluate the integral on the right-hand side of Eq. (10.59). To this end we consider the behavior of $T_1(t_r)$ at small intervals of the argument change. After expanding $T_1(t_r)$ into the Taylor series

$$T_1 = T_{1.0} + \left(\frac{dT_1}{dt_r}\right)_0 dt_r + \ldots \tag{10.60}$$

and substituting the derivative $(dT_1/dt_r)_0 = (q/\rho_1 c_{p_1}) W(T_{1.0})$ found from Eq. (10.14) we obtain:

$$\theta_1 = \frac{Da}{\gamma}\bar{W}(\beta)d\bar{t}_r \tag{10.61}$$

Since the gaseous and liquid phase temperatures are approximately equal in the vicinity of t_r=0 (T_2~$T_{1.0}$), the integral in Eq. (10.59) may be represented as

$$\int_0^{\bar{t}_{ind}} (\theta_1 - \theta_r)d\bar{t}_r \approx \frac{Da}{\gamma}\bar{W}(\beta)\bar{t}_{ind}^2 \tag{10.62}$$

In the activated combustion processes the chemical reaction rate may be represented in the form

$$W(T_1) = z\rho_{a.0} f(\eta)\exp\left(-\frac{E}{RT_1}\right), \tag{10.63}$$

where $f(\eta)$ is the function of completeness of combustion: $f(\eta)=1$ when $t<t_{ind}$; $f(\eta)=0$ at $t>t_{ind}$.

Using this expression for the chemical reaction we rewrite Eq. (10.59) in the form

$$\theta_{1m} = \frac{1-\sigma\omega^2}{\gamma\sigma 2\bar{\chi}\bar{\Omega}} \tag{10.64}$$

where $\bar{\chi} = t_z/(t_1 F)$, $t_z = z^{-1}\exp(\beta)$.

B. Parametric study of bubbly combustion reactor states. We now address the analysis of possible states of bubbly combustion reactors. With this aim we consider the behavior of the solution of the thermal balance equation under parameter variation. We also consider the effect of different parameters on the extension of the domains corresponding to the existence of stationary and non stationary combustion regimes.

To describe these states we use Eqs. (10.50) and (10.51). Using the expression for $W(T_1)$, we can present the average heat release function as

$$<Q_1> = q\int_0^{\infty} \varphi(t_r)\rho_{a.0} f(\eta)\exp\left[-\frac{E}{RT_1(t_r)}\right]dt_r \tag{10.65}$$

To evaluate the latter integral we can use the expressions (10.56) and (10.57) determining temperature variation in the gas–vapor mixture. This approach is, however, very complicated. To simplify the calculations, we approximate the chemical reaction rate taking into account that in real systems, chemical reactions are activated. In this case the reactions are localized in the high temperature domain where most of the mass of reacting material is converted. Hence we can represent chemical reaction rate in reactive bubble as

$$W = W(t_r)\delta(t_r - t_{ind}) \tag{10.66}$$

where $\delta(t_r - t_{ind})$ is the delta function.

Using Eq. (10.66) we find the following approximate expression for the average heat release function

$$< Q_1 >= \rho_1 c_{p1} T_{1.0} j_o (\frac{\beta}{\sigma\gamma \bar{t}_1}) \exp(-\omega) \tag{10.67}$$

Rearranging Eq. (10.51) by using Eq. (10.52), we arrive at the following average heat removal function

$$< Q_2 >= A_4 (1 + \beta < \theta_1 >) - A_5 \tag{10.68}$$

where $$A_4 = \rho_1 c_{p1} j_o T_{1.0} \left[1 + F^{-1} \left(\rho_{2.1} c_{p2.1} + \frac{Da}{Se_w \gamma} \right) \right]$$

and

$$A_5 = \rho_1 c_{p_1} j_o T_{1.0} \left\{ 1 + \left(\rho_{2.1} c_{p2.1} + \frac{Da\gamma}{Se_w} \right) \right.$$

$$\left. \times \left[1 - F^{-1} \left(\frac{Se^o_{1-2}}{Da} c_{p2.1} \gamma + \frac{Se^o_{1-2}}{Se_w} \cdot \frac{Tw}{T_{1.0}} \right) \right] + \frac{Da}{Se_w} \gamma \left(\frac{Tw}{T_{1.0}} - 1 \right) \right\}$$

Substituting expressions (10.67) and (10.68) in the equation $<Q_1>=<Q_2>$, we determine stationary states of the reactor, which yields (at $\bar{q}_e \ll 1$) the following parametrical equation

$$f + \frac{\sigma\omega}{2\bar{\chi}\bar{\Omega}^2} \Lambda(\bar{\Omega}\Sigma + \frac{F-1}{F}\Lambda) = \frac{1-\Lambda}{1+\bar{\Omega}} + F(\Sigma - 1) + \tag{10.69}$$

$$\alpha_* \theta_w \sigma\gamma\bar{\Omega}^{-1} \frac{\Lambda + \bar{\Omega}}{1+\hat{\Omega}} e^{\omega} \left[\Sigma \left(1 - uf\bar{\Omega} \right) - \Lambda(1+uf) \right]$$

where $\Sigma = \bar{t}_1 / j_*$

C. Stationary and critical states. Equation (10.69) defines the domains of existence of stationary and non stationary states of bubbly combustion reactors. It contains eight parameters accounting for the hydrodynamic, thermal and kinetic effects. In the multidimensional parametric space ($\omega, \Lambda, F, \bar{\Omega}, \Sigma, \alpha_*, \theta_w$ and $\bar{\chi}$) the limiting values of these parameters correspond to the surface, subdividing the parametric space into domains corresponding to various regimes of bubbly combustion reactors.

The solution of Eq. (10.69) may be represented as 56 spatial (or 28 planar) diagrams which establish the correlation between any three (or two, for the planar graphs) parameters when all the other parameters are fixed. These diagrams allow us to outline the domains of existence of different states in the quasi-stationary approximation and to determine the values of regime parameters corresponding to the critical states.

Consider some of these diagrams, which are of great importance for the understanding of the general properties of the combustion process in bubbly reactors. First we examine a simple case when the initial temperature of the liquid reagent is equal to that of the reactor walls (i.e. $\theta_w = 0$) and the parameter $\Sigma = 1$. Then Eq. (10.69) takes the form

$$f + \frac{\sigma\omega^2}{2\bar{\chi}\bar{\Omega}^2}\Lambda(\bar{\Omega} + \frac{F-1}{F}\Lambda) = \frac{1-\Lambda}{1+\bar{\Omega}} \tag{10.70}$$

From Eq. (10.70) we see that there are parametric spaces $\omega - \bar{\Omega} - \Lambda$, $\bar{\chi} - \bar{\Omega} - \Lambda$ etc. (in which stationary states are possible), which are bounded by the limiting plane $\Lambda = 1$ (Fig. 10.9) where the space $\omega - \bar{\Omega} - \Lambda$ is shown. Indeed, since the parameters $\bar{\Omega}, \bar{\chi}, \omega$ and Λ are larger than or equal to zero and $F \geq 1$, and $f > 0$ for any physically realistic values of these parameters, the first two terms on the left in Eq. (10.7) are positive. Therefore, the term on the right in Eq. (10.70) should also be positive and hence $\Lambda \leq 1$. On the one hand, from the physical point of view, the value of the parameter Λ should be limited (from above) because the amount of heat spent in heating the bubbly medium and compensating for the losses to the walls and convection, changes as Λ varies. An increase in Λ (or decrease of heat losses) results in the heating of the system that prevents the emergence of the stationary regimes. On the other hand, an increase in Λ (when the heat losses are fixed) is accompanied by an increase in the specific heat of the medium. In this case large values of Λ correspond to lower temperatures when stationary states cannot appear. Note that in real reactive bubbly media the product $\rho_{2.1}c_{p2.1}$ is of the order 10^3. Therefore, in the given case, the condition corresponding to stationary states of the reactor ($\Lambda < 1$) may be satisfied when the parameter F has the same order of magnitude, i.e. when the specific flow rate of the mixture or heat removal to the reactor walls are large enough. It is clear that the combustion process in an adiabatic reactor without flow is absolutely unstable.

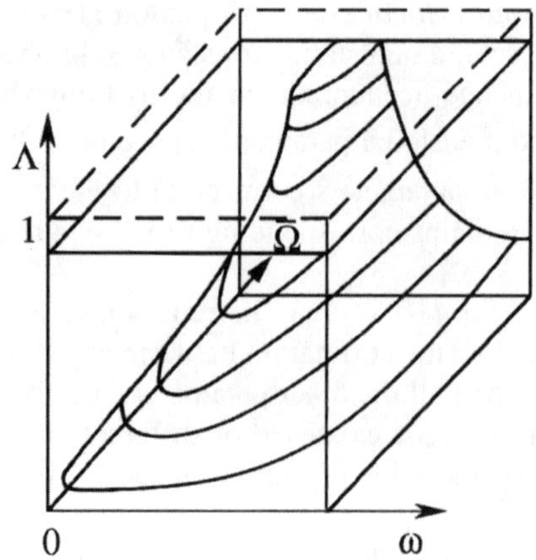

Fig. 10.9 Bubbly combustion reactor. The surface corresponding to the stationary states $(\Sigma = 1,\ \theta_w = 0)$, $\Lambda < 1$ corresponding to the domain of stationary states; $\Lambda > 1$ corresponding to the domain of nonstationary states; dashed line corresponds to the limiting plane $\Lambda = 1$. Reprinted from Yarin and Hetsroni (1995), with permission

Consider in more detail the general case when the parameter Σ is arbitrary. We assume as before that $\theta_w = 0$ in Eq. (10.69) and solve this equation with respect to Λ. As a result, we obtain the following equation

$$\Lambda^2 + K_1\Lambda + K_2 = 0 \qquad (10.71)$$

Its solution has the form

$$\Lambda = \frac{1}{2}\left(-K_1 \pm \sqrt{K_1^2 - 4K_2}\right) \qquad (10.72)$$

where

$$K_1 = \left[\frac{\sigma\omega^2\Sigma}{2\chi\bar{\Omega}} + \frac{1}{1+\bar{\Omega}}\right] / \frac{\sigma\omega^2}{2\chi\bar{\Omega}} \cdot \frac{F-1}{F}$$

$$K_2 = [\Phi - F(\Sigma - 1)] / \frac{\sigma\omega^2}{2\chi\bar{\Omega}} \cdot \frac{F-1}{F},$$

$$\phi = f - (1+\bar{\Omega})^{-1}$$

Since the parameters $\sigma, \omega, \bar{\Omega}, \bar{\chi}$ and Σ are positive and $F \geq 1$ (accordingly, $K_1 > 0$), we find that the physically plausible solution corresponds to the plus sign before the radical. In this case K_2 should be negative (or equal to zero) and consequently the difference $\Phi - F(\Sigma - 1)$ is smaller than or equal to zero. The value of the function $\Phi(\bar{\Omega},\omega)$ belongs to the interval $-1 \le \Phi \le 0$ at any values of ω and $\bar{\Omega}$. Thus we have the following estimate of Σ

$$\Sigma \ge (1+\frac{\Phi}{F}) \tag{10.73}$$

Since the right-hand side of Eq. (10.69) is positive, we have

$$\Lambda \le 1+F(\Sigma-1)(1+\bar{\Omega}) \tag{10.74}$$

In an activated reaction the induction time is very small compared to the characteristic residence time. Assuming $\omega \ll 1$ we transform Eq. (10.69) to the form

$$\frac{\sigma\omega^2}{2\chi\bar{\Omega}}\Lambda(\Sigma\bar{\Omega}+\frac{F-1}{F}\Lambda) = -\frac{\Lambda}{1+\bar{\Omega}}+F(\bar{\Omega}-1)+\alpha_*\theta_w\sigma\gamma\frac{\lambda(\Sigma-\Lambda-\bar{\Omega}+1)}{\bar{\Omega}(1+\bar{\Omega})} \tag{10.75}$$

Since the right hand side of Eq. (10.75) is positive, we obtain the estimate

$$\alpha_*\theta_w \ge \frac{\Lambda-F(\Sigma-1)(1+\bar{\Omega})}{(\Sigma-1)-(\Lambda+\Omega)}\cdot\frac{\bar{\Omega}}{\Lambda}\cdot(\sigma\gamma)^{-1} \tag{10.76}$$

The equality in Eq. (10.76) corresponds to the surface subdividing the parametric space in the domains with stationary and non stationary states. The dependence $\alpha_*\theta_w(\Sigma)$ is plotted in Fig. 10.10. It is seen that an increase in F (with the other parameters fixed) leads to an increase of the domains corresponding to the stationary states for both $\theta_w > 0$ and $\theta_w < 0$.

To determine a relationship between the parameters corresponding to stationary states of bubbly combustion reactors in the case of unequal initial temperatures of reactive mixtures and reactors walls, we solve Eq. (10.69) with respect to Λ

$$\Lambda = \frac{1}{2}(-K_3 \pm \sqrt{K_3^2+4K_4}) \tag{10.77}$$

where

$$K_3 = \left\{\frac{\sigma\omega^2\Sigma}{2\chi\bar{\Omega}}+\frac{1}{1+\bar{\Omega}}-\alpha_*\theta_w\sigma\gamma\right\}\cdot\left\{\frac{\sigma\omega^2}{2\chi\bar{\Omega}^2}\;\frac{F-1}{F}+\frac{\alpha_*\theta_w\sigma\gamma}{\bar{\Omega}(1+\bar{\Omega})}\right\}^{-1},$$

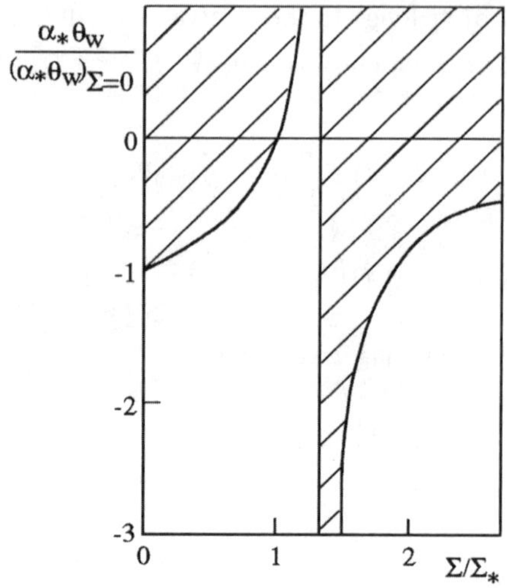

Fig. 10.10 Bubbly combustion reactor. Dependence of $\alpha_* \theta_w$ on Σ. The domains corresponding to the existence of stationary states are shaded. Reprinted from Yarin and Hetsroni (1995), with permission. $\Sigma_* = 1 + [\Lambda/(F(1+\bar{\Omega}))]$

$$K_4 = F(\Sigma - 1) \cdot \left\{ \frac{\sigma\omega^2}{2\bar{\chi}\bar{\Omega}} \quad \frac{F-1}{F} + \frac{\alpha_* \theta_w}{\bar{\Omega}(1+\bar{\Omega})} \sigma\gamma \right\}^{-1}$$

Since Λ should be positive, the physically realistic solution of Eq. (10.77) is possible only under the following conditions:

i. $K_3 > 0$, $K_4 > 0$, plus sign before the radical

ii.

$$K_3 < 0 \quad \begin{cases} a - K_4 > 0, \text{ plus sign before the radical} \\ b - K_4 < 0, \text{ plus sign before the radical } (K_3^2 > 4K_4) \end{cases} \tag{10.78}$$

Consider in detail these conditions, restricting the analysis to physically realistic situations when $\theta_w > 0$. We estimate the range of possible variation of the parameters when the initial temperature of the reactive mixture is lower than that of the walls.

When K_3 and K_4 in Eq. (10.77) are positive [the first case in (10.78)], Σ should be greater than $(1+\bar{\Omega})$. The parameters ω, $\bar{\Omega}$, $\bar{\chi}$ and F satisfy the following inequality

$$\frac{\sigma\omega^2}{2\bar{\chi}\bar{\Omega}}\left[\Sigma-\frac{F-1}{F}(\Sigma-\bar{\Omega}-1)\right]+(1+\bar{\Omega})^{-1}>0 \tag{10.79}$$

for any values of the parameters χ, ω and F. Accordingly, we can write the conditions of the existence of stationary solution of Eq. (10.75) in the form $\Sigma>(1+\bar{\Omega})$, $\omega>0$, $\bar{\chi}>0$, and $F\geq 1$.

When K_3 and K_4 have different signs, [the second case in (10.78)], the stationary solution of Eq. (10.75) exists if the following inequalities are satisfied

$$\Sigma>1, \quad \alpha_*\theta_w>\left(\frac{\sigma\omega^2}{2\bar{\chi}\bar{\Omega}}+\frac{1}{1+\bar{\Omega}}\right)\frac{\bar{\Omega}(1+\bar{\Omega})}{\Sigma-\bar{\Omega}-1}(\sigma\gamma)^{-1} \tag{10.80}$$

When K_3 and K_4 are negative [the second case in (10.78)] stationary solutions are absent. Indeed, from the expression for K_4 it follows that $\Sigma<1$, which is incompatible with the condition $K_3<0$.

The effect of the parameter F on the average temperature of the gaseous phase is illustrated in Fig. 10.11, where $<\theta_1>$ is plotted versus F for several values of Λ. An increase in Λ with a fixed flow rate of reactive mixture leads to a monotonic (practically linear) decrease in average temperature of the gaseous phase due to the growth in the heat capacity of bubbly medium. An increase in the flow rate of the reactive mixture also leads to a decrease in the average temperature of the gaseous phase. The latter results from an increase in convective heat losses at fixed heat capacity of the reactive medium and the heat release in it. The dependence of the average temperature of the liquid phase on the parameters the F and Λ is very weak.

The dependence $\omega(\bar{\Omega})$ corresponding to the stationary states of the bubbly reactor is shown in Fig. 10.12. In this graph the critical points are also marked. It is seen that on the high temperature branch of the curve in Fig. 10.12, there is a minimum (at the point M) corresponding to the regime with maximal heat release. The existence of the extremum in the curve $\omega(\bar{\Omega})$ corresponds to the possibility of stationary regimes with equal intensities of heat release at different values of the parameter $\bar{\Omega}$. This effect manifests itself when two or more parameters, which govern convective interfacial heat flux and heat removal to the walls, vary.

In particular, in order to change from the regime with the maximal heat release to the regime $\omega_1(\Omega')$ we must decrease the intensity of convection or heat removal to the walls (while to change to the regime $\omega_1(\Omega'')$ the said intensity must be increased) as follows from the condition $F=\text{const}$. In this case the intensity of heat release in the first and the second states is the same at $\Omega'\neq\Omega''$.

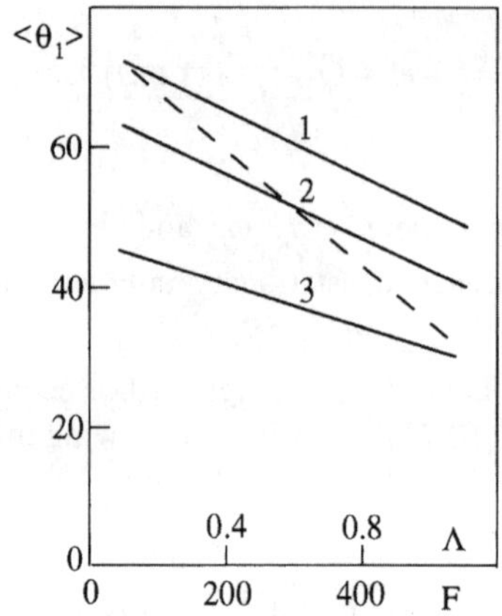

Fig. 10.11 Bubbly combustion reactor. Dependence of the average temperature of the gaseous phase on the parameter F. Reprinted from Yarin and Hetsroni (1995), with permission. Curve 1, $\Lambda = 0.25$; curve 2, $\Lambda = 0.5$; curve 3, $\Lambda = 0.75$; the dashed line shows the dependence of $< \theta_1 >$ on Λ at F=300

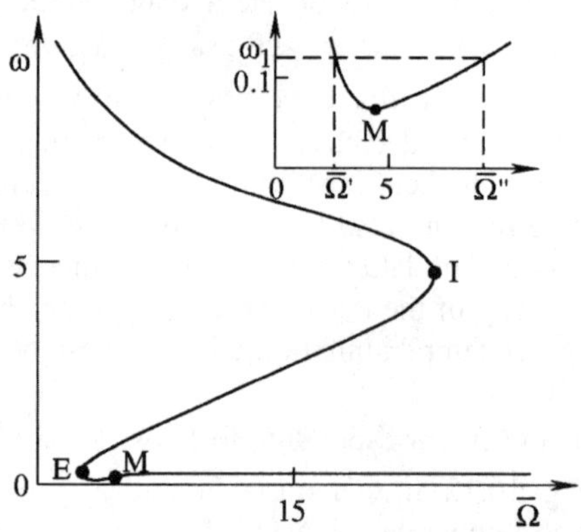

Fig. 10.12 Bubbly combustion reactor. Dependence $\omega(\bar{\Omega})$. I is the point of ignition; E is the point of extinction; M is the point of the maximum heat release. Reprinted from Yarin and Hetsroni (1995), with permission

The shape of the $\omega(\Omega)$ curves corresponding to the solution of Eq. (10.75) at different values of χ and at F=const, Λ=const, $\theta_w = \text{const}$ is shown in Fig. 10.13. The points corresponding to ignition and extinction of a bubbly reactor. They separate regions corresponding to high-temperature (a–b), low temperature (c–d) and intermediate (unstable) (b–c) states. The curves of the critical states separate the parametric plane $\omega - \bar{\Omega}$ into several of domains corresponding to high-temperature I, low-temperature II and intermediate (unstable) regimes.

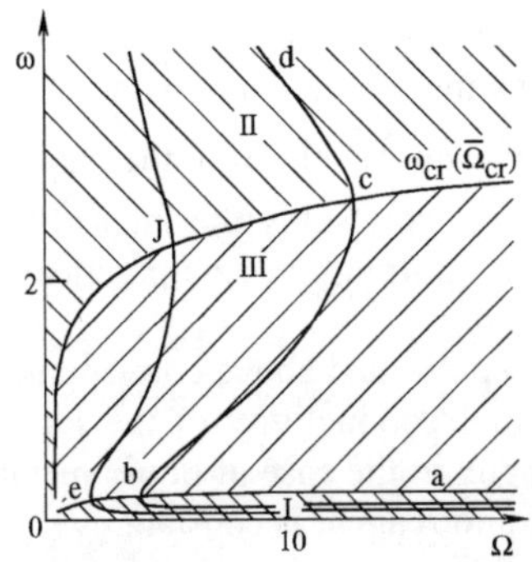

Fig. 10.13 Bubbly combustion reactor. Reprinted from Likhachev et al. (1991), with permission. 1 - $\bar{\chi}=20$, $2-\bar{\chi}=8$; $F=201$; $\Lambda=0.06$; $\theta=9\cdot10^{-4}$. I, High temperature states; II, low temperature states; III, intermediate (unstable) states

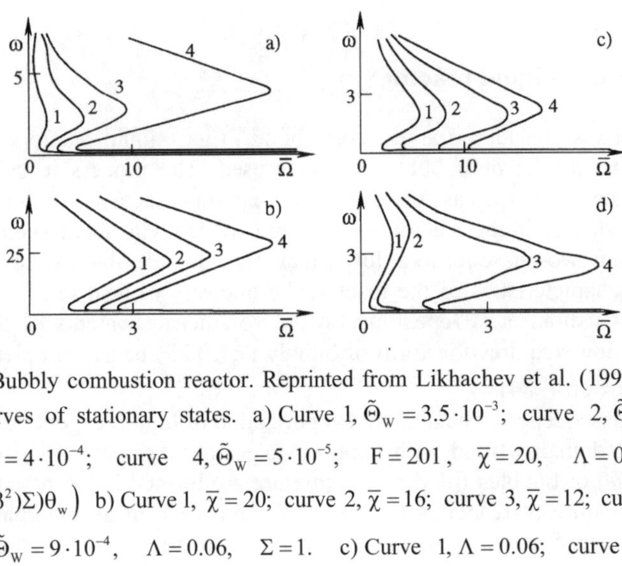

Fig. 10.14 Bubbly combustion reactor. Reprinted from Likhachev et al. (1991), with permission. curves of stationary states. a) Curve 1, $\tilde{\Theta}_W=3.5\cdot10^{-3}$; curve 2, $\tilde{\Theta}_W=9\cdot10^{-4}$; curve 3, $\tilde{\Theta}_W=4\cdot10^{-4}$; curve 4, $\tilde{\Theta}_W=5\cdot10^{-5}$; $F=201$, $\bar{\chi}=20$, $\Lambda=0.06$, $\Sigma=1$, $\left(\tilde{\Theta}_W=((\gamma/\beta^2)\Sigma)\theta_w\right)$ b) Curve 1, $\bar{\chi}=20$; curve 2, $\bar{\chi}=16$; curve 3, $\bar{\chi}=12$; curve 4, $\bar{\chi}=8$; $F=201$, $\tilde{\Theta}_W=9\cdot10^{-4}$, $\Lambda=0.06$, $\Sigma=1$. c) Curve 1, $\Lambda=0.06$; curve 2, $\Lambda=0.08$; curve 3, $\Lambda=0.1$; curve 4, $\Lambda=0.16$; $\bar{\chi}=20$, $F=201$, $\theta=9\cdot10^{-4}$, $\Sigma=1$

Calculations show that, as the parameter $\bar{\chi}$ decreases, the domain of the stationary states of the bubbly reactor expands as a consequence of an increase in the reactivity of the vapor–gas mixture. A decrease in F (reduction of the intensity of convection and heat removal to the walls) expands the domain of the intermediate state noticeably. The latter is due to the fact that under the constant intensity of the interfacial heat flux the relationship between the time of thermal relaxation and residence time of the bubbles in the reactor also changes as F decreases. At a sufficiently small F (large $\bar{\Omega}$) when the time of thermal relaxation is greater than the residence time, the hot bubbles transmit only a part of their heat to the liquid and hence, to the cold bubbles, thus hindering the establishment of thermal equilibrium. Figure 10.14 shows the effect of the parameters θ_w, Σ, Λ and $\overline{\chi}$ on the shapes and positions of the $\omega(\overline{\Omega})$ curves in the parametric plane $\omega - \overline{\Omega}$. It is seen that those parameters affect noticeably the intensity of heat release and the temperature corresponding to ignition but have practically no influence on those values in the extinction regime.

Note that the results presented above correspond to the combustion of bubbly media, having low thermal conductivity of the liquid reagent. At large thermal conductivities (for example, in the case of liquid metal) it is natural to assume that the temperature of the liquid reagent is constant over the whole bulk of the reactor. Therefore, the bubbly medium may be represented as the one-temperature liquid continuum that contains gas inclusions with different temperatures. In the framework of this model, the general approach to the description of the thermal regime of bubbly stirred combustion reactors is identical to the approach developed above for combustion of liquid with low thermal conductivity.

3.10.5 Jet gas–liquid reactor

The results considered above correspond to gas–liquid reactors in which a premixed gas–droplet or bubbly mixture is used. The process in such reactors changes noticeably in the case of separately supplying reactants when their mixing occurs immediately in the reactor volume (Fig. 10.15). Gas jets issued into liquid reagent form a two-phase jet-like flow which, to a considerable extent, determines the general characteristics of the reactor: the intensity of heat release, completeness of combustion, etc. Depending on the volumetric contents of the gaseous phase such flow acquires the form of bubbly (ξ >0.25) or gas–droplet (ξ <0.25) jets (Abramovich 1984).

As a consequence of heating and vaporization of liquid reagent due to chemical reaction, domains filled with vapor–oxidizer mixture containing droplets of reactive liquid or bubbles filled by the mixture are formed in the reactor. At high initial temperature of reactants (or due to the presence of an external igniter) a high-temperature zone—the diffusion flame—is formed in the two-phase mixture. In the case when combustion products are gaseous, the diffusion flame is located in an open cavity (Fig. 10.16a); when combustion products are condensed, the diffusion flame is located in a closed gas cavity (Fig. 10.16b). Its size depends not only on the physico-chemical properties of reactants and intensity of their mixing, but also on the boiling temperature of combustion products that determine the position of the condensation zone.

Sukhov and Yarin (1981, 1983), and Yarin and Sukhov (1987) developed the theory of laminar immersed flames for the case when combustion products are gaseous. Using a number of simplifying assumptions (e.g. that the rate of chemical reaction is infinitely large, the thermal conductivity and diffusion coefficients are constant, the effect of buoyancy force is negligible, etc.) they reduced the problem to solving the system of equations

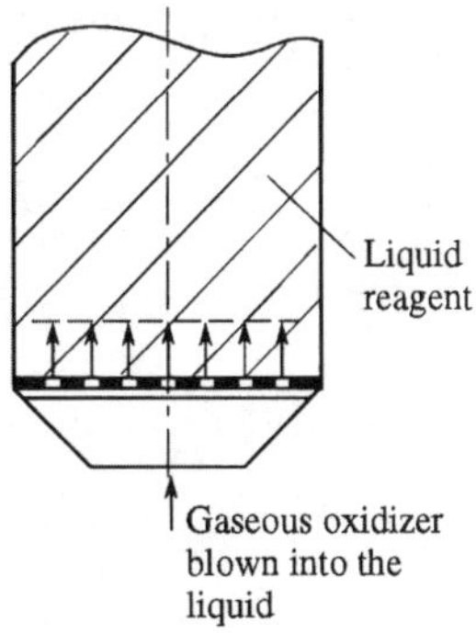

Fig. 10.15 Scheme of jet gas–liquid reactor

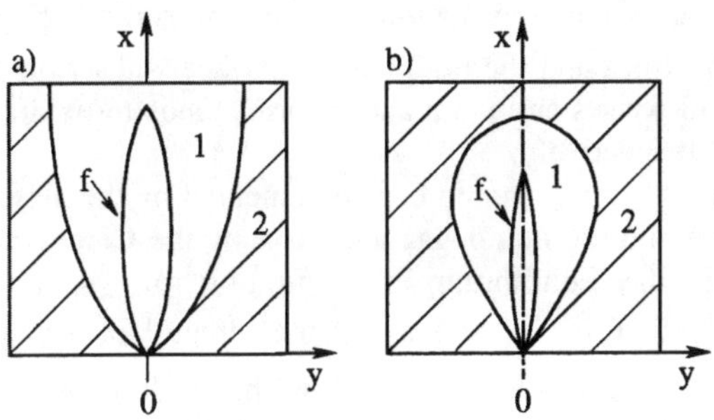

Fig. 10.16 Scheme of immersion flame. a) Flame with gaseous combustion products. b) Flame with condensed combustion products

$$\begin{aligned}
& u_i \frac{\partial u_i}{\partial x} + v_i \frac{\partial u_i}{\partial y} = \nu_i y^{-k} \frac{\partial}{\partial y}\left(y^k \frac{\partial u_i}{\partial y} \right) \\
& \frac{\partial}{\partial x}(y^k u_i) + \frac{\partial}{\partial y}(y^k v_i) = 0 \\
& u_1 \frac{\partial C_j}{\partial x} + v_1 \frac{\partial C_j}{\partial y} = D_j y^{-k} \frac{\partial}{\partial y}\left(y^k \frac{\partial C_j}{\partial y} \right) \\
& u_i \frac{\partial T_i}{\partial x} + v_i \frac{\partial T_i}{\partial y} = \tilde{\alpha}_i y^{-k} \frac{\partial}{\partial y}\left(y^k \frac{\partial T_i}{\partial y} \right) \\
& \sum_j C_j = 1
\end{aligned} \qquad (10.81)$$

In the system of Eqs. (10.81) the first, third and fourth equations are the momentum species and energy equations in the boundary layer approximation; the second one is the continuity equation.

The solution of Eqs. (10.81) is subject to the following boundary conditions

$$
\begin{aligned}
&y = 0: && \frac{\partial u_1}{\partial y} = 0 \quad v_1 = 0, \quad \frac{\partial C_a}{\partial y} = \frac{\partial C_c}{\partial y} = 0, \quad C_{b_1} = 0 \\
&y = y_\phi(x): && C_a = C_{b_1} = 0, \quad C_c = 1, \qquad T_{1'} = T_\phi \\
&y = y_*(x): && u_1 = u_2 = u_*, \quad \frac{\partial u_1}{\partial y} = \rho_{2.1} v_{2.1} \frac{\partial u_2}{\partial y}, \quad C_a = 0, \ T_1 = T_2 = T_* \\
&y \to \infty: && u_2 \to 0, \quad T_2 \to T_\infty
\end{aligned}
\qquad (10.82)
$$

where u and v are the longitudinal and transverse velocity components, ν, $\tilde{\alpha}$ and D are the kinematic viscosity, thermal diffusivity, and diffusivity (D_j=D), respectively, k=0 or 1 for plane or axisymmetric flow, subscripts ϕ and * correspond to the combustion front and the liquid–gas interface, subscripts i=1and 2 refer to gaseous and liquid phases and j=a,b and c corresponds to oxidizer, fuel and combustion products, respectively.

The conditions (10.82) should be supplemented by the mass and thermal balances at the flame and the liquid–gas interface, by the Clapeyron–Clausius equation determining the equilibrium concentration of vapor at the interface $C_{b1*} = \chi \exp\left[-q_e/(R_b T_*)\right]$, as well as by the integral invariants to find a nontrivial solution of the corresponding jet flow. In the framework of the boundary layer theory the slope of the flame front to the flow axis is small enough $((\partial/\partial n)_\phi \approx (\partial/\partial y)_\phi$, n is the normal to the flame front). Then assuming the Lewis number Le=1, we obtain the above-mentioned additional conditions at the flame front and the liquid–gas interface, as well as the integral invariants in the following form

$$
\left(\frac{\partial C_a}{\partial y}\right)_\phi + \sigma\left(\frac{\partial C_{b1}}{\partial y}\right)_\phi = 0 \qquad (10.83)
$$

$$
\left(\frac{\partial T_1}{\partial y}\right)_{b.\phi} - \left(\frac{\partial T_1}{\partial y}\right)_{a.\phi} + \frac{q}{c_p}\left(\frac{\partial C_{b1}}{\partial y}\right)_\phi = 0 \qquad (10.84)
$$

$$
v_{1*} C_{c*} + D\left(\frac{\partial C_c}{\partial y}\right)_* = 0 \qquad (10.85)
$$

$$
\rho_2 v_{2*} - \rho_1 v_{1*} C_{b_{1*}} - \rho_1 D\left(\frac{\partial C_{b1}}{\partial y}\right)_* = 0 \qquad (10.86)
$$

$$\lambda_1\left(\frac{\partial T_1}{\partial y}\right)_* - \lambda_2\left(\frac{\partial T_2}{\partial y}\right)_* + \rho_2 v_{2*} q_e = 0 \tag{10.87}$$

$$\int_0^{y_*} u_1^2 y^k dy + \rho_{2.1}\int_{y_*}^{\infty} u_2^2 y^k dy = \frac{I_0}{\rho_1} = \text{const} \tag{10.88}$$

$$\int_0^{y_*} u_1 G y^k dy = \frac{M_0}{\rho_1} = \text{const} \tag{10.89}$$

where $G=(1+\Delta C)$, $\Delta C=(C_a/\sigma)-C_b$ is the Schvab–Zel'dovich variable, I_0 and M_0 are contants.

Assuming self-similar velocity distributions of the liquid and gaseous phases, we assume the following form of them:

$$\bar{u}_i = A_i \bar{x}^{\alpha_i} F_i'(\varphi)\varphi^k \tag{10.90}$$

where $\bar{u}_i = u_i/u_0$, $\bar{x}_i = x/L$, $u_0 = I_0/M_0$, $L=\left(M_0^2/(\rho_1 I_0)\right)^{1/(k+1)}$, $\varphi = y/y_*(x)$, $\bar{y}_* = B\bar{x}^{\beta}$, A, B, α and β are constants.

Then we obtain from Eqs.(10.81) the following equation for the function F_i

$$\left(\frac{F_i'}{\varphi}\right)^2 + \frac{F_i}{\varphi^k}\left(\frac{F_i'}{\varphi^k}\right) + \frac{1}{\varphi^k}\left[\varphi^k\left(\frac{F_i'}{\varphi^k}\right)'\right]'\frac{3-k}{(k+1)(6-5k)} = 0 \tag{10.91}$$

The boundary conditions for Eq. (10.91) are obtained from Eq. (10.82) in the form

$$\varphi = 0,\quad \left(\frac{F_1'}{\varphi^k}\right)' = 0,\quad \frac{F_1'}{\varphi^k} = 1,\quad F_1 = 0$$
$$\varphi = 1,\quad F_1' = \nu_{2.1} F_2',\quad \left(\frac{F_1'}{\varphi^k}\right)' = \rho_{2.1}\nu_{2.1}\left(\frac{F_2'}{\varphi^k}\right)' \tag{10.92}$$
$$\varphi \to \infty\quad \frac{F_2'}{\varphi^k} \to 0,\quad \left(\frac{F_2'}{\varphi^k}\right)' \to 0$$

Integrating Eq. (10.91) we obtain

$$F_1 = \mathrm{th}\varphi; \quad F_2 = a_1 \frac{\exp(a_1\varphi) - a_2 \exp(-a_1\varphi)}{\exp(a_1\varphi) + a_2 \exp(-a_1\varphi)} \tag{10.93}$$

for the plane flow, and

$$F_1 = \frac{4\varphi^2}{8+\varphi^2}, \quad F_2 = \frac{(2+a_5)\varphi^{a_5} - (2-a_5)a_3}{\varphi^{a_5} - a_3} \tag{10.94}$$

for the axisymmetric flame

In Eqs. (10.90), (10.93) and (10.94)

$$A_1 = (I_1 + \rho_{2.1} v_{2.1}^2 I_2)^{-\frac{2}{3-k}} \left[\frac{Re_1}{6-5k}\right]^{\frac{1+k}{3-k}}, \qquad A_2 = v_{2.1} A_1$$

$$B = (I_1 + \rho_{2.1} v_{2.1}^2 I_2)^{\frac{1}{3-k}} \left[\frac{Re_1}{6-5k}\right]^{-\frac{2}{3-k}},$$

$$\alpha_i = -\frac{k+1}{3-k}; \qquad \beta = \frac{2}{3-k},$$

$$I_1 = \int_0^1 \frac{(F_1')^2}{\varphi^k} d\varphi, \qquad I_2 = \int_1^\infty \frac{(F_2')^2}{\varphi^k} d\varphi,$$

$$a_1 = \frac{\sqrt{v_{1.2} + [v_{1.2} \sin h(1)/\rho_{2.1}]^2}}{\cos h(1)},$$

$$a_2 = \frac{a_1 - v_{1.2} \tan h(1)/\rho_{2.1}}{a_1 + v_{1.2} \tan h(1)/\rho_{2.1}} \exp(2a_1), \qquad a_3 = \frac{4 - 9\rho_{2.1} v_{2.1}(a_5+2)}{4 + 9\rho_{2.1} v_{2.1}(a_5-2)},$$

$$a_4 = \frac{8(1 + 8v_{2.1}\rho_{2.1}^2 - 9v_{2.1}\rho_{2.1}}{81(v_{2.1}\rho_{2.1})^2}, \qquad a_5 = \sqrt{4+2a_4}, \quad Re_1 = \frac{I_0^{\frac{1}{2}}}{\rho_1^{\frac{1}{2}} v_1}$$

Note that Eq. (10.91) has three solutions (for the liquid domain of the axisymmetric flow) corresponding to different values of constant a_4: $a_4 > -2$, $a_4 = -2$, and $a_4 < -2$; the physical meaning has only the solution corresponding to $a_4 > -2$.

In order to find the concentration and temperature distribution we use the integral method of calculation as well as the hypothesis of local similarity. The profiles G and θ_i are expressed as (Sukhov and Yarin 1981).

$$G(\bar{x}, \bar{y}) = G_m (1-\varphi)^2 + G_* \varphi^2 \tag{10.95}$$

$$\theta_1 = \theta_m + \frac{G_m - G}{G_m - 1}(\theta_\phi - \theta_m); \quad \text{for} \quad 0 < \varphi \le \varphi_\phi \quad \text{and} \tag{10.96}$$

$$\theta_1 = \theta_* + \frac{(G - G_*)}{(1 - G_*)}(\theta_\varphi - \theta_*) \quad \text{for} \quad \varphi_\phi \le \varphi \le 1 \tag{10.97}$$

$$\theta_2 = \theta_* - \frac{(\theta_* - 1)}{I_4}\int_o^\varphi \varphi^{-k} \exp[-nI(\varphi)]d\varphi \tag{10.98}$$

where

$$\theta = \frac{T}{T_\infty}, \; G_m = \frac{1}{F_1(1) - I_3}\left(\frac{\overline{x}^\alpha}{A_1 B^{k+1}} - G_* I_3\right), \; \vartheta = \frac{q}{c_p T_\infty}$$

$$\theta_\phi = \theta_* + [(\theta_m + \theta_*) + (G_m - 1)\vartheta]\frac{1 - G_*}{G_m - G_*}, \; I(\varphi) = \int_0^\varphi \varphi^{-k} F_2 d\varphi, \; I_3 = \int_0^1 \varphi^2 F_1' d\varphi$$

$$I_4 = \int_0^\infty \varphi^{-k} \exp[-nI(\varphi)]d\varphi, \; n = \frac{6 - 5k}{(k+1)(3-k)} Pr$$

The shape and length of the flame are expressed as follow:

$$\overline{y}_\phi = \overline{y}_*(\overline{x})\left[\frac{G_m(\overline{x}) - 1}{G_m(\overline{x}) - G_*}\right]^{\frac{1}{2}} \tag{10.99}$$

$$\ell_\phi = \frac{Re_1}{6 - 5k}\left(I_1 + \rho_{2.1} v_{2.1}^2 I_2\right)^{\frac{1-k}{1+k}} [F_1(1) - I_3(1 - G_*)]^{\frac{k-3}{k+1}} \tag{10.100}$$

The effect of various parameters on the characteristics of the immersed laminar flame is clearly visible via the dependence of its length on them, namely on the reactants, temperature, issue velocity, etc. An increase of the reagents, temperature, as well as preheating of the reactive mixture lead to shortening of the diffusion flame length. This results from growth of the vapor concentration at the interfacial surface and, accordingly from growth of the rate of gaseous oxidizer consumption per unit of flame length. An opposite effect takes place on increasing the latent heat of evaporation of the liquid reagent. As in the other types of diffusion flames (Vulis and Yarin (1978), the characteristics the immersed flame depend on the low geometry. For example, the lengths of the axisymmetric and plane flames ℓ_ϕ follow different laws, namely, they are proportional to the first and the third powers of C_{b*}, respectively. In both cases ℓ_ϕ, as well as the intensity of heat release in the flame are proportional to the initial velocity of the gas jet. The composition of the gas jet (in particular, the presence of inert admixture) also affects the flame characteristics, Fig. 10.17 (Sukhov and Yarin 1983). An increase

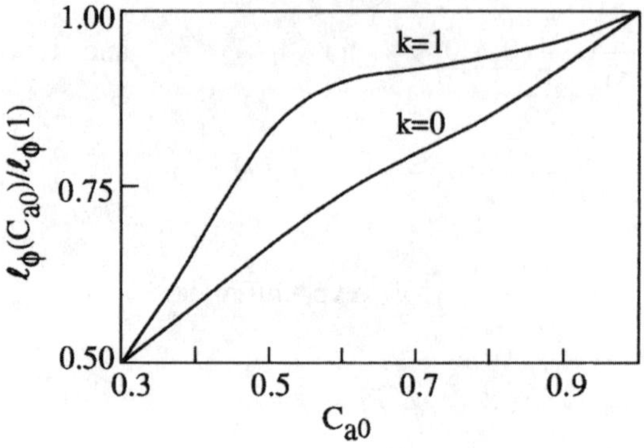

Fig. 10.17 The effect of inert admixture on the length of a plane and axisymmetrical immersion flame. $C_{a.0}$ is the oxidizer concentration in initial mixture. Reprinted from Yarin and Sukhov (1987), with permission

in content of an inert admixture in gaseous oxidizer is accompanied by a noticeable decrease in the lengths of the axisymmetric and plane immersed flames.

Avery and Faeth (1974) considered an axisymmetric immersed turbulent flame resulting from an issue of gaseous oxidizer into a liquid metal. To study the process characteristics in the bubbly flow regime they used the integral method of calculation. Accordingly, considering the bubbly jet as a gas jet with a variable density ρ, velocity u and enthalpy i they presented the distribution of flow parameters as

$$\frac{\rho u}{(\rho_m u_m)} = F_1(\eta); \quad \frac{\rho u^2}{\rho_m u_m^2} = F_2(\eta)$$
$$\frac{\rho u \Delta i}{\rho_m u_m \Delta i_m} = F_3(\eta); \quad \frac{\rho u \tilde{C}}{\rho_m u_m \tilde{C}_m} = F_4(\eta) \tag{10.101}$$

where $\eta = y/\delta(x)$, $\Delta i_m = i_m - i_\infty$, $\tilde{C} = (C_a/C_{a\infty})\sigma - (C_b/C_{b\infty}) + 1$, $\delta(x)$ is the jet cross-sectional radius, and subscript m corresponds to the jet axis.

The axial values of the parameters $\rho_m, u_m, \Delta i_m, \tilde{C}_m$, as well as $\delta(x)$ are found from the integral relations

$$\frac{d}{dx}\int_0^\infty \rho u y dy = -(y\rho v)_{y\to\infty} \tag{10.102}$$

Equation (10.102) characterizes ejection features of turbulent jet due to the liquid and inflow from surrounding fluid.

$$2\pi\int_0^\infty \rho u^2 y dy = I_0 \tag{10.103}$$

$$2\pi\int_0^\infty \rho u \Delta i y dy = Q_0 \tag{10.104}$$

$$2\pi\int_0^\infty \rho u \tilde{C} y dy = G_0 \tag{10.105}$$

where I_0, Q_0 and G_0 are the integral invariants of the problem $I = I_0 + I_x$, $I_0 = m_0 u_0$, $I_x = (P_0 - P_\infty)((\pi d^2)/4)$, $Q_0 = m_0 \Delta i_0$, $G_0 = m_0 \tilde{G}_0$, $m_0 = 2\pi\int_0^\infty \rho u y dy$ at x=0, u_0 is the initial velocity of the gas jet, $\Delta i_0 = i_\infty - i_0$, and d is diameter of the nozzle.

Assuming that the transversal mass flux at the external boundary of the bubbly jet is proportional to $u_m \delta(x)$

$$-(\rho v y)_{y\to\infty} = \varepsilon \rho_\infty b(x) \cdot u_m \tag{10.106}$$

and using Eqs. (10.102)–(10.105) and the conditions $\tilde{C}_m = \varepsilon_1$ at $x = \ell_\phi$ (in diffusion flame $\varepsilon_1 = 1$), we find the expression for the flame length

$$\frac{\ell_\phi}{d_0} = \frac{m_0 I^{\frac{3}{2}}}{\varepsilon \;\; \varepsilon_1 d_0 B_1 I_4 \sqrt{2\pi\rho_\infty I}} \tag{10.107}$$

where $b(x) = \delta(x)\left(\rho_m/\rho_\infty\right)^{1/2}$, ε an empirical constant, $B_1 = (\tilde{C}_0 - \varepsilon_1 (I_4/I_1))^{-1}$, $I_i = \int_0^\infty F_i(\eta) d\eta$.

The comparison of experimental data on the length of diffusion flames of various types with the theoretical predictions of Avery and Faeth (1974) is presented in Fig. 10.18. It is seen that there is a fairly good agreement between the data of measurements and the theoretical results. The latter shows that in spite of a number of simplifications used in the Avery and Faeth (1974) theory, it accounts for the general trends of the complicated process.

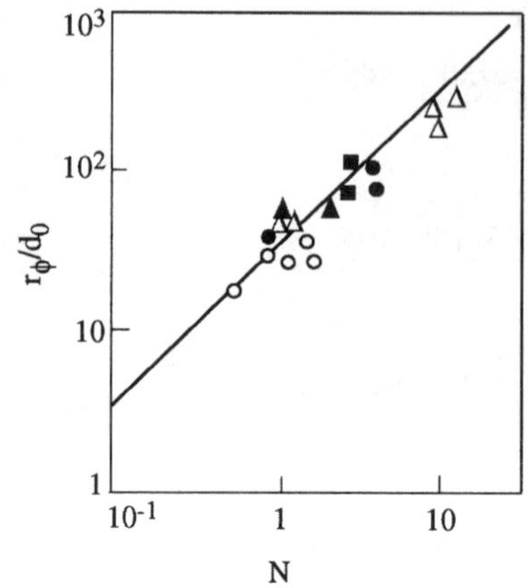

Fig. 10.18 Dependence the length of turbulent diffusion flames on parameter $N = m_0/(d_0 B_1 \sqrt{\pi\rho_\infty})$. Reprinted from Avery and Faeth (1974), with permission. ○ , immersed flame (system $Na + NaCl + Cl_2$), ●, H_2; ■, urban gas; ▲, CO; Δ , $CH_3CH_2CH_4$: free turbulent flame (oxidizer=air). Straight line, Avery and Faeth theory

3.10.6 Gas–solid particle reactor

The models developed in the previous sections were extended to the other types of ideally-stirred combustion reactors, in particular, to the ideally-stirred combustion reactor that converts a solid reactant. The stationary states of such a reactor were investigated by Genkin et al. (1981). Consider the thermal regime of an ideally-stirred gas–particle combustion reactor following this work.

The equations employed in the analysis of the gas–liquid reactor were obtained in Section 3.10.2. They express the mass and energy conservation in a two-phase multicomponent reactive medium: Eqs. (10.23–10.27). In addition the following simplifying assumptions will be used in the present section: (i) a monodispersed gas–particle reactive mixture with low volumetric content of condensed phase is supplied to the reactor inlet, (ii) the particles are spherical, (iii) the residence times of all components of the two-phase mixture in the reactor (particle, oxidizer, combustion product) are identical, (iv) the heat transfer intensity in the carrier gas is infinitely high, so that the gaseous phase temperature in different continua is the same. Assuming that the average parameters of the reactive mixture are independent of time ($d < \Pi >/dt = 0$, Π is the considered parameter, for example, $\Pi = \rho_\ell$; $\rho_{1.2}$, i...), we present the energy and mass balance equations in the form

$$\int_0^\infty \eta\varphi(t_r)dt_r - \sum_{i=1}^{3} \Delta H_i = 0 \tag{10.108}$$

$$\int_0^\infty \frac{d\eta}{dt}\varphi(t_r)dt_r - j_0\int_0^\infty \eta\varphi(t_r)dt_r = 0 \qquad (10.109)$$

where q is the heat of reaction, j_0 is the specific volumetric flow rate of the mixture fed into the reactor, $\eta = 1-(\rho_1/\rho_{1.0})$ is the completeness on chemical transformation, $\varphi(t_r)$ is the distribution function of particle residence time in the reactor. Also

$$\Delta H_i = H_i - H_{i.0};\quad H_1 = \rho_{1.0}j_0c_1T_{1.0}\int_0^\infty (1-\eta)\bar{T}_1\varphi(t_r)dt_r$$

$$H_2 = \rho_{2.0}j_0c_2T_{2.0}(1-\sigma\int_0^\infty \eta\varphi(t_r)dt_r)$$

$$H_3 = \rho_{3.0}j_oc_3T_{3.0}(1+\sigma)\int_0^\infty \eta T_3\varphi(t_r)dt_r,\quad \bar{T}_i = \frac{T_i}{T_{1.0}};\quad T_{1.0} = T_{2.0} = T_{3.0}$$

where c_I is the specific heat of the i th phase; subscripts 1, 2 and 3 correspond to solid reagent, carrier gas, and combustion products, respectively, 0 to the parameters at the reactor inlet.

The system of Eqs. (10.108) and (10.109) should be supplemented by the energy and mass balance equations for a single particle determining its instantaneous temperature and completeness of combustion. The first of these equations is

$$m_1c_1\frac{dT_1}{dt} = q_*\rho_*\frac{d\delta}{dt}S + h(T_\infty - T_1)S + \tilde{\sigma}\varepsilon(T_\infty^4 - T_1^4)S \qquad (10.110)$$

where m_1 and c_1 are the mass and specific heat of a particle, q_*, ρ_*, are the heat release with respect to the combustion product, the density of combustion product, δ is thickness of the oxide film on the particle surface, S is the particle surface, h is the heat transfer coefficient; $\tilde{\sigma}$ and ε are the Stefan–Bolzmann constant and emissivity of the oxide film.

Equation (10.110) accounts for the convective and radiative heat transfer under the assumption of optical transparency of gas and ideal reflection of radiation from the reactor walls, the temperature of which is assumed to be equal to the gas temperature.

Assuming, for example, that the reactive mixture contains metallic particles, we employ the simplest kinetic law determining such particle oxidation (see Chapter 1.3).

$$\frac{d\delta}{dt} = \frac{k_nC_s^m}{\delta^n}\exp\left[-\frac{E}{RT_1}\right] \qquad (10.111)$$

where k_n is the pre-exponential coefficient in the corresponding kinetic rate law, m is the order of reaction, E is the activation energy, R is the universal gas constant and C_s is the oxidizer concentration on the particle surface.

Then Eq. (10.110) can be presented as

$$\frac{dT_1}{dt} = \left(\frac{q}{c_1} + aT_1\right)\frac{3z}{r\delta^n}\exp\left(-\frac{E}{RT_1}\right) + \frac{3h}{c_1\rho_1 r}\left[(T_\infty - T_1) + \chi(T_\infty^4 - T_1^4)\right] \tag{10.112}$$

where q is the heat release with respect to solid reagent, $z = k_n C_s^m$, $a = -(\sigma c_3)/c_1$, and $\chi = (\sigma\varepsilon)/h$.

Taking into account characteristics of the activated process, we use the following approximation of the functions $T_1(t)$ and $\eta(t)$: $T_l = \text{const}$, $\eta = 0$ when $t < t_{ind}$ ($T_{1.0} \le T_l \le T_i$, T_i is the ignition temperature, t_h is the heating time up to particle ignition); $T_1=T_3=\text{const}$, $\eta = 1$ at $t \ge t_{ind}$ ($T_2 \le T \le T_{ad}$, T_{ad} is the adiabatic temperature of combustion). Note that the last assumption is rigorously fulfilled for gaseous combustion product and is approximately fulfilled for finely dispersed combustion products.

In the case of a small change in T_1 and T_2 during particle heating it is possible to use a linear approximation of the radiative heat transfer and of the rate of heat release due to chemical reaction, as well as to neglect the change in the particle size. Using these simplifications the integration of Eqs. (10.108) and (10.109) yields the expression

$$\omega = \omega_* \ln\left(\frac{\bar{T}_2 - K_I}{\bar{T}_2 - K_{II}\bar{T}_i}\right) \tag{10.113}$$

where $\omega = t_h \cdot j_0$, $\omega_* = t_0 j_0$, $\bar{T}_2 = T_2/T_{1.0}$, $t_0 = [(c_1\rho_1 r)/(3h_\Sigma)]\left(1-(h/(h_\Sigma\gamma Se))\right)^{-1}$, $K_I = 1-(h/h_\Sigma)\gamma Se(1+(\beta/\gamma))$, $K_{II} = 1-(h/h_\Sigma)Se(1+(\beta/\gamma))$, $h_\Sigma = h + h_R$, h_R is the radiative heat transfer coefficient, r is the particle radius.

The ignition temperature T_i is determined from Eq. (10.110) and the condition $d^2T_l/dt^2 = 0$. The relation for the calculation T_i has the form

$$\frac{a}{\delta^n}\gamma S_e \exp\left[-\frac{1}{\beta}\left(\frac{\beta}{\theta_i} - 1\right)\right] - \frac{1 + 4\chi\left(\dfrac{\theta_i}{\beta}\right)^2}{1 + \dfrac{1}{\beta} + \dfrac{1}{a\gamma\beta^2\theta_i^2}} = 0 \tag{10.114}$$

where $\theta = (RT)/E$.

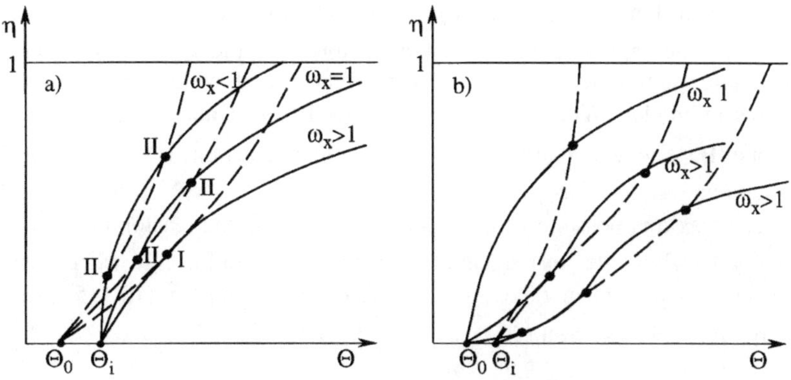

Fig. 10.19 The dependences $\eta_I(\Theta)$ dashed line (a/b) and $\eta_{II}(\Theta)$ solid line (a/b). Reprinted from Genkin et al. (1981)

Using the assumption that the thermal effect of the reaction does not depend on temperature, which means that $\sigma c_2 + c_1 = (1+\sigma)c_3$ we transform Eqs. (10.108) and (10.109) to the form

$$\eta_I = G\left(\frac{\Theta - \Theta_0}{\Theta_* - \Theta}\right) \tag{10.115}$$

$$\eta_{II} = \left(\frac{\Theta - \Theta_i}{\Theta - 1}\right)^{\omega} \tag{10.116}$$

where

$$\Theta = \frac{\Theta_2}{\beta K_I}, \quad \Theta_0 = \frac{\theta_{2.0}}{K_I \beta} + c_{1.2}\rho_{1.2} \cdot \frac{1}{K_I}(1 - \frac{\theta_1}{\beta}); \quad \Theta_i = \frac{K_{II}}{K_{II}} \; \frac{\theta_i}{\beta}, \quad \Theta_* = \frac{1}{K_I}\left(\frac{\theta}{\beta} + \frac{\beta}{\gamma}\right)$$

$G = c_{2.1}\rho_{2.1}$, $h_{I,II} = \int_0^\infty \eta\varphi(t_r)dt_r$ is total completeness of combustion of solid reagent, subscripts I and II correspond to Eqs. (10.108) and (10.109), respectively.

The system of equations (10.115) and (10.116) determines stationary values of gas temperature Θ and total completeness of combustion η on parameters G, ω_*, Θ_i, Θ_0 and Θ_*. The dependences $\eta_I(\Theta)$ and $\eta_{II}(\Theta)$ calculated using Eqs. (10.115) and (10.116) are shown in Fig. 10.19. The points of intersection of curves $\eta_I(\Theta)$ and $\eta_{II}(\Theta)$ correspond to steady states of a gas–solid particle reactor which are characterized by steady values of the temperature Θ_{st} and completeness of combustion η_{st}. When the temperature of the carrier gas entering the reactor is less than ignition temperature, i.e. $\Theta_0 < \Theta_i$ (Fig. 10.19a) one (I) or two (II) stationary states are possible. Note that in this case there is still another low-temperature state of the reactor, which cannot be found within the framework of the present model. When $\Theta_0 > \Theta_i$ (see Fig. 10.19b) and $\omega_* \le 1$ there exists only one (I) stationary state and when $\omega_* > 1$ there exist one (I), two (II) or three (III) stationary states.

The maximal possible value of the temperature in the reactor Θ_{max} is determined by the flow rates of reagent at the reactor inlet. In the case when the gaseous reagent is present in excess Θ_{max} can be found from Eq. (10.115) provided that $\eta = \eta_{max} = 1$. The result is

$$\Theta_{max} = \frac{\Theta_* + G\Theta_0}{1+G} \tag{10.117}$$

On the other hand in the case of an excess of the solid phase $(A \le 1)$, when $\eta = \eta_{max} = A$, by relation

$$\Theta_{max} = \frac{K\Theta_* + \Theta_0}{1+K} \tag{10.118}$$

where $K = c_{1.2}(\xi/\sigma)$.

Since the operating regimes of a two-phase gas–particle reactor are determined by five parameters, the stationary states of such a reactor can be presented in three-dimensional parametric space in the form of ten diagrams. Each of these diagrams corresponds to certain values of the governing parameters. As an analysis shows, the parameters ω_*, G, Θ_* and the relationship between Θ_i and Θ_0 have the most substantial effect on the operational regimes of the gas–particle combustion reactor. In Fig. 10.20 the regions of the existence of stationary states are shown in the space of the parameters ω_*, G and Θ_* (at $\Theta_0 > \Theta_i$). The characteristic shapes of several sections of the parametric space by planes $\Theta_* = \text{const}$ and $\omega = \text{constant}$ is shown in Fig. 10.21 ($\Theta_i = 3$, $\Theta_0 = 4$, $K = 1$). In Fig. 10.20 it is seen that in the general case, the parametric space $\omega_* - G - \Theta_*$ is separated by four surfaces F_a (1, 2, 3, 4), F_b (5, 6, 7, 8), F_e (9. 10,. 12) and F_d (9, 10, 13), determined by the conditions: $\Theta_{max} = \Theta_0 - (F_a)$; η = A, $\eta_I^{'} = \eta_{II}^{'} - (F_c, F_d)$. The surface subdivided the parametrical space into five domains corresponding to one (II), one or two (IV) and three (III) stationary states of the reactor, or to the ab sence of solutions (I, IV). There are also domains where stationary states are impossible (I, IV).

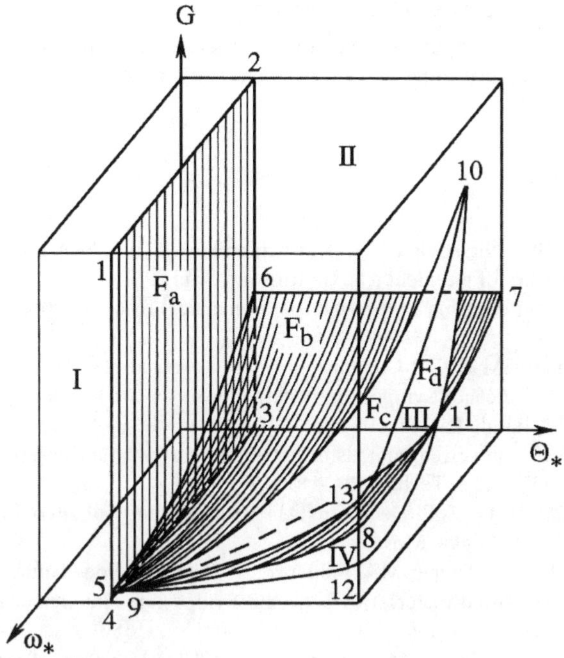

Fig. 10.20 Domains of stationary states of a two-phase gas–particle reactor. Reprinted from Genkin et al. (1981), with permission

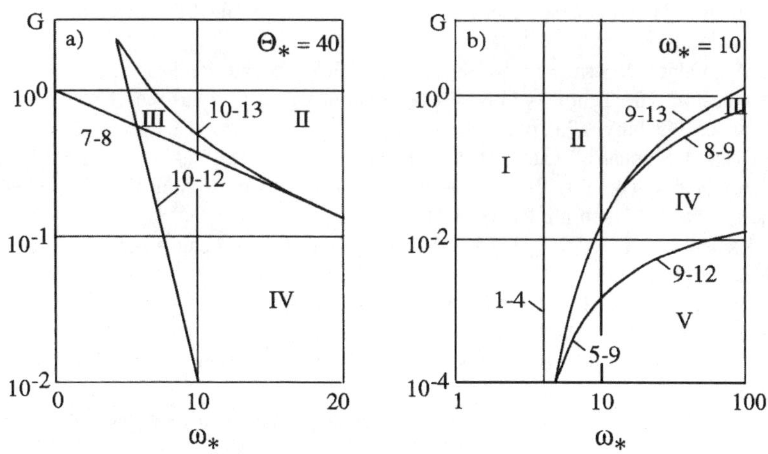

Fig. 10.21 Sections of parametrical space $\omega_* - G - \Theta_*$. Reprinted from Genkin et al. (1981). a) Dependence $G(\omega_*)$. b) Dependence $G(\Theta_*)$

Similarly it is possible to construct a diagram for $\Theta_0 < \Theta_i$. In this case the parametric space is separated into four characteristic domains which correspond to one or two stationary states or to the absence of solutions.

References

Abramovich GN, Girshovich TA, Krasheninnikov SYu, Sekundov AN, Smirnova, IP (1984) Theory of turbulent jets. (in Russian) Nauka, Moscow

Aris R (1965) Introduction to the analysis of chemical reactors. Prentice-Hall, Englewood Cliffs. N. J.

Avery JF, Faeth GM (1974) Combustion of a submerged gaseous oxidizer jet in a liquid metal. The Fifteenth Symposium (International) on Combustion. The Combustion Institute, Pittsburgh, Pa., pp. 501–512

Barat RB (1992) Jet-stirred combustor behavior near blowout: observations and implications. Combust. Sci. Technol. 84: 187–197

Barichnikov NV, Gager VE, Denisov ND (1979) Metallurgy of zirconium and hafnium. (in Russian) Metallurgya, Moscow

Bradley D, Chin SB, Draper MS, Hankinson G (1967). Aerodynamic and flame structure within a jet-stirred rector. The Sixteenth Symposium (International) on Combustion. The Combustion Institute, Pittsburgh, Pa., pp. 1571–1580

Buyevich YuA, Korolyova NA, Natalukha IA (1993a) Modeling of unsteady combustion regimes for polydispersed fuels. 1 - instability and auto-oscillations. Int. J. Heat Mass Transfer 36: 2223–2231

Buyevich YuA, Korolyova NA, Natalukha IA (1993b) Modeling of unsteady combustion regimes for polydispersed fuels. 2 - parametrically controlled combustion. Int. J. Heat Mass Transfer 36: 2233–2238

Clarke AE, Odgers J, Stringer FW, Harrison AJ (1965) Combustion processes in a spherical combustor. The Tenth Symposium (International) on Combustion. The Combustion Institute, Pittsburgh, Pa., pp. 1151–1166

Evangelista JJ, Shinnar R, Katz S (1969) The effect of imperfect mixing on stirred combustion reactors. The Twelfth Symposium (International) on Combustion. The Combustion Institute, Pittsburgh, Pa., pp. 901–912

Frank-Kamenetskii DA (1969) Diffusion and heat transfer in chemical kinetics - 2nd edn., Plenum, New York

Fridman NB, Kitain MM, Shteinberg AS, Merzhanov AG (1981) The mechanics of bubble ignition. Sov. Phys. Dokl. 258: 961–965 (in Russian)

Garmata BA, Gulyanitskii, B.S. 1968. The metallurgy of titanium. (in Russian) Metallurgiya, Moscow

Genkin AL, Gusika PL, Yarin LP (1981) Stationary states of a two-phase flow reactor. Combust. Explos. Shock Waves 17: 553–558

Hottel, HC, Williams GC, Nerheim NM, Schnieder GR (1965) Kinetic studies in stirred reactors: combustion of carbon monoxide and propane. The Tenth Symposium (International) on Combustion. The Combustion Institute, Pittsburgh, Pa., pp. 111–121

Labowsky M (1980) Calculation of the burning rates of interacting fuel droplets. Combust. Sci. Technol. 22: 217–226

Lignola PG, Reverchon E (1988) A jetstirred reactor for combustion studies: design and characterization. Combust. Sci. Technol. 60: 319–333

Likhachev VN, Sukhov GS, Yarin LP (1989) Towards a theory of high-temperature gas–liquid ideally stirred reactors. Sov. Phys. Dokl. 309: 914–917

Likhachev VN, Sukhov GS, Yarin LP (1991) The theory of bubble combustion reactors. Combust. Explos. Shock Waves 27: 191–199

Longwell JP, Weiss MA (1955) High-temperature reaction rates in hydrocarbon combustion. Ind. Eng. Chem. 47: 1634–1643

Merzhanov AG 1973. The problem of technological combustion. In: Merzhanov AG (ed) Combustion Process in Chenical Technology and Metallurgy. AN SSSR, Chernogolovka, pp. 15–28 (in Russian)

Merzhanov AG, Abramov BG (1976). The thermal regimes of the exothermal processes in ideally stirred flowing reactor. Preprint. (in Russian) Institute of Chemical Physics, Chernogolovka

Merzhanov AG, Abramov BG (1977). Thermal regimes of the exothermic processes in continous stirred tank reactors. Chem. Eng. Sci. 32: 475–481

Nigmatulin RI (1991) Dynamics of multiphase media. vols. 1 and 2. Hemisphere, London

Perlmutter DD (1972) Stability of chemical reactors. Prentice-Hall, Englewood Cliffs, N.J.

Pushkin VN, Sukhov GS, Yarin LP (1993) Thermal conditions of ideal-mixing gas–drop reactor. Combust. Explos. Shock Waves 29: 50–56

Sangiovanni JJ, Kesten AS (1975) Effect of droplet interaction on ignition in monodispersed droplet streams. The Sixteenth Symposium (International) on Combustion. The Combustion Institute, Pittsburgh, Pa., pp. 577–592

Sukhov GS, Yarin LP, (1981) Combustion of a jet of immiscible fluids. Combust. Explos. Shock Waves 17: 146–151

Sukhov GS, Yarin LP (1983) Calculating the characteristics of immersion burning. Combust. Explos. Shock Waves 19: 155–158

Thornton MM, Malte PC, Crittenden AL (1987) A well-stirred reactor for the study of pyrolysis and oxidation kinetics: carbon monoxide and n-pentane oxidation. Combust. Sci. Technol. 54: 275–297

Tsai JS, Sterling AM 1990. The combustion of linear droplet arrays. The Twenty-third Symposium (International) on Combustion. The Combustion Institute, Pittsburgh, Pa., pp. 1405–1411

Van Heerden (1953) Autothermic Process. Properties and Reaction Design. Ind. Eng. Chem. 46: 1242–1252

Vaughn CB, Sun WH, Howard JB, Longwell JP (1991) Measurements and modeling of light hydrocarbons in rich C_2H_4 combustion in a jet-stirred reactor. Combust. Flame 84: 38–46

Vulis LA (1961) Thermal regimes of combustion. McGraw-Hill, New York

Vulis LA, Yarin LP (1978) Aerodynamics of a torch. (in Russian) Energia, Leningrad

Williams FA (1985). Combustion theory. 2nd edn., The Benjamin/Cummings, Menlo Park, Calif.

Williams GC, Hottel HC, Morgan AS (1969) The combustion of methane in a jet-mixed reactor. The Twelfth Symposium (International) on Combustion. The Combustion Institute, Pittsburgh, Pa., pp. 913–924

Yarin LP, (1990) Thermal regime of combustion of bubbly media. Arch. Combust. 10: 185–200

Yarin LP, Hetsroni G (1995) On gas–liquid combustion reactor theory (ideally stirred reactor). Combust. Sci. Technol. 109: 93–120

Yarin LP, Sukhov GS (1987) Fundamentals of combustion theory of two-phase media. (in Russian) Energoatomizdat, Leningrad

Zel'dovich YaB, Barendlatt GI, Librovich VB, Makhviladze GM (1985) Mathematical theory of combustion and explosion. Plenum, New York

3.11 Displacement reactor

3.11.1 The kinematic balance method

Displacement combustion reactors are intended to achieve a large number of technological processes that are related to solid fuel gasification and combustion, high temperature synthesis of refractory materials, etc. As usual, they are operated as straight channels in which reactive material (one or two phase mixture of reactants) is entered and combustion products are removed (Fig. 11.1).

The operation of displacement combustion reactors depends on various factors including the thermal, diffusive and kinetic properties of reagents, the rate of supply of the reactive substance, as well as the intensity of heat transfer to the reactor walls. At finite rates of heat and mass transfer, the temperature and concentration fields in displacement reactors are essentially inhomogeneous. In the case of activated reaction, the conversion of initial reagents into combustion products occurs, mainly, within a thin region of flow which is identified with the combustion (flame) front. Thus, the development of the combustion process in displacement reactors is, in a sense, similar to that in reactive flow. In both cases the process is accompanied by the formation of a combustion wave which propagates over the reactive medium in the opposite direction to the flow velocity. This circumstance allows the application of the methods of combustion theory in studying the characteristics of displacement reactors (Merzhanov and Filonenko 1963, Merzhanov 1967, Butakov et al. 1978, Sukhov and Yarin 1978). Significant simplification of the problem is achieved by using a one-dimensional model of the displacement combustion reactor. In the frame of such an approximation the problem of the adiabatic displacement reactor is reduced to the known problem of the development of an exothermic reaction in a gas flow with distributed parameters (Zaidel and Zel'dovich 1962, Khaikin and Rumanov 1975). A number of important results related to displacement combustion reactors were obtained by Merzhanov and Filonenko (1963) and Butakov et al. (1978). It was shown that regimes of such reactors are determined by the ratio of the rate of the combustion wave corresponding to the initial temperature $u_f(T_0)$ to the supply velocity of reactive mixture at the reactor inlet u_*. On this basis, the classification of possible regimes of displacement reactors was formulated as follows: (i) separationless combustion regime at $u_f > u_*$ (flame front is located close to the inlet), (ii) self-ignition or induction regime at $u_f << u_*$ (flame front is located far from the inlet) and (iii) intermediate regime at $u_f < u_*$ (Fig. 11.2).

Sukhov and Yarin (1982) suggested a kinematic approach for studying stationary states of displacement combustion reactors. This approach takes into consideration a high inhomogeneity of flow parameters over a channel length that is due to localization of the reaction in a thin zone of flow and the flame front's capability between self-propagation over a substance. The state in the reactor channel is obviously determined in this case by a competition two factors: the supply

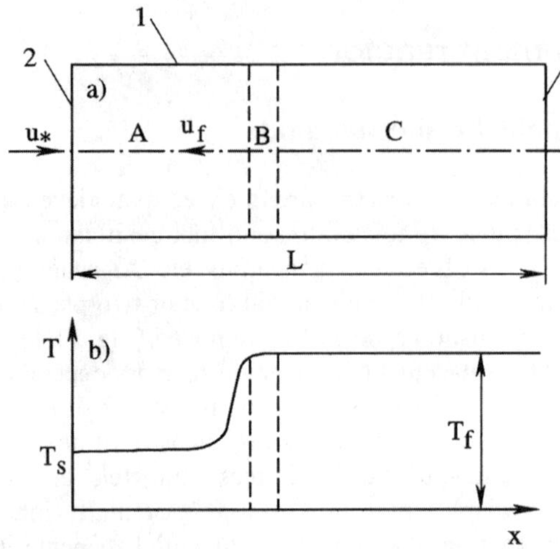

Fig. 11.1 A model of a displacement combustion reactor. a) 1, reactor channel; 2, 3, inlet and outlet cross-sections; A, reactive mixture zone, B, reaction zone (flame front), C, combustion products zone. b) Temperature distribution

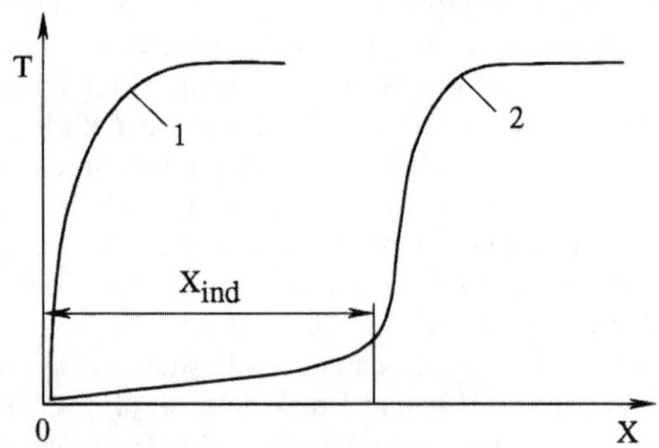

Fig. 11.2 Temperature distribution in a displacement combustion reactor. 1, Separationless combustion regime; 2, separation regime

of initial substance with a given rate u_* and transformation of the reactants into the final product in a combustion wave propagating over a substance in the opposite direction with the rate u_f. The rate u_f is unknown a priori, but it is determined by the set of physico-chemical characteristics of a medium and by the regime conditions of the process. Under these circumstances the theory describing the combustion reactor's operation should obviously be based on the kinematic principle of comparing the local values of u_f, calculated over the channel length, with the rate of supply of the reactive mixture u_*. The difference between these values in each cross-section of a channel allows for a judgement whether the process is stationary or not, and in which direction the flame front propagates in the channel.

The equality of these rates determines the position of a stationary combustion zone and its principal characteristics. As a matter of fact, the kinematic balance method rests upon a kinematic interpretation of the energy balance of a displacement reactor, in which the heat release and heat transfer intensities are proportional, respectively, to the flame propagation rate u_f and to the substance supply rate u_*.

The possibilities of the kinematic balance method are illustrated below by the particular example of a homogeneous displacement combustion reactor (Sukhov and Yarin 1982). In the case when the temperature and concentration fields are similar (Le=1) the governing system of equations describing the steady states of the displacement reactor takes the form

$$\rho c_p u_f \frac{dT}{dx} = \lambda \frac{d^2T}{dx^2} + qW \tag{11.1}$$

$$u_f = u_* \tag{11.2}$$

where

$$W(T) = z\left[\frac{c_p(T_f - T)}{q}\right]^n \exp\left(-\frac{E}{RT}\right)$$

z and n are the kinetic constants; E is the activation energy, R is the universal gas constant, q is the heat of combustion, λ is the thermal conductivity, c_p is specific heat, subscript f corresponds to the flame front.

The boundary conditions of the problem are

$$x = -x_f, \quad T = T_0 \tag{11.3}$$

$$x = L - x_f, \quad \frac{dT}{dx} = 0 \tag{11.4}$$

where L is the reactor length, x_f is the distance from the inlet of reactor to the flame front, and subscript zero corresponds to the initial state.

The boundary condition (11.3) corresponds to a reactor with a cooling inlet. The condition (11.4) is satisfied automatically in a reactor with a motionless layer of catalyst and in a reactor without a catalyst at certain values of the Peclet (Pe) and the Damkohler (Da) numbers, when the reactive mixture fully burns out within the reactor channel. The range of permissible Pe and Da values may be found by means of parametric analysis of the system of governing equations (Butakov et al. 1978).

In the general case corresponding to separated combustion regime when the flame front is located far from the reactor inlet, the integration of the energy equation (11.1) yields (using the Zel'dovich–Frank-Kamenetskii approximation) the following rate of the combustion wave:

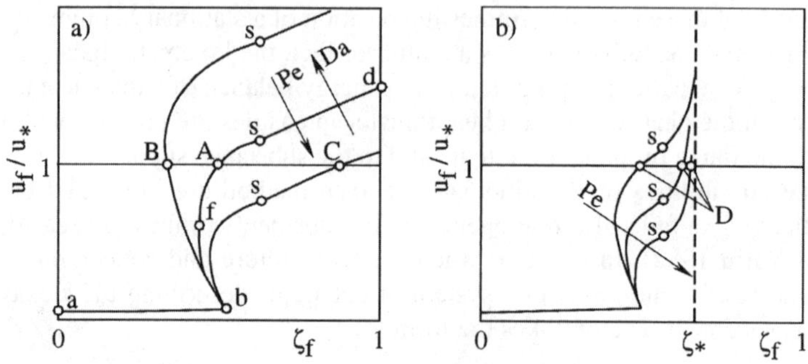

Fig. 11.3 The dependence $u_f/u_* = f(\zeta_f)$. Reprinted from Sukhov and Yarin (1982). a) regimes without self-ignition. b) Regimes with self-ignition

$$\left(\frac{u_f}{u_*}\right)^2 = \frac{z\gamma^{n-1}Da(1+\beta\theta_f)^{2(n+1)}\Gamma(n+1)}{Pe[\theta_m - \theta_0'(\zeta_f)]}\exp\left(\frac{\theta_f}{1+\beta\theta_f}\right) \tag{11.5}$$

where

$$\beta = \frac{RT_0}{E},\ \gamma = \frac{c_p RT_0^2}{C_o qE},\ \theta = \frac{E}{RT_0^2}(T - T_0),\ Da = \frac{z}{u_*}C_o^{n+1}L\exp\left[-\frac{E}{RT_0}\right]$$

is the Damkohler number, $Pe = (u_* L)/\alpha$, is the Peclet number, α is the thermal diffusivity, $\theta_m = 1/\gamma$ is the adiabatic temperature of combustion, θ_f is the final temperature, θ_0' is the temperature at the end of the induction zone, C_o is the reactive substance concentration in the initial mixture, $\zeta_f = x_f/L$, and $\Gamma(n+1)$ is Gamma function.

As has already been mentioned, the final temperature in a displacement combustion reactor depends on the position of the flame front in the reactor channel. When the distance from the reactor inlet to the flame front significantly exceeds the width of the heating zone (separate regime of combustion) the final temperature is close to the adiabatic one. In this case the rate of the combustion wave always increases with increasing ζ_f. When the combustion wave is broader at the reactor inlet (induction heating of the mixture is absent) another process occurs. The final temperature is determined in this case by integrating Eq. (11.1) over the width of the heated zone without taking into account the chemical source

$$\theta_{fin} = \theta_m\left[1 - \exp\left(-\frac{u_f}{u_*}Pe\zeta_f\right)\right] \tag{11.6}$$

where θ_m is the maximal temperature ($\theta_m \approx \theta_{ad}$, θ_{ad} is the adiabatic combustion temperature).

The dependence $u_f/u_* = f(\zeta_f)$ has the characteristic S-shape, see curve abfs in Fig. ll.3a. The stationary states of a displacement combustion reactor correspond to the intersection points of the curve determining the rate of combustion (Eq. 11.5) with the line $u_f/u_* = 1$ (Fig. 11.3).

The rate of the combustion wave depends essentially on the values of the Peclet and the Damkohler numbers. An increase of Pe, as well as decrease of Da, displaces the curve $u_f/u_* = (\zeta_f)$ downwards (Fig. 11.3a). When Pe is large enough (Da is small) only one stationary state of the displacement reactor, corresponding either to the combustion regime (point A) or to the separated regime (point C) is possible. A small Pe corresponds to an unsteady state (point B at the section bs of the curve $u_f/u_* = (\zeta_f)$ with negative slope). At very large Pe a high temperature stationary process is impossible, since within the range of $0 \le \zeta_f \le 1$ the ratio u_f/u_* is less than one. This result is valid only for the states at which the Damkohler number does not exceed some critical value Da_{cr}. At $Da > Da_{cr}$ the reactive mixture is ignited by means of induction heating, and high temperature states become possible at any values of Pe (point D in Fig. 11.3b).

The kinematic balance method permits many possibilities, which allow for judgement about the number of possible stationary states and their stability. In parallel, it is possible to distinguish the domains of existence of a kinematic hysteresis, to predict self-sustained oscillation combustion regimes, the extinction crises, etc. or, in other words, to study, in essence, all the complex phenomena related to the high-temperature operation of a displacement reactor. In particular, it was successfully used studying the thermal regime of heterogeneous combustion reactors (Sukhov and Yarin 1988 a, b, Yarin and Sukhov 1992, Stolyarova et al. 1980, 1981, Guzhiev et al. 1986, Likhachev et al. 1991, 1992). Some of the results of the above mentioned works are presented below.

3.11.2 Bubble displacement reactor

The reactor model. The principal scheme of a displacement bubble combustion reactor is shown in Fig. 11.4. It consists of a mixer (1), a straight insulated tube (2) in which high temperature conversion of a reactive bubble mixture occurs, and a hopper (3) for accumulation of the combustion products. The liquid and gaseous reagents enter the mixer where a uniform monodisperse bubble mixture is formed. At large enough residence time of such a mixture in the mixer, its components are under conditions of thermal equilibrium. In this case the temperatures of the gaseous and liquid reagents at the entrance cross-section of the reactor channel are equal to each other. At low temperatures of the two-phase mixture the rate of chemical reaction is negligible. That allows for the assumption that the concentrations of gaseous and liquid reagents are almost fixed and the completeness of their conversion is close to zero. Thus, the conditions corresponding to the state of the reactive mixture at the entrance cross-section of the reactor channel are

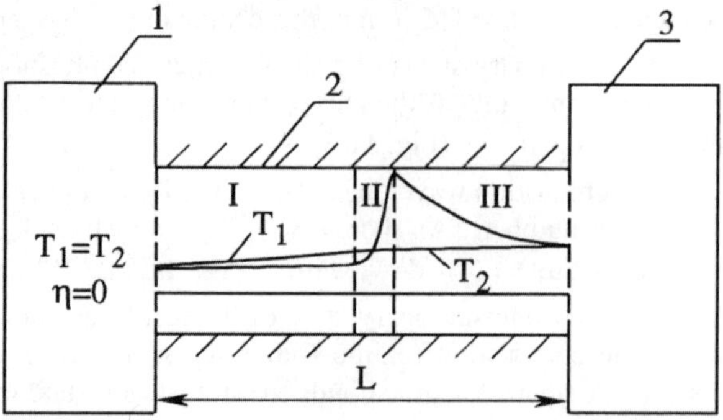

Fig. 11.4 Bubble displacement reactor model. Reprinted from Likhachev et al. (1992) with permission. 1, Tank in which the initial mixture is formed; 2, straight insulated tube; 3, hopper for product collection

$$x = -x_f \quad T_1 = T_2 = T_0, \quad \eta = 0 \tag{11.7}$$

where x_f is the distance from the flame to the entrance of the reactor channel, $\eta = (\rho_{a0} - \rho_a)/\rho_{a0}$ is the completeness of combustion of the gaseous oxidizer, ρ is the effective density, subscript 0 refers to the initial state of reactive mixture, subscripts 1 and 2 correspond to gaseous and liquid phases and subscript a corresponds to gaseous oxidizer.

In order to study the thermal regimes of a displacement bubble combustion reactor, one uses the previously developed (see Chap. 2.7) phenomenological approach to the description of combustion wave propagation in a reactive bubbly medium. In accordance with this approach the bubbly medium is represented as an ensemble of continua with effective density $\rho_i = \xi_i \rho_i^o$, where ρ_i^o and ρ_i are the physical and the effective densities, respectively, ξ_i is the volumetric content of the i th phase. Taking into account that the expansion of bubbles does not lead to qualitatively new results for the combustion wave propagation (see Chap. 2.7), we use the model of a medium with frozen (non-deformable) bubbles. In this case the system of mass and energy conservation equations (in the frame of reference associated with the combustion front) is

$$\frac{\rho_{a0} u_f}{\sigma} \frac{d\eta}{dx} = W(\eta, T_1) \tag{11.8}$$

$$c_1 \rho_1 u_f \frac{dT_1}{dx} = qW(\eta, T_1) - hs(T_1 - T_2) \tag{11.9}$$

$$c_2 \rho_2 (1 - \xi_1) u_f \frac{dT_2}{dx} = \frac{d}{dx}\left[(1 - \xi_1)\lambda_2 \frac{dT_2}{dx}\right] + hs(T_1 - T_2) \tag{11.10}$$

where c is the specific heat, λ is the thermal conductivity, u_f is the combustion wave rate, σ is the stoichiometric oxidizer-to-fuel mass ratio, q is the heat of combustion, h is the heat transfer coefficient, $s = (36\pi N\xi^2)^{1/3}$ is the specific interfacial area, N is the numerical concentration of bubbles, and $W(\eta, T_1)$ is the volumetric rate of chemical reaction.

Assuming that the n-th order reaction with respect to oxidizer leads to complete burning-out of the oxidizer and the formation of gaseous combustion products, we present the kinetic function $W(\eta, T)$ in the form

$$W(\eta, T_1) = \xi_1 z f(\eta) \exp\left(-\frac{E}{RT_1}\right) \tag{11.11}$$

where $f(\eta) = (1-\eta)^n$ is the burning out function.

At low pressure when the ratio $\rho_1^\circ / \rho_2^\circ \ll 1$, the volumetric content of the gaseous phase and its density change only slightly during the combustion process. That allows for the assumption that $\xi_1 \approx \xi_0$, $\rho_1 \approx \rho_{10}$. Accordingly the energy equation for the reactive mixture as a whole is obtained by combining Eqs. (11.9)–(11.11). It is

$$c_1\rho_1 u_f \frac{dT_1}{dx} + c_2\rho_2^\circ(1-\xi_0)u_f \frac{dT_2}{dx} = \frac{d}{dx}\left[(1-\xi_0)\frac{dT_2}{dx}\right] + \frac{\rho_{a0} q}{\sigma} u_f \frac{d\eta}{dx} \tag{11.12}$$

The boundary condition at $x=L-x_f$ (L is the length of the reactor channel see Fig. 11.4) is

$$x = L - x_f, \quad \frac{dT_1}{dx} = \frac{dT_2}{dx} = 0, \quad \eta = \eta_{fin} \tag{11.13}$$

where subscript fin refers to the exit cross-section of the reactor channel.

It should be noted that the boundary conditions (11.7) and (11.13) correspond to the displacement combustion reactor with a cooled inlet, adiabatic channel and uniform state of the material in the exit hopper, which is achieved by intensive mixing of the products. Other boundary conditions are possible, depending on the construction of the reactor and its operating regimes.

Integrating Eq. (11.12), we obtain the following expression for the final temperature:

$$\overline{T}_{fin} = 1 + \frac{\vartheta \overline{u}_f \eta_{fin} - \rho^o_{21} c_{21} \dfrac{(1-\xi_0)}{\xi_0} \left(\dfrac{d\overline{T}_2}{d\zeta} \right)_{-\zeta_f}}{\overline{u}_f \left[1 + \rho^o_{21} c_{21} \dfrac{1-\xi_0}{\xi_0} \right]} \tag{11.14}$$

where $\overline{T} = T/T_0$, $\overline{u}_f = u_f/u_0$, $u_0 = \alpha_2/L$, $\alpha_2 = \lambda_2/\rho^o_2 c_2$, $c_{2.1} = c_2/c_1$, $\rho^o_{21} = \rho^o_2/\rho^o_1$, $\vartheta = q/(\sigma c_1 T_0)$ and $\zeta = x/L$.

Excluding the variable x from Eqs. (11.8) and (11.9), we obtain

$$\frac{d\eta}{dT} = c_1 \sigma \frac{W(\eta, T_1)}{qW(\eta, T_1) - hs(T_1 - T_2)} \tag{11.15}$$

By integrating Eq. (11.15) we find the expression the completeness of combustion of the gaseous oxidizer

$$\eta_{fin} = \sigma \int_1^{\overline{T}_{fin}} \frac{\overline{W}(\eta, T_1)}{\vartheta \overline{W}(\eta, T_1) - (\overline{T}_1 - \overline{T}_2)} d\overline{T}_1 \tag{11.16}$$

where $\overline{W}(\eta, T_1) = W(\eta, T_1)/W_o$ and $W_o = hs/c_1$.

Rate of combustion wave propagation. To find the rate of propagation of the combustion wave, we use the condition by Sukhov and Yarin (1981 a, b) for the rate of a combustion wave in a reactive bubbly medium

$$qz\xi_0 \exp\left(-\frac{E}{RT_{1*}} \right) = hs(T_{2*} - T_{1*}) \tag{11.17}$$

where T_{1*} and T_{2*} are the temperatures of the gas and liquid phases at the cross-section $x = -x_*$, which separates the heating and reaction zones.

In order to calculate the characteristic temperatures T_{1*} and T_{2*} corresponding to the borderline between the heating and combustion domains, consider the structure of the combustion wave. Bearing in mind the characteristics of the process, it is possible to represent the combustion wave as a succession of three zones: (I) heating, (II) reaction, (III) thermal relaxation (Fig. 11.4).

Heating zone. Since within the heating zone $T_1 \approx T_2$, it is possible to omit the term containing $(T_1 - T_2)$ in Eq. (11.10). Then integration of Eq. (11.10) subject to the boundary conditions $x = -x_f$, $T_2 = T_0$ and $x = -x_*$, $T_2 = T_{2*}$, leads to the following expression for the liquid phase temperature

$$\overline{T}_2 = 1 + (\overline{T}_{2*} - 1)\frac{\exp(\overline{u}_f\zeta) - \exp(-\overline{u}_f\zeta_f)}{\exp(-\overline{u}_f\zeta_*) - \exp(-u_f\zeta_f)} \tag{11.18}$$

In accordance with Eq. (11.18) the derivative $\left(d\overline{T}_2/d\zeta\right)_{-\zeta_f}$ is expressed as

$$\left(\frac{d\overline{T}_2}{d\zeta}\right)_{-\zeta_f} = \overline{u}_f(\overline{T}_{2*} - 1)\frac{\exp(-\overline{u}_f\zeta_f)}{\exp(-\overline{u}_f\zeta_*) - \exp(-\overline{u}_f\zeta_f)} \tag{11.19}$$

Taking into account Eq. (11.18) and for the estimate of the chemical reaction rate of $-\infty < \zeta < -\zeta_*$, we can integrate Eq. (11.19). The resulting distribution of the gaseous phase temperature is

$$\overline{T}_1 = 1 + (\overline{T}_{2*} - 1)\times \tag{11.20}$$

$$\frac{\varepsilon\dfrac{\exp(\overline{u}_f\zeta_*)}{\overline{u}_f^2 + \varepsilon} - \exp(-\overline{u}_f\zeta_f)\left[\exp\left(-\varepsilon\dfrac{\zeta_*}{\overline{u}_f}\right) - \exp\left(-\dfrac{\zeta_f(\overline{u}_f^2 + \varepsilon)}{\overline{u}_f}\right)\dfrac{\overline{u}_f^2}{\overline{u}_f^2 + \varepsilon}\right]}{\exp(-\overline{u}_f\zeta_*) - \exp(-\overline{u}_f\zeta_f)}$$

where $\varepsilon = (hsL)/(c_1\rho_1 u_0)$.

Since the thickness of the reaction zone is very small, and thus x_* is close to zero, it is possible to take $T_1 = T_{1*}$ at $x = -x_* \sim 0$. In this case we find from Eq. (11.20) the expression for T_{1*}

$$\overline{T}_{1*} = 1 + (\overline{T}_{2*} - 1)\frac{\dfrac{\varepsilon}{\overline{u}_f^2 + \varepsilon} - \exp(-\overline{u}_f\zeta_f)\left[1 - \exp\left(-\dfrac{\zeta_f(\overline{u}_f^2 + \varepsilon)}{\overline{u}_f}\right)\dfrac{\overline{u}_f^2}{\overline{u}_f^2 + \varepsilon}\right]}{1 - \exp(-\overline{u}_f\zeta_f)} \tag{11.21}$$

Relaxation zone. Taking into account the fact that the temperature of the liquid reagent changes only slightly in a bubble suspension $(T_2 \simeq T_f)$ and within the relaxation zone $W(\eta, T_1) = 0$, we obtain the solution of Eq. (11.9) in the form

$$\overline{T}_1 = \overline{T}_{fin} + (\overline{T}_{1.max} - \overline{T}_{fin})\exp(-\frac{\varepsilon\zeta}{\overline{u}_f}) \tag{11.22}$$

where $T_{1.max}$ is the maximal gaseous phase temperature.

In order to determine the temperature of the liquid reagent, we integrate Eq. (11.10) taking into account Eq. (11.22). Bearing in mind the condition $\zeta = 1-\zeta_f$, $\overline{T}_2 = \overline{T}_{fin}$, $d\overline{T}_2/d\zeta = 0$ we obtain the following expression for the liquid phase temperature:

$$\overline{T}_2 = \overline{T}_{fin} - \frac{\varepsilon}{c_{21}\rho_{21}} \frac{(\overline{T}_{1max} - \overline{T}_{fin})}{(k_2 + \frac{\varepsilon}{\overline{u}_f})(k_1 + \frac{\varepsilon}{\overline{u}_f})} \exp(-\zeta\varepsilon\overline{u}_f) + N_1 \exp(k_1\zeta) + N_2 \exp(k_2\zeta) \tag{11.23}$$

where

$$k_{1.2} = \frac{1}{2}\left[\overline{u}_f \pm \sqrt{\overline{u}_f^2 + 4\varepsilon c_{21}\rho_{21}}\right]$$

$$N_1 = \frac{1}{k_2 - k_1} \exp[-k_1(1-\zeta_f)] \frac{\varepsilon(\overline{T}_{1max} - \overline{T}_{fin})}{c_{21}\rho_{21}(k_1 + \frac{\varepsilon}{\overline{u}_f})} \exp[-\varepsilon\frac{1-\xi_f}{\overline{u}_f}]$$

$$N_2 = \frac{1}{k_1 - k_2} \exp[-k_2(1-\zeta_f)] \frac{\varepsilon(\overline{T}_{1max} - \overline{T}_{fin})}{c_{21}\rho_{21}(k_1 + \frac{\varepsilon}{\overline{u}_f})} \exp[-\varepsilon\frac{1-\xi_f}{\overline{u}_f}]$$

At $\zeta = \zeta_*$ Eq. (11.23) yields the following expression for the temperature of the liquid reactant at the ignition point:

$$\overline{T}_{2*} = \overline{T}_{fin} - \frac{\varepsilon(\overline{T}_{1max} - \overline{T}_{fin})}{c_{21}\rho_{21}(k_2 + \frac{\varepsilon}{\overline{u}_f})(k_1 + \frac{\varepsilon}{\overline{u}_f})} + N_1 + N_2 \tag{11.24}$$

To determine the maximal temperature of the gaseous phase, we examine the nature of the variation of T within the combustion wave. Considering that in zone I in Fig.11.4 T_2=T_0, in zone II $c_1\rho_1 u_f (dT_1/ax) = qW(\eta, T_1) \gg hs(T_1 - T_2)$ and in zone III $c_1\rho_1 u_f (dT_1/dx) \approx hs(T_1 - T_2) \gg qW(\eta, T_1)$, we find the temperature values at the boundaries of the zones. Combining Eqs. (11.8) and (11.9) and taking into account the above conditions, we obtain the following equation which describes the temperature distribution in the gaseous phase in zone I in Fig. 11.4:

$$\frac{d\overline{T}_1}{d\zeta} = \vartheta\frac{d\eta}{d\zeta} - \frac{\varepsilon}{\overline{u}_f}(\overline{T}_1 - 1) \tag{11.25}$$

Since heating of the bubbles in zone I is insignificant (Sukhov and Yarin 1981 a, b), the following linear approximation is acceptable in describing the gas temperature:

$$\overline{T}_1 = 1 + \left(\frac{d\overline{T}_1}{d\zeta}\right)_{-\zeta_f} (\zeta + \zeta_f) \tag{11.26}$$

Substituting the expression (11.26) into Eq. (11.25), we obtain

$$\frac{d\overline{T}_1}{d\zeta} = \vartheta\frac{d\eta}{d\zeta} - \frac{\varepsilon}{\overline{u}_f}\left(\frac{d\overline{T}}{d\zeta}\right)_{-\zeta_f} (\zeta + \zeta_f) \tag{11.27}$$

Integrating Eq. (11.27) over ζ from $-\zeta_f$ to $-\zeta_*$ and assuming a narrow heating zone (ξ_* close to zero), we find

$$\overline{T}_{1*} = 1 + \vartheta\eta_* - \frac{\varepsilon}{\overline{u}_f}\left(\frac{d\overline{T}_1}{d\zeta}\right)_{-\zeta_f} \frac{\zeta_f^2}{2} \tag{11.28}$$

The temperature distribution in zone II satisfies the equation

$$\frac{d\overline{T}_1}{d\zeta} = \vartheta\frac{d\eta}{d\zeta} \tag{11.29}$$

The integral of Eq. (11.29) is

$$\overline{T}_{1\max} - \overline{T}_{1*} = (\eta_{\max} - \eta_*)\vartheta \tag{11.30}$$

Equations (11.28) and (11.30) yield

$$\eta_{\max} = \frac{1}{\vartheta}\left[(\overline{T}_{1\max} - 1) + \frac{\varepsilon}{\overline{u}_f}\left(\frac{d\overline{T}_1}{d\zeta}\right)_{-\zeta_f} \frac{\zeta_f^2}{2}\right] \tag{11.31}$$

Determining the rate of chemical reaction $W(\eta, T_1)$ from Eq. (11.11) (for n=1) and using the approximation $T_{2\max}=T_0$, we obtain Eq. (11.9) at the maximum position, where $d\overline{T}_1/d\overline{x} = 0$ in the following form:

$$qz\xi_0(1 - \eta_{\max})\exp\left(-\frac{E}{RT_{1\max}}\right) = hs(T_{1\max} - T_0) \tag{11.32}$$

Substitution of η_{max} from Eq. (11.31) into Eq. (11.32) results in the equation which determines $\overline{T}_{1max}$ as a function of ζ_f, $\overline{u}_f\left(d\overline{T}_1/d\zeta\right)_{-\zeta_f}$ and the other parameters.

The expressions (11.24) and (11.28) relating to the temperatures T_{1*} and T_{2*} of phases at the ignition point , as well as the maximal temperature of the gaseous phase $T_{1.max}$ take the dimensionless forms

$$\mathrm{Se}\exp\left(\frac{1-\overline{T}_{1*}^{-1}}{\beta}\right)=\frac{\overline{T}_{2*}-\overline{T}_{1*}}{\beta} \tag{11.33}$$

$$\mathrm{Se}\exp\left(\frac{1-\overline{T}_{1max}^{-1}}{\beta}\right)(1-\eta_{max})=\frac{\overline{T}_{1max}-1}{\beta} \tag{11.34}$$

where $\beta=(RT_0)/E$ and $\mathrm{Se}=[(z\xi_0 qE)/(hsRT_0^2)]\exp(-1/\beta)$.

The system of equations (11.33), (11.34) and (11.14), (11.16), (11.21), (11.24) and (11.31) determines the characteristic temperatures and completeness of combustion, as well as the rate of the combustion wave and its position in the reactor channel.

Reactor regimes. Figure 11.5 shows the results of a calculation of the combustion wave speed as a function of its position in the reactor channel for various values of T_0, of the reactivity of the bubble mixture, and of the rate of interfacial heat exchange. The line $\overline{u}=\overline{u}_f=$ const., which characterizes the feed rate of the mixture into the reactor, is also drawn in Fig. 11.5. According to the kinematic balance method, the intersection points of the curves $\overline{u}_f(\zeta_f)$ and $\overline{u}_f=$ const. correspond to the steady states of the reactor. The curve $\overline{u}_f(\zeta_f)$ has a non-monotonic character and in the general case it intersects the line $\overline{u}_f$=const. at points 1 and 2. These points correspond, respectively, to the position of the combustion front near the reactor entrance or in a region which is quite far from it. The first of these corresponds to a regime in which heat removal in the intake plays a significant role, and the second to combustion under practically adiabatic conditions. With an increase in the initial temperature and reactivity of the mixture, and also with a decrease in the interfacial heat exchange rate, points 1 and 2 approach each other and for certain critical values of T_0, Se, and ε, the curves $\overline{u}_f(\zeta_f)$ and $\overline{u}_f=$ const. lose their common points. This means that for $T_0>T_{cr}$ ($\mathrm{Se}>\mathrm{Se}_{cr}$, $\varepsilon<\varepsilon_{cr}$, subscript cr corresponding to the critical state of the reactor), there is no steady state, due to considerable reaction acceleration and self-ignition in the system (thermal explosion).

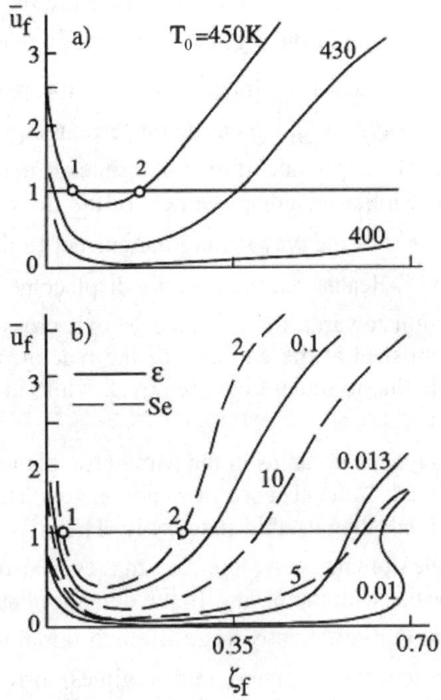

Fig. 11.5 Kinetic curve. Reprinted from Likhachev et al. (1992) with permission. a) $u_f(\zeta_f)$ with T_0 being the parameter, while the other parameters being fixed. b) $u_f(\zeta_f)$ with Se and ε are parameters, while the other parameters being fixed; points 1 and 2 correspond to the position of the flame front near the reactor inlet or to the region which is quite far from it

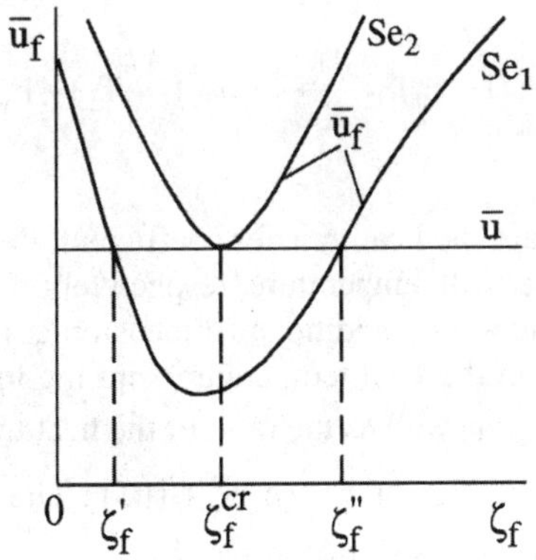

Fig. 11.6 Steady regimes of bubble combustion reactor Reprinted from Likhachev et al. (1992) with permission. ζ_f'' corresponds to a stable state and ζ_f' to an unstable state

By using a method similar to that of the Semenov diagram, one can show that steady states are unstable in the region $0 < \zeta_f < \zeta_f^{cr}$, while they are stable in the region $\zeta_f^{cr} < \zeta_f < 1$ (Fig. 11.6). Indeed, if for some reason the combustion front is deflected to the right of ζ_f'', then the corresponding increase in the combustion rate relative to the supply rate of the mixture leads to a displacement of the wave to its initial state. Similarly, a displacement of the front to the left of ζ_f'' will be accompanied by a drift of the wave along the current to the point ζ_f''. Deflection from the position ζ_f' leads to a progressive displacement of the combustion wave along the direction towards the entrance or exit cross-section. In the first case, the wave is established at the entrance of the reactor, where a combustion explosion is impossible due to intense heat removal, while in the second case, the wave is shifted towards ζ_f''.

Figure 11.7 displays the domains in the parametric plane $Se - \varepsilon$ in which the various states are realized: I, the absence of steady states; II, the presence of stable and unstable states; III, the unstable states only. The effect of the parameter β on the boundaries of the domains corresponding to these various regimes is shown in Fig. 11.8. It is seen that with increasing β the domain of steady (stable and unstable) states contracts. For sufficiently large β (high initial temperature), the domain of steady states degenerates. Here, only regimes corresponding to thermal explosion and to combustion due to intense heat release at the entrance domain are possible, respectively.

Effect of heat losses. The existence of heat losses to the reactor wall significantly affects the extension of the domains of stable and unstable states of a displacement reactor. The losses can be incorporated in the energy equation of the carrier fluid which in the given case takes the form

$$c_2\rho_2^o(1-\xi_0)u_f\frac{dT_2}{dx} = \frac{d}{dx}\left[(1-\xi_0)\lambda_2\frac{dT_2}{dx}\right] + hs(T_1 - T_2) - h_w s_w(T_2 - T_w) \tag{11.35}$$

where h_w, s_w and T_w are the heat transfer coefficient, the specific area of the reactor wall surface and the wall temperature, respectively.

The problem includes two additional dimensionless parameters accounting for the difference between the wall temperature and the initial temperature of the carrier fluid, $\overline{T}_w = T_w/T_0$, as well as the ratio of the heat transfer coefficient to the interfacial heat transfer coefficient $\overline{h} = (h_w S_w)/(hS)$. The effect of these parameters on the domain boundaries is shown in Fig. 11.7.

The possible regimes of displacement combustion reactor with heat losses are illustrated in Fig. 11.9 in which curves $\zeta_f(Se)$ are plotted in the parametrical

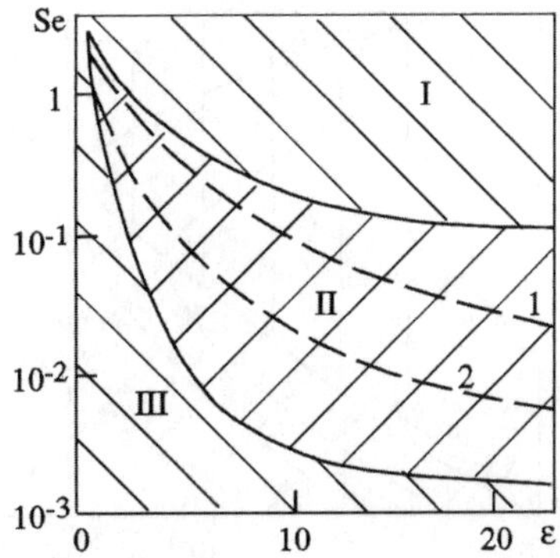

Fig. 11.7 Steady regime domains, and the effect of heat removal to the channel walls. Reprinted from Likhachev et al. (1992) with permission. I, The absence of steady states; II, the presence of stable and unstable states; III, unstable states only. Curve 1, $\overline{T}_w = 0.5$; curve 2, T_w=1.5

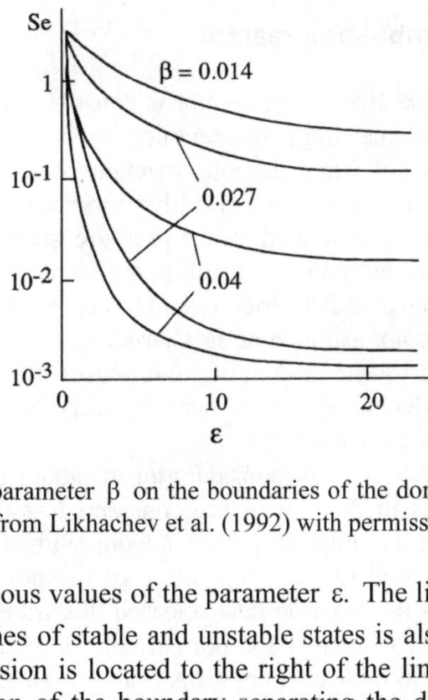

Fig. 11.8 The effect of parameter β on the boundaries of the domains of various combustion regimes. Reprinted from Likhachev et al. (1992) with permission

plane $\zeta_f - \mathrm{Se}$ for various values of the parameter ε. The limiting curve $\zeta_{\ell im}(\mathrm{Se})$ which separates regimes of stable and unstable states is also shown here. The region of thermal explosion is located to the right of the line Se=Se_{cr}. Figure 11.9 also shows the position of the boundary separating the domain of steady states from that in which such states cannot be attained for any values of $\overline{T}_w$ (the dashed line). An increase in the wall temperature leads to a contraction of the domain of steady states. This is related to the growth of the reaction rate with heat intake from outside. An increase of ratio $(h_w s_0)/(hs)$ is accompanied by expansion of the region of stable states.

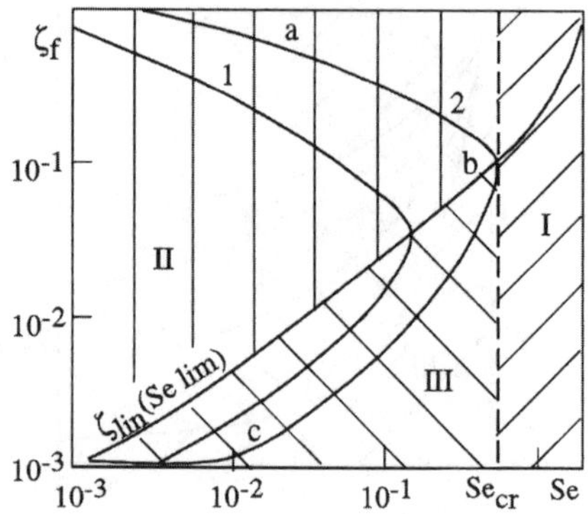

Fig. 11.9 The domains of possible regimes of displacement combustion reactor in the ζ_f - Se plane. Reprinted from Likhachev et al. (1992) with permission. Curve 1, ζ_f(Se) for $\varepsilon = 20$; curve 2, $\varepsilon = 10$; ab, stable state; bc, unstable state; I, stationary states cannot be realized for any values of $\overline{T}_w$; II, region of stable states; III, region of unstable state

3.11.3 Filtration combustion reactor

Physical model of reactor. The reactor is constructed as a straight cylindrical tube inside which the high temperature exothermal reaction proceeds (Fig. 11.10). Dispersed solid and gaseous reactants enter the combustion zone through the cooled inlet of the reactor. The filtration motion of gas in the channel obeys the Darcy law and is generated by the pressure gradient arising as a consequence of either oxidizer burning out at the flame front (natural filtration) or the action of a suction device at the channel outlet (forced filtration). In the latter case the oxidizer enters the combustion zone in a larger quantity than is required from the reaction stoichiometry. The motion of solid phases (the reactant and the solid product) may occur under the effect of a gravity force, a pressure differential in the channel or some other reasons.

The analysis is restricted by consideration of some principal factors only. Therefore, the processes in the reactor are considered assuming a uniform gas–particle temperature which implies porous media with a developed interface, where gas is rapidly heated to the temperature of the porous medium skeleton. Also, it is assumed that the reaction heat released, the transport coefficients, and the specific heat capacity of phases and porosity of the medium are constant. The absence of heat transfer into the channel walls is also assumed.

Governing equations. To describe possible stationary and non stationary reactor states we shall make use of heat- and mass-transfer equations with sources in the reference frame associated with the combustion front. In the quasi-stationary approximation these equations are as follows:

$$\frac{d(\rho_g u)}{dx} = -\sigma W \tag{11.36}$$

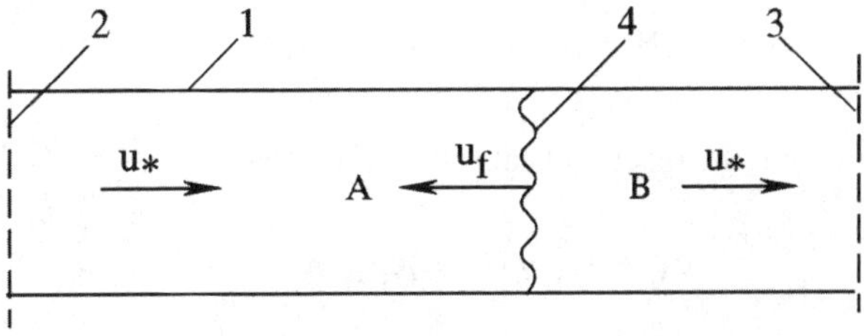

Fig. 11.10 Sketch of the filtration combustion reactor. Reprinted from Yarin and Sukhov (1992) with permission. 1, the reactor channel; 2, 3, the inlet and outlet cross-sections; 4, the flame front; A, solid reactant and gaseous oxidizer zone; B, condensed product zone

$$\frac{dP}{dx} = -\frac{u - u_f}{k} \tag{11.37}$$

$$\rho_{M0} u_f \frac{d\eta}{dx} = W \tag{11.38}$$

$$[c_v \rho_g u + (c_M \rho_M + c_p \rho_p) u_f] \frac{dT}{dx} = \lambda \frac{d^2T}{dx^2} + qW \tag{11.39}$$

Here x is the longitudinal coordinate in the reactor, σ is the stoichiometric oxidizer-to-solid reactant mass ratio, λ is the thermal conductivity, P is the pressure, T is the temperature, ρ is the density, u is the gas velocity, $\eta = (\rho_{Mo} - \rho_M)/\rho_{Mo}$ is the completeness of combustion of the solid reactant, q is the reaction heat released, k is the permeability coefficient, c_M, c_p are the specific heats of the solid reactant and the product, and c_v is the specific isochoric heat of the gas. Subscripts denote: f, flame; g, gas phase; M, solid reactant; p, product. In the following subscripts 0, 1, 2 are the states at the channel input, before and after the front.

It is emphasized, that the quasi-stationary approximation used here implies the absence of partial time derivatives in the transport equations. The validity of such an approximation in problems of nonstationary combustion of condensed media is justified by the smallness of the combustion wave propagation rate in these systems compared to the gaseous media.

We supplement the system of Eqs. (11.36) – (11.39) by the gas equation of state

$$mP = \rho_g R_g T \tag{11.40}$$

by stoichiometric relations

$$\rho_M = \rho_{M0}(1-\eta), \quad \rho_p = (1+\sigma)\rho_{M0}\eta \tag{11.41}$$

by the condition that the reaction heat released is constant

$$c_M + \sigma c_v = (1+\sigma)c_p \tag{11.42}$$

by an alternative condition of cessation of combustion according to Aldushin et al. (1974)

$$\rho_2(1-\eta_2) = 0 \tag{11.43}$$

and by a macrokinetic law expressing the characteristics of a solid-phase interaction of reactants through the oxide layer under high activation energy $(\beta = (RT_o)/E \ll 1)$.

$$W = zmf(\eta)P^{\nu}\exp\left[-\frac{E}{RT}\right] \tag{11.44}$$

where m is the medium's porosity, $f(\eta)$ is the burning-out function; $f(\eta)=1$ for $0 \le \eta < 1$, $f(1)=0$, z and ν are the kinetic coefficients $(0 \le \nu \le 2)$. In the case of strong inhibition of the chemical reaction by the growth of an oxide layer, the burning-out function may be presented in the form (Aldushin et al. 1974, Yarin and Sukhov 1992)

$$f(\eta) = (\eta_o + \eta)^{-n}, \quad f(1) = 0 \tag{11.45}$$

where $\eta_o \ll 1$, and n is the kinetic coefficient. Equation (11.45) allows for the application of the present approach to the case when there is incomplete conversion $\eta < 1$ in the kinetic combustion models of the second kind (Merzhanov 1980).

The condition (11.43) indicates two possible cases of flame propagation: when the oxidizer filtration does not provide complete burning-out of a solid reactant in the flame zone $(P_2 = 0,\ \eta_2 < 1$, the filtration regime) and when the oxidizer enters the combustion zone in excess $(P_2 > 0,\ \eta_2 = 1$, the kinetic regime). In the general case the flame propagation rate is not necessarily equal to the supply rate. The equality is established only at stationary states. To describe these states the system of Eqs. (11.36)–(11.44) should be supplemented by the kinematic relation

$$u_f = u_* \tag{11.46}$$

where u_* is the rate of supply of the solid reagent.

The boundary conditions to the problem may be different depending on the construction features and design of the reactor. When specified in the form

$$\begin{aligned} &x = -x_f, \quad T = T_0, \quad P = P_0, \quad \eta = 0 \\ &x = L - x_f, \quad \frac{dT}{dx} = 0, \quad \rho_f u = g \end{aligned} \tag{11.47}$$

these conditions correspond to a permeable solid (with respect to oxidizer) in a channel adiabatic reactor with a cooled inlet and a suction device at the outlet. Here L is the channel length, x_f is the distance from the reaction zone to the channel inlet, g is the oxidizer mass flux in the product zone. Thus, when the problem is solved in the quasi-stationary approximation, the time dependence of a process is implicit, that is, it manifests itself via the parameter $x_f(t)$ that appears in the boundary conditions. Its value is determined by the equation

$$\frac{dx_f}{dt} = u_* - u_f \tag{11.48}$$

with the initial condition that defines the place of combustion wave initiation in the channel:

$$t = 0 \qquad x_f = x_f(0) \tag{11.49}$$

Integral relations. The solution of the problem will be sought in the form of a quasi-stationary combustion wave propagating over a substance with a speed $u_f(x_f)$. We shall use the concepts of combustion in a flow of substance advanced by Zaidel and Zel'dovich (1962), Khaikin and Rumanov (1975), as well as Butakov et al. (1978). According to these concepts, the mixture of reactants, passes through some specific zones along the channel. We shall integrate approximately Eqs. (11.36)–(11.39) within the range of each of these zones. The solutions obtained, along with the matching conditions at the boundaries of the above-mentioned zones, determine the combustion front velocity u_f along the channel in the form of a function $u_f/u_* = F(x_f)$. As shown below, by comparing this function with the line $u_f/u_* = 1$, it is possible not only to determine the number of possible stationary states, but also to investigate all the fundamental properties of the processes in the reactor channel.

Comparing the combustion wave propagation rate with the supply rate it is necessary to distinguish two limiting regimes. In the case of $u_f(x_f)>u_*$ the wave propagates in the direction opposite to the flow direction and is stabilized near the inlet (the separationless combustion regime). Such a state is characterized by considerable conductive heat transfer from the reactor zone to a cooled inlet. This results in a decrease in the flame temperature. For $u_f(x_f)<u_*$ the convective heat transfer in the channel prevails over the conductive one, and the self-heating of a moving active mixture becomes dominant. In the case of long reactor channels this process is terminated by self-ignition and complete burning out of one of the reactants (the self-ignition, or the induction regime). It is similar to the processes that occur in the flow of active gaseous mixtures (Zaidel and Zel'dovich 1962, Khaikin and Rumanov 1975, Butakov et al. 1978).

Along with the limiting cases mentioned above an intermediate case is also possible, when the induction process occurring in the channel zone close to the inlet is terminated by burning out of the mixture at the flame front, rather than by self-ignition (combustion with flame separation). Unlike separationless combustion, in this region the wave propagates in the channel over a mixture of reactants, which are heated in the induction zone to a temperature $T_o' > T_o$. A similar case was distinguished by Butakov et al. (1978) in their investigation of a gas-phase combustion reactor. The induction zone, within the range of which convective heat transfer is dominant, serves as a kind of heat insulation that prevents heat transfer from the flame to the cooled inlet. In this case the temperature of a final product is close to the adiabatic one. The combustion with flame separation, which incorporates the elements of the limiting regimes mentioned above (the induction process and the flame front), obviously corresponds to the general case of combustion reactor operation (Fig. 11.11).

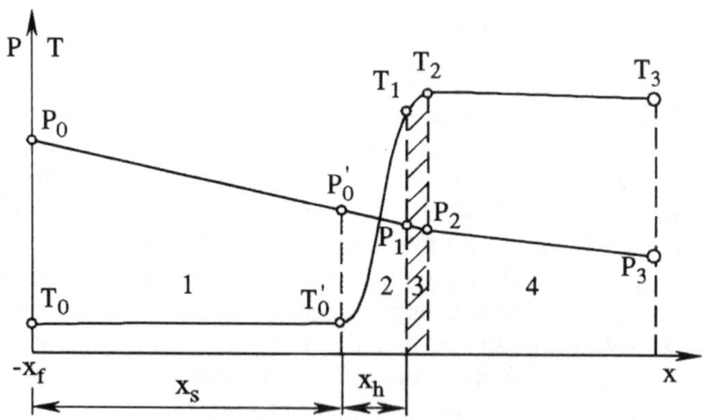

Fig. 11.11 Combustion structure in the flame separation regime. Reprinted from Yarin and Sukhov (1992) with permission. Characteristic regions: 1, induction heating-up; 2, hearing-up in a combustion wave; 3, flame; 4, hot product

Taking into account the above considerations, we determine the flame propagation rate in the reactor channel. Using the boundary conditions (11.47), we find the first integral of Eqs. (11.36) and (11.39) as

$$\rho_g u = g + \sigma \rho_{Mo} u_f (\eta_2 - \eta) \tag{11.50}$$

$$\lambda \frac{dT}{dx} = \left[c_v g + c_M \rho_{Mo} u_f (1 + \tau \eta_2)\right](T - T_2) + g \rho_{Mo} u_f (\eta_2 - \eta) \tag{11.51}$$

where $\tau = (\sigma c_v)/c_M$.

Eliminating the gas velocity from the Darcy law (11.37) by using (11.50), we obtain the filtration equation for the case of intensive blowing ($g \gg \rho_g u_f$) in the form

$$\frac{dP^2}{dx} = -2R_g T \frac{g + \sigma \rho_{Mo} u_f (\eta_2 - \eta)}{mk} \tag{11.52}$$

In the further analysis it is convenient to use the temperature coordinate in Eqs. (11.38) and (11.52). Applying Eq. (11.51) for this purpose, we obtain

$$\rho_{Mo} u_f \frac{d\eta}{dT} = \frac{z P^{\nu}(T) \lambda \exp\left(-\dfrac{E}{RT}\right)}{[c_v g + c_M \rho_{Mo} u_f (1 + \tau \eta_2)](T - T_2) + g \rho_{Mo} u_f (\eta_2 - \eta)} \tag{11.53}$$

$$\frac{dp^2}{dT} = -\frac{2\lambda R_g T[g + \sigma \rho_{Mo} u_f (\eta_2 - \eta)]}{mk\{[c_v g + c_M \rho_{Mo} u_f (1 + \tau \eta_2)](T - T_2) + g \rho_M u_f (\eta_2 - \eta)\}} \tag{11.54}$$

After separating variables, Eq. (11.53) can be approximately integrated using the standard high-activation-energy asymptotic approach (Zel'dovich et al. 1985) taking into account the features of this sort of combustion (Aldushin et al. 1974). In particular, the function P(T) is approximated by the linear dependence $P^2(T) = P_2^2 + \left(dP^2/dT\right)(T - T_2)$, where thc derivative is defined by Eq. (11.54). The integral of the Eq. (11.53) determines the flame propagation speed as a function of the dimensionless variables

$$\left(\frac{u_f}{u_*}\right)^2 = \frac{2\gamma Da (1 + \beta\theta_2)^2 \exp\left(\Pi \pi_2^2 + \dfrac{\theta_2}{1 + \beta\theta_2}\right)}{Pe \cdot \Pi^{\frac{\nu}{2}} (\eta_2 - \eta_o^{'})^2} \Gamma(1 + \frac{\nu}{2};\ \Pi \pi_2^2) \tag{11.55}$$

Here $\theta = \frac{E(T - T_0)}{RT_0^2}$, $\pi = \frac{P}{P_0}$, $\beta = \frac{RT_0}{E}$, $\gamma = \frac{c_M RT_0^2}{qE}$,

$\Pi = \frac{mkP_0^2 \rho_{Mo} u_f \eta_2}{2\lambda R_g RT_2^3 (g + \sigma\rho_{Mo} u_f \eta_2)}$, $Da = zP_0^{\nu} L \frac{\exp\left(-\frac{E}{RT_0}\right)}{\rho_{Mo} u_*}$ is the Damkohler number, $Pe = \frac{\rho_{Mo} u_* c_M L}{\lambda}$ is the Peclet number, $\Gamma(1 + \frac{\nu}{2}; \Pi\pi_2^2) = \int_{\Pi\pi_2^2}^{\infty} t^{\nu/2} \exp(-t)dt$ is the incomplete gamma-function; η_o' is the completeness of chemical transformation at the end of the induction zone.

Equation (11.55) is the basic one in studying the filtration combustion reactor. It includes a set of the a priori unknown parameters $\pi_2, \theta_2, \eta_2, \eta_o'$, whose values are determined by the integrals of Eqs. (11.39) and (11.52). The latter were obtained by Stolyarova et al. (1980) using some simplifying assumptions (the absence of conductive heat transfer in the induction zone, the negligible filtration resistance of the combustion wave, etc.) in the following form:

$$\pi_2^2 = 1 - \frac{u_f \eta_2}{u_* N}\left(1 + \frac{G u_*}{u_f \eta_2}\right)\zeta_s \tag{11.56}$$

$$\eta_o' = \gamma[\tau G + (1 + \tau\eta_2)\frac{u_f}{u_*}]\theta_o' \tag{11.57}$$

$$\theta_o' = -\ln\left\{1 - \frac{2NDa(1 - \pi_2^{2+\nu})}{\gamma(2+\nu)\left(G + \eta\frac{u_f}{u_*}\right)\left[\tau G + (1 + \tau u_2)\frac{u_f}{u_*}\right]}\right\} \tag{11.58}$$

$$\theta_2 = \theta_{ad} = \eta_2 \gamma \frac{u_f}{u_*}\left[\tau G + (1 + \tau\eta_2)\frac{u_f}{u_*}\right] \tag{11.59}$$

Here $\zeta = x/L$, $G = g/(\sigma\rho_{Mo} u_*)$, $N = (mkP_o^2)/(2\sigma\rho_{Mo} u_f T_o L)$ is the filtration parameter, θ_{ad} is the adiabatic flame temperature, and ζ_s is the induction zone length (the flame separation zone).

The system of the integrals (11.55)–(11.59) should be supplemented by the relationships for determining the distance from the flame front to the channel inlet

$$\zeta_f = \zeta_s + \zeta_h \tag{11.60}$$

and for calculating the heating-up zone length ζ_h in the combustion wave structure, which appears in (11.60). This value is found from the temperature profile in the combustion wave

$$T = T_0 + (T_{ad} - T_0)\exp\left\{[c_v g + c_M \rho_{Mo} u_f (1 + \tau\eta_2)]\frac{x}{\lambda}\right\} \tag{11.61}$$

as a distance from the flame front to the cross-section, where the difference $(T - T_0)$ is two orders of magnitude lower than $(T_{ad} - T_0)$, that is

$$\zeta_h = 4.6\left\{\mathrm{Pe}\left(\tau G + (1 + \tau\eta_2)\frac{u_f}{u_*}\right)\right\}^{-1} \tag{11.62}$$

Equation (11.56) is of principal importance among the relations obtained, since it defines, along with Eq. (11.43), the domains of existence of the kinetic and filtration combustion regimes. The analysis of these relations indicates that the process in the channel is terminated by complete transformation of a solid reactant if

$$\zeta_s < N\frac{u_*}{u_f}, \qquad \pi = 1 - \zeta_s N\frac{u_f}{u_*}, \quad \eta_2 = 1 \tag{11.63}$$

and by incomplete transformation if

$$N\frac{u_*}{u_f} < \zeta_s \le 1, \; \pi_2 = 0, \; \eta_2 = \zeta_s N\frac{u_*}{u_f} < 1 \tag{11.64}$$

The change of the regimes occurs under the condition that

$$\zeta_s = \zeta^o = N\frac{u_*}{u_f}, \quad \pi_2 = 0, \quad \eta_2 = 1 \tag{11.65}$$

For stationary states $\zeta^o = N$. If the change of the regimes occurs at the reactor outlet, then $\zeta^o = N = 1$. The improvement of the oxidizer filtration conditions $(\zeta^o = N > 1)$ allows for sustaining the kinetic combustion regime throughout the reactor length (the "short" channel). An increase of the filtration resistance $(\zeta^o = N < 1)$ is accompanied by the appearance of a zone of incomplete burning out of a solid reactant in the reactor (the "long" channel).

The relations (11.55)–(11.64) (called the equations of group I hereinafter) allow for calculation of the spectrum of local values of the combustion wave propagation speed in the form a function $u_f/u_* = F(\xi_f)$ in the quasi-stationary approximation. The relations of group I are suitable, however, for calculating the separated flame rate only, i.e. they are applicable only in the case where the induction zone exists between the combustion front and the channel inlet. This zone does not exist in the case of separationless combustion, and the process proceeds under the conditions of intensive heat transfer to a cooled inlet and at a considerable decrease of the flame temperature $(T_2<T_{ad})$. For calculating the temperature T_2 in this case one should use, instead of Eq. (11.59), the expression

$$\theta_2 = \theta_{ad}\left\{1 - \exp\left(-\left[\tau G + (1+\tau\eta_2)\frac{u_f}{u_*}\right]Pe \cdot \zeta_f\right)\right\} \tag{11.66}$$

obtained by integrating the energy equation over the heating-up zone in the combustion wave taking into account the heat transfer to the reactor inlet.

Since the induction heating-up of a mixture is absent in the separationless combustion $(\zeta_s = 0)$, according to the Eq. (11.60) the position of the flame front in the channel is determined by the relation

$$\zeta_f = \zeta_h \tag{11.67}$$

One should remember that the value ζ_h cannot be calculated in this case via Eq. (11.62), which is suitable for the adiabatic flame only. It follows from Eqs. (11.43) and (11.56) that in the separationless combustion $\eta_2 = \pi_2 = 1$.

The expressions (11.55), (11.66) and (11.67) with $\eta_2 = \pi_2 = 1$, form the group II of equations, allowing for the calculation of the function $u_f/u_* = F(\zeta_f)$ in close proximity of the inlet cross-section of the reactor channel.

Thus, the relations found make it possible to calculate the local values of the flame propagation speed at any cross-section of the reactor and to plot the dependence of u_f/u_* on ζ_f. According to Eq. (11.55), the distribution of the flame rate is determined by the combustion kinetics. Therefore, the dependence $u_f/u_* = f(\zeta_f)$ will be called the kinetic curve hereinafter. This curve is constructed according to the following rules:

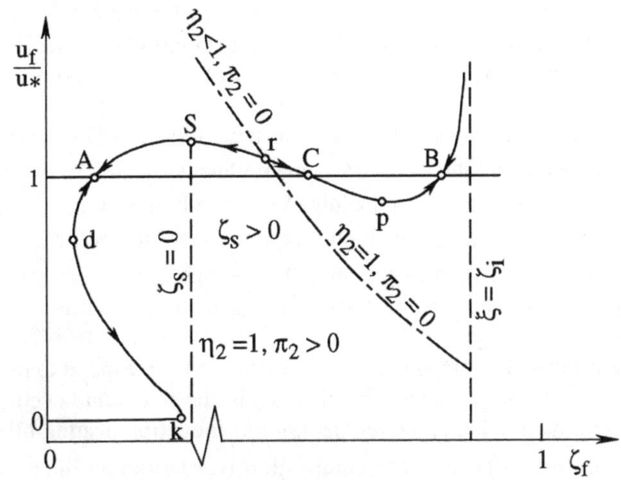

Fig. 11.12 Configuration of the kinetic curve. The arrows show the direction of combustion wave propagation in the channel. Reprinted from Yarin and Sukhov (1992) with permission

- Specifying ζ_s within the range from 1 to 0, one calculates ζ_f and the section of a kinetic curve corresponding to the combustion with flame separation, from the equations of group I.
- In the vicinity of the inlet cross-section, where $\zeta_s = 0$ (separationless combustion regime), the calculation is performed using the equations of group II; the value ζ_h is specified as increasing from the value $\zeta_h = 0$; the "matching" (intersection) of the kinetic curve sections occurs at point S (Fig. 11.12) in the vicinity of the cross-section, where $\zeta_s = 0$.

The $u_f/u_* = F(\zeta_f)$ dependence found according to the above procedure, makes it possible to solve the problem (i.e. to determine the non stationary characteristics of the process). Indeed, with the $u_f/u_* = F(\zeta_f)$ curve known and for the initial conditions (11.49) the law of the combustion wave propagation in the channel $\zeta_f(t)$ will be determined by the integral of Eq. (11.48)

$$\bar{t} = \int_{\zeta_f(0)}^{\zeta_f} \frac{d\zeta}{1-F(\zeta)} \tag{11.68}$$

where $\bar{t} = (tu_*)/L$.

Because all the parameters of the problem can be determined in the form of functions of the variable ζ_f, the integral (11.68) describes their variation in time.

The intermediate result, the kinetic curve, is of fundamental importance, because, as will be shown below, once this curve is determined, one can investigate the whole complex of properties of the combustion reactor in both stationary and nonstationary regimes.

In the general case the kinetic curve has a complicated Ω -like shape (Fig. 11.12) and intersects the line of the stationary states $u_f/u_* = 1$ at three points corresponding to separationless combustion (A) and to combustion with flame separation (B and C). The absence of high-temperature solutions for $\zeta_f \ll 1$ (the Ok branch on the kinetic curve corresponds to low-temperature oxidation) suggests of extinction of the flame because of heat losses near the cooled inlet of the reactor channel.

Different rates of combustion wave propagation correspond to different positions of the wave in the channel. The changes in the rate value occur due to several factors. On the one hand, the temperature of induction heating of the medium T_0' grows with the distance of the combustion wave from an inlet. On the other hand, simultaneously, the filtration zone length before the wave increases and pressure P_2 at the wave front decreases. When the wave propagates to a considerable distance into the channel, the transition to the filtration combustion regime becomes possible (P_2=0, $\eta_2 < 1$), which is followed by lowering of the final temperature T_2 due to incomplete burning out of a solid reactant. The induction heating of a mixture accelerates the combustion wave according to Eqs. (11.55) and (11.57). The growth of filtration resistance of the induction zone and the decrease of the final temperature, in contrast decelerate the wave. Accordingly, different sections of the kinetic curve reflect a prevailing influence of any of the above-mentioned factors. For example, the decrease of the flame propagation rate on the S–r branch in Fig. 11.12 is due to an oxidizer pressure drop at the flame front, and a similar decrease of the rate on the r–p branch is due to a reduction in final temperature caused by incomplete burning out (r is the point of transition from the kinetic combustion regime to the filtration regime). Near the $\zeta = \zeta_f$ cross-section, where the self-ignition of a heterogeneous mixture is possible, the self-heating process in the induction zone is activated (T_0' increases drastically). Accordingly, Eq. (11.55) results in an increase in u_f and, respectively, in transformation of the process to the rising branch p–B of a kinetic curve.

The adiabatic ignition induction length ζ_i is determined by integrating the energy equation (11.39) taking into account the high activation energy $E/(RT_0) \gg 1$ and assuming λ=0.

$$\zeta_i = \frac{N}{G+\eta_2}\left\{1-\left[1-\frac{(2+\nu)(G+\eta_2)\eta_2[1-\exp(-\theta_{ad})}{2NDa\theta_{ad}}\right]^{\frac{2}{2+\nu}}\right\} \tag{11.69}$$

As the Peclet and Damkohler numbers vary, the kinetic curve varies and the stationary states also change. Then, various limiting cases are possible, where the flame front just coincides with the channel outlet ($\zeta_s = 1$), or with the boundary of the zones of complete and incomplete burning out ($\zeta_s = N$), or with the cross-section corresponding to the transition from separationless combustion to combustion with flame separation ($\zeta_s = 0$). Transitions to these states correspond to functional dependences of the $Pe = \varphi(Da)$ type, which are determined by the system of equations of group I under the condition $u_f / u_* = 1$ and for ζ_s values listed above. These dependences form a foundation basis for the classification of reactor regimes in the Pe–Da parametric plane by subdividing the latter into domains with different character of proceeding processes.

Stationary reactor regimes. We shall consider some typical regimes of a filtration combustion reactor. For this purpose we shall analyze first of all stationary states of the reactor with complete consumption of oxidizer (the "long" channel: G=0, N<1).

The position of the combustion wave at the cross-section $\zeta_s = 0$ corresponds to the condition

$$Pe = 2\gamma \Pi^{-\frac{\nu}{2}} Da(1+\beta\theta_{ad})^2 \exp\left[\Pi + \frac{\theta_{ad}}{1+\beta_{ad}}\right]\Gamma(1+\frac{\nu}{2};\Pi). \tag{11.70}$$

where

$$\Pi = N \cdot \frac{Pe}{\gamma(1+\beta\theta_{ad})^3}, \quad \Gamma(1+\frac{\nu}{2};\Pi) = \int_{\Pi}^{\infty} t^{\frac{\nu}{2}} \exp(-t)dt$$

is the incomplete gamma-function. The dependence (11.70) is represented by line 1 on the parametric plane Pe–Da (Fig. 11.13).

The combustion wave localized at the cross-section $\zeta_s = N$, where the kinetic combustion regime is transformed into the filtration regime, is characterized by the relation

$$Pe = 2\gamma \Pi^{-\frac{\nu}{2}} Da(1+\beta\theta_{ad})^2 (1-\eta_o')^{-2} \exp\left[\frac{\theta_{ad}}{1+\beta\theta_{ad}}\right]\Gamma(1+\frac{\nu}{2}) \tag{11.71}$$

where

$$\Gamma(1+\frac{\nu}{2}) = \int_0^{\infty} t^{\frac{\nu}{2}} \exp(-t)dt$$

is the gamma-function. The dependence (11.71) is shown by curve 2 on the diagram in Fig. 11.13.

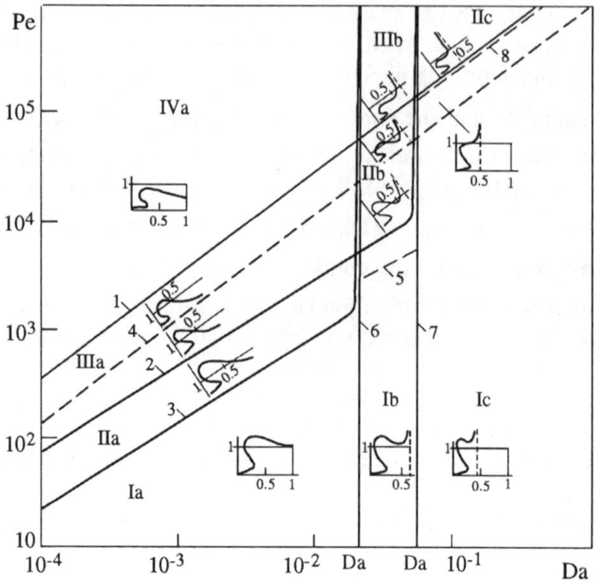

Fig. 11.13 Parametric diagram of states of a "long" reactor. Reprinted from Yarin and Sukhov (1992) with permission. Insets to the graph illustrate the shape of kinetic curves corresponding to different states of a long reactor in the parametric plane $u_f - \zeta_f$. N=0.5, $\tau = 1$, $\nu = 0.5$, $\gamma = 0.01$, $\beta = 0.04$

The state of the reactor with the combustion wave at the outlet cross-section of the channel $(\zeta_s = 1)$ is described by the relation

$$\mathrm{Pe} = 2\gamma \mathrm{Da}(1+\beta\theta_{ad})^2 \Pi^{-\frac{\nu}{2}} (\mathrm{N} - \eta_o^{'})^2 \exp\left(\frac{\theta_{ad}}{1+\beta\theta_{ad}}\right) \Gamma(1+\frac{\nu}{2}) \tag{11.72}$$

(curve 3 in Fig. 11.13).

The spontaneous attainment of high-temperature states of the reactor is possible only in the case where the self-ignition of a mixture occurs during its passage through the channel, i.e. if the adiabatic length of ignition induction does not exceed the reactor channel length $(x_i \le L)$. Therefore, the critical condition of self-ignition in the reactor should be written as

$$x_i = L \quad \text{or} \quad \zeta_i = 1 \tag{11.73}$$

Taking into account that in this case $\pi_2 = 0$ and $\eta_2 = N$, one determines from relation (11.69) the critical Damkohler number corresponding to the given state:

$$\mathrm{Da}_{\mathrm{cr.1}} = \frac{1}{2}(2+\nu)\mathrm{N}\frac{1-\exp(-\theta_{\mathrm{ad}})}{\theta_{\mathrm{ad}}} \tag{11.74}$$

In a similar manner it is possible to determines the other critical Damkohler number corresponding to self-ignition at the intermediate cross-section ($\zeta_i = \mathrm{N}$, $\pi_2 = 0$, $\eta_2 = 1$)

$$\mathrm{Da}_{\mathrm{cr.2}} = \frac{1}{2}(2+\nu)\frac{1-\exp(-\theta_{\mathrm{ad}})}{\mathrm{N}\theta_{\mathrm{ad}}} \tag{11.75}$$

The dependences (11.74) and (11.75) are shown on the parametric diagram in Fig. 11.13 by straight lines 6 and 7 which are the asymptotes of the curves 3 and 2, respectively. Lines 6 and 7 separate the parametric space into domains, where the self-ignition either is absent ($\mathrm{Da}_{\mathrm{cr.1}} > \mathrm{Da}$; $\zeta_i > 1$) or does exist and is terminated by the complete ($\mathrm{Da} > \mathrm{Da}_{\mathrm{cr.2}}$; $\zeta_i < \mathrm{N}$) or by incomplete ($\mathrm{Da}_{\mathrm{cr.1}} < \mathrm{Da} < \mathrm{Da}_{\mathrm{cr.2}}$, $\mathrm{N} < \xi_i < 1$) burning out of solid reactant. Along with the lines 1–3 these boundaries subdivide the Pe–Da plane into a set of domains with different process laws.

We consider these laws in more detail by analyzing the form of the kinetic curves and the character of their intersection with the line of stationary states in different domains of the parametric plane Pe–Da. For clarity, the corresponding results were plotted on reactor state diagram directly (Fig. 11.13). In region Ia, confined by lines 3 and 6, the separationless combustion regime holds. Zone IIa between lines 2 and 3 corresponds to the regimes of separationless combustion ($\eta_2 = 1$) and flame separation combustion with incomplete burning out ($\eta_2 < 1$). In zone IIIa between lines 1 and 2 the separationless and flame-separation combustion processes are terminated by the complete chemical transformation of solid reactant. And, finally, in region IVa situated above line 1 in Fig. 11.13 the supply rate everywhere exceeds the flame propagation rate. As a result, the stabilization of the combustion wave in the channel is impossible, and the wave is carried away by a substance flow outside its limits. The pressing out of the wave from the channel occurs by a jump, during the transition through boundary 1 from the domain IIIa (the kinematic flame blow-off). The absence of the induction self-ignition at $\mathrm{Da} < \mathrm{D}_{\mathrm{cr.1}}$ implies that the reactor cannot operate in a high-temperature regime in the domain IVa.

In domain IIb which is bounded by the lines 1, 5, 6, and 7 high-temperature processes proceeding with complete and incomplete chemical transformation of a solid reactant at the flame front localized near the inlet ($0 < \zeta_f < \mathrm{N}$) and at some intermediate channel cross-section ($0 < \zeta_f < 1$) are possible. The transfer from IIb into the adjacent domains IIIb and Ib corresponds, respectively, either to a decrease or to an increase of the flame propagation rate according to Eq. (11.55). In this case the flame front is either stabilized near the self-ignition point (domain

IIIb, $\zeta_f > N$, $\eta_2 < 1$), or occupies a position near the channel inlet (domain Ib, $\eta_2 = 1$).

In domains Ic and IIc, confined by the lines 1 and 7, the high-temperature stationary processes proceed with the complete chemical transformation of a solid reactant at the flame front near the channel inlet (Ic) or in an intermediate channel cross-section (IIc), respectively.

As noted above, at $Da > Da_{cr.1}$ self-ignition is possible in the reactor channel. As a result, the high-temperature states are established in the corresponding part of the parametric space, as the active reactants enter the channel. In contrast, the high-temperature regimes in domains Ia, IIa and IIIa should be considered only as potentially possible ones, which can be implemented due to the presence of a special igniter in the channel.

The reactor with a "short" channel (N>1) is characterized by a lower number of possible stationary states due to the absence of incomplete burning out of a solid reactant in it. The parametric diagram of this reactor (Fig. 11.14) is simplified, accordingly. It includes only three principal confining lines corresponding to the limiting stationary states of the reactor with a combustion wave at the channel inlet line (1), at the channel outlet line (3) and to substance self-ignition at the exit cross-section line (6). The formula for the boundaries 1 and 3 is obtained from the equations of group 1 in which, along with the assumptions $u_f/u_* = 1$ and $\eta_2 = 1$, one should put $\zeta_s = 0$ or $\zeta_s = 1$. The critical Damkohler number corresponding to boundary 6 is determined by Eq. (11.69) with $\zeta_i = 1$ and $\eta_2 = 1$.

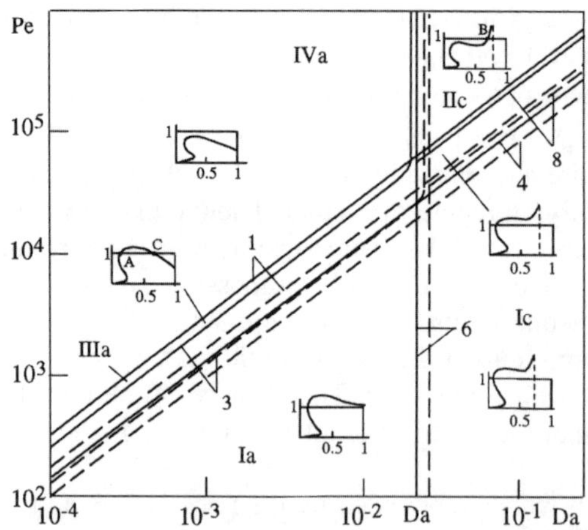

Fig. 11.14 Parametric diagram of states of a "short" reactor. Reprinted from Yarin and Sukhov (1992) with permission. Insets to the graph illustrate the shape of kinetic curves corresponding to different states of a short reactor in the parametric plane $u_f - \zeta_f$. N=2, $\nu = 0.5$, $\gamma = 0.01$, and $\beta = 0.04$. Solid lines, G=0; dashed line, G=0.2

Stability of stationary states. The analysis of the parametric diagrams Pe–Da shows that within a rather wide range of the Peclet and Damkohler numbers the multiplicity of stationary states exist. The non uniqueness of solutions, naturally, raises the problem of their attainment, i.e., the stability problem. The problem of flame stability in the channel of finite length has two aspects which were discussed earlier by Barenblatt et al. (1962).

The stability is determined, on the one hand, by the ratio of the intensities of local heat- and mass- transfer processes in the closest vicinity of the flame front (the local stability) and, on the other hand, by the characteristics of implementation of the combustion process in an apparatus (the supply and removal of a solid phase at a rate u_*, the maintenance of specific conditions at the channel inlet and outlet, the channel length, the induction heating of substance before the flame front, the heat transfer towards the cooled inlet, etc.). In the first approximation, it is possible to assume that both factors affecting stability act independently. In this case to study the effect of the local heat and mass transfer it is possible to apply the method of small perturbations as was done to evaluate the stability of a filtration combustion front in an unbounded medium by Sukhov and Yarin (1980) and by Aldushin and Kasparyan (1981). The effect of the construction of the apparatus is elucidated by the analysis of the kinetic curves in the vicinity of points of their intersection with the line of stationary states of a reactor. Obviously, only those stationary states can be attained which are stable with respect to both of the above-mentioned factors.

Leaving aside the problem of local stability of the flame (it can be achieved by a proper choice of the temperature T_0 and pressure P_0), we shall consider the effect of characteristics of the channel reactor on the stability of its stationary regimes. Referring to the kinetic curve mentioned above (Fig. 11.12), one can easily conclude that among the stationary states corresponding to points B, A, and C the stable states correspond to the positive slope of the kinetic curve $du_f/d\zeta_f$ and the unstable states to the negative slope. Indeed, random deviations of the combustion wave to the left or the right from points A and B, where $du_f/d\zeta_f > 0$, transfer the wave into the domains where $u_f<u_*$ and $u_f>u_*$, respectively. In the first case the wave will be carried away by the flow into the initial state, whereas in the second case it also returns into the initial state, following the flow upwards. In a similar way it is possible to show that the stationary state corresponding to point C in Fig. 11.12, where $du_f/d\zeta_f < 0$, is unstable.

Taking into account the above considerations, one can find the stable states on the parametric diagram in the domains with a multiplicity of stationary regimes. For example, at $Da<Da_{cr.1}$ (Fig. 11.13) below line 1 only the state corresponding to the stable point A is attained, which corresponds to a separationless combustion. This case is illustrated by the kinetic curve for Pe=8,000 in Fig. 11.15. A decrease of the value of Pe results in a rise in the upper branch of the kinetic curve. The positive slope of the kinetic curve at the point of its intersection with the line of stationary states changes to a negative one, and the state corresponding to point A in Fig. 11.12 loses its stability. The transition to the instability corresponds to Pe and Da values at which the limiting point on the kinetic curve

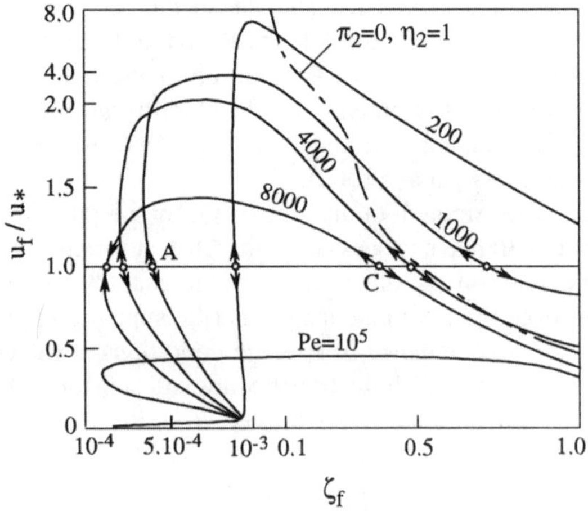

Fig. 11.15 The family of the kinetic curves. Reprinted from Yarin and Sukhov (1992) with permission. N=0.5, Da=0.005 ($200 \le Pe \le 10^5$)

with the slope $du_f/d\zeta_f = \infty$ is located on the line $u_f/u_* = 1$. On the parametric diagram in Fig. 11 13 the set of such Pe and Da values forms the boundary 4.

For $Da_{cr.1} < Da < Da_{cr.2}$ in domain IIIb (Fig. 11.13) only the stationary state corresponding to point B is stable; in the domain IIb among three formally possible states A, B and C only states A and B are stable. In transition through line 4 state A loses its stability, and at boundary 5 the stable state B is lost.

For $Da > Da_{cr.2}$ the stable state corresponding to point B is attained in the domain Iic (Fig. 11.13). After transition into domain Ic in the area between the lines 1 and 8 only two states, A and B, are stable among three stationary states. As the Peclet number decreases, the state corresponding to point B is lost on line 8, and on line 4 the state corresponding to point A becomes unstable.

Thus, the stable high-temperature stationary process in the combustion reactor channel is possible only in that part of the parametric space that corresponds to $Da < Da_{cr.1}$, and is confined from below by lines 5 and 4 in Fig. 11.13. The upper limits are absent in this case, because of the unbounded growth of the Peclet number that is accompanied by degeneration of stable regimes with a combustion wave and their transformation into thermal explosion (self-ignition) rather than by blowing off of the flame.

Hysteresis phenomena. The evolution of the kinetic curves in transition from domain Ic into domain IIc in Fig. 11.13 suggests the existence of a specific hysteresis phenomenon (kinematic flame hysteresis). Its characterized by a jumpwise change of the flame front coordinate ζ_f with variation of the Peclet number (Fig. 11.16). The movements of the front, which are inadequate with increasing and decreasing Pe, are also accompanied by a change of combustion temperature from $\theta_2 < \theta_{ad}$ on the lower branch of the hysteresis curve in Fig. 11.16 (separationless combustion) to $\theta_2 = \theta_{ad}$ on the upper branch (flame separation). A similar effect can take place when the other parameters are varied for instance, the Damkohler number. The domain of existence of the hysteresis on the parametric diagram lies between lines 1 and 8 (combustion with complete burning out of solid reactant) and between the lines 1 and 5 (incomplete burning out); see Fig. 11.13.

Combustion regimes with self-sustained oscillations. The flame instability in the separationless combustion regimes (the state corresponding to point A in Fig. 11.12) can give rise to the development of a self-sustained oscillation process in the reactor. The self-sustained oscillations arising in the part of the parametric space confined by line 4 from above in Fig. 11.13 develop differently depending on the whether the self-ignition of the substance is possible or not within the limits of the reactor channel. Therefore, when the self-sustained oscillation mechanism is considered, it is worthwhile to distinguish the cases $Da<Da_{cr.1}$ and $Da>Da_{cr.1}$.

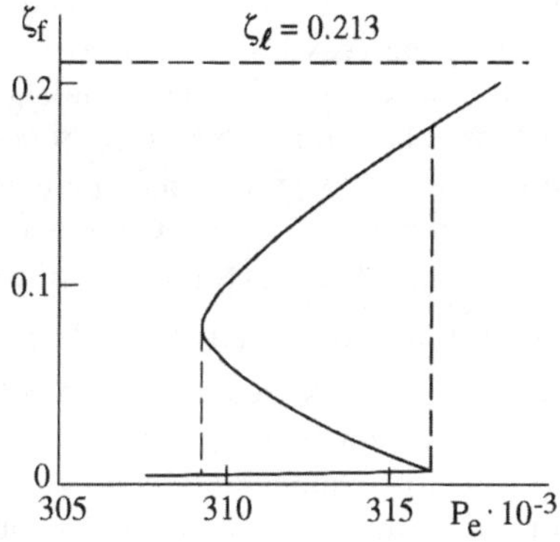

Fig. 11.16 Kinetic flame hysteresis. Reprinted from Yarin and Sukhov (1992) with permission N=0.5, Da=0.1, and $\eta_2 = 1$.

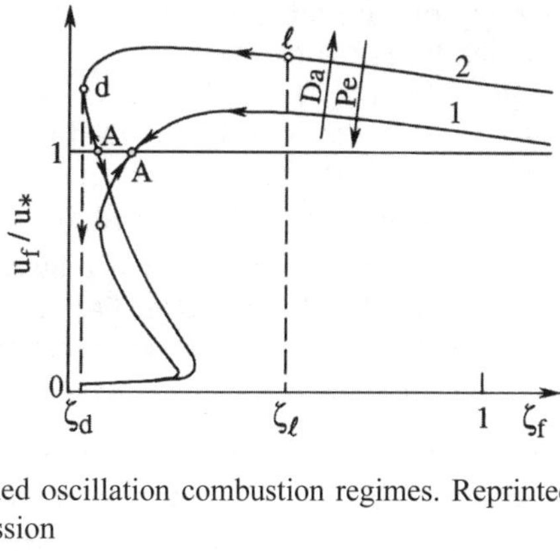

Fig. 11.17 Self-sustained oscillation combustion regimes. Reprinted from Yarin and Sukhov (1992) with permission

A. *Subcritical states* ($Da<Da_{cr.1}$). As noted above, stable combustion is attained in the domain IIIa of Fig. 11.13 near the channel inlet (see Fig. 11.17, curve 2, point A). Under these conditions the ignition of an active mixture from an external high-temperature source in the channel inevitably results in the transition of the combustion front formed into a stable state corresponding to point A in Fig. 11.12. A decrease of the supply rate u_* (a decrease of Pe with a simultaneous increase of Da) leads to a rise in the kinetic curve and to a loss of the combustion stability (see curve 1, point A in Fig. 11. 17). In this case the limiting point d in Fig. 11.17, at which $du_f/d\zeta_f = \infty$, is situated above the line of the stationary states.

The combustion wave, upon reaching the limiting point, leaves it, because at point d the condition $u_f > u_*$ is satisfied. In this case the process transforms into the low-temperature oxidation state ($T_2 \approx T_o$) corresponding to the lower branch of the kinetic curve in Fig. 11.12. This transition, marked by a dashed line in the Fig. 11.17, essentially implies the cessation of combustion corresponding to the thermal losses caused by the proximity of a cooled inlet. When the fresh active mixture enters the channel, it is heated at the surface of contact with hot reaction products. After some time elapses, this process may be terminated by the ignition of a solid reactant and the formation of a new combustion wave at the channel cross-section with coordinate ζ_f. The wave formed moves towards the channel inlet and ceases its existence at the cross-section ζ_d. Then the process is repeated again. Thus, in the domain of the parametric space below line 4 in Fig. 11.13 combustion in a self-sustained oscillation regime is possible with alternation of high-temperature (flashes) and low-temperature (depressions) stages. The duration of a depression is determined by the ignition time of mixtures t_i, and amplitude by displacement of a solid reactant during the depression time $u_* t_i$. If the ignition time is large enough ($u_* t_i > L$), the process in the channel has no time to come out of the depression state, and the self-sustained oscillation regime is not attained. In this case the loss of stability of the state corresponding to point A in Fig. 11.12 means the cessation of a high-temperature process.

B. *Supercritical states* ($Da>Da_{cr.1}$). For $Da>Da_{cr.2}$ self-ignition of the substance in the channel always gives rise to self-sustained oscillations in transition through boundary 4 in Fig. 11.13. In this case the amplitude of the oscillatory process depends on the ignition time and is determined by the value of $u_* t_i$.

For $Da_{cr.2}>Da>Da_{cr.1}$ the existence of stable combustion with incomplete burning out below line 4 in Fig. 11.13 (state B in Fig. 11.12) complicates the process of transition to self-sustained oscillations. The latter arises only in the case where the ignition time is small enough and the condition $u_* t_i < \zeta_c$ is fulfilled (here ζ_c is the front coordinate corresponding to the unstable state C in Fig. 11.12). In the opposite case, when $u_* t_i > \zeta_c$, the cessation of combustion near the inlet leads to the appearance of stable combustion inside the channel (state B in Fig. 11.12). However, during the transition through line 5 in Fig. 11.13 into the domains where this state is absent, the self-sustained oscillations become the only possible form of existence of a high-temperature process in the reactor.

The self-sustained oscillation process considered above is a feasible fundamental property of high-temperature plug flow reactors. This process was observed by Butakov and Shkadinskii (1978) in experiments with styrene polymerization in a liquid-phase tubular reactor, as well as by Hlavacek and Hofmann (1970). This process is known as a vibrational combustion in the operation of gas burners and combustion chambers of jet engines.

Combustion in a permeable (with respect to oxidizer) reactor channel. For $G \neq 0$ the combustion develops under the conditions of flame ballasting by an excess quantity of oxidizer supplied from the channel by filtration through the hot product layer. The flame ballasting by gas results in quantitative and qualitative changes in the flow pattern. The calculations have shown that the increase in the capability of the channel to support the oxidizer flow leads only to displacement of all borderlines on the reactor's parametric diagram (Fig. 11.14) to the right, whereas all the principal features inherent in the processes at $G=0$ continue to exist. A further increase of ballasting gives rise to a combustion crisis, when it reaches the critical value G_*. This crisis is formally manifested in the absence of high-temperature solutions corresponding to a combustion wave at $G>G_*$. On the reactor's parametric diagram lines 1, 3, 4 and 8 disappear. Only one boundary 6 continues to exist, which corresponds to self-ignition at the reactor outlet. This boundary separates the parametric space into two domains, where the process proceeds in a low-temperature regime with some insignificant chemical transformation of a solid reactant ($Da<Da_{cr}$) and where the induction process in the channel is terminated by ignition of the substance, which then completely burns out ($Da>Da_{cr}$).

One should note that the value G_* is not constant, but depends on characteristics of the medium and regime conditions. For example, $G_*=0.2$ for $N=2$, $Pe=120{,}000$ and $Da=0.04$.

References

Aldushin AL, Merzhanov AG, Khaikin BI (1974) Regimes of layer-by-layer filtration combustion of porous metals. Sov. Phys. Dokl. 215: 616–612

Aldushin AP, Kasparyan SG (1981) Stability of stationary filtrational combustion waves. Combust. Explos. Shock Waves 17: 615–625

Barenblatt GI, Zel'dovich YaB, Istratov AG (1962) On the diffusional-thermal stability of a laminar flame. Zh. Prikl. Mekh. Tekh. Fiz. 4: 21–26 (in Russian)

Butakov AA, Shkadinsky KG (1978) A self-sustained oscillation regime of exothermal reaction proceeding in a tubular reactor. Sov. Phys. Dokl. 238: 166–169

Butakov AA, Maksimov EI, Shkadinskii GK (1978) Theory of chemical displacement reactors. Combust Explos. Shock Waves 14: 48–54

Guzhiev AV, Soldatkina NN, Sukhov GS (1986) Operating states of a filtrational reactor. Combust. Explos. Shock Waves 22: 207–214

Hlavacek H, Hofmann H (1970) Modeling of chemical reactors—xix. Transient axial heat and mass transfer in tubular reactors. The stabillity considerations. Chem. Eng. 25: 1517–1526

Khaikin BI, Rumanov EN (1975). Exothermic reaction regimes in a one-dimensional flow. Combust. Explos. Shock Waves 11: 573–578

Likhachev VN, Sukhov GS, Yarin LP (1991) The theory of bubble combustion reactors. Combust. Explos. Shock Waves 27: 191–199

Likhachev VN, Sukhov GS, Yarin LP (1992) Towards a theory of bubble reactor combustion (displacement reactors). Combust. Explos. Shock Waves 28: 129–136

Merzhanov AG (1967) Combustion processes in chemical engineering. Pre-print. Branch of Inst. Chem. Phys. (in Russian) AN SSSR, Chernogolovka

Merzhanov AG (1980) SHS-process: combustion theory and practice. Arch. Combust. 1: 23–48

Merzhanov AG, Filonenko AK (1963) About the thermal self-ignition of homogeneous gaseous mixture in flow. Sov. Phys. Dokl. 152, 1: 143–146

Stolyarova NN, Sukhov GS, Yarin LP (1980) Theory of a filtration reactor with a stabilized combustion front. Combust. Explos. Shock Waves 16: 174–180

Stolyarova NN, Sukhov GS, Yarin LP (1981) Steady conditions in a filtrational reactor. Combust. Explos. Shock Waves 17: 642–646

Sukhov GS, Yarin LP (1978) Towards the theory of displacement filtration reactors. Sov. Phys. Dokl. 234: 1442–1444

Sukhov GS, Yarin LP (1979) Steady conditions of filtration combustion. Combust. Explos. Shock Waves 15: 1–7

Sukhov GS, Yarin LP (1980) Two-dimensional instability of the combustion of porous substances in a gaseous oxidizer. Combust. Explos. Shock Waves 16: 275–280

Sukhov GS, Yarin LP (1981a) Combustion waves in bubbly media. Sov. Phys. Dokl.. 256: 376–380

Sukhov GS, Yarin LP (1981b) Laws of combustion of bubbled media. Combust. Explos. Shock Waves 17: 251–257

Sukhov GS, Yarin LP (1982) Towards the analysis of steady states of displacement reactors. Found. Chem. Technol. 16: 391–394

Sukhov GS, Yarin LP (1988a) Combustion-reactor theory: the dynamic-balance method. Combust. Explos. Shock Waves 24: 1–6

Sukhov GS, Yarin LP (1988b) Operating conditions of combustion reactors. Combust. Explos. Shock Waves 24: 263–268

Yarin LP, Sukhov GS (1992) On filtration combustion reactor theory. Combust. Sci. Technol. 84: 15–32

Zaidel RM, Zel'dovich YaB (1962) On possible stationary combustion regimes. Zh. Prikl. Mekh. Tekh. Fiz. 4: 27–32 (in Russian)

Zel'dovich YaB, Barenblatt GI, Librovich VB, Makhviladze GM (1985) Mathematical theory of combustion and explosion. Plenum, New York

Nomenclature

A_c	acceleration parameter
a	transverse semi-axes
B	Spalding transfer number
Bi	Biot number
b	longitudinal semi-axes
C	concentration
C_f	fuel concentration
C_o	oxidizer concentration
C_v	vapor concentration
$C_\ell(C_s)$	concentration of the ℓ-th (s-th) component of two-phase mixture
$C_{o\infty}$	ambient oxidizer concentration
C_d	drag coefficient
C_d	total drag coefficient
C_{dAC}	total drag coefficient of accelerated particle
$C_{d.c}$	drag coefficient for creeping flow at $Re \to 0$
$C_{d.st}$	Stokes drag coefficient
$C_{d.u}$	drag coefficient in uniform flow
C_H	historical drag coefficient
C_m	virtual mass drag coefficient
C_ℓ	lift force coefficient
c	specific heat
c_i	specific heat of the i-th phase
c_p	specific heat at constant pressure; specific heat of combustion product
c_v	specific heat at constant volume
c_m	specific heat of solid reagent
$c_{2.1}$	dispersed phase to continuous phase specific heat ratio
$c_{(*)}$	root-mean square of molecules' velocities
D	diffusivity
D_s	diffusion coefficient of the s-th component
Da	Damkohler number
Da_{cr}	critical Damkohler number
d	characteristic size
d	particle, droplet, bubble diameter
d_{eq}	volume equivalent sphere diameter

d_f	fluid element diameter
d_N	nozzle diameter
d_0	initial droplet diameter
E	activation energy
E_o	Eotvos number
$\mathbf{e}_x$	unit vector in x direction
$\mathbf{F}$	force exerted by a viscous fluid on a particle
$\mathbf{f}$	body force
$\mathbf{f}_B$	Basset force
$\mathbf{f}_d$	total drag force
$\mathbf{f}_e$	drag force due to mass transfer
$\mathbf{f}_f$	viscous drag force
$\mathbf{f}_g$	gravity force
$\mathbf{f}_\ell$	lift force
$\mathbf{f}'_\ell$	lift force due to velocity gradient
$\mathbf{f}_m$	virtual mass drag force
$\mathbf{f}_p$	pressure drag force; drag force due to pressure gradient
Fo	Fourier number
Fr	Froud number
Fz	coal ignition index
f	frequency of particle collision
f_i	cross-section surface area particle of the i-th fraction
$f(\eta)$	burning-out function
G	specific mass flux; droplet mass flux; ratio of heat release due to chemical reaction to heat losses from bubble to surrounding liquid reagent
$G_{1.0}$	oxidizer mass flux at nozzle exit
$G_{2.0}$	fuel mass flux at nozzle exit
$\bar{G}$	specific total interface flux
$\tilde{G}_\ell^{(i)}$	specific total mass flux due to a distinction of gaseous component concentrations in the i-th and j-th continuum
G	volumetric flow rate
$G_\ell^{(i)}$	specific volumetric flow rate of a substance that is extracted from a combustion reactor
$G_{\ell.0}^{(i)}$	specific volumetric flow rate of a substance that is fed into a combustion reactor
G_{fl}	mass flow rate of combustion product in the tip of the torch
G_v	volumetric flow rate of reactive mixture
G_x	interfacial mass flow rate in a two-phase jet
G_i	specific rate of change of the i-th phase due to chemical reaction
G_1	specific rate of gaseous phase conversion due to chemical reaction; mass expense of gaseous reagent; mass flow rate of gaseous reagent
G_2	specific rate of solid phase conversion due to chemical reaction; mass expense of solid reagent, mass flow rate of solid reagent
G_s	specific rate of change of the s-th component due to chemical reaction

Gr	Grashof number
g	gravity acceleration; mass flux
g_{ij}	specific mass flux from the i-th to the j-th continuum
H	total enthalpy; equilibrium dissolution constant
H_i	rate of the i-th phase enthalpy change due to heat exchange with other phases of a heterogeneous mixture
h	heat transfer coefficient
h_f	flame heat transfer coefficient
h_{ij}	heat transfer coefficient for contact heat transfer between the i-th and the j-th phases
$h_{ij.1}$	heat transfer coefficient corresponding to heat transfer from the i-th to the j-th gaseous continuum
$h_{ij.2}$	heat transfer coefficient corresponding to heat transfer from the i-th to the j-th liquid continuum
$h_{1\text{-}2}$	interfacial heat transfer coefficient
$h_{w.1}$	heat transfer coefficient corresponding to heat transfer from gaseous phase to reactor wall
$h_{w.2}$	heat transfer coefficient corresponding to heat transfer from liquid phase to reactor wall
h_m	mass transfer coefficient
h_{ij}	mass transfer coefficient corresponding to mass transfer from the i-th to the j-th gaseous continuum
I	total diffusion flux; radiative heat flux
I_{abs}	absorbed energy flux
I_0	radiative heat flux from external source of radiation; jet momentum
i	entalphy
i_n	unit vector normal to droplet surface
J_0	initial momentum of fluid element
j	specific volumetric flow rate of reactor exit
j_0	specific volumetric flow rate of reactor inlet
K	Phase interaction coefficient; surface curvature; burning rate constant
k	Boltzman constant; chemical rate constant; evaporation rate constant; pyrolysis rate constant; permiability coefficient; number of particles
k_{ef}	effective constant of chemical reaction
k_f	filtration coefficient
k_0	pre-exponential factor
k_s	pre-exponential in Arrhenius law
k_*	rate of oxidation constant
L	length of reactor; length of core of constant velocity in jet; extension of bubble suspension; integral scale of turbulence
$\bar{L}$	dimensionless length of constant velocity core in two-phase jet
ℓ	characteristic length; mixing length; distance between particles centers; particle mean free path; scale of turbulence
ℓ_h	length of heating zone
ℓ_f	flame length; length of combustion zone

ℓ_{far}	characteristic length of far field in coal dust flame
ℓ_{near}	characteristic length of near field flow in coal dust flame
ℓ_{react}	length of reaction zone
ℓ_R	dynamic relaxation length
ℓ_w	characteristic wake length
$\ell_{(*)}$	mean free-path of molecules; particle mean free path
$\overline{\ell}$	dimensionless scale of turbulence
M	molecular weight; heterogeneity parameter; mass of fluid element; current mass of droplet
M_p	total mass of particles in fluid element
M_{pi}	total mass of particles in the i-th fraction
m	particle mass; porosity; ratio of mass content of large to small particles with fixed total mass content; total mass flow rate; mass of species involved in reaction, order of reaction
m_i	mass of particles of i-th fraction
N	total number of particles in reactor volume; numerical concentration of droplets or bubbles
Nu	Nuselt number
$\overline{Nu}$	average Nuselt number
$\overline{Nu}_N$	average Nuselt number for natural convection
$\overline{Nu}_F$	average Nuselt number for forced convection
$\mathbf{n}$	outer unit normal vector at particle surface
n	order of reaction; number of particles in unit volume; number of fluid cells in continuum; ratio of particle to carrier gas velocities at nozzle exit; kinetic coefficient
n_m	mass concentration
n_{**}	numerical concentration
n_v	volumetric content of condensed phase
$n_{(*)}$	numerical concentration of molecules
P	pressure
Pe	Peclet number
Pe_*	diffusional Peclet number
P_i	pressure of i-th phase
Pr	Prandtl number
P_s	pressure at particle surface
P_{sat}	saturation pressure
$P(\delta)$	particle-size distribution
P_∞	ambient pressure
Q_I	heat released
Q_{II}	heat losses
Q_i	rate of i-th phase enthalpy change due to phase transition
Q_ℓ	heat losses due to completeness of particle burning

$Q_{1-2}^{(i)}$	specific interfacial heat flux from the i-th gaseous to the i-th liquid continuum
$Q_{w.1}^{(i)}$	specific interfacial heat flux from the i-th gaseous continuum to reactor's wall
$Q_{w.2}^{(i)}$	specific interfacial heat flux from the i-th liquid continuum to reactor's wall
$Q_{ij.1}^{(i)}$	specific interfacial heat flux from the i-th to the j-th gaseous continuum
$Q_{ij.2}^{(i)}$	specific interfacial heat flux from the i-th to the j-th liquid continuum
q	heat of reaction; heat flux
q_e	latent heat of evaporation
q_i	specific heat flux due to the i-th phase thermal conductivity
q_{py}	latent heat of pyrolysis
q_R	radiant heat flux
q^2	specific turbulent kinetic energy
q_t^2	overall energy of fluctuation in a wave of a single particle
q_v^2	energy of fluctuation generated by particles in volume of two-phase mixture
q_Σ^2	total turbulent energy
R	universal gas constant
Ra	Rayleigh's number
Re	Reynolds number
Re_τ	turbulent Reynolds number
Re_ω	rotational Reynolds number
R_g	gas constant
R_v	vapor gas constant
r_f	standoff radius of a diffusion flame
r_0	initial particle radius
r, φ, θ	spherical coordinates, associated with particle center
$\dot{r}$	regression rate of droplet surface
r_s	particle radius
$r_{1.2}$	principal radius
r_f/r_s	standoff ratio
S	particle surface area; interception cross-section
S_{eq}	volume equivalent sphere surface area
S_f	particle cross-section surface area; flame surface
S_w	reactor wall surface area
S_{1-2}	interfacial surface area
s	specific surface area
s_w	specific reactor wall surface area
s_{1-2}	specific interfacial surface area
S_{ij}	specific interfacial surface area
S_{ij}^R	specific interfacial surface area for radiative heat transfer

Sc	Schmidt number
Se	Semenov number
Se_{cr}	critical value of the Semenov number
Sh	Sherwood number
S_r	shear rate
St	Stefan number
T	temperature
T_{ad}	adiabatic combustion temperature
T_b	boiling temperature
T_f	flame temperature
T_{fin}	final temperature
T_i	temperature of i-th phase
T_I	ignition temperature
T_{i*}	temperature corresponding to boundary of heating and combustion zones
T_p	particle temperature
T_{p0}	initial particle temperature
T_s	particle surface temperature; surface decomposition temperature
T_{w-b}	wet-bulb temperature
T_w	wall temperature
T_0	initial temperature
T_∞	ambient temperature
T'	fluid temperature fluctuation
T'_p	particle temperature fluctuation
T_u	turbulent intensity
T_ω	torque
t	time
t_b	particle burning time; complete oxidation time
t_h	hydrodynamic characteristic time
t_{ind}	induction time
t_{ox}	time needed to start oxidation of particle surface
t_{py}	time of particle pyrolysis
t_r	viscous relaxation time; resident time
t_R	chemical reaction characteristic time
t_t	total time of droplet vaporization
t_w	characteristic time of combustion wave propagation
t_*	thermal relaxation time; characteristic time; characteristic time of bubble harmonic oscillations
t^*	particle and carrier fluid interaction time
t_*^R	radiative heat transfer relaxation time
U	dimensionless speed of combustion wave
u	longitudinal component of flow velocity
u_f	combustion wave speed; rate of pyrolysis front
u_{fT}	turbulent flame velocity

u_m	mean velocity at jet axis
u_n	combustion wave speed in vapor–oxidizer mixture
u_δ	velocity of undisturbed flow
$u_{1.0}$	carrier gas velocity at nozzle exit
$u_{1.m}$	maximal longitudinal velocity in jet cross-section
$\langle u_{1.m} \rangle$	average velocity within reaction zone on torch
u'	carrier fluid fluctuating velocity
$\tilde{u}'$	root-mean square of turbulent fluctuations of carrier fluid
u_*	supply velocity of reactive mixture at reactor inlet; velocity characteristic scale
u_∞	ambient velocity
u_c	particle burning rate
u_s	mass burning rate
V	volume; total volatile material evolved up to time t
V^*	limiting value of V at $t \to \infty$
$\mathbf{v}$	velocity vector with components u, v and w directed along x,y, and z axes
v_r, v_φ, v_θ	velocity components in spherical coordinates r, φ and θ associated with particle center
v_f	filtration velocity
v_{fn}	normal component of filtration velocity
v_{in}	normal component of velocity
v'	carrier fluid fluctuating velocity
v'_p	particle fluctuating velocity
$\bar{v}'$	carrier fluid dimensionless fluctuating velocity
$\bar{v}'_p$	particle dimensionless fluctuating velocity
$v'_{1.m}$	particle mean velocity during interaction time
v'_{mon}	particle fluctuating velocity in monodisperse system
v'_{pol}	particle fluctuating velocity in polydisperse two-phase flow
v_{py}	speed of pyrolysis front propagation
$\tilde{v}$	relative velocity between carrier fluid and particle
v_∞	velocity of undisturbed flow
v_*	characteristic velocity scale
W	specific chemical reaction rate
W_ℓ	conversion rate of the l-th component
W_{py}	pyrolysis rate
$\overline{W(T)}$	average reaction rate
$W(\bar{T})$	reaction rate corresponding to average temperature
W_s	rate of heterogeneous reaction
W_e	evaporation rate
$\bar{W}$	dimensionless chemical reaction rate

We	Weber number
w	vapor velocity relative to the inertial system of coordinates
w_s	diffusion velocity
x	arclength coordinate along generatrix
x, y, z	Cartesian coordinates
x_f	distance from reactor inlet to flame front
y_δ	torch thickness
Z	pre-exponential factor, coordinate normal to surface

GREEK SYMBOLS

α	thermal diffusivity; mass change reaction coefficient
$\alpha_{2.1}$	dispersed to continuous phase thermal diffusivity ratio
β	thermal expansion coefficient; dimensionless initial temperature; coupling function; stoichiometric factor
γ	total mass content of admixture; Euler number; dimensionless angular velocity; ratio of density of interior to density of exterior fluids
γ_{ox}	excess of oxidizer
γ_i	total mass content of particles of the i-th fraction
δ	laminar layer thickness; heating zone thickness; reaction zone thickness; oxide film thickness; jet (wave) width; particle diameter
$\delta_{(*)}$	molecular diameter
ε	emissivity; mean dissipation rate; eddy thermal diffusivity; porosity; piezoconductivity
ε_T	temperature fluctuation intensity
ζ	dimensionless longitudinal coordinate associated with combustion front
ζ_f	dimensionless distance from reactor inlet to flame front
ζ_h	dimensionless heating zone length
ζ_s	induction zone length
η	completeness of combustion; dimensionless transverse coordinate measured from jet center line
θ, Θ	dimensionless temperature
θ_{ad}	adiabatic combustion temperature
θ_E	extinction temperature
θ_f	flame temperature
θ_I	ignition temperature
θ_{pI}	particle ignition temperature
θ_{pcr}	critical particle temperature

θ_{p0}	initial particle temperature
θ_∞	ambient temperature
$\theta_{\infty I}$	ignition ambient temperature
ϑ	dimensionless heat released
χ	transverse to longitudinal semi-axes ratio; large to small semi-axes ratio; average reaction rate to reaction rate corresponding to average temperature ratio
χ_0	fuel to oxidizer mass flux ratio at nozzle exit
λ	thermal conductivity
λ^c	thermal conductivity for conductive heat transfer
λ^R	thermal conductivity for radiative heat transfer
$\lambda_{2.1}$	dispersed to continuous phase thermal conductivity ratio
Λ	correction factor to Stokes drag
λ_ω	rotation parameter
μ	dynamic viscosity; molecular weight
μ_∞	ambient dynamic viscosity
$\mu_{2.1}$	dispersed to continuous phase dynamic viscosity ratio
ν	kinematic viscosity; stoichiometric coefficient
ν_T	turbulent viscosity
$\nu_{2.1}$	dispersed to continuous phase kinematic viscosity ratio
ν_∞	ambient kinematic viscosity
ξ	volumetric content
ξ_i	volumetric content of the i-th phase
ξ_ℓ	volumetric content of the ℓ-th component
ξ_0	initial volumetric content
ξ, η, φ	bipolar coordinates
Π	normal stress vector exterior to droplet
ρ	density; effective density; particle density
ρ_e	coke density
ρ_i	effective density of the i-th phase
ρ_g	gas density
ρ_ℓ	melt density; effective density of the ℓ-th component
ρ_m	solid reagent density
ρ_{met}	metaplast density
ρ_p	combustion product density; particle density
ρ_v	vapor density
$\rho_{2.1}$	particle to carrier fluid density ratio

ρ_∞	ambient density
ρ^0	physical density
ρ, z, φ	cylindrical coordinates
σ	stoichiometric oxidizer-to-fuel mass ratio; enhancement coefficient; standard deviation
σ_{ik}	tensor of viscous tension
σ_t	surface tension
$\tilde{\sigma}$	Stefan–Boltzmann constant
τ	characteristic time
τ_c	particle burning time
τ_d	diffusion time
τ_{ev}	droplet evaporation time
τ_h	hydrodynamic time; heating time
τ_k	kinetic time
τ_R	chemical reaction time
$\tau_{R.f}$	particle residence time in far flow field of torch
$\tau_{R.n}$	particle residence time in near flow field of torch
τ_s	tangent viscous traction on particle surface
φ	distribution function; increment of fluctuations
$\varphi(\eta)$	burning out function
Φ	fuel-to-air mass ratio; shape factor
ψ	particle to ambient temperature ratio; steam function; frequency of fluctuations
ω	angular speed of particle rotation; particle-size ratio; rate of volatile production due to pyrolysis
Ω	heating parameter; complex frequency

SUBSCRIPTS

a	inert component
ad	adiabatic condition
c	conductive heat transfer
cr	critical state
d	drag
e	evaporation
F	forced convection
f	fuel; flame front
fin	final state
g	gas phase
i	ignition

ind	induction
ℓ	$=a, b_1, b_2, c, d$ correspond to gaseous oxidizer, vapor of reactive liquid, liquid reagent, combustion product and gaseous inert admixture, respectively
M	solid reagent
m	maximum
n	normal to particle surface
N	natural convection
O	oxidizer
p	particle
py	pyrolysis
r	relaxation
R	radiant heat transfer
s	saturating state; particle surface
v	vapor
w	surrounding wall
0	initial state
∞	ambient
1,2	continuous and dispersed phases
*,**	small and large particles
< >	average over ensemble of particles

Subject Index

Author Index

Printing: Saladruck Berlin
Binding Stein+Lehmann, Berlin